MANPRINT

AN APPROACH TO SYSTEMS INTEGRATION

MANPRINT

AN APPROACH TO SYSTEMS INTEGRATION

Edited by

Harold R. Booher
United States Department of the Army

VNR VAN NOSTRAND REINHOLD
New York

Library of Congress Catalog Number 90-12016
ISBN 0-442-00383-8

Printed in the United States of America

Van Nostrand Reinhold
115 Fifth Avenue
New York, New York 10003

Van Nostrand Reinhold International Company Limited
11 New Fetter Lane
London EC4P 4EE, England

Van Nostrand Reinhold
480 La Trobe Street
Melbourne, Victoria 3000, Australia

Nelson Canada
1120 Birchmount Road
Scarborough, Ontario M1K 5G4, Canada

16 15 14 13 12 11 10 9 8 7 6 5 4 3 2

Library of Congress Cataloging in Publication Data

MANPRINT, an approach to systems integration / edited by Harold R.
Booher.
p. cm.
Includes bibliographical references.
ISBN 0-442-00383-8
1. Human engineering. 2. Systems engineering. I. Booher, Harold
R.
T59.7.M34 1990
620.8'2—dc20

90-12016
CIP

CONTENTS

FOREWORD

PERSPECTIVE

This book is important to everyone concerned with the design and development of people-oriented systems. The Manpower and Personnel Integration (MANPRINT) program is a major military system procurement initiative adopted by the Army to focus on the needs and capabilities of the soldier. This program is unique in that it integrates six areas of user concerns which include human factors engineering, manpower, personnel, training, health hazards, and system safety throughout the development cycle of Army materiel.

Even though MANPRINT was developed for Army systems, the philosophy and techniques used in this program extend well beyond military systems used by soldiers. It can be applied to all products and systems used by people such as automobiles, airplanes, boats, control rooms, automated manufacturing, telecommunications, computers, and medical equipment.

Interestingly, the impetus for MANPRINT came from the senior managers who buy these systems. During the early and mid-1980s, two Army generals, M. R. Thurman and R. M. Elton, who served successively as the Deputy Chief of Staff for Personnel, were instrumental in fostering MANPRINT development. By the end of the 1980s, this program was integrated throughout the standard procurement system of the Army. The formal statement of acquisition policy is contained in Army Regulation 602-2. Some aspects of the MANPRINT approach have been initiated in each of the other services and in the Office of the Secretary of Defense. A recent Department of Defense Directive (Number 5000.53), for example, provides guidance for all military departments in addressing the domains of MANPRINT in the system acquisition process. The concepts of MANPRINT are already spreading to other government agencies (e.g., the Federal Aviation Administration) and to other countries (e.g., the United Kingdom and West Germany).

The focus of this book is on the philosophy, not the techniques, of MANPRINT. This philosophy is organized around four major considerations dealing with the *Organization/Management Context, User-Centered Design Advances, Systems Integration Methodologies, and Sources of User-Centered Technology* in Parts I through IV, respectively. This book is appropriate for a variety of readers. Whether you are a user, designer, researcher, or manager of people-oriented systems, you should be interested in this philosophy.

Users. Each of us use complex systems in our daily lives, and we are often frustrated with the improper design of equipment in our homes and offices. The philosophy of MANPRINT deals with these design problems by providing a procedure to keep the needs and requirements of the user central to the design and development of the products they will ultimately use.

Part II and Part IV of this book which deal with user-centered design advances and technology sources are particularly appropriate to the user. These sections of the book address a general approach to providing information about user needs as well as highlight specific considerations, methods, and sources of information that are central to incorporating user requirements in the design process.

Designers. Integration is key to the proper design of systems. People-oriented design involves the selection of appropriate users, the appropriate training of the user, proper operation by the user, and adequate maintenance of the system by the user. The design philosophy of this book is unique in two aspects. First, there is an emphasis on the user of the system early in design which continues throughout the design process. Second, the MANPRINT design philosophy integrates the various system acquisition functions throughout design. For example, trade-offs are often required between design and training, and these two functions must be considered simultaneously. Often design and training are viewed as totally independent in which training of the user is considered after the design is determined. By trading off these two considerations, total system cost and performance can be improved.

Part III of this book which deals with systems integration methodologies is particularly appropriate for designers. In addition, the methods described in Parts II and IV should be used in the design process in order to make designs more user-centered.

Researchers. This book also suggests several opportunities to the research community. In order to support the MANPRINT design philosophy, a variety of design techniques need to be developed. Methodological research needs to be goal oriented to provide direct support of these design procedures. Although many techniques are already available for use in the MANPRINT process, additional research is still needed.

Examples of design-oriented techniques such as computer-based ergonomic design tools, workload analysis methods, and computer modeling procedures are provided in Parts II and III. In addition, the need for national human performance information centers is described in Part IV. Research disciplines dealing with human factors engineering, computer science, operations research, management science, and cognitive psychology should each be interested in research to augment the MANPRINT design philosophy.

Managers. This book provides an overall orientation for managing the entire system acquisition cycle. This orientation is centered on the ultimate user of the system. Managers of both the buyers and the builders of these systems need to work cooperatively in order to produce truly usable products.

All four parts of this book deal directly with the functions and process of this management process. Specifically, Part I addresses the overall orientation and management context needed for enhancement of user-centered design. More specifically, the acquisition decision process described in Part III provides an overall description of the management analysis and decision processes of MANPRINT in the acquisition of military systems, but commercial applications are noted as well.

CHALLENGE

Read this book from your perspective as either a user, designer, researcher, or manager. Consider this people-oriented design philosophy carefully. As an academic, I read the MANPRINT philosophy and discovered a variety of implications for graduate training and research in human factors engineering.

MANPRINT offers us an exciting opportunity to keep the ultimate user of the system central in the design process. Many of the lessons learned from the Army experience are now starting to appear. In the future, we should expect to have more case studies and examples of improved system acquisition and performance which can be related directly to the use of MANPRINT. In addition, I hope we will develop new methods and augment the training of the professionals needed to support this process.

As users, designers, researchers, and managers of these systems, we must meet the challenge of MANPRINT to improve the design and utilization of our products. Perhaps we can make the decade of the 1990s the decade of user-oriented design. The philosophy of MANPRINT can help us reach that goal.

Robert C. Williges
Professor
Virginia Polytechnic Institute and
State University

PREFACE

Global competition, demographic trends, and high-risk technology are conspiring to bring both military and commercial industry from "equipment" domination to a more "people-equipment" orientation. The purpose of this book is to describe a new systems integration philosophy known as MANPRINT (Manpower and Personnel Integration) which is involved in this change. Through a series of edited works which speak to the concepts and complexities of integrating people, machines, and organizations, it is written to those who want to know more about the MANPRINT role in a changing industrial culture.

In a kickoff meeting before we actually started to write, I asked each author to focus their piece on an advanced technical or managerial concept that could help enhance the overall MANPRINT approach as it was being practiced by the Department of the Army. We set as a broad goal for each chapter that it describe the state of the art of each concept and explain its relevance to MANPRINT as visualized in the symbol shown on the cover of this book. The symbol, by showing primary colors combining to become white, emphasizes the MANPRINT focus on the integration process. More specifically, we asked each contributor to reflect on the complex relationships among (1) people as designers, users, and repairers of machines, (2) government and industrial organizations which regulate, design, manufacture, and/or operate machines, and (3) methods and processes for design, production, and operation of machines. Although this goal was much more difficult to achieve than any of us envisioned at the outset, I believe the reader will see not only a fairly comprehensive treatment of specific topics, but also how the broader relevance of the topic to complex systems is appreciated by the contributors.

I had in mind then, and even more so now that the book is complete, that human factors as an engineering discipline is fully ready to be an inherent activity within systems integration and that the MANPRINT approach can be beneficial in that regard. But just as surely, I am convinced that fairly wide sweeping organizational and managerial changes will be necessary for any organization to adopt MANPRINT or any similar human factors approach as routine in its systems project management. As a grass roots philosophy, the idea of human factors technology has great appeal. It seems only common sense that we should design things so they are safe and easy to use, break seldom, but are easy to fix if they do break. Moreover, the technology of human factors is sufficiently mature that if applied, I would estimate that 90 percent of the problems users have with machines could be avoided. Yet, in my 20-plus years with the profession, little has changed on a national level. Human factors continues to be rediscovered nearly every time there is a well publicized disaster in which "human error" is involved. Major human factors programs were introduced in each branch of the military and the Department of Transportation in the 1960s, yet the nuclear industry knew literally nothing of the discipline until Three Mile Island drew it to their attention. Those in the discipline have always been "crying in the wilderness." It seems so obvious to design-in human technology early to avoid later problems. Why hasn't the engineering world recognized its importance?

Charles Perrow, coming with a macro-view of human factors and its role in our society, has probably come closest to putting his finger on the problem ("The Organizational Context of Human Factors Engineering," *Administrative Science Quarterly*, 1983; *Normal Accidents*, Basic Books, 1984). If we are to understand why "the principles of human factors engineers are so neglected" in most military and industrial organizations, we need to have a better appreciation of the relationship between equipment design and organizational structure. After reviewing a wide variety of organizational systems that involve high risk technologies – "nuclear power plants, chemical plants, aircraft and air traffic control, ships, dams, nuclear weapons, space missions and genetic engineering" – Perrow asked, "Who bears the consequence of poor design in high-technology systems?" Except in highly publicized catastrophes, it is usually no one at all in the organization which produced the equipment. It is rather the operator or the repairman who, if not a victim himself, is at least tagged with "human error."

Often those who rise to the defense of the operator or other users tend to blame the design engineers. That is perhaps understandable since there is so much human factors technology that the engineers seem to ignore. But, Perrow forces us to search deeper for insights that can lead to more profitable approaches. It is time to stop blaming the design engineer and take into account the pervasive social casual factors inherent in organizations which make and operate our machines. Perrow describes the situation from an organizational analyst's point of view. He reminds us that managers and professionals respond to "rewards and sanctions and prevailing belief systems of top management." There is nothing to prevent top management, if it wishes, from informing designers about human factors principles. Furthermore, it is top management who "can require that these principles be utilized." They alone "can structure the reward system so that it encourages designers to take these principles into account." The issues of rewards and consequences are at the very heart of the problem and suggest a solution.

In the procurement of military equipment, we have often reflected on the reward/consequences system for program managers. They are rewarded for meeting near term costs and schedules and are usually far removed to another job should disastrous consequences occur later in the hands of the soldier, sailor, or airman. MANPRINT in the Army has not totally changed the reward/consequences system, but attitudes of top management have changed and government program managers and industry personnel have experienced either career enhancements or career setbacks directly attributed to MANPRINT.

MANPRINT developed independently of Perrow's insights, but his ideas come the closest to explaining why MANPRINT has been steadily gaining stature in government and industry. Perrow's work identifies the problem, but this book is an attempt at a solution. In spite of the difficulty of assigning blame, it seems nearly inconceivable, however, that an approach to systems integration which focuses on people and organizations has not been written years ago. But, based upon reflection of my experiences in both defense and commercial environments, it is perhaps more timely now than before. There are three overriding reasons for this. First, the western industrial world, particularly the United States, has not been open to the kind of radical changes MANPRINT envisions. So long as both systems decision makers and designers on the one hand and consumers on the other hand have defined technological progress in hardware (and software) terms only, human factors technology and people integration philosophies were secondary to all mainline decisions. Now, high procurement costs, poorly performing equipment, current economic conditions stemming from worldwide competition, and the changing demographics of the '90s are forcing a new look at engineering design and manufacturing processes. There is also the ever increasing awareness of the potentially disastrous consequences to public health, safety, and environment that arise from manufacturing and operating complex technology. These are some of the reasons government policy makers and industrial leaders now seem ready to hear about quality improvement concepts like MANPRINT.

Second, the domains of MANPRINT (manpower, personnel, training, human factors engineering, health hazards, safety) have been focused too low and too narrowly to enable wide sweeping changes in systems integration. While perhaps not *sufficient*, two critical managerial ingredients are at least *necessary* if there is to be any hope of lasting success. One ingredient already mentioned above is to have high visibility and support from the very top of an organization. If the highest interest is mid-level management or lower, the MANPRINT concept will not be as successful. The Army MANPRINT program started with the personal involvement and support of both the military leadership (General Maxwell Thurman, when Vice Chief of Staff, Army; now Commander, U.S. Army Southern Command and leader of the Panama campaign) and the civilian leadership (Mr. Delbert Spurlock, when he was Assistant Secretary of the Army, Manpower and Reserve Affairs).

Bringing together as many human factors related disciplines as possible into systems integration is the other critical ingredient. In addition to areas like human factors engineering, health hazards, and safety, which are traditionally closer to equipment design, successful MANPRINT insists that the disciplines of manpower, personnel, and training also be included. Only in this way can those who make equipment design decisions be meaningfully linked to those who can be specific about the large costs associated with people in systems.

Third, the technology for people integration into systems conceptualization and systems design has only recently become available. Part III of the book describes many of the relevant concepts, but this is not a "how-to" book. This book introduces the concepts, points out where we are, and what needs to be done. Many more books are needed on how to do systems integration the MANPRINT way. In fact, the extreme difficulty of putting together Part III convinced me more than ever of the infancy of this technology. MANPRINT to date has only begun to identify the enormous potential that exists for future research and development in this field. This, of course, does not lessen the immediate improvements possible from using existing technology.

There is a word used extensively throughout the entire book which is extremely hard to define. That word is *integration*. Its definition, I believe, will become more apparent with the reading of each chapter. When properly applied, the concept of integration is an extremely powerful one. Some of the specific Army program examples will point this out. Poorly executed, it is just another word with little meaning. One warning though – *Integration is never easy*. But organizations can make it easier to achieve integration by better utilizing the capabilities of individuals and disciplines that exist throughout the organization. Individuals in these organizations should find many of the principles, techniques, and technologies described both thought provoking and practicable. And for those interested in research and education, nothing would please us more than for this book to become their primer in exploring entirely new careers in systems integration.

As the Army responds to a world of changing priorities, the continuing vision and support for MANPRINT is much appreciated. The Army commitment is still highly visible under current military leaders to include, General Robert W. RisCassi,Vice Chief of Staff, Army; General William G. T. Tuttle, Jr., Commander, Army Materiel Command; General John W. Foss, Commander, Training and Doctrine Command; and under the new civilian leadership, The Honorable G. Kim Wincup, Assistant Secretary of the Army (Manpower & Reserve Affairs); and The Honorable Stephen K. Conver, Assistant Secretary of the Army (Research, Development and Acquisition).

We are especially grateful to Hay Systems, Inc., for conducting the planning workshop and producing the entire manuscript and to the National Security Industrial Association for introducing the completed volume as part of their proceedings during the 11th Annual Conference of the Manpower and Training Committee on "Integrating People, Machines and Organizations: The Winning Approach to Systems Acquisition."

ACKNOWLEDGMENTS

This book is the result of the dedication of many individuals from widely varying backgrounds contributing a broad range of roles, from technical expertise, to fiscal and moral support, to editorial and secretarial perseverance. Deeply appreciated are the efforts of each person who assisted in the project and who in one way or another provided stimulation, encouragement, expertise, and selfless labor needed to conceive and execute an extraordinarily complicated project and unusually rapid publication schedule. From the day in August 1988, when Dr. Joyce Shields originated the idea to the date when Dr. William Rouse and Dr. Kenneth Boff agreed to share their experience and expertise only a few months elapsed. Shortly thereafter in December 1988, a small group workshop met to plan the book and by April 1989 manuscripts started to arrive for review. In less than one year every manuscript was written, reviewed, made camera ready, printed and distributed in time for the National Security Industrial Association Proceedings. The editor wishes to thank each of the chapter authors and members of the advisory committee for their tireless dedication to meeting the tight publication schedule. Special recognition is given to Joyce, Bill, and Ken whose expertise, unique insights, and constant encouragement guided the entire process and saw it to completion.

In addition, there were important contributions from numerous other participants. The help provided at the workshop by Dr. Oscar Grusky, Dr. Lawrence Hanser, Dr. Edgar Johnson, Mr. John Miles, Dr. Charles Overbey, CPT Timothy Schroth, Dr. Paul Thurmond, Dr. Richard Vestewig, Dr. Richard Christ, and Ms. Nancy Holup was very beneficial. The consensus of who was most valuable to this project being completed "on time" and "on schedule" is without question due to Gilda Weisskopf of Hay Systems, Inc. She indefatigably, most pleasantly, successfully "integrated" all editorial, administrative, and secretarial activities. There were several others whose assistance was exemplary. Robin Walthour of Hay Systems, Inc., appeared at the eleventh hour and placed the entire manuscript into camera ready format. The guidance and assistance of the Van Nostrand Reinhold staff was critical in meeting the schedule without sacrificing quality. Marjorie Spencer and Joy Aquilino were particularly helpful. Within the Directorate of MANPRINT, the assistance of Barbara Frank is especially appreciated. She coordinated all the governmental administration of the project. Also, Daniel Pieloch did an incredible amount of daily chores ranging from library research to typing and hand carrying manuscripts. There were also those who had to shoulder the bulk of the directorate work while the rest of us concentrated on the project. Our gratitude to Margaret Simmons, Nelson Laughton, Marjorie Zelko, and Harry Chipman for cheerfully carrying the added burdens.

The environment provided by the Deputy Chief of Staff for Personnel, LTG Allen K. Ono, and the Assistant Deputy Chief of Staff for Personnel, MG Larry D. Budge, for the undertaking of a project of this nature is greatly appreciated. Gratitude is also expressed to those of the Office of the Deputy Chief of Staff for Personnel who have to fight the daily fires resulting from past oversights in attending to the needs of the user, and who have encouraged the project; especially associates Mr. Gary Purdum, Director of Manpower, Mr. Raymond Sumser, Director of Civilian Personnel, BG Theodore Stroup, Director for Military Personnel Management, and MG Robert Ord, III, Commanding General, U.S. Total Army Personnel Command. COL David Fabian as the executive officer has been very understanding of shifting workload priorities as a result of effort put to this project.

The vital encouragement and support given by many in meeting the challenge of institutionalizing MANPRINT is deeply appreciated. These include GEN Maxwell Thurman, Mr. Delbert Spurlock, and LTG Robert Elton (USA, Ret.), who created the MANPRINT senior executive position. Dr. Jay Sculley, GEN Arthur Brown (USA, Ret.), and LTG Donald Pihl, who solidified a materiel acquisition environment that focused on the solider; LTG Donald Jones and Dr. Earl Alluisi, who expanded the concepts to the Office of the Secretary of Defense; BG Jack Pellicci, as the first Deputy Director of MANPRINT; Dr. Robin Keesee and Dr. John Weisz, as the providers of human performance expertise; Mr. Joseph Cribbins, for his logistics expertise and personal encouragement; Mr. George Singley, for providing leadership in future technology; MG Donald Eckelbarger, for his concerns for soldier safety; MG Stephen Woods, for raising the voice of the user; and MG Ronald Andreson, for managing the leading success story.

Acknowledged also are those who provided additional vision, leadership, and support as the program started to blossom. Foremost have been GEN Richard Thompson (USA, Ret.), GEN Louis Wagner, Jr. (USA, Ret.), LTG Benjamin Register, Jr. (USA, Ret.), Mr. Walter Hollis, Mr. George Dausman, Mr. Richard Vitali, MG Richard Beltson, MG Wilson Shoffner, MG James Drummond (USA, Ret.), Mr. Hunter Woodall, Jr., and MG David Maddox.

Finally, this project incurred considerable sacrifice on the part of the families of the many contributors. The Army recognizes the importance of the family to any successful mission. A special heartfelt thanks to those many families, and to Anne for her patience and understanding. This book is dedicated to the soldiers and civilians of tomorrow, who like Catherine, Alice, Susanna, John, and Emily will inherit the technological decisions of today.

CONTRIBUTORS

William O. Blackwood is a principal staff member with Advanced Technology Incorporated. He left the Army in 1989 with over 20 years of military service. He was one of the principal architects of the Army's MANPRINT program. Currently, he is the program manager for a United States Coast Guard project on the impact of technology on the work force.

Robert E. Blanchard is Chief Scientist of BehaviorMetrics, a research and consulting firm located in San Diego, California. He is a fellow of the Human Factors Society and an Adjunct Professor of Industrial and Organizational Psychology at the California School of Professional Psychology, San Diego.

Kenneth R. Boff is Director of Design Technology at the Human Engineering Division of the Armstrong Aerospace Medical Research Laboratory (AAMRL) at Wright-Patterson Air Force Base. He is Editor of the two-volume *Handbook of Perception and Performance* (Wiley, 1986) and the four-volume *Human Engineering Data Compendium* (AAMRL, 1988) for display and control designers.

Harold R. Booher is Director for MANPRINT in the Office of the Deputy Chief for Personnel, U.S. Department of the Army. Author of over 70 publications and internal presentations, he has written on topics in addition to MANPRINT covering nuclear power safety regulations, transportation safety research, military job performance aiding, and the history and philosophy of science. He is Chairman of the Human Factors Society Public Interest Committee.

Dennis D. Collins is a senior operations research analyst with the MANPRINT Directorate. He directs the development of the MANPRINT Combat Model Program and the development of an overall approach to MANPRINT Analysis. He was study team leader for the Close Combat-Light Mission Area Analysis published in 1982 and has developed analyses for numerous military systems.

Keith M. Fender is a program analyst with the Department of Defense. Originally a member of the Army Headquarters staff that was instrumental in establishing MANPRINT in 1986, he is currently involved in examining major defense programs and their integration processes.

Paul M. Haas is President of Concord Associates, Inc., of Knoxville, Tennessee, a consulting firm offering services in human factors, reliability, and systems safety. Previously, he initiated the human factors research and development program at Oak Ridge National Laboratory and led the development of a Department of Energy national reliability data center for advanced nuclear reactors.

Sandra G. Hart is the Chief of the Rotorcraft Human Factors Branch in the Aerospace Human Factors Research Division at the National Aeronautics and Space Administration, Ames Research Center. With long-term research interest in pilot workload, she has recently directed a research program in human factors for the design and operation of current and advanced civil and military helicopters.

Robert T. Hennessy is Vice President of Monterey Technologies, Inc., a human factors research and consulting firm in Carmel, California. He is a fellow of the American Psychological Association and the American Association for the Advancement of Science.

Glen M. Hewitt is Chief of Research and Analysis with the Department of the Army's MANPRINT Directorate (Office of the Deputy Chief of Staff for Personnel). He is a Certified Systems Professional and recently completed an Arroyo Center Research Fellowship at the RAND Corporation.

Mark A. Hofmann is the Associate Director of the Human Engineering Laboratory which is a laboratory under the United States Army Laboratory Command. He holds memberships in a number of professional organizations and serves on a number of national and international committees.

Kenneth M. Johnson, Hay Systems, Inc., is a senior instructor at the MANPRINT Staff Officer and Senior Training Courses and program manager for the Air Force's Integrated Manpower, Personnel, and Comprehensive Training and Safety (IMPACTS) training development program. He has over 20 years military and civilian experience in the areas of manpower, personnel, training, and safety.

Rudy Laine is currently Chief of the Policy Branch for the United States Army Manpower and Personnel Integration (MANPRINT) Directorate at Headquarters, Department of the Army. He has been deeply involved in the development of MANPRINT associated data bases at the Department of Defense Training Performance and Data Center and the United States Army Materiel Readiness Support Center.

Joe W. McDaniel is a scientist/engineer at the Armstrong Aerospace Medical

Research Laboratory. He performs ergonomics research and develops computer tools to aid crew system designers, including COMBIMAN (Computerized Biomechanical Man-model), a 3-D computer-aided design model of a pilot, and CREW CHIEF, a model of a maintenance technician.

Elliott Mittler is Assistant Professor of Systems Management at the University of Southern California. He specializes in the management of complex situations including project management, management of research and development, and the public policy process dealing with natural hazards. He is author of *Natural Hazards Policy Setting*.

Frederick A. Muckler is an independent consultant working on training requirements estimation, medical product labeling, work restructuring, and MANPRINT enhancement. An Adjunct Research Professor at the Center for Effective Organizations (CEO), San Diego State University, he has taught human factors and related courses at the University of Southern California; University of California at Los Angeles; California State University, Northridge; and California State University, Los Angeles.

Judy A. Oneal is President of Oneal Brooks Associates, a support contractor to the Department of Defense and industry. Her work in past 20 years has focused on the development and evaluation of personnel and training systems.

Sam Parry is an Associate Professor of Operations Research at the Naval Postgraduate School. His primary teaching and research interests are in the development of air-land combat models for the Army.

Harold E. (Smoke) Price is a Vice President of Essex Corporation and Director of the Human Factors and Training Group. He has been working in human factors and systems engineering for more than 30 years and is a fellow of the Human Factors Society and a member of the Institute of Electrical and Electronics Engineers (IEEE).

Robert N. Riviello is Program Director for MANPRINT Training and Testing for Hay Systems, Inc. A consultant to numerous Department of Defense agencies and industries building military equipment, his research and analysis have encompassed all aspects of the weapon system acquisition program from a human performance and resource requirement perspective.

William B. Rouse is Chairman and Chief Scientist at Search Technology, Inc., a firm specializing in contract research and development and engineering services in decision support and training systems. His professional interests are in human decision making and problem solving, human-computer interaction, and design of information systems. He has published several books and a large number of technical articles in these areas.

Sally A. Seven, a human factors consultant with a military command and information systems background, is currently working on work restructuring, operational workload measurement and prediction, and early training systems requirements estimation. She has been a member of the faculty at Harvey Mudd College, Scripps College, and Claremont Men's College.

John B. Shafer is a Human Factors Engineer in the IBM Systems Integration Division and has been contributing to major weapon systems development for over 30 years. He is a recipient of an IBM Outstanding Achievement Award for implementing a corporate wide usability emphasis. He also has served four terms as Mayor of the Village of Owego.

Joyce L. Shields is President of Hay Systems, Inc., specializing in forecasting how to best integrate people, technology, and organizations. She has directed and conducted research across a broad range of government and industrial sectors with a focus on human resource planning and development. She is a fellow of the American Psychological Association, a member of the Army Science Board, and on the editorial board of the *Human Performance Journal*.

Harold P. Van Cott is Senior Staff Officer and Study Director for the Committee on Human Factors of the National Academy of Sciences/National Research Council. His principal interest is the integration of principles and data about human performance into the system design process.

Sally J. Van Nostrand is Chief of Concepts Analysis and Planning, U.S. Army Laboratory Command. She is responsible for wargaming and other analyses of future Army high technologies. She is a Certified Systems Professional and member of the Board of Directors, Military Operations Research Society.

Charles A. Vehlow, employed by McDonnell Douglas Helicopter Company, is the Director of Supportability for the Light Helicopter Experimental (LHX) SuperTeam located in Mesa, Arizona. Having served on active duty in the United States Army as a helicopter and fixed wing pilot, he has also worked to determine aviation requirements as a member of the United States Army Reserve. He has written numerous articles on the man-machine interface requirements of emerging weapons systems.

Christopher D. Wickens is a Professor of Psychology and Head of the Aviation Research Laboratory at the University of Illinois. His research interests are in attention, visual displays and models of pilot performance. He is the author of a textbook in *Engineering Psychology* and is a fellow of the Human Factors Society.

MANPRINT

AN APPROACH TO SYSTEMS INTEGRATION

CHAPTER 1

INTRODUCTION: THE MANPRINT PHILOSOPHY

Harold R. Booher

BACKGROUND

On March 28, 1979, the central control room operators were alerted to a loss of feedwater to the pressurized water reactor (PWR) of the Three Mile Island Unit 2. The safety injection pumps came on automatically to pump in auxiliary water to the reactor vessel. The indicators for the pressurized vessel, however, showed a dangerous "water-solid" condition for the pressurized vessel. In an attempt to reduce the pressurizer water level, the operators turned off the safety injection pump. Consequently, the reactor core water cover became depleted, leading to a near meltdown. As a result, the nuclear industry radically changed its management and organizational approaches to nuclear safety, but has never fully recovered from the effects of the accident or alleviated the public concerns with nuclear energy.

In February 1987, Northwest Flight 255 crashed on take off at Detroit, killing 156 people. The cause was attributed to pilot failure to recognize flaps in the wrong position for take off. The backups of automatic computer indication and copilot checkoff of critical flap position had failed as well.

In the Persian Gulf on July 3, 1988, the U.S. Navy crew of the Vincinnes shot down an Iranian commercial jetliner killing all 290 civilians on board.

These are only a few of the more dramatic examples of the effects on society from failures in technology at the interface of people and machines. These highly visible incidents are international. Remote areas like Bhopal, India, and Chernobyl, USSR, are now part of household vocabularies because of the tragedies there. The Challenger showed us even space is not too remote, and although safety is constantly improving in some technologies, it is still the leading cause of death in others. More U.S. people die in automobile accidents every year than were killed in the entire Vietnam war.

The loss of lives is not the only cost. The waste in productivity every year is astronomical. The Three Mile Island accident, where no lives were lost, cost General Public Utilities Nuclear over $1.3 billion in radioactive waste cleanup and borrowed electrical energy expense. The costs every year to the American industry from failure, rework, and waste resulting from substandard manufacturing has been estimated at over $600 billion[1] – enough to totally eliminate the national debt in less than five years.

Although technology is constantly improving, the number of catastrophic incidents can be expected to continually rise if for no other reason than the opportunities for both human and machine failures and the consequences of human and/or machine failure continue to rise with rapidly developing technologies involving greater and greater operational complexity.

Solutions to this modern societal problem of unnecessary losses in lives, productivity, and quality of life are extremely complex. People are both the cause and the solution. People are the benefactors and the victims. Through human error in design, operation, or repair of machines, others are hurt, killed, made unhappy or, at the least, inconvenienced. On the other hand, it is through human intelligence and unique human skills that equipment, organizations, and knowledge enhancing products are designed and operated effectively, efficiently, and safely. The quality of any product produced by any industrial organization depends ultimately on the interplay of several primary factors, all under the control of people themselves.

Focus on Quality

It is a fundamental belief of the authors of this book that it is possible to achieve both (a) dramatic waste and victim reductions on the debit side of society's ledger, and (b) dramatic increases in system performance and productivity on the credit side. One word which pervades such a belief is "quality." Initially providing and then continually improving the quality of products for human use requires looking at all aspects of a higher goal. That goal must be to maximize human benefits while minimizing waste and harm on as global a scale as possible. Another fundamental belief is that positive changes can be achieved by any organization independent of the forces from outside the organization. For example, greater total fiscal resources need not be invested to see net benefits. In general, however, the larger and hierarchically higher the organization is defined, the greater the potential for immediate, dramatic, and lasting improvements in quality.

The secret behind achieving any of the promises of increased productivity and increased corporate profits, while at the same time reducing the complexity and risk to the consumer, lies in recognizing that our "equipment" oriented culture needs to change to one that is "people" oriented. While the overall benefits would be greatest if the entire country's culture should change, any independent organization can change dramatically on its own. Any organization which wishes to achieve these promises can do so; however, only if the top decision makers for the organization decide that a "people" orientation is desired. If top management does decide on such an orientation, it is critical they be encouraged to invest in a system integration approach which will (1) take advantage of existing systems integration expertise, decision aids, and technology, and/or (2) develop the necessary systems integration expertise, decision aids, and technology. A system integration approach developed by the Army provides all the ingredients for the proper change in orientation.

Army MANPRINT Program

The Army MANPRINT program is currently the most ambitious attempt in the U.S. to implement major portions of the new integration philosophy described in this book. For the Army, MANPRINT is a management and technical program designed to improve its weapon systems and units performance. Its leaders have adopted the idea that it is necessary to change the focus of equipment developers away from "equipment-only" toward a "total system" view – one that considers soldier performance and equipment reliability together as a system. The program is extremely broad and includes all Army management, technical processes, products, and related information covering the domains (see Table 1-1) of manpower, personnel, training, human factors engineering, system safety, and health hazards.

The most unique aspect of the program is effective integration of human factors into the mainstream of materiel development, acquisition, and fielding. Organizationally, these functions are spread throughout the Army, with major roles being performed by Army Materiel Command, Training and Doctrine Command, Office of the Surgeon General, Army Safety Center, Army Research Institute, and Human Engineering Laboratory. Responsibility for integrating these varied human factors functions into the materiel acquisition process lies with the Deputy Chief of Staff for Personnel on the Department of Army staff. The policy that lays out the formal definition and various roles and responsibilities is presented in Army Regulation 602-2, Manpower and Personnel Integration (MANPRINT) in the Materiel Acquisition Process (May 1990).

Philosophy vs. Program

For purposes of this book, which addresses the MANPRINT philosophy rather than details of the Army MANPRINT program, it may be helpful to point out some features which distinguish "The Program" from "The Philosophy." The MANPRINT program applies to the Army only. It is rapidly becoming part of the institution's way of doing business. Consequently, its policy documentation has already changed. Decision makers' attitudes have

Table 1-1
MANPRINT Domains

MANPOWER

The number of human resources, both men and women, military and civilian, required and available to operate and maintain Army systems.

PERSONNEL

The aptitudes, experience, and other human characteristics necessary to achieve optimal system performance.

TRAINING

The requisite knowledge, skills, and abilities needed by the available personnel to operate and maintain systems under operational conditions.

HUMAN FACTORS ENGINEERING

The comprehensive integration of human characteristics into system definition, design, development, and evaluation to optimize the performance of human-machine combinations.

SYSTEM SAFETY

The inherent ability of the system to be used, operated, and maintained without accidental injury to personnel.

HEALTH HAZARDS

Inherent conditions in the operation or use of a system (e.g., shock, recoil, vibration, toxic fumes, radiation, noise) that can cause death, injury, illness, disability, or reduce job performance of personnel.

changed. Specific systems are being affected in a specific way. A high degree of routine process is in effect.

The MANPRINT philosophy, on the other hand, applies to the rest of Department of Defense and beyond – both nationally and even internationally. It promotes a radical change in an institution's way of doing business. It is visionary, operating with ideas and concepts in various stages of maturity. It is evolving, not static or routine. The Army, of course, started with the philosophy, made several radical changes, and now fosters an ongoing program. However, the Army also looks to the MANPRINT philosophy for still more radical changes in the future, particularly as it affects the fundamentals of the technology base in its future research and technology investments.

SYSTEMS INTEGRATION PHILOSOPHY

The purpose of the book is to describe several of the advanced technical and managerial concepts utilized by MANPRINT as a systems integration philosophy. The description includes the state of the art of various concepts and their relevance to the complex relationships among (1) people as designers, users, and repairers of machines, (2) government and industrial organizations which regulate, design, manufacturer, and/or operate machines, and (3) methods and processes for design, production, and operation of equipment. It also addresses specific multidisciplinary challenges for the researchers, educators, and practitioners resulting from MANPRINT.

Recognition of the status and relevance of these advances, both technically and managerially, is considered critical to how successful institutions will be in attempts to make radical changes from an industrial culture that is oriented toward technology of machine to one that is oriented toward technology for people who use machines.

Attempts to make wide, sweeping changes in government organizations that either design and operate or influence the design and operation of machines are not new. Major human factors programs were initiated by the Air Force, Navy, and Department of Transportation as early as the mid-1960s (Air Force, 1967; Fucigna, 1968; Little, 1966). More recently, the Nuclear Regulatory Commission has recognized the need for a comprehensive human factors program (Hopkins et al., 1982; Moray & Huey, 1988).

Without exception, however (although numerous specific examples of positive human factors influence can be cited), it is fair to conclude that all past attempts to incorporate human factors as a primary consideration in government policy for the procurement or regulation of the nation's technology have been marginal at best. Human factors continue to be viewed as a contributor to or supporter of design and operations, but it is not on equal footing with engineering or operations disciplines.

Unique Aspects

Table 1-2 lists unique aspects of the MANPRINT philosophy which illustrates why organizations which adopt a people-oriented concept in their systems integration approaches to product design and manufacture have a much better chance for success in the 1990s. To begin, the benefits of a common focus of management is provided through the power and authority

Table 1-2
MANPRINT Philosophy Unique Aspects

- High level visibility of people-oriented concepts
- Focus throughout total organization on competence and motivation
- Top-down approach rather than bottom-up
- Multidisciplinary views of design
- Quantification of people variables
- Systematizes early warning of human error consequences
- Provides trade-off techniques early in design
- Pushes technology and aids engineering advances
- Inherent part of system – not just supporting role
- Communicates in decision maker's language
- Encourages resources redirection rather than net increases
- Educates all people in the process
- Reduces demand for manpower, personnel, and training

that a top-down approach allows. By giving high level visibility to people-oriented concepts, the desired wide-sweeping changes have a realistic environment in which to grow. Understanding the concepts at the very top of an organization can bring focus on people throughout and set up a reward system which instills competence and motivation in its employees. The chapters in Part I, *Organization/Management Context*, provide considerable detail for readers to evaluate how well their organizations understand and support certain crucial concepts.

For example, the strengths and weaknesses of various disciplines are recognized. It may be wise to let the human factors engineer take the lead for integrating several other disciplines into any specific system design. But MANPRINT would immediately remind the systems integrator of military systems, for example, that it is the enormous unnecessary manpower, personnel, and training costs resulting from poor design that influence the minds of government decision makers, all the way up to Congress.

In MANPRINT, decision makers and facilitators take advantage of technological developments in system integration. Inherent in several of these advances is quantification of people variables. This is important because it allows system trade-offs to be made with people variables on the same footing as product variables. These newer methods also allow better decisions to be made early in the design and development process where changes are relatively inexpensive to make.

MANPRINT subscribes to the idea that investment in the front end on human factors will provide paybacks tenfold in the long term. But it goes beyond this to promise more immediate benefits as well. In the past, one of the problems of long-term payback is that the long-term rewards for the front-end decision makers do not accrue for them personally. MANPRINT looks for more immediate productivity and cost avoidance measures first and, then as a bonus, points out the long-term advantages.

MANPRINT forces product technology to become more innovative. A company who recently adopted MANPRINT into its military helicopter engine design found out that it had also produced an engine more competitive for commercial purposes. It did so at *no added cost* over its original design plan. Routinely it is being found that MANPRINT provides the needed incentive to make the product not only user-friendly, but more reliable and cheaper to produce.

A fundamental concept of MANPRINT is that people are considered part of any system being developed. Chapter 10 by Shields, Johnson, and Riviello on *The Acquisition Decision Process* describes this aspect of MANPRINT in step-by-step detail.

The other chapters on *System Integration Methodologies* in Part III all recognize how crucial it is to describe issues and recommended actions for decision makers in their own language. Generally, these are in terms of mission, resource, product, and/or process information.

Once introduced into an organization, a challenge for MANPRINT is to remain viable during periods of budget restraint. It is easier to introduce new ideas when resources are increasing. But ideas considered as frills during good times are vulnerable to extinction in bad times. MANPRINT, therefore, encourages resource redirection rather than looking for a portion of a total net resource increase. Funds allocated to MANPRINT and the disciplines it integrates rise and fall proportionately with the health and ills of the organization. In bad times, MANPRINT may, however, be one of the few areas where increased investment is warranted. Where else can it be logically expected to really produce more with less.

Education is absolutely essential to MANPRINT. All people involved in the process from the top to the bottom must understand the concepts. Specialists are needed in certain areas, but to be successful MANPRINT cannot rely solely on human factors specialists. Muckler and Seven in Chapter 18 on *National Education and Training* address this problem and describe the various strategies that are being considered to provide MANPRINT expertise to meet the increasing demand.

Finally, an aspect which will be receiving more and more attention as the availability decreases (and consequently costs rise) for higher skilled manpower, is emphasis on simpler design. Products which can be operated and repaired by fewer people, by lesser skilled people, and/or people with lesser training will be in greater demand. It is not an impractical expectation for the military, and probably in many commercial areas as well, to demand MANPRINT designs which will allow cost reductions in all three areas – manpower, personnel, and training (MPT) – together. Too often in the past, cost reductions in one of these areas has merely been shifted to one or both of the others. For example, when the Army wanted to reduce total number of people, costs went up in recruiting higher skills. If it now tries to reduce cost of recruiting and retaining high quality individuals, it will find increased training will be needed to maintain performance effectiveness. Improvements in equipment design is the most practical way to get major net reductions in MPT costs.

Levels of Sophistication

The MANPRINT approach to systems integration inspires considerably different effects on total organizational culture depending on the level of sophistication employed. The first level, one which has been in existence for at least 30 years in military systems development, is MANPRINT as an "ility." By this is meant human factors concerns are addressed in a supporting role to the design engineering role in a manner similar to reliability, maintainability, etc. There is much that can be done in such a role, but too often, as in the concern for safety in the nuclear industry, human factors effects on design are viewed as a reaction to or at best a prevention of future, similar deficiencies. This is an important aspect to consider if equipment has already been developed and is operational. The Army, for example, is reducing injuries and repair costs and saving soliders' lives each year by addition of seat belts and rollover bars to its standard jeeps. It would have been better in terms of the net cost per vehicle had the seat belts and rollbars been there from the start, but this approach is far better than never making the improvement.

The second level (the present emphasis on MANPRINT in the Army) is to have people and equipment be given equal consideration in the early stages of design. MANPRINT is thought of as emphasizing total system performance, where total system means the soldier is included as part of the system. But for the greatest impact from MANPRINT, systems integration needs to be seen as a way to get more out of technology. This is the future level of sophistication. This concept has already been discussed above, but much more detail is provided in a number of the chapters. Parry, Collins, and Van Nostrand in Chapter 11, *Complex Environment Models in Systems Integration*, for example, show a model for stages of implementation where MANPRINT can play a primary role in the direction technology itself is to take.

PAST:	MANPRINT as an "ility"
PRESENT:	MANPRINT as total system performance
FUTURE:	MANPRINT as a technology driver

Beyond High Technology

The idea of the human user driving technology advances has already caught on in Japan. Yamada (1989) reports that Japanese businesses are well into the process of changing their paradigms of the 1980s with its focus on high technology (primarily electronics) to what they call "human technology" for the 1990s. Ranging from the use of neural networks in "emotional computers" to the "science of comfort" in homes, Yamada defines human technology as "advanced technology that enhances performance for humans in such areas as comfort, enjoyment, and usability." Figure 1-1 illustrates this shift in product and service developmental interest. According to Yamada, the technological forefront is "human oriented improvements" in which "the most advanced sciences and technology are being used" to command the market lead.

MANPRINT PRINCIPLES

Table 1-3 lists eight fundamental MANPRINT principles for organizations which procure products. Each of these is described by illustrating the progress the Army has made in applying these principles.

Top Level Decision Positions

An active seat representing MANPRINT exists on the Army's top-level materiel acquisition decision board. This ensures that critical human performance issues are at least raised and acknowledged before a final decision on design, procurement, testing, or fielding is made.

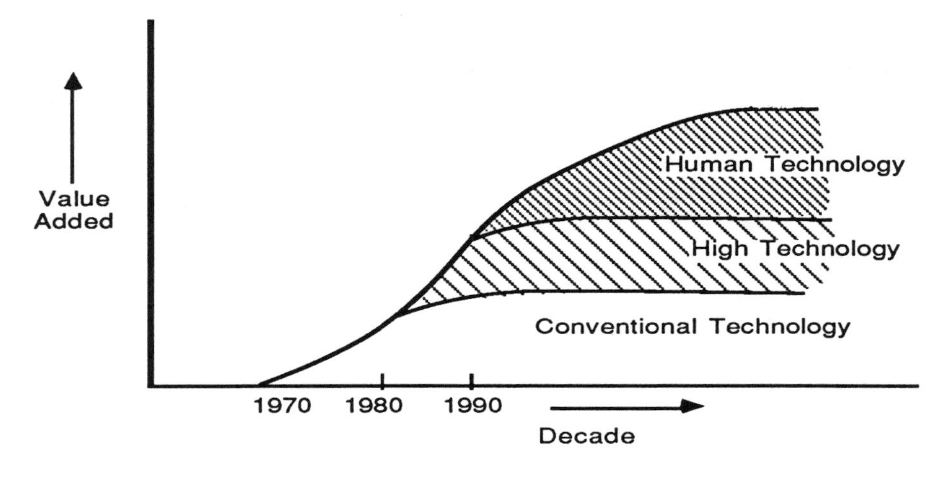

Figure 1-1
Japanese Shifts in Product and Service Development from Improvements in Conventional Technology in the 1970s to Electronic "High" Technology in the 1980s to Human Technology in the 1990s
(Adapted from Yamada, 1989)

Table 1-3
MANPRINT Priniciples for Product Procuring Organizations

1. Knowledgeable person(s) occupy top level positions providing voice, authority, and support.
2. Continuing education and training are provided.
3. User requirements are defined and included in product requirements documents.
4. User requirements applicable to procurement documents are clearly specified.
5. People considerations are provided in primary decision making trade-off models.
6. High visibility of people factors is included in source selection evaluations.
7. Human performance requirements are included in test plans, and tests are conducted with user defined as part of system.
8. Knowledge, skills, and abilities represented by all human factors disciplines are fully utilized.

Continuing Education and Training

A program exists for the continuing education and training of Army civilian and military personnel at all levels in the MANPRINT process. Courses are also open to other government employees and industry. As of March 1990, over 2,750 individuals have been trained in MANPRINT. Significant also is the importance that military leaders have placed on MANPRINT training. In the summer of 1987, General Maxwell R. Thurman as commander of the Training and Doctrine Command (TRADOC) and General (Ret.) Louis C. Wagner, Jr., as commander of Army Materiel Command (AMC) both conducted one-day seminars for their general officers and senior executive civilians. On September 15, 1988, and January 11, 1989, General (Ret.) Arthur E. Brown, while Vice Chief of Staff of the Army, chaired government/industry seminars on MANPRINT. In attendance by special invitation were more than 100 industry vice presidents, Army general officers, and senior executive civilians. New leaders since have continued these seminars, e.g., General John W. Foss (Commander, TRADOC), November 27, 1989, and General William G. T. Tuttle, Jr. (Commander, AMC) March 20, 1990.

User Requirements

Human factors requirements are documented in the Army user requirements process for new or improved equipment. At the earliest possible milestone, MANPRINT acts to ensure that a Target Audience Description (TAD) is produced as part of the System MANPRINT Management Plan (SMMP) and incorporated into the TRADOC Required Operational Capability (ROC) statement. The TAD specifically defines the personnel (in terms of skill level, numbers, anthropometric ranges, and training needs) for whom the system is to be designed.

Procurement Documents

All human factors requirements applicable to procurement documents (e.g., Requests for Proposals [RFPs]) are clearly specified. This area can often be fraught with inconsistencies and is most difficult to monitor especially with nonmajor procurements. Several guides for program managers exist (see Bogner, Kibbe, & Laine, 1990; and Chapter 12 by Booher & Hewitt). Pressure must be continually exerted to improve the quality and quantity of MANPRINT requirements in RFPs.

Trade-off Models

MANPRINT considerations are required in the Army's primary trade-off models for decisions on the development or production of equipment. For MANPRINT to become institutionalized, human performance data will need to be included in underlying combat models and better human performance methodologies for interfacing the combat models with the primary decision-making tools will need to be developed. This is a complex issue that is unlikely to be resolved satisfactorily without further research and development. An increased emphasis is currently being placed by the Army on this problem area as described in Parry, Collins, and Van Nostrand in Chapter 11.

High Visibility

Efforts to assure high visibility of MANPRINT in source selection evaluations is crucial. To the degree human factors can make the difference between whether contracts are won or lost is the single most reliable indicator to industry that the Army is serious about its expressed commitment to human factors. Several major Army programs can be pointed to in which MANPRINT made a visible difference in the source selection outcome. These include the Pedestal-Mounted Stinger (now called AVENGER) missile system; the Line of Sight-Forward Heavy (LOS-F-H), an air defense artillery system; the improved Howitzer Program (HIP); and the new helicopter T-800 Engine.

Testing

Human performance requirements are included in test plans. The Army system being procured can fail the test because soldiers cannot perform adequately just as readily as if the hardware or software fails. The total system with "soldier in the loop" is what is tested for acceptance. Progress here is excellent for those systems actually being tested. For example, improved performance that can be directly attributed to MANPRINT has been recognized in the LOS-F-H, the Airborne Target Hand Over System (ATHS) (an air-to-ground and air-to-air communications systems for target location), and the T-800 Helicopter Engine. One system, the AQUILA, was canceled at least partially because of human factors deficiencies. Many systems are procured, of course, that do not receive formal tests. It is a continually difficult goal to achieve, but user tests should be conducted on all equipment before fielding. A steady increase in this area is a true measure of MANPRINT long-term effectiveness.

Interdisciplinary

Recognize strength in union. Too often the subareas of human factors are forced to compete against one another for limited funds: human factors engineers versus safety engineers; manpower and personnel versus training; human factors versus integrated logistics support. Perhaps more than anything else, MANPRINT has sought to combine the strengths and minimize the weaknesses of the varied people-oriented subdisciplines to provide an integrated front to those who make people/hardware trade-off decisions.

CONCEPTS OVERVIEW

The chapters of the book are loosely categorized within four parts as illustrated in Figure 1-2. This is not meant to be restrictive in any sense, since any one of the chapters could with a different orientation provide valuable insight into each of the other categories. But the book organization does have a certain logic which is meant to aid the reader in understanding why the chapters are placed where they are.

Part I, *Organization/Management Context*, reminds us that except for totally new organizations, any organization in which MANPRINT is to play a role already has a process for producing its product. Most of the decision makers like things the way they are. If a new systems integration philosophy is to be introduced into any organization, major changes will need to take place. In order for that to happen, top management has to understand the concept and has to be motivated to implement the change required. To be motivated, payoffs to the organization leaders must be demonstrated.

What MANPRINT adds with its advanced concepts are facilitators to help the decision makers make the changes needed and to help them run an organization which will in fact produce user-centered products. Facilitation most succinctly comes in the form of systems integration expertise, in-house education and training, and decision aids.

Part II, *User-Centered Design Advances*, recognizes that organizations build things a piece at a time and then combine them in various ways to build more and more complex systems. But each level of design and production can add something to the total process if the most recent advances for integrating user requirements are included. Because of their familiarity with the engineering aspects of the organization, it is often the human factors engineer who will play a leading role in identifying and exercising the advances of Part II.

Part III, *Systems Integration Methodologies*, brings the high cost drivers – manpower, personnel, and training – fully into the systems integration process. These methods which include analytical tools, simulators, and other techniques to help decision makers are essential in demonstrating the

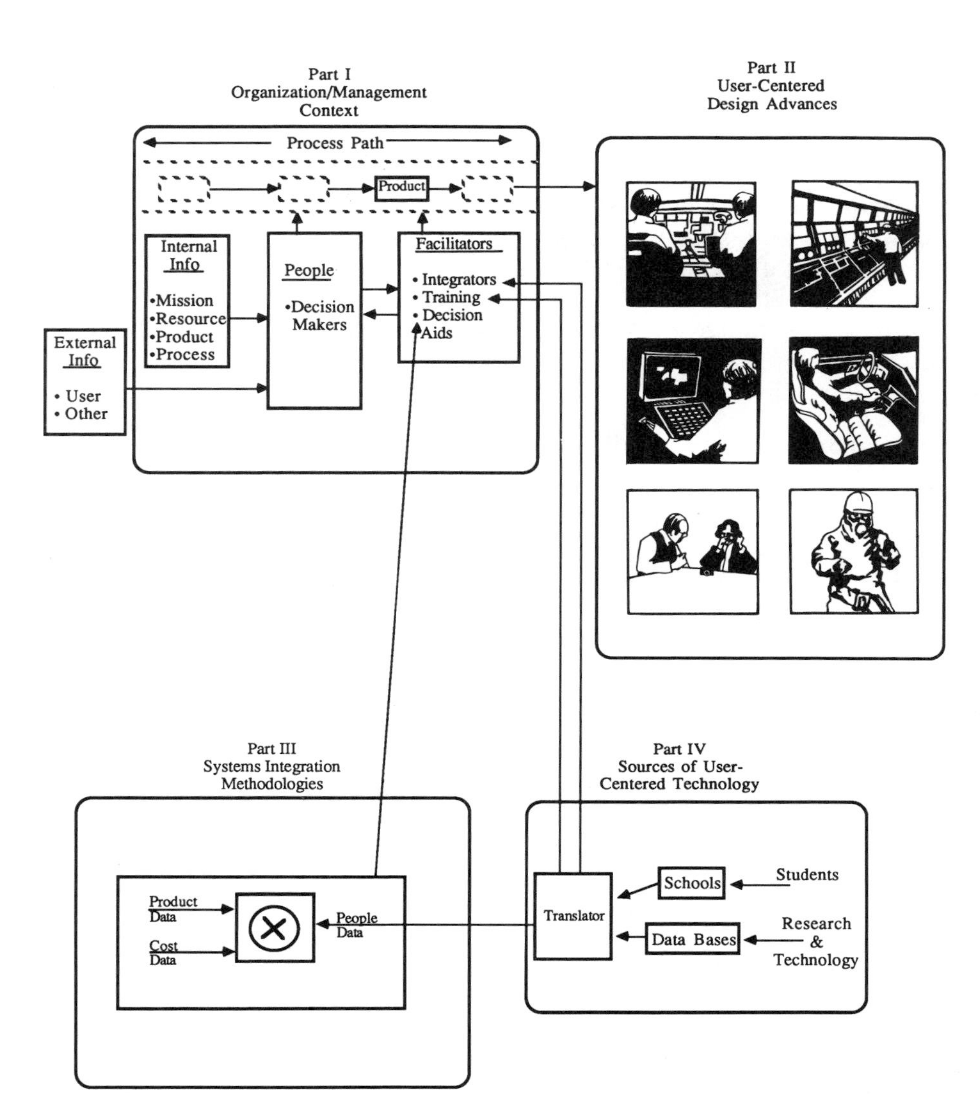

Figure 1-2
Overview of MANPRINT Systems Integration Concepts
(Source for Part II Illustration: Human Factors Society[2])

entire picture of cost effectiveness to the decision makers. These methodologies in conjunction with the other facilitators will allow the decision makers to make "people-equipment" oriented decisions that are consistent with their interests in "mission, resource, product and process."

Finally Part IV, *Sources of User-Centered Technology*, answers many of the fundamental questions asked by those facilitators who take on the responsibilities of integrator, educator, or trainer. For example:

- Where do you find the relevant human performance data?
- Where do you find and how do you evaluate the qualifications of MANPRINT experts?
- What MANPRINT techniques are relevant and available

Taken together or separately, each of the chapters fills an important role in the MANPRINT philosophy. A further more detailed introduction to each of the chapters is provided at the beginning of each part.

NOTES

[1]Quality experts frequently state 20 to 40 percent of payroll costs can be associated with waste, failure, and rework. The $600 billion estimate is based on 20 percent of labor costs per annum, around $3 trillion (1989).

[2]From the *Human Factors: Designing for Human Use*, 1988. Copyright 1988 by The Human Factors Society, Inc., and reproduced/adapted by permission.

REFERENCES

Air Force Systems Command, Headquarters (1967). *Handbook of instructions for aerospace personnel subsystems design* (AFSCM 80-3). Washington, DC: Andrews Air Force Base.

Bogner, M. S., Kibbe, M., & Laine, R. (1990). *Directory of design support methods*. Developed for Department of Defense Human Factors Engineering Technical Group (Designing for the User Subgroup). Washington, DC: Headquarters, Department of the Army, Office of the Deputy Chief of Staff for Personnel, MANPRINT Directorate.

Fucigna, J. (1968). *Human performance development program for Naval Material Command.* Darien, CT: Dunlap and Associates.

Hopkins, C. O., et al. (1982). *Critical human factors issues in nuclear power regulation and recommended comprehensive long range plans* (3 vols.; Technical Report NUREG CR-2833). Washington, DC: Nuclear Regulatory Commission.

Little, A. D. (1966). *The state of the art of traffic safety: A critical review and analysis of the technical information in factors affecting traffic safety.* Washington, DC: Automobile Manufacturers Association.

Moray, N., & Huey, B. (1988). *Human factors research and nuclear safety.* Washington, DC: National Academy Press.

Yamada, S. (1989, June 1). Paradigm shift in product/service development. *Japanese business strategies at a turning point.* New York: NRI Forum International.

PART I

ORGANIZATION/MANAGEMENT CONTEXT

Organizations vary considerably to the degree they have top level managers who a) understand and b) are capable and willing to introduce the MANPRINT philosophy into the organizational process. Chapter 2, *Total Quality Management (TQM) and MANPRINT*, by Booher and Fender describes a new management approach currently being examined by numerous U.S. industries and the Department of Defense. A central theme of TQM is the importance of the voice of the user in product design and manufacture. Feedback from the product user becomes an inherent part of the institution's process. This is illustrated in Figure 1 for Part I by user feedback passing through the decision makers to modulate the process. Organizations which adhere to the TQM process provide a favorable atmosphere for MANPRINT implementation. In this chapter the authors provide clues for evaluating how applicable TQM is to their organization.

In Chapter 3, *Change Management Process*, Blanchard and Blackwood describe the various principles central to carrying out any major institutional change. In large organizations, especially, a drastic change from "business as usual" is generally necessary to truly implement a philosophy as wide-sweeping as MANPRINT. Institutional change can be brought about both from inside forces and outside forces. Change by inside forces can only be introduced by the top decision makers but successful change implementation requires far more than written policy changes. Successful change ultimately requires strong and favorable attitudinal changes throughout the organization. Outside forces such as Government procurement policy or regulatory measures levied on industry or competitive forces within industry can also be stimuli for institutional change. Blanchard and Blackwood introduce a theoretical model for change management which highlights the role of change agents, policy making authority, communication paths and the importance of education, training and individual awards. They also illustrate several change intervention tactics and approaches found useful in applying the MANPRINT philosophy to the Army's Materiel Acquisition Process.

Chapter 4, *Management Integration Methods and Principles,* by Mittler, Hewitt, and Vehlow discusses the difficult problem of managing the integration of various disciplines, processes, and products and the history leading up to modern project management approaches. From this, the authors describe eight principles which play important roles in effective management integration. Finally, after evaluating several Army programs which have recently applied MANPRINT, they describe a data base management system which can act as a total integration methodology for MANPRINT projects in industry.

In Chapter 5, *MANPRINT in a Systems Engineering Organization*, John Shafer describes how IBM made a major institutional change in their systems engineering organization to incorporate many of the tenets of the MANPRINT philosophy. He discusses in some detail the policy changes and the various tools and courses used to train its system engineers.

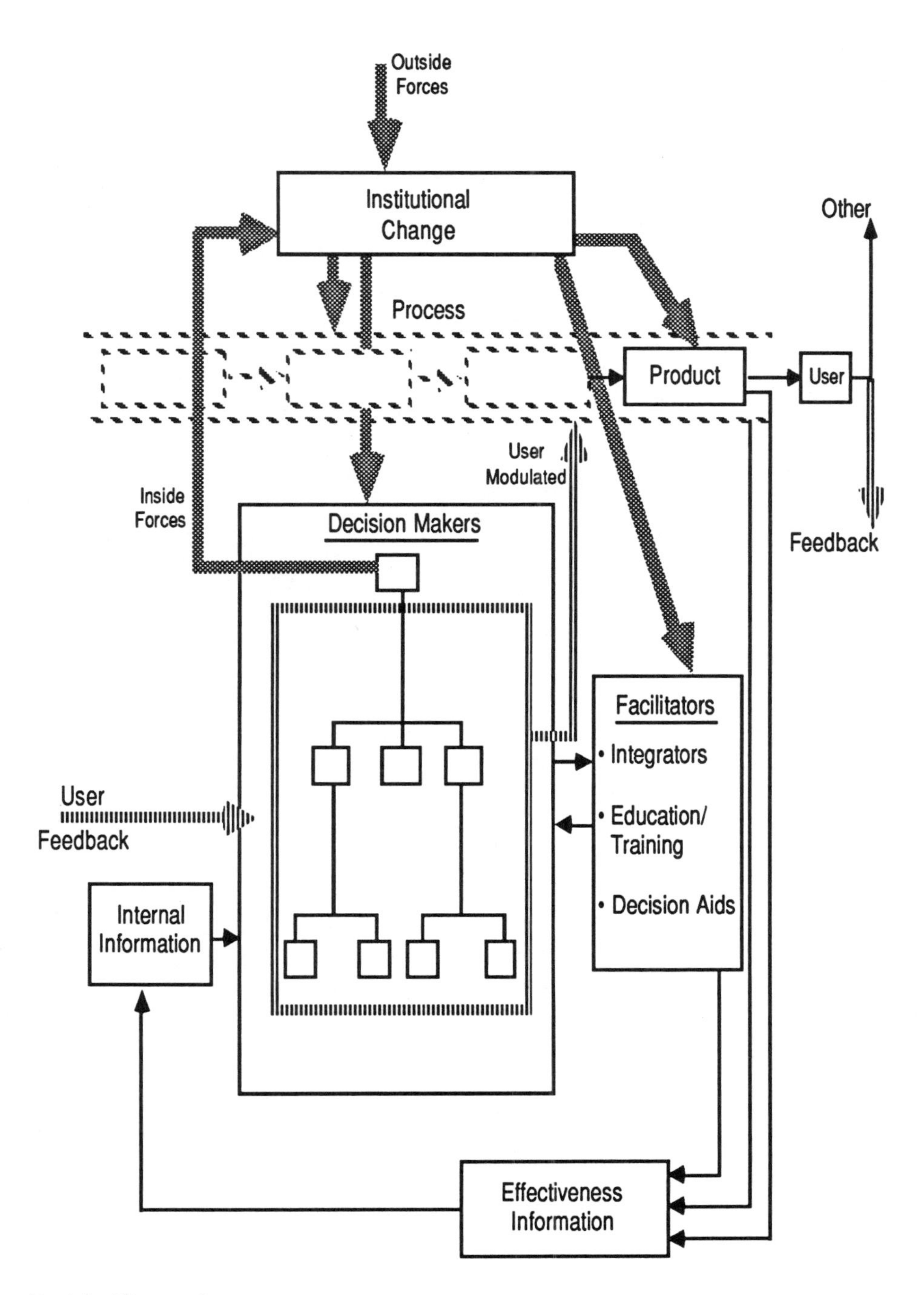

Part I, Figure 1
Organization/Management Context

CHAPTER **2**

TOTAL QUALITY MANAGEMENT AND MANPRINT

Harold R. Booher
Keith Fender

ABSTRACT

General national concerns with America's loss of market share and weakening of the industrial base along with specific Department of Defense concerns about the effectiveness of its weapons procurement process have led to renewed United States interest in methods to improve the quality of its products. American management for both commercial and military products is becoming especially intrigued with the statistical process improvement methodology offered by Total Quality Management (TQM) philosophers such as W. E. Deming. This chapter describes the Total Quality Management approach and points out the similarities and differences to the MANPRINT management approach. The two management philosophies although different in scope are complementary. They have much in common and need similar management expertise. This is brought out, for example, in the language used to describe TQM and MANPRINT principles, e.g., the voice of the user, focus on process, organizational "total" involvement. Implementation of the programs can be quite different, however. Depending on the "quality" of the TQM approach in any particular organization, MANPRINT can take on significantly different roles. Guidelines are given, therefore, to aid the readers in (1) assessing the status of TQM in their organization and (2) taking actions as MANPRINT proponents depending on whether the TQM environment is *mature, open-minded, or closed-minded.*

INTRODUCTION

It has been well recognized beginning in the '70s and progressing through the '80s that America's industrial base has not kept pace with other highly industrial nations, especially Japan. In the late '80s, the United States faced an accelerating balance of trade deficits with most of its trading partners, while the performance and reputation of United States goods and services continued to decrease. Although improving, the predicted deficit for 1989 was still over $100 billion ("Is this the moment," 1989). This has created a favorable attitude in American industry in general, and within the Department of Defense (DoD) and its major contractors in particular, for a reemphasis on approaches which offer significant improvements in productivity. A most promising approach sponsored by the DoD for major systems acquisition is Total Quality Management (TQM). During this same period, DoD has become increasingly aware of the rising operations and support costs, especially the manpower and personnel bill which is over 50 percent of the DoD budget. The affordability of systems already being procured is still yet another concern. To help offset these potentially disastrous fiscal trends, the MANPRINT program in the Army and similar initiatives throughout DoD were spawned. Because of the natural relationship and many common goals of the two management approaches, it is important to better understand how they can work together. The purpose of this chapter, therefore, is to describe the relevance of MANPRINT to both the philosophy and method of TQM. The remainder of the book can perhaps then be better appreciated within an overall "quality" context.

Definitions

Total quality management has been defined as the application of management methods and human resources to control *all processes* with the objective of achieving *continuous improvements* in quality (Defense Systems Management College, 1989). Organizations which adopt this approach emphasize "Total" in front of "Quality Management" because it is unlikely that anything less than full commitment throughout the entire organization can accomplish the direct dramatic increases in productivity. The enormous differences in products resulting from companies in Japan which use the TQM approach and the United States companies which do not is illustrated in Figure 2-1, for example. By making much greater quality effort early in the development stage, the Japanese company is able to nearly eliminate the need for engineering changes (problem solving) well before production.

For the purposes of this chapter, MANPRINT can be defined as the application of management methods to control the *acquisition and design processes* with the objective of enhancing *total* system performance and supportability. As applied in the Army and described in Chapter 1, the scope includes all management, technical processes and products and related information under the six labels of human factors engineering, manpower, personnel, training, system safety, and health hazards. Shields, Johnson and Riviello (Chapter 10) describe the DoD acquisition process in which both the TQM and MANPRINT programs are receiving primary attention. As is discussed there and elsewhere (Booher, 1988), the most unique aspect of the program is effective integration of human factors into the mainstream of materiel development, acquisition, and fielding.

Organization/Management Model

Part I, Figure 1 provides a conceptual model applicable to both TQM and MANPRINT. In order to assume the maximum benefits from either the Total Quality Management or MANPRINT approach, the entire organization must change from its current culture to a new one. The degree of transition difficulty depends on how able and willing top management is to make the *institutional change* and how well the factors of the change process itself are understood and implemented (see Chapter 3, Blanchard & Blackwood). Before the organization can make the cultural change, basic quality concepts must be understood and adopted as the vision for organization's management.

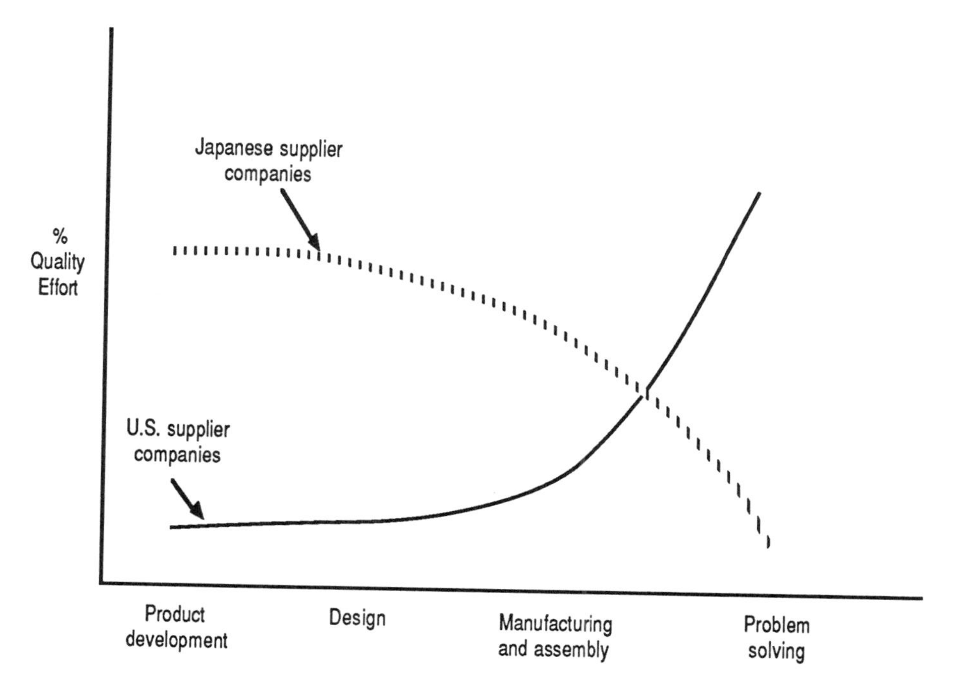

Figure 2-1
U.S. and Japanese Comparison
Quality Effort by Activity
(Source: Sullivan, 1986)[1]

Processes influenced by TQM and MANPRINT are *user modulated.* Whether they are commercial or military customers, the *voice of the user* is the primary consideration in product design. The voice of the user takes on special meaning, however, if it is understood that the user is not necessarily the buyer. For commercial equipment like automobiles, VCRs, radios, refrigerators, the user is often also the buyer. But in most military systems, the buyer is the United States government bureaucracy while the ultimate user is the soldier, sailor or airman. *Quality Function Deployment* (QFD) methods of TQM recognize this distinction. See Figure 2-2 for a recent example on a large government procurement. Similarly, most commercial airplanes are bought by commercial airlines, not the ultimate customer who is the passenger. This situation is further complicated by the pilot also being a user. MANPRINT makes it clear that user means first and foremost the operator or maintainer, i.e., the person who actually uses or repairs the machine. Both the buying organization and the product developer/ manufacturer must strive to "delight" the *real user* if TQM is to be successful.

Recognition of the real users is important because they are the ones who will ultimately recognize and expect quality in their products. Once recognized it is then possible to learn how to listen to the user and even take the next step of helping them better identify and articulate their needs. Scholtes and Hacquebord (1987) conclude you should be aware not only of problems resulting directly from defects in your product but of related problems experienced by customers even when your product is functioning properly. To accomplish this they note effective and creative customer feedback systems are vital. The feedback system should function so well that the quality organization actually leads customers into the future by exploring technological possibilities so distant that a demand is created for products and services not yet imagined. The model (Part I, Figure 1) shows *effectiveness information* provided internal to the organization is needed to predict user acceptance. This may be accomplished through models, simulations, small scale experiments, the use of experts, and/or various analytical techniques. What is important is that the results can translate user needs into both system performance (MANPRINT) and executive language (TQM).

The next section describes in some detail the TQM philosophy and methodology. If the reader is already familiar with TQM, he/she may wish to skip to TQM/MANPRINT Comparison later in this chapter which compares the two management approaches.

TQM: A SHORT COURSE

The Japanese emphasis on quality did not naturally appear simply as an offshoot of their culture. They feel particularly indebted to W. Edward

RELATIONSHIP MATRIX Customer Needs & Desires	IMPORTANCE TO USER						
	Pilot	Maintainer	Unit Commander	Materiel Developer	Other Services/ Foreign Services	Congress	Mean
Warranty	0	1	0	5_2	3	5_2	2.33
Operational Effectiveness	5_3	1	5_1	5_1	5_2	5_1	4.33
Affordable							
International Collaboration for Affordable Program	0	0	2	3	5_1	5_4	2.50
Cost Reduction Through Competition	0	0	0	4	1	5_3	1.67
Cost Reduction Through Product Development	0	0	0	4	3	2	1.50
Great Supportability							
Effective Training System	4	4	5_3	5_9	5_7	2	4.17
Breakthrough Integrated Logistics Support (ILS) System	2	5_3	4	4	4	1	3.33
Simplified Maintenance	3	5_1	5_6	5_5	5_3	1	4.00
Fit Aircraft to Soldier	4	4	5_8	5_{10}	5_{10}	2	4.17
Maximum Availability	3	5_2	5_2	4	4	5_5	4.33
Highly Capable Air Vehicle that Satisfied all Mission Requirements							
Powerful Man-Machine Capability	5_6	2	4	4	4	3	3.67
Highly Survivable Weapon System	5_1	4	5_5	5_3	5_5	4	4.67
High Performance Airframe	5_4	1	4	5_6	5_6	3	3.83
Lightweight Rugged Airframe	4	3	4	5_8	5_9	3	4.00
Highly Effective Mission Equipment Package (MEP)	5_5	4	5_7	5_4	5_8	5_6	4.83
Lethal Effective Armament System	5_2	2	5_4	5_7	5_4	5_7	4.50
Mean	3.12	2.56	3.62	4.56	4.32	3.50	3.61

This diagram is an example of an exercise by decision makers of a large aircraft corporation on a large helicopter design project. The chart is the result of several hundred speculated customer needs and desires. The vertical list of 16 items were the final list of what were considered important to the customers.

Horizontally is each type of user considered. In such a complex procurement, the company recognized it had many users ranging all the way from the pilot and maintainer to Congress. The cells were filled in with numbers ranging from 0 to 5 (least to most important) from the particular user's point of view. For example, the pilot and unit commander do not consider the warranty on aircraft delivery important at all. For each of the users, more than one of the items were considered most important. In order to set priorities on the most important items, the number 5 items were further ranked for each user. For example, the pilots number 1 concern is a highly survivable weapon system. This is indicated by the subscript "1" next to the 5.

Of the six items which were all extremely important to the pilot, powerful man-machine capability was sixth, thus the subscript "6."

When all the users were considered together, the most important item was not a high performance or light weight rugged airframe, but a highly effective mission equipment package. The least important items to the real users were any cost reduction desires.

Figure 2-2
Quality Function Deployment Diagram

Deming. Building on the work of Walter Shewhart, the father of process control at Bell Telephone Laboratories in the thirties, Deming introduced the Japanese to statistics in the early fifties. The importance of statistical process control was not appreciated in the United States, where invented, but was readily accepted in Japan because the Japanese were desperate for change. Deming's biographer, Nancy Mann, explains, "why this miracle happened in Japan lies in the coming together of all the necessary factors for a transformation. Japanese industry, after the war, experienced a 'bottoming out.' W. Edwards Deming came upon the scene and learned to appreciate the Japanese personality and culture. He had an awareness of what needed to be done, and he saw to it that the message was communicated to the people with the ability to take action." She goes on to conclude: "During the war when the Western mentality would have been more conducive to receiving a message explaining how to improve productivity and competitive position, there was no one in top industrial management who had ever thought that statistical quality control could become important. Besides, no one understood at that time the responsibilities of management. The attitude of working for short-term profits and blaming business failures on fiscal policies and the hourly workers was a deeply ingrained habit, particularly in the U.S."[2] (Mann, 1989, p. 25).

The current resurgence of interest in statistical quality control and TQM in the United States is for the same reason as the Japanese in the '50s. Top management recognizes they are in deep trouble (Tribus, 1984). Ford Motor Company, for example, has taken a total new approach to quality consciousness as an integral part of corporate culture primarily because of a traumatic 1.8 billion dollar loss in one year (Strickland, 1988). The costs of waste for businesses who do not manage quality is estimated at anywhere from 20-40 percent of net sales (Crosby, 1979, p. 38; Harrington, 1987, p. 46). The opportunity for enhanced productivity is tremendous if quality management can but reliably identify the specific problem sources. For example, well managed companies have reduced the cost of quality down to 2.5 percent or even lower (Crosby, 1979, p. 39; Harrington, 1987, p. 46).

Deming (1982, 1986) warns that a whole new way of thinking needs to take place in American management if it is to compete with the Japanese. Some of the traditional approaches to fighting the problem (e.g., throwing dollars at it, purchasing new technology gadgetry, blaming the workers, or even trying harder) will not work. American industry and major buyers like the DoD have to find a smarter way to produce better quality products for lower costs. Total Quality Management is believed by many to be the way to awake the sleeping giant and turn it into a leader once again.

Total Quality Management is not free, however. There are considerable front- end investment costs, not in tooling or facilities, but in education and training of employees. There is also the recognized need to stay in business while undergoing transition. Further, Rouse (1989) raises a problem generally overlooked by TQM proponents. Since TQM investment strategy is to develop employee expertise, Rouse asks what is to offset employee demand for increased wages which, if granted, could drive up prices? Moreover, although there are strong arguments to the contrary, TQM Japanese style has numerous cultural limitations for application in the United States. Japan, for example, prohibits business takeovers by law (Tribus, 1984). Also, although limited to the largest companies, a large portion of Japanese workers work all their life for one company and, of course, the country's culture has always fostered an ethic of team work and cooperativeness, whereas the United States is much more individual oriented. These differences, along with Japanese fiscal and banking policies which encourage long term investment, make TQM far more difficult to implement in United States companies. In 1985 the Japanese saved 17 percent of their disposable income compared to the United States 3 percent that year (Hall & Hall, 1987, p. 88). Finally, we cannot be sure that much of the Japanese success is not simply from working harder and longer. The average Japanese student spends 30 percent more time in school, and the average Japanese businessman spends far less time with his family than the United States businessman does (Hall & Hall, 1987). Nevertheless, with all these limitations, TQM does seem to offer great opportunities to bring the United States "Out of the Crisis" (Deming, 1986).

Continuous Quality Improvement

The executive's language is "money," while the workers language is "things." Since quality must be expressed in the language of the executive, the mid-level manager must be "bilingual." Product oriented disciplines (e.g., engineering, logistics support, human factors) have not done well at translating from "things" to the language of the executives. They will need to build on the work of Juran, for example, by elaborating how management for quality is no different than that for finance. The *Juran Trilogy* (Juran, 1987) is useful in recognizing that managing for quality is carried out using the same three processes as managing for finance.

> *Financial Planning* sets out business goals, develops the actions and resources needed to meet goals, translates goals into money, and summarizes them into the financial plan, or budget.
>
> *Financial Control* consists of three steps: evaluate actual performance; compare actual performance to goals; take action on the difference. It takes such forms as cost control or expense control.
>
> *Financial Improvement* aims at doing better than the past. It takes such forms as cost reduction, purchase of new facilities to raise productivity, mergers, acquisitions, etc.[3]

A survey by Juran showed that 80 percent of managers are satisfied with their companies' performance of *Quality Control*, but around half are dissatisfied with their *Quality Planning* and *Quality Improvement* (Juran, 1987, p. 2). His trilogy model (Figure 2-3) illustrates that attending only to Quality Control results routinely in large amounts of chronic waste and from time to time unpredicted sporadic spikes outside the zone of Quality Control. By utilizing Quality Planning and Quality Improvement, however, whole new zones of Quality Control can be realized. Through feedback of lessons learned back to Quality Planning, continual improvement in quality becomes realizable.

Crosby (1979) describes Quality Improvement in terms of *Quality Management Maturity Stages* (Figure 2-4) starting with "uncertainty," where the reported cost of quality is not known, but conservative estimates of United States companies are around 20 percent of sales. Quality Improvement continues through the stages of "awakening," "enlightenment," "wisdom," and finally to "certainty" where reported and actual costs of quality are equal and small. The importance of *long-term focus* on Quality Improvement is shown in the example of Company A vs. Company B (Figure 2-5). Company A may be well ahead at a fixed point in

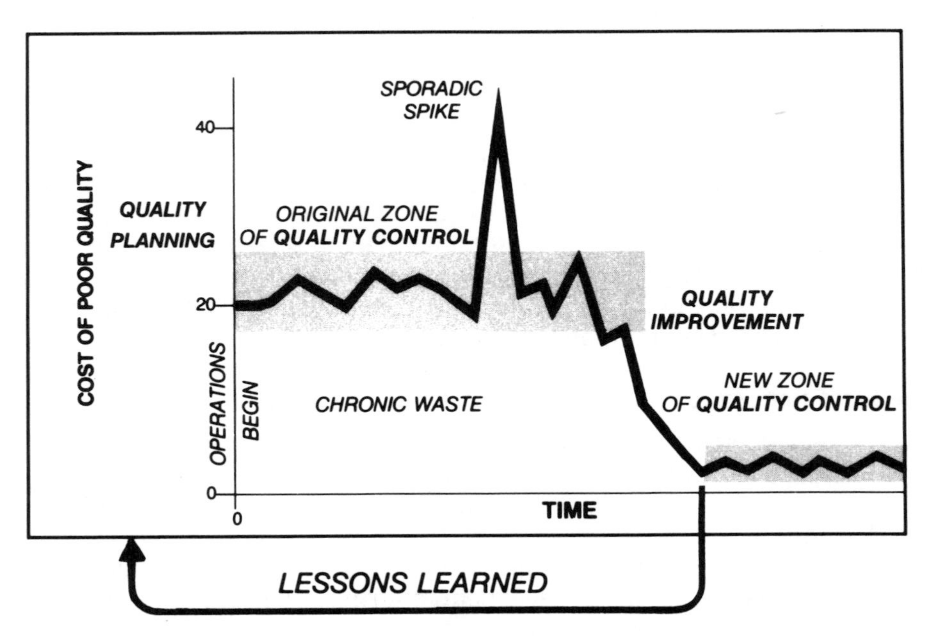

Figure 2-3
The Juran Trilogy for the Quality Process
The Juran Trilogy is a registered trademark of Juran Institute, Inc.
(Source: Juran, 1987)[4]

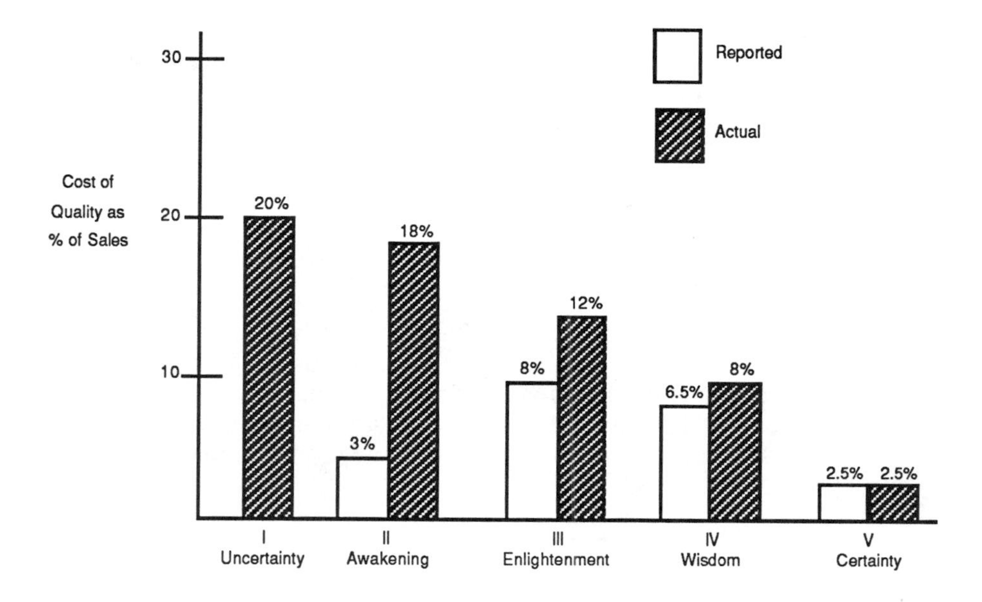

Figure 2-4
Quality Management Maturation Stages
(Adapted from Crosby, 1979, pp. 38-39)

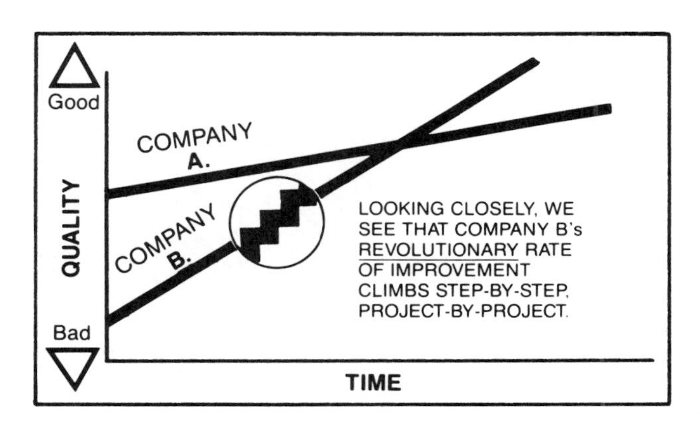

Figure 2-5
Comparison of Company A and Company B Quality Improvement Rates
(Source: Juran, 1987)[5]

time, but if the rate of improvement is less than Company B, Company A will soon be overtaken. It is best to start Quality Improvement by addressing problem areas where the costs are great but fixes are relatively inexpensive. Particularly relevant to MANPRINT is the recognition that the earlier errors are detected in the development process, the less expensive for the company to correct them (Figure 2-6).

Process Variation and Control

The central scientific concept underlying the success of the TQM approach of Deming and others (Feigenbaum, Ishikawa, Taguchi, Tribus, Willoughby) is the recognition that variation is present in all processes. The behavioral sciences have long understood variation in human behavior, but engineering science has tended to treat management approaches to the design process with a similar fixity as experienced in natural chemical or physical processes.

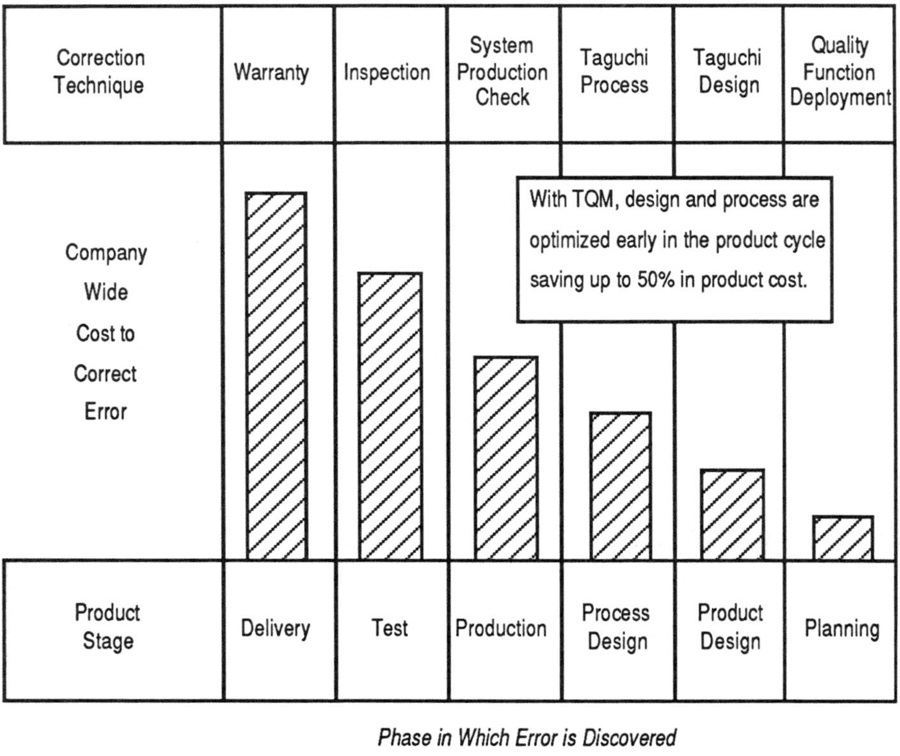

Correction Technique	Warranty	Inspection	System Production Check	Taguchi Process	Taguchi Design	Quality Function Deployment
Company Wide Cost to Correct Error						
Product Stage	Delivery	Test	Production	Process Design	Product Design	Planning

With TQM, design and process are optimized early in the product cycle saving up to 50% in product cost.

Phase in Which Error is Discovered

Figure 2-6
Cost Reduction Potential with TQM
(Source: Stuelpnagel, 1988)[6]

The two approaches to handling variation are:

(1) The *engineering or traditional approach* which has the object of just meeting specifications and tends to result in products with maximum allowable variation (Figure 2-7a).

(2) The *statistical approach* which has the object of process consistency and incidentally results in consistent products.

The problem with the traditional approach is that many products which should work, too often do not (returned by user, or lost customer) and those which should work, might if they had not been scrapped (waste good parts). If attempts are made to tighten the specification without considering the system process, waste simply increases (Figure 2-7b). Concern for Quality

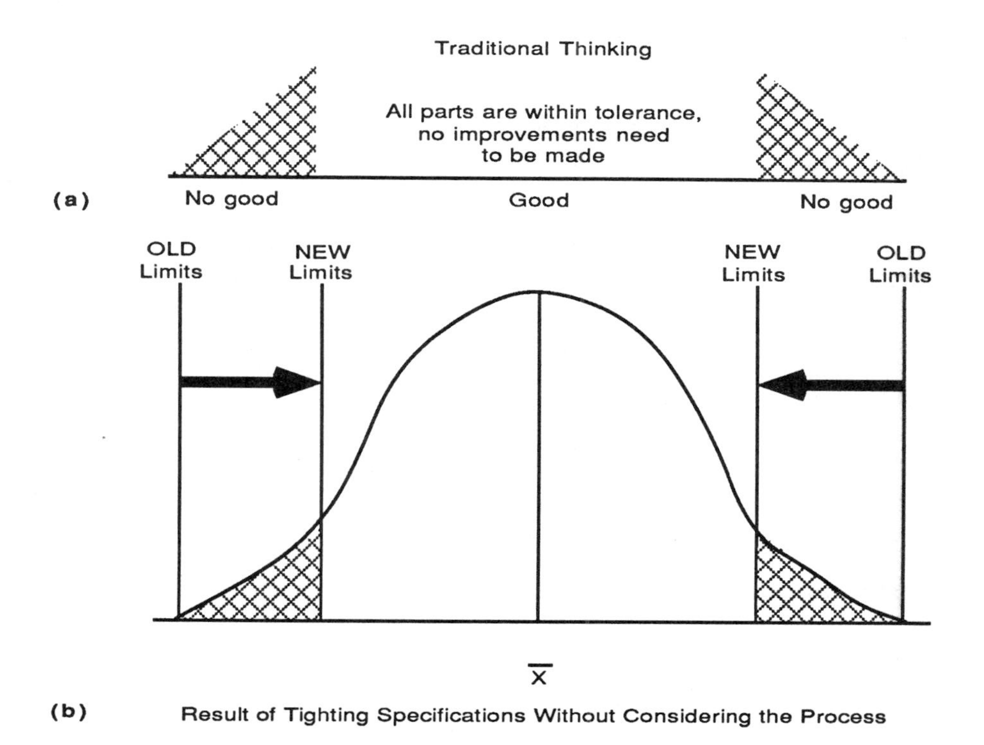

Figure 2-7
Traditional Method of Controlling Variations

in the United States has concentrated on inspection and quality assurance techniques (traditional approach). Sullivan (1986) notes that this accomplishes only about 40 percent of the commitment needed to justify putting "total" into an organization's TQM approach. Deming (1986, p. 327) shows with his "red beads" experiment that inspection has no effect on improving productivity. In fact, using Monte Carlo experiments with a funnel, he concludes any attempt to obtain improvement by adjusting a stable process, results in an output that is worse than if the process had been left alone.

The statistical approach allows the organization to continually improve the process by reducing the variability (Figure 2-8a) and with special mathematical management tools pinpoint where the problems are in the process. Eighty-five percent of the time, the problem is in the process, not the people (Mann, 1989, p.7). Further, problems in the process are often measured against such artificial standards as "user dissatisfaction" rather than internal measures of productivity. The Taguchi method (e.g., Figure 2-8b) allows management to calculate its loss to the customer in relation to the product's deviation from its target value (which is a flawless product). These methods also allow management to fairly easily calculate overall corporate productivity in meaningful terms.

Figure 2-9 illustrates the difference between controlled and uncontrolled variation. Since the object of TQM is to control variation, it is necessary to distinguish between that variation that is relatively easy to identify (*assignable*) and that which is difficult (*random*). Fortunately, Deming teaches, assignable variation is the worst (unexpected and large) but most often economically correctable in most organizational processes. TQM directs its attention, therefore, to assignable variation. For human factors it is important to recognize that this is the area to concentrate effort to systematically eliminate error due to the design process. The fact that random error will still occur in design or can occur in operator performance may best be considered through the concept described by Rouse (Chapter 8). Table 2-1 lists what TQM leaders generally agree is experimentally verifiable in the work place (Deming, 1986; Ishikawa, 1985; Scherkenbach, 1988; Taguchi, 1986).

Scientific Approach

Deming and others stress that the scientific approach is necessary. Quantitative data, observation and experiment must be used in planning, problem solving and decision making. Guesswork, passing the blame and giving people ulcers are poor management techniques for quality management. The best methods for accomplishing work are determined by the careful gathering and analysis of data on that work (Scholtes & Hacquebord, 1987).

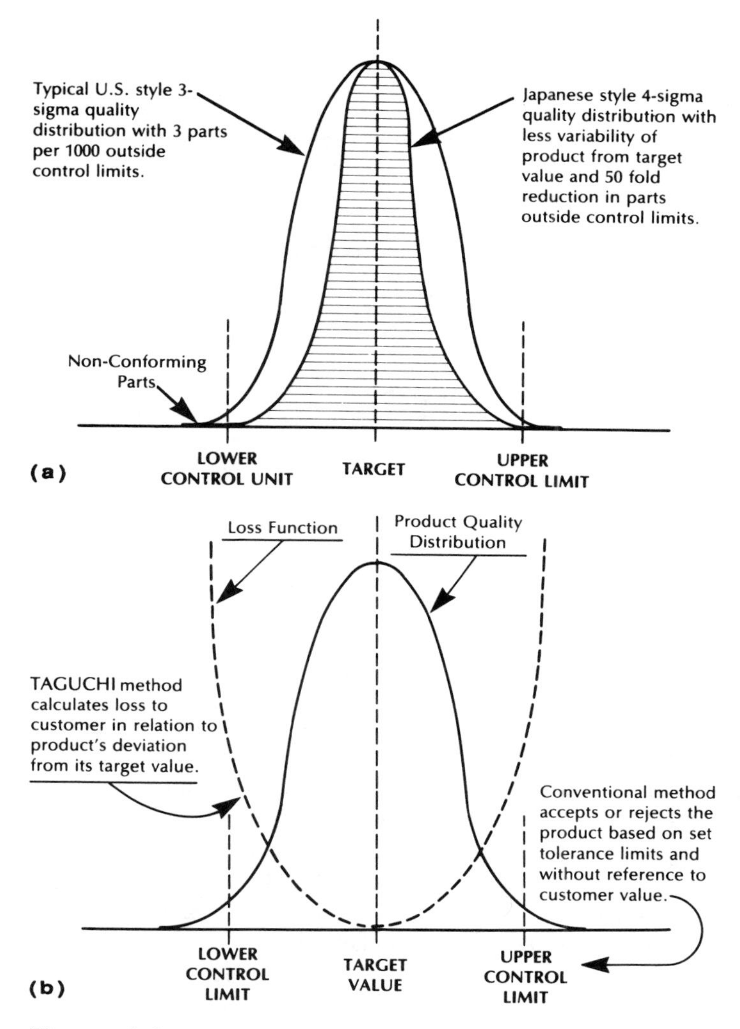

Figure 2-8
U.S. and Japanese Product Quality Distributions:
a. Variability Reduction; b. Quality Loss Function
(Source: Stuelpnagel, 1988)[7]

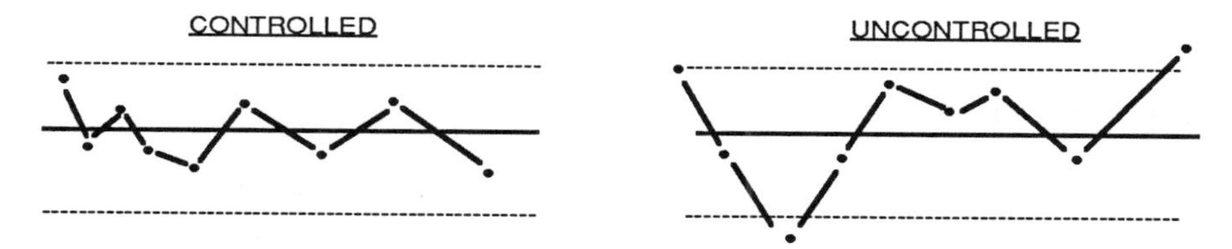

Figure 2-9
Controlled and Uncontrolled Variation

Table 2-1
Fundamentals of Process Variability Control

1. All processes have variability.
2. The control of quality is largely the control of variation.
3. Causes of variability are either chance or assignable (i.e., common or special).
4. Assignable causes can be found and eliminated. This is the responsibility of the local work force.
5. The future can be predicted in terms of past behavior only for processes in a state of statistical control.
6. The only economical way to improve a process in statistical control is to change the organizational system, which is the responsibility of management.

Most of the tools and methods used by TQM are variations on standard statistical data collection and analysis techniques well known to behavioral scientists and specialists in systems integration. Also, many of the methods of display, histograms, pie charts, trend lines, etc., are well known to executives. What appears new to TQM is the blending of these two disciplines so that valid statistical data about the product and the process are presented in common formats readily understandable by managers. Especially useful is the Deming's Plan, Do, Check, Act (P.D.C.A.) cycle (sometimes called the Deming Cycle or the Shewhart Cycle) as a method for continuous improvement. It allows everyone in the company to follow a simple, but useful, common approach to attacking and describing problems. This and other TQM tools and methods that have been helpful in making this translation are illustrated in Figure 2-10. Recommended readings for explanations of how to apply these various techniques are Ishikawa (1986), Mizuno (1988), and Joiner (1986).

Education and Training

TQM places great emphasis in total personnel involvement and speaks of people as its largest and most valuable investment. What this means in MANPRINT philosophy is development of competent and motivated people. The principal way TQM accomplishes this is through continual

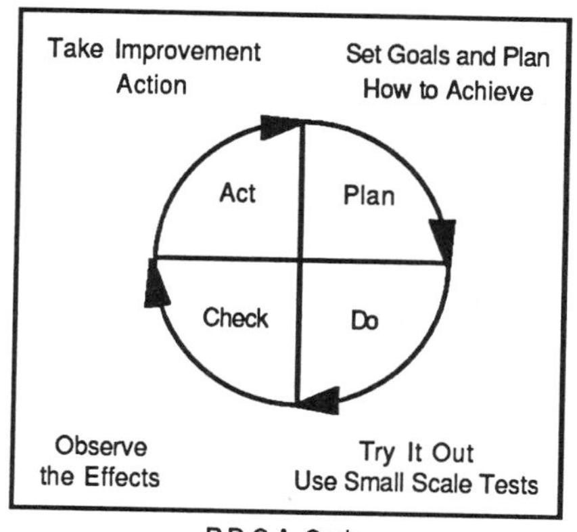

P.D.C.A. Cycle

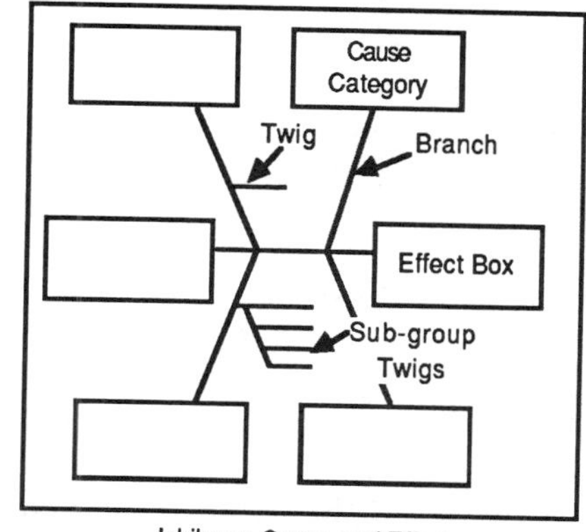

Ishikawa Cause and Effect
or "Fishbone" Diagram

Taguchi Method
Parameter Optimization

Define the Problem
↓
Brainstorm
(Select characteristics with good additivity)
↓
Design of Experiments
↓
Experimentation/Simulation
↓
Response Table
↓
Plot Averages for Strong Effects
↓
Optimization Prediction
↓
Confirmation Run
↓
Conclusion

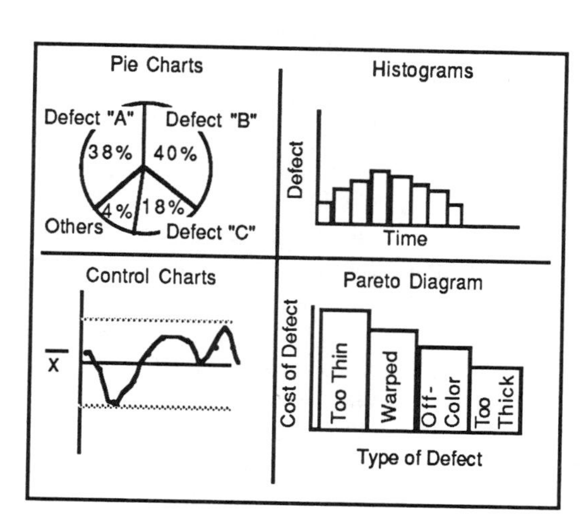

Figure 2-10
TQM Tools and Methods

education and training. The MANPRINT model looks at the training of high level decision makers (top and mid-management), facilitators (mid-managers for system engineering and support activities) and product decision makers (design engineers). TQM looks at the very top and near top as leaders, and everyone else as workers.

Decision Makers (Top and Mid-Managers)

Studies seeking to define generic characteristics of leadership have been inconclusive, primarily because of the extreme variability of leader characteristics. Bennis and Nanus (1985), however, concluded from their study of 200 leaders from various walks of life, that leaders do tend to pursue four basic strategies. These are *visioning, communication, positioning,* and *deployment of self.* Visioning sets the focus for the organization; communication tells the members specifically what the goals and rewards are and gives feedback; positioning gives constancy of direction and specifies actions to implement the vision; deployment of self means the leaders personality flavors that of all interpersonal relationships of the group being led. These strategies obviously can be employed in ways that either move or impede the organization toward quality. TQM stresses, therefore, the importance of the education and training for leaders.

The discussion of Scholtes and Hacquebord (1987) is most helpful in translating the TQM philosophy to organizational application. They recommend, for example, three areas that leaders should study from the beginning and understand so well that they feel them "in their bones." These are first of all Deming's concepts and methods (see especially Deming, 1986; Scherkenbach, 1988; Gitlow & Gitlow, 1987; and Mann, 1989) including attending a Deming four-day seminar. Second, leaders need to understand the nature of variation (Scherkenbach, 1988, Chapter 5, is recommended). Third, decision makers need to understand statistical thinking and use of data. In addition to the above references, Ishikawa (1985) and Taguchi (1986) are recommended for decision makers.

Managers, key staff, and supervisors at every level will need help in understanding the concepts. For TQM, they will need to understand their new jobs as redefined within the new view of the organization and act to lead through the transformation. They will need to learn new skills: planning, group and meeting management, inquiry skills. For MANPRINT, facilitators will take on the primary job of guiding this effort.

Designers and Other Non-Management Employees

TQM recommendations for these employees will aid in the MANPRINT approach to systems integration. Scholtes and Hacquebord (1987, pp. 24-

25) summarize the type and extent of training recommended:

a. *Technical training related to specific job skills*. Everyone needs to fully understand their job, acquire and maintain competence so variation between workers is minimum.

b. *Systems orientation* for all individuals and groups to provide the overall focus needed for team work.

c. *New technical and maintenance skills*. TQM seeks to elevate everyones' level of technical competence.

d. *Basic orientation to quality*.

e. *Technical advisor training*. This is provided to a network of personnel (facilitators) capable of providing consultation and technical assistance to others. Especially important for these individuals are the basic tools of the scientific approach, skill in project planning and management, abilities in team development, and communication with management.

f. *Basic improvement skills*. The goal of TQM is to gradually have everyone learn basic skills in planning and managing projects, working in groups, gathering and analyzing data. Scholtes and Hacquebord do not recommend mass training in improvement skills, but rather to provide training "as needed" in the judgment of technical advisors.

Good basic references for education and training include Deming (1986, pp. 52-54, p. 86), Juran (1988, Section 17), Gitlow and Gitlow (1987, Chapters 6 & 13), Scherkenbach (1988, Chapters 11 & 12), and Harrington (1987, pp. 98ff).

Understanding Jobs

Even though an organization has made the commitment to people outside and inside the organization as an approach to total quality, TQM recognizes that the organization cannot completely change overnight. It must make a transformation while still doing business. Change to TQM is not easy. From the point of view of the executive it costs money, takes time, and takes dedication. Even with this investment, it could fail if the organization is unable to create true systems integration. Scholtes and Hacquebord note (1987, p. 3) that in a quality organization "the work of one step of a process must be the perfect antecedent to the next step of the process. The work of one division must reinforce the work of another division."[8]

In order for all the systems and processes to work in a consistent, coordinated, complementary manner, everyone must work in concert. This translates to a spirit of teamwork with common dedication to "delighting the customer, understanding the organization's systems and processes and a shared commitment to their on-going improvement"[9] (Scholtes & Hacquebord, 1987, p. 3). But everyone still has a job to do. It is most critical that everyone know exactly what their job is. Scholtes and Hacquebord (1987, pp. 3-4) describe what each person should know about his/her job:

> • Understanding where my work fits into the various larger systems and processes of which I am part. What and who precedes me and follows me in the sequence of activities. How my work relates to the final product and ultimate user or consumer.
> • Knowing my internal customers: those who receive my work and spend on what I do to accomplish their own work. Knowing what they want and don't want. Knowing what would satisfy and delight them.
> • Mastering the information and skills necessary to perform tasks related to my work. Constantly reviewing and upgrading my knowledge and skills.
> • Understanding the process or technology with which I work: how it functions, what are its capabilities, what causes variation and breakdown. Constantly getting to know it better and learning how to improve its performance.[10]

They conclude this level of understanding of one's job can be achieved only with continuous education and regular feedback between each employee and his or her external and internal customers.

TQM/MANPRINT COMPARISON

Figure 2-11 lists ten factors which are important descriptions of both TQM and MANPRINT, but which often mean something somewhat different to each. For example, both claim to focus on improving the process, but TQM means establishing a work environment which continually strives to improve all processes, while MANPRINT has a much narrower focus. Its focus is toward influencing existing decision processes. These are primarily design and development in corporations and program management and planning in the government. Recognizing the similarities and differences in the two programs can help improve communication between and about the programs. To the degree that an organization adopts the novel ideas of TQM and organizes itself for effective communication, the atmosphere for MANPRINT management will be quite favorable.

Novel Ideas for Management

Deming's 14 Points (Table 2-2) reflect what is generally considered the most completely developed philosophy for TQM. The DoD, after examining

FACTORS	TQM	MANPRINT
1. Competitive Advantage	Better customer satisfaction gives market advantage	Better performance in hands of users wins more contract awards
2. "TOTAL"	"Better" results from Total Quality throughout all corporate processes	"Better" results from total systems performance including limitations imposed by human considerations
3. Source of Awareness	Started as response to national and DoD level concerns	Started as Army Program and has expanded
4. Process Focus	Directed toward continually improving all processes	Directed toward inserting human consideration into existing decision processes. Design process is continually improved.
5. Variability	Uses analytical tools to reduce variability and error due to process; primarily a management probe	Recognizes error and variability largely driven by human beings, correctable by management and designers using analytical tools on process and product
6. User Focus	Primarily on buyer whose acceptance of product determines organizational success	Primarily on end user, who is clearly defined by data tools like a "Target Audience Description"
7. Top-Down Implementation	Depends on commitment to quality at top and follow down to design and production floor	In government, depends on commitment at the top. Requirements passed from government to industry, but may not be highly visible to top management in industry.
8. Integration Process	Integrates quality around constant emphasis on meeting user needs. Measures results through extensive analytical monitoring of process variance.	Focus on integration process within the design process. Measures results through testing with end users.
9. Change	Must be tolerant to changes because users need is constantly changing due to competition - either commercial or enemy action	Same as TQM
10. Distribution Function	Permeates entire organization and influences all aspects of organization including personnel staffing, promotion, and compensation decisions.	Primarily influences product and process design and development decisions.

Figure 2-11
Comparison of Corporate Factors

Deming and others (Feigenbaum, 1983; Tribus, 1984; Ishikawa, 1985), developed a modified but similar list for its TQM instruction at the Defense Systems Management College. As discussed more fully later in the chapter, an organization's implementation of TQM ideas will also give a favorable atmosphere for MANPRINT. Some of the most novel ideas behind the TQM approach, which also have relevance to MANPRINT, can be summarized as follows:

Table 2-2
Deming's 14 Points
(Source: Deming, 1986, cover)[11]

1. Create constancy of purpose for improvement of product and service.
2. Adopt the new philosophy.
3. Cease dependence on inspection to achieve quality.
4. End the practice of awarding business on the basis of price tag alone. Instead, minimize total cost by working with a single supplier.
5. Improve constantly and forever every process for planning, production, and service.
6. Institute training on the job.
7. Adopt and institute leadership.
8. Drive out fear.
9. Break down barriers between staff areas.
10. Eliminate slogans, exhortations, and targets for the work force.
11. Eliminate numerical quotas for the work force and numerical goals for management.
12. Remove barriers that rob people of pride of workmanship. Eliminate the annual rating or merit system.
13. Institute a vigorous program of education and self-improvement for everyone.
14. Put everybody in the company to work to accomplish the transformation.

TQM is Fully People Oriented

An organization which signs up for TQM is saying that it believes investment in people is its primary target of opportunity. It is really convinced that profits will be greatest if it considers *delighting* customers and employee *job satisfaction* as its foremost goals.

TQM is Meaningless Without Quality in Leadership

Starting at the very top and moving down as far as possible throughout the organization, Quality Leadership must start with *competence* and *ethical values*, progress to *understanding*, and then to *commitment*. Without ethics no honest concern exists for customer or employee, and there is no long-term focus. Without competence, understanding is unlikely; and without understanding, commitment is meaningless. Fortunately many major organizations do have competent and ethical leaders at the very top. If investment in people can be demonstrated to improve quality which can produce profits, they are capable of understanding the economics. Commitment then is all that is needed to effect change. Long-term profits must be balanced with near-term profits. The company must be able to stay in business while trying to do better.

TQM Can Remove Barriers to Communications

If the organization is committed to long-term quality, knows what delights the customer, and has provided an environment where its employees are competent and happy, the single greatest impediment remaining is often ineffective communication. TQM, properly implemented, rewards teamwork, drives out fear and selfishness which stifles communication, and provides methods for rapid communication up management levels and across functional areas. Rapid communication down management levels is seldom a problem.

TQM Does It Early

Both TQM and MANPRINT stress "the earlier the better." The earlier errors are discovered, the lower the cost to the organization for process correction (TQM) and the more rapidly and inexpensively it is to make system design changes for specific products (MANPRINT). In developing complex systems, it is especially important that TQM and MANPRINT are involved early enough to make critical trade-off decisions (Figure 2-12). Considering that from 75 percent to 90 percent of the manufacturing costs are

		Alternatives				
Criteria	Weight	1	2	3	4	5
TECHNICAL	35%					
Weight						
Survivability						
Susceptibility						
Flight Performance						
COST	20%					
Procurement						
Operations and Support						
RELIABILITY/AVAILABILITY/	20%					
MAINTAINABILITY (RAM) &						
INTEGRATED LOGISTICS						
SUPPORT (ILS)						
Reliability						
Logistics Support						
Battle Damage Repair						
MANPRINT	20%					
Training						
Personnel						
Soldier/System Interface						
PRODUCIBILITY	5%					
SCORE						

A trade-off sheet from an Air Frame design team.

In a complex design, several alternatives will be considered. A team will evaluate the alternatives by considering the criteria with predetermined weighting factors. Note that MANPRINT and RAM/ILS are given 40 percent of the design alternative weight. This is an extremely different design concept from past where human factors, training and ILS were considered support considerations, not equipment design considerations.

Figure 2-12
Critical Trade-off Decisions

determined by the design (Daetz, 1987, p.66), it is extremely critical that the design engineer is aware and motivated to adopt these principles.

Innovation in Processes, Products, and Services

The DoD has added this TQM principle. Innovation may mean new technology or it may simply mean innovative thinking which can eliminate *non-value-adding* operations and administrative barriers to improvement. It is sometimes overlooked in TQM discussions that innovation in new technology is still in the organization's interest. If the four principles above are followed, especially with a view toward MANPRINT in product design, technology can often advance at little or no added overall financial investment.

TQM Success Is Measurable

The contribution of TQM to corporate profits can be shown at fairly significant levels within a reasonable period of time, i.e., three to four years. If it is not shown in this time, it will likely be discarded. TQM done correctly should show pockets of improvement within one to two years. If the organization also applies MANPRINT principles, measures of success in cost and performance terms are possible within the first year of implementation. Figure 2-13 shows a theoretical comparison of various approaches and their potential effects on productivity. The traditional approach (Company A) still

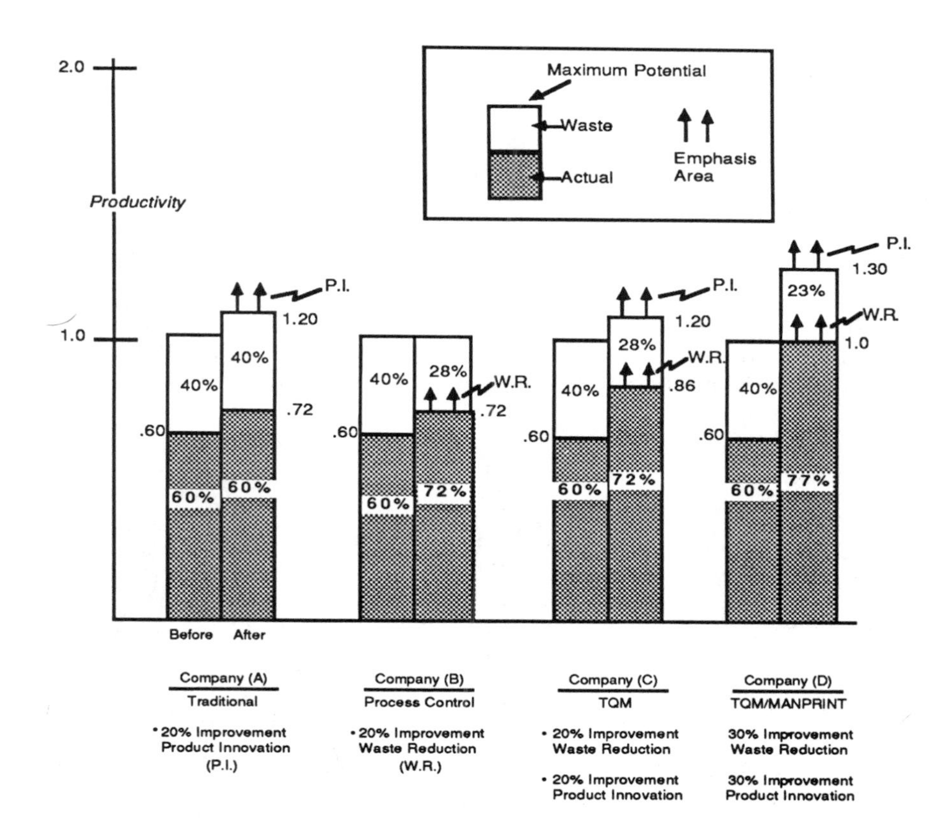

Figure 2-13
Productivity Potential with Varying Improvement Approaches

has 40 percent waste even with 20 percent increase in sales because of product innovation. Company B using process control alone with no increase in sales would be more profitable. The TQM approach (Company C) increases sales and cuts down waste. MANPRINT amplifies the entire process (Company D) by precise user definition, low front-end start-up costs, and added efficiencies in process control.

Organization For Effective Communication

Institutional change is extremely difficult to achieve from within an organization. Outside forces are usually most effective. If American organizations (industry and government) are desperate enough, change can possibly occur without replacing top decision makers, but only if independent proponents for change exist at the very top and throughout the organization. Successful Japanese companies have accomplished this for management quality by starting at the very top at the level of the Board of Directors. At this level they have a "company wide quality control committee" along with finance and other highly respected committees. The president reporting to the board also has a vice president for quality at the same level as sales, manufacturing finance, etc.

This top level visibility and authority is necessary for both TQM and MANPRINT. Authority and accountability is needed at the top but a new approach to management for communication is also needed. Figure 2-14 shows how many companies are structured. Kerzner (1984, p. 4) describes the management problem: "There are always 'class or prestige' gaps between various levels of management. There are also functional gaps

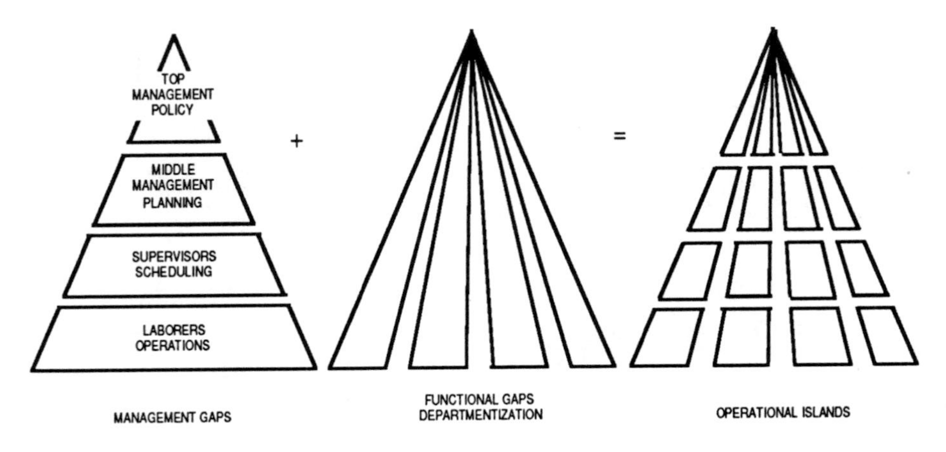

Figure 2-14
Organizational Structure Leading to Operational Islands
(Source: Kerzner, 1984)[13]

between working units of the organization. If we superimpose the management gaps on top of the functional gaps, we find that companies are made up of small *operational islands* that refuse to communicate with one another for fear that giving up information may strengthen their opponents."[12] In the design or development of any product, the manager's responsibility, at both top and mid levels, is to get these islands to communicate cross-functionally toward common goals and objectives. This suggests that in addition to traditional chain of command ways are also needed to describe the organization as a flow of processes, with interdependence of functional areas and group accountability. Deming provides a functional systems way to view an organization (Figure 2-15).

Because of its unconventional presentation, Deming's intent is not readily apparent from the Figure. Deming's purpose with this diagram was to get the viewer to visualize an organization in an entirely new manner. In particular, he sought to depict several factors and dynamic relationships usually missed from conventional organization charts.

- The interdependency of organizational processes.
- The primacy of the customer (consumer).
- The impact of customer feedback (consumer research).
- Continuous improvement based on customer feedback.
- The importance of suppliers.
- The network of internal/supplier/customer relationships.

(Scholtes & Hacquebord, 1987)

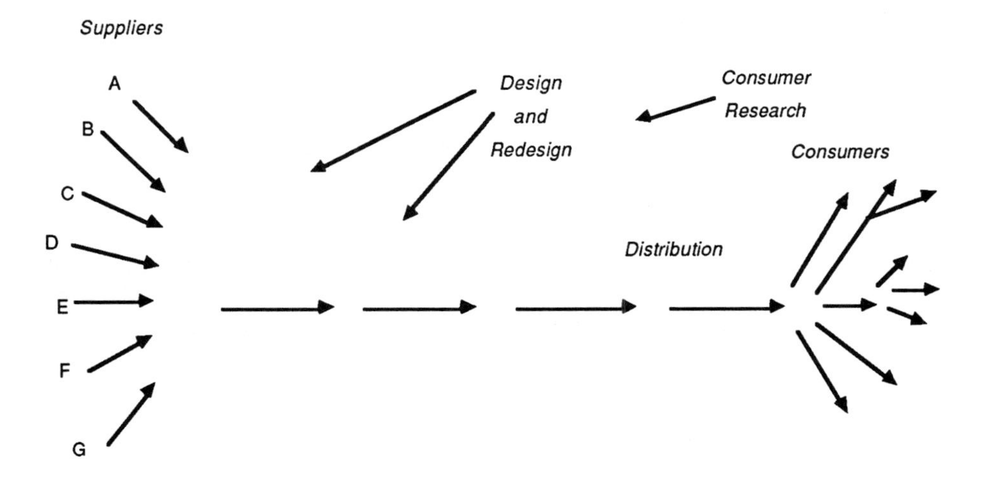

Figure 2-15
Organization as a System (Deming, 1986)[14]

IMPLEMENTING MANPRINT IN TQM ENVIRONMENTS

Scholtes and Hacquebord suggest establishing a series of improvement projects, carefully selected and guided by managers (exemplary in TQM themselves), conducted by cross-divisional teams using the scientific approach and coached by technical advisors. They note also that improvement projects should be "anchored" by surrounding individuals or groups involved in the projects with a network of other similar activities. It appears that adopting the MANPRINT philosophy on a project which designs a new product would be ready made for a TQM demonstration. What is exciting about MANPRINT, in large industrial organizations especially, is that MANPRINT does not require major additional start-up costs or special time away from the job. Underutilized people often exist within the organization with expertise *ready to go*. The Army has initiated several programs (e.g., the Light Helicopter Experimental (LHX), the T800 engine, and the Advanced Antitank Weapon System-Medium) where companies planning on developing these systems have already made the connection of MANPRINT to TQM. Mittler, Hewitt, and Vehlow (Chapter 4) provide some specific examples of MANPRINT applications that fit well with the TQM approach.

Those wishing to apply MANPRINT principles and concepts within their own organization need to first assess the overall corporate values, competence, understanding, and commitment to quality improvement. From the assessment, the organizational status of TQM can be classified as *mature, open-minded or closed-minded.* Knowing the status of organizational TQM allows MANPRINT managers to better plan and implement their activities.

Assess Status

Figure 2-16 is an evaluation form which allows readers to make an assessment of how well their organization compares with a fully mature TQM organization. To make the evaluation, the reader should read the description for each clue and then make a check in the most appropriate column for the organization being assessed. Score 1 point for each "not true," 2 points for "can't tell," 3 points for "generally true," and 4 points for "very true." Sum up the scores of each column and add together for a total score. Compare your score with the following table to determine which category is most applicable to your organization.

TQM CATEGORY	SCORE
MATURE	31 - 40
OPEN-MINDED	21 - 30
CLOSED-MINDED	10 - 20

CLUES	DESCRIPTION	NOT TRUE	CAN'T TELL	GENL'Y TRUE	VERY TRUE
		1	2	3	4
1. User's Voice	Top level management knows who the user is. Operators and/or maintainers of product are among the users. Management already sees natural link of MANPRINT to TQM.				
2. Proponent Level	The CEO personally claims responsibility in well distributed company policy, and several other corporate and executive level individuals show personal interest in TQM.				
3. Organized for Integration	Training across functional lines is rewarded. Decision making appears generally to be done by consensus at all levels of the organization. Physical separation of team members does not impede daily communication on design decisions.				
4. Funds Dedicated	Function and operating budgets appear adequate for non-mainline engineering functions, e.g., RAM, ILS, human engineering, training, value engineering, could-cost, as well as quality management. These functions combined are budgeted about equal to engineering design in all areas, including travel and training.				
5. Training	Quality improvement courses are available for all levels of management. Technical course work is encouraged by mid to lower management. Courses are valued by "movers" and "shakers." Course quality is outstanding.				
6. Integration in Operation	Quality improvement approaches appear evident in several organization processes, not only in those to meet specific contract requirements. If you fail to produce a report somebody outside your functional area would notice. Design engineers frequently request assistance. Support personnel can and do provide useful design information.				
7. Decision History	Records of process and product variation, measures of "goodness" and decision history are kept and are accessible for evaluation.				
8. Expertise	Highly qualified individuals are placed in quality improvement positions at all levels of management. The number of people being hired for non-mainline engineering functions is about equal to traditional engineering positions.				
9. Career Prestige	Practitioners' salaries and promotion rates are all about equal across functional areas, e.g., senior human factors engineers are about the same as senior design engineers.				
10. Data Base	A TQM data base is operational, is used frequently and contains MANPRINT data.				
	COLUMN TOTALS SCORE______				

Figure 2-16
Evaluation of TQM Status Form

Open-Minded, Closed-Minded, or Mature

After making a realistic assessment of the organizational TQM environment, a plan should be devised to correct deficiencies without endangering current successes. Realism must be the rule. First ensure that you and your immediate organization have a good understanding of MANPRINT, the theory, resources, tools and data bases and how the six domains of MANPRINT (Chapters 1 and 4) are integrated. The concepts described in this book are relevant to good MANPRINT regardless of the TQM status, but the manager's approach may vary considerably depending on the atmosphere generated by top management's approach to quality.

The Open-Minded Company

If your company is open-minded, then you have many more opportunities to formalize TQM and have MANPRINT take its place along side TQM. On the other hand, the company could start to move toward a closed-minded environment if quality does not appear to be paying off or if an area like MANPRINT is seen to be a cost not contributing to quality. In either case, building on the successes that you have already found from your assessment, do the following:

User's Voice: Ensure that whenever management looks at "users" they clearly differentiate between the decision makers and operators. MANPRINT can provide top management with precise, immediate information which defines the operator/maintainer and the relationship of human considerations to performance specifications and evaluation criteria.

Proponent Level: Find out which executives support TQM and how they evaluate quality. Focus your effort to succeed by their criteria. They are your "user." Look for *their* management by objective statements in *their* job evaluation criteria and in other status reports.

Look for the real leadership of the company, the "shakers" and "movers" who lead the company. A few of these influential people supporting a change will attract more and more support. Try to join the programs that will show your success and will be seen.

Organizing for Integration: Identify the informal organizations and build your ties there first for the quickest payoff. Find the real leaders, the people who really do the work and help them succeed. Then, giving appropriate credit, take this success public for formal recognition. Make extensive use of cross training and distributed operations. Loan your best people to others to solve problems. Make your shop a training site for management trainees in systems and organizational integration. You will make valuable allies and quickly increase your "network."

Remember that as you extend integration throughout the organization it will cause change and resistance. Allow people the opportunity to plan for the change and give them a chance to be included. Institutional change must have the full support of the top management, and TQM will give you that needed protection while change is taking place.

Funds Dedicated: In order to get more funds, you must make the case that you are worth it. There is always enough money to do what CEO's want to do and they almost always want to make more profit. Be able to show the *value added* of your operation. TQM records of variability, historical documents, engineering and financial records will help. Be able to show the influence of MANPRINT on the products of the process and then show, in dollars, how your operation accounted for cost savings or cost avoidance (the language of management). If you can make the case that what you do adds value, it will be much easier to talk to program managers and to executive managers. Try to make the case that integration and quality doesn't cost, it pays!

Training: Insist on top quality training for your people. The MANPRINT integration course should become the corporate centerpiece for all integration training for all disciplines. This course should also be an important step in networking the key leaders.

Integration in Operations: Focus on the informal organizations and their needs. The design engineers have immediate and specific needs. Satisfy these "users" and integration and quality will follow. A record of this informal assistance will help overcome any formal resistance later. Define the "users" who pay the bills for the company and actively support them by showing MANPRINT cost savings or avoidance. Finally, work to make MANPRINT the model for other integration efforts.

Decision History: TQM and MANPRINT both use information to improve product and process. You must be able to show decision makers where you made a difference. Historical records should show not only the difference but the degree of success. Some integration efforts have to be better than others. Reinforce success. If you cannot tell who or what is better or not so good, your group may be seen as part of the quality problem rather than the solution.

Expertise/Career Prestige: If you have the record of variation/decision history, the record of *value added* from the integration process, and can articulate success in cost terms, you will be able to hire and promote the talented people that you need. As these people take on more responsibility, you will be able to fill key jobs from the TQM/MANPRINT integration team, increasing the amount of networking and influence.

Data Bases: The record of variation and decision history have all assumed usable data bases. The criterion of success is not the size of the data base. How often it is used and who uses it are better criteria. MANPRINT managers should be sure they have a detailed understanding of who the organization believes its users are and what their needs are. Reviewing the use of the data bases is an excellent measure of TQM and how well it is working for the customers.

The Closed-Minded Company

If you have a closed-minded company whose top managers have not been effective quality leaders, all is not lost. MANPRINT can still contribute as the "petunia in the onion patch." Adapt the "ONION PATCH STRATEGY" which is "think big but stay close to your roots" (Scholtes & Hacquebord, 1987, p. 14). Build on the successes of your organization. From this foundation actively look for opportunities to contribute to resolving other visible problems. Build a network of supporters while you improve the quality of your product for all your new customers. The opportunity may first arise when a company has its first project where MANPRINT is a requirement.

Be ready to make full use of fleeting opportunities when dealing with senior management. When the issue of quality comes up, be ready to quickly show how you can help. There is no substitute for a book of successes backed by pictorial illustrations and numerical data to make a convincing case.

Out of desperation, it is possible that the closed-minded company could change to an open-minded one for a short time. Be prepared, therefore, for applying the suggestions given above.

The Mature Company

The mature company will tend to recognize good MANPRINT and support it. MANPRINT itself may not be mature, however, so again the suggestion for the open- minded company may be applicable.

SUMMARY AND RECOMMENDATIONS

Two major new management philosophies, Total Quality Management and MANPRINT, have much in common and need similar management expertise. Both are being looked at by the Department of Defense acquisition community because of the potential they offer in improving effectiveness of the weapons procurement process. Both offer American management reasonable expectation to improve product competitiveness with the foreign market. The two programs are fundamentally similar in two ways: (1) They are both people oriented. TQM looks both to the external customer and to the internal worker as the overriding difference in making quality products. MANPRINT looks at total system performance with the user being a critical performance factor in the system; and (2) They both rely on methods which rise above simple philosophies to reliably evaluate success. TQM and MANPRINT are complementary here in that TQM uses a new measurement and reporting system for executives which is not just financial, while MANPRINT can show its performance improvement in financial language. By the provision of a common vision, i.e., long-term focus on people and verification of success with scientific tools, measurable progress is possible.

The central scientific concept underlying TQM is the recognition that variation is present in all processes. TQM argues that the statistical approach (object of process consistency) is the way to handle variation and, thereby, control quality. MANPRINT recognizes that system performance variation is frequently attributed to people, and the way to handle the problem is to recognize these sources of variation early in system design stages.

The comparison of key management concepts of both philosophies reveals that a Total Quality Management Approach in an organization will tend to vary from one that is mature to one that is closed-minded. This can be evaluated as an organization's propensity for change and integration. MANPRINT proponents should assess TQM in their organization to determine actions to take within the organization. MANPRINT has a scope smaller than TQM, but success can be more easily and quickly determined. If an organization is open to TQM, a MANPRINT project can be a good candidate for TQM demonstration. By showing value added, TQM approach can be strengthened. If the TQM environment is closed-minded, MANPRINT will likely succeed only as a project-by-project basis, i.e., MANPRINT can succeed but only in a specific project which adheres to MANPRINT principles. Successful MANPRINT could, however, help TQM become open-minded if the link is recognized.

NOTES

[1]Sullivan, L. P. (1986). The seven stages in company-wide quality control. *Quality Progress, Vol. 19(5)*, pp. 77-88. Reprinted with permission by L. P. Sullivan, ASI.

[2]The Keys to Excellence, Third Edition, 1989, by Nancy R. Mann, Prestwick Books, Los Angeles.

[3]Juran, J. (1987). *On quality leadership.* Wilton, CT: Juran Institute, Inc.

[4]Ibid.

[5]Ibid.

[6]Stuelpnagel, Thomas R. (1988). Total quality management. *National Defense, Vol. 73(442)*, Figure 9. Copyright American Defense Preparedness Association.

[7]Ibid., Figures 4 and 5.

[8]Scholtes, P., & Hacquebord, H. *A practical approach to quality*, p. 3. Copyright 1987 Joiner Associates, Inc.

[9]Ibid.

[10]Ibid., pp. 3-4.

[11]Reprinted from *Out of the Crisis* by W. Edward Deming by permission of MIT and W. Edward Deming. Published by MIT, Center for Advanced

Engineering Study, Cambridge, MA 02139. Copyright 1986 by W. Edward Deming.

[12]Kerzner, H. (1984). *Project management: A systems approach to planning, scheduling, and controlling*. New York: Van Nostrand Reinhold.

[13]Ibid.

[14]Reprinted from *Out of the Crisis* by W. Edward Deming by permission of MIT and W. Edward Deming. Published by MIT, Center for Advanced Engineering Study, Cambridge, MA 02139. Copyright 1986 by W. Edward Deming.

REFERENCES

Bennis, W., & Nanus, B. (1985). *Leaders*. New York: Harper & Row.

Booher, H. R. (1988). Progress of MANPRINT - The Army's human factors program. *Human Factors Bulletin, Vol. 31 (12)*, pp.1-3.

Crosby, P. B. (1979). *Quality is free*. New York: McGraw-Hill.

Daetz, D. (1987, June). The effect of product design on product quality and product cost. *Quality Progress*, pp. 63-67.

Deming, W. E. (1982). *Quality, productivity and competitive position*. Cambridge, MA: Massachusetts Institute of Technology, Center for Advanced Engineering Study.

Deming, W. E. (1986). *Out of the crisis*. Cambridge, MA: Massachusetts Institute of Technology, Center for Advanced Engineering Study.

Defense Systems Management College (1989, February). *Total quality management*, fact sheet, Program Manager's Notebook.

Feigenbaum, A. (1983). *Total quality control*. New York: McGraw-Hill

Gitlow, H. S., & Gitlow, S. J. (1987). *The Deming guide to quality and competitive position*. Englewood Cliffs, NJ: Prentice-Hall.

Hall, E. T., & Hall, M. R. (1987). *Hidden differences*. New York: Anchor Press/Doubleday.

Harrington, H. J. (1987). *The improvement process*. New York: McGraw-Hill.

Is this the moment Washington has been waiting for? (1989, September 4). *Business Week*, pp. 46-47.

Ishikawa, K. (1985). *What is total quality control?* Englewood Cliffs, NJ: Prentice-Hall.

Ishikawa, K. (1986). *Guide to quality control*. White Plains, NY: Asian Productivity Organization Quality Resources.

Joiner, B. (1986). Using statistics to help transform industry in America. *Quality Progress, Vol. 19(5)*, pp. 46-50.

Juran, J. (1988). *Quality control handbook*. New York: McGraw-Hill.

Juran, J. (1987). *On quality leadership*. Wilton, CT: Juran Institute, Inc.

Kerzner, H. (1984). *Project management: A systems approach to planning,*

scheduling, and controlling. New York: Van Nostrand Reinhold.
Mann, N. R. (1989). *The keys to excellence.* Los Angeles, CA: Prestwick Books.
Mizuno, S. (Ed.) (1988). *Management for quality improvement - The 7 new quality control tools.* Cambridge, MA: Productivity Press.
Rouse, W. (1990). Human resource issues in system design. In N. P. Moray, W. R. Ferrell, & W. B. Rouse (Eds.), *Robotics, control, and society.* London: Taylor & Francis.
Scherkenbach, W. W. (1988). *The Deming route to quality and productivity.* Rockville, MD: Mercury Press.
Scholtes, P., & Hacquebord, H. (1987). *A practical approach to quality.* Madison, WI: Joiner Associates, Inc.
Strickland, J. C. (1988, June 18). Total quality management, Institutionalization in DoD Acquisition. *First Annual National Minority Contracting Forum OASD(P&L),* Pentagon, Washington, DC.
Stuelpnagel, Thomas R. (1988). Total quality Management. *National Defense, Vol. 73(442),* pp. 57-62.
Sullivan, L. P. (1986, May). The seven stages in company-wide quality control. *Quality Progress, Vol. 19(5),* pp. 77-83.
Taguchi, G. (1986). *Introduction to quality engineering.* White Plains, NY: Asian Productivity Organization Quality Resources.
Tribus, M., (1984, June). *Reducing Deming's 14 points to practice.* Cambridge, MA: Massachusetts Institute of Technology, Center for Advanced Engineering Study.

CHAPTER 3

CHANGE MANAGEMENT PROCESS

Robert E. Blanchard
William O. Blackwood

ABSTRACT

Implementing the MANPRINT philosophy requires changes to organizational structures, policies, and procedures and to the values and norms of groups and individuals. Furthermore, such changes must be "institutionalized" to protect against the organization reverting to its previous ways of doing business. This chapter describes some primary concepts important to a unified model for organizational change, discusses the processes underlying organizational change management, and critiques the theoretical base of current change management technology. For illustration, the Army experience with various change factors is summarized. Finally, a Change Management Model, for use in implementing and assessing progress in adopting the MANPRINT philosophy, is proposed.

BACKGROUND OF CHANGE MANAGEMENT TECHNOLOGY

An organization which adopts the MANPRINT philosophy does so to increase assurance that the system produced will be able to meet operational requirements while minimizing costly personnel and material resources. But implementing the MANPRINT philosophy requires changes to organizational structure, policies, procedures and to the values and norms of the people who compose those organizations. It is to be expected that individuals (or groups) will resist any changes which appear to threaten their self interests. Such resistance must be overcome if desired changes are to become part of the organization's institution.

There have been numerous attempts to make major changes in government organizations that design and develop human-machine systems (see Chapter 1). Unfortunately, these programs were largely unsuccessful. In a seminal article, Perrow (1983) suggested that a major problem with past efforts was that human factors issues had not been considered from an organizational change viewpoint. He maintained that if human factors was to achieve its due regard in system acquisition, it would have to be incorporated as part of the organizational structure and culture of those institutions responsible for the conception, design, and development of human-machine systems.

Change Management and MANPRINT

Consistent with Perrow's admonition, the technology for implementing and managing organizational change is an inherent part of the MANPRINT philosophy. It is acknowledged that individuals and groups in the system design and development community need to accept the policies and procedures of MANPRINT and incorporate them as a part of the normal way of doing business. Furthermore, such changes must be "institutionalized" to protect against the organization reverting to its previous way of doing business. There is constant pressure from numerous forces to revert to the previous status quo; to return to the "good old days." These forces must be recognized and understood and properly countered if the changes installed are to be lasting ones.

Institutionalization is a continuing process whereby new norms, values and beliefs become incorporated within the existing organization. In the instance of MANPRINT, these norms, values and beliefs reflect on how people are considered within system conception and design. The "systems concept" is important to the notion of organizational change and to insuring the persistence of such change. Perceiving the organization as an operating system with subsystems, internal and external environments, missions and interactions is useful to understanding change management technology which is discussed in this chapter in some detail.

Need for Systems Model

The organizational change literature is diverse and uneven in the treatment of factors, variables and processes which appear to influence human behavior in organizational change situations. Huse (1975, p. 330) commented on this "patchwork of techniques" which seems to exist quite independently from one another.

Most of the techniques reported for implementing change are directed at the individual within the organization's internal environment rather than at the organization as a system. Too often system properties are disregarded and changes in individual variables (rewards, advancement, job satisfaction) are equated with changes in the organization (Katz & Kahn, 1966, p. 425).

Most contributors to the literature acknowledge the need for a systems viewpoint in organizational design. Schein (1980, p. 241) admonishes the practitioner to develop a model of "how the system works" as a prerequisite to developing change interventions; however, no insights are provided as to how that is to be accomplished. Schein further suggests that a systems model is needed to filter, assimilate and integrate the many seemingly unrelated notions, paths, lists of principles, guidelines and appeals which seem to characterize this literature.

Huse (1975, p. 329) argues that the systems approach calls for greater emphasis on the appropriate level of intervention considering system implications rather than placing undue reliance on the use of a particular technique. He notes that the process should be system-driven (needs) rather than technique-driven (tools). Such an approach would generate linkages with other individuals within an organization who can have a profound impact on change persistence. Architects and building planners, wage and salary administrators, workplace and workarea designers, or computer applications specialists are illustrations of the type of individuals who can strengthen change.

PRIMARY CONCEPTS

This section describes how current organizational change theory and techniques have been organized into a model or approach which can be used in conjunction with the MANPRINT philosophy. Due to the diversity described above, only the better established factors and precepts which underlie the organizational change process are included. For a more detailed discussion, see Blanchard (1988).

The primary concepts important to a unified model of organizational change management for MANPRINT can be classified under:

- Organizational Structure
- Systems Concept
- Power and Authority

Organizational Structure

Although MANPRINT has the potential for broad application, it originated within the structure of a governmental bureaucracy. The characteristics of this type of organization impact directly on the design of change interventions, particularly from the standpoint of forces resisting change, authority conflict, turf controversies, and leader (manager) turnover. The politicized nature of this structure must be considered to fully grasp change strategy development, implementation and the critical problem of institutionalizing change.

By their very nature, bureaucracies resist change. Typically, they are very conservative, highly segmented organizations with heavy regimentation, strong hierarchical (vertical) relationships and entrenched procedures and regulations governing the way the system is to be operated (Hall, 1977, p. 346). Bureaucratic characteristics are often associated with government, but they are not confined there. Many large corporations in the private sector exhibit bureaucratic organizational characteristics. The public bureaucracy differs from business firms, labor unions, universities and other institutions by its public nature. This involves exposure to public scrutiny and concern with the "public interest" (Bernstein, 1958, p. 201).

By contrast, organizations which are less mechanistic and not dominated by tall hierarchies have better communication not only above and below the various units but also laterally. Such organizations typically do not suffer from what Kanter (1983, p. 75) calls "information pathologies" as do the more structured organizations. Healthy information flow is considered the life blood of innovation and organizational change. Chapter 4 (Mittler, Hewitt, & Vehlow) describes management and organizational structures in more detail.

System Concept in Organizations

As discussed earlier, a systems viewpoint is needed to deal with the change management process, particularly for changes as extensive as that suggested by MANPRINT or Total Quality Management (TQM). Without a systems viewpoint, we tend to see problems and their solutions in isolation without full understanding of their impact on the total organization. "System" is meant as an entity comprising a set of operational elements (or subsystems) which function together in a collaborative manner to achieve a common output or goal. Organizations, therefore, can be viewed as social systems made up of interacting parts or subsystems functioning in operational environments (with boundaries and cultures) to achieve a common output or goal.

Social Systems

Social systems are viewed as open, input-output systems with organizational environments and boundaries. Systems import energy from an external environment, transform the energy into something else, and export some form of product back to the external environment. The product then generates new energy by which the cycle is repeated. Figure 3-1 illustrates the open system process.

Social systems are neither fully closed nor fully open (Robbins, 1983, p.11). All systems have some interaction with their environments if they are to survive. Degree of openness varies with the extent to which the organization increases or decreases its awareness or sensitivity to its environment. As Schein (1980, p. 192) points out, open systems are not merely adaptive to whatever the environment provides, but are actively engaged in interacting with their environments.

Entropy is a law of nature which states that all forms of organization within a closed system move toward disorganization and failure. However, unlike the closed system, the open social system can import more energy from its environment than it expends and thereby can achieve "negative entropy" which allows it to survive or even grow. Social organizations which cease to import more energy than they use (cease to effectively sense their external environment) tend to go out of business in the private sector or cease being funded and are disestablished in the public sector. (See Katz & Kahn, 1966, pp. 19-26, for a discussion of open social systems). Theoretically, an open social system (organization) which effectively senses its external environment (and imports more energy than it uses) can adapt to the need for change in an effective manner and can survive indefinitely. An organization is more amenable to changes such as embodied in the MANPRINT philosophy to the extent it is open and responsive to stimuli from its operating environment.

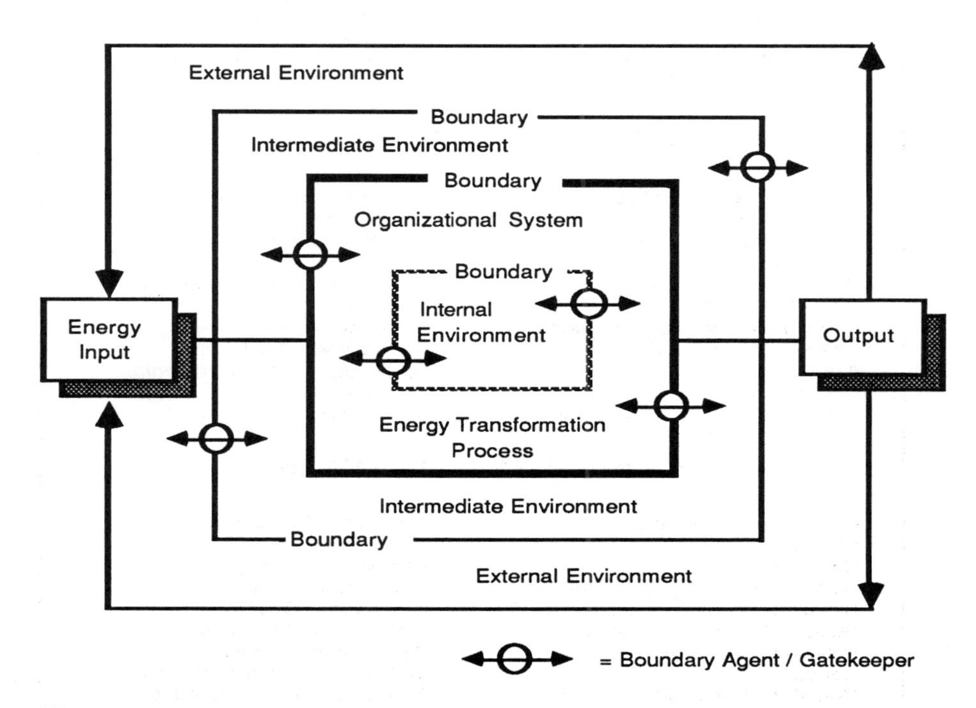

Figure 3-1
Open System with Operating Environments and Boundaries

Organizational Environments and Boundaries

For purposes of managing organizational change, environments can be classified into three types: (a) *External environment* which refers to the economic, political, cultural and technological context within which most, if not all, organizations are embedded; (b) *Intermediate environment* composed of organizations, agencies, units and individuals with which an organization is in direct interaction but lie outside the focal organization's direct authority chain; and (c) *Internal environment* which refers to the structure, technology, procedures, policies, processes and people that make up the focal organization and are subject to its direct authority. Table 3-1 lists sample sets of factors for MANPRINT's environmental domain.

Table 3-1
Sample Factors Composing MANPRINT's Environment Domain

EXTERNAL ENVIRONMENTAL FACTORS	
International Geopolitics	Advances by Potential Adversaries
Political Parties/Priorities	System Obsolescence
Public Attitudes Toward Military	Technological Advancements
Global Economy	Mass Media Emphasis
Foreign Nations and Alliances	Human Resource Availability
National Defense Posture	Goals/Objectives - National

INTERMEDIATE ENVIRONMENTAL FACTORS	
Civilian Personnel Office	Army Materiel Command
General Accounting Office	Forces Command
Defense Contracting Agencies	Combat Commands
Comptroller (Army)	Technical Laboratories
Budgets/Funding Levels	Science Advisory Boards
Congressional Committees	Competitors (Navy/Air Force)
Office of the Secretary of Defense	Consumer Action Groups
Secretary of the Army	Private Study Institutes
Assistant Secretaries (Army)	Technological Advancements
Chiefs of Staff	Universities
Training and Doctrine Command	Goals/Objectives - Army

INTERNAL ENVIRONMENTAL FACTORS	
Organizational Structure	Personnel Ceiling/Staffing
Process/Procedures	Media (Immediate)
Contractors	Line/Staff Harmony
Human Resources Administration	Leader Effectiveness
Research & Development Resources	Organizational Climate
Budgets/Control/Management	Staff Morale/Commitment
Policies/Procedures	Personnel Capabilities
Consultants	MANPRINT - Service/Support
Goals/Objectives - MANPRINT	Goals/Objectives - Individual

Between each environment there exists a *boundary* which is defined by the characteristics above. Boundaries may differ depending upon the nature of the problem and individuals involved. *Boundary roles* are linkages or interfaces which are created to interact with a particular environmental sector to sense developments and to reduce the threat of uncertainty. Boundary roles typically exist with external and intermediate environments and might include congressional liaison officers, memberships on joint planning groups, memberships in professional groups, and so forth. *Boundary agents* perform boundary roles and are individuals who scan for information pertinent to the well-being of the focal organization and provide a buffering or *gatekeeping* function. It is through the vigilance of boundary agents that the focal organization is able to scan and monitor its environments and to sense developments which might require a response such as organizational change. In open system's theory, the boundary agent is one approach to insuring negative entropy and the continued existence of the organization.

It is to the organization's advantage to attempt to control its level of dependence on external and intermediate environments and to reduce the level of uncertainty. Various strategies useful in managing environments include:

(a) contracting to obtain information and technical expertise to reduce uncertainty and obtain support;

(b) co-opting by including an individual or unit from a conflicting environment in the decision making process to reduce their threat or level of criticism;

(c) combining with one or more other organizational entity (merging or coalescing) for purposes of controlling common environments;

(d) use of third parties to solicit support for the focal organization (lobbyists, trade associations, regulatory agencies);

(e) advertising the focal organization's position to promote products, ideas or viewpoints; and

(f) shifting the focal organization's domain of interest to align with a more favorable (less uncertain) environment; for example, a research agency might strive to broaden its mission to include more direct operational support in a period of declining basic research funding.

Organizational Culture

How well an organization can adapt to change depends on its culture. Organizational culture is complex and often subtle, but it is extremely important that the change manager understands it in order to determine an effective approach to change intervention.

An organization's culture is a system of shared values and beliefs which concern its people, structures, and control systems and which produce the norms for the organization's behavior. This system determines what is important, how things work, and the manner in which the organization's business is conducted. Schein (1983, p. 14) suggests that culture comprises a set of "rules" or "guidelines" that a given group has evolved or discovered in the process of stresses and challenges. These rules determine: (a) the patterns of organizational behavior; (b) policy; (c) flow of communications; (d) rewards, practices, procedures; (e) what is discussed, remembered, slogans, war stories; (f) normative behavior, dress, office decor, facilities, traditions.

One of the first steps in planning for change is to determine the organization's cultural readiness for such a change. Changes which are at odds with the cultural makeup of the organization have little chance of success. Therefore, a prospective change manager may first wish to explore the following questions, suggested by Leibowitz, Farren and Kaye (1986, p. 48):

(a) To what extent are upper and middle management aware of the forces in the organization's external and intermediate environments which may be creating pressures for change? Is this awareness shared by the body of employees?

(b) Are the mission objectives and goals of the organization clearly articulated? To what extent are these goals understood and shared by the employees or staff? Is there congruence between current organizational unit goals and those that may be incorporated in the planned change?

(c) What are the main messages from top management as to what is important? Are these messages consistent with organizational goals and the reality of the organization's operating environments?

(d) How effective are vertical and lateral information networks within the organization? Does information tend to flow freely or are there impediments?

(e) Which norms, patterns or values appear to be sacrosanct? Can planned change be implemented without challenging these values, at least initially?

The paradox in dealing with organizational culture is that the existing culture must be respected while at the same time certain cultural patterns must be altered if change efforts are to be successful. Leibowitz et al. (1986, p. 57) suggests a strategy in which the organization's culture is generally accepted while at the same time effort is made to slowly change one, particularly restrictive cultural pattern. By this approach, insight can be gained into the organization's culture which will be helpful in designing effective follow-on strategies for change interventions.

Power and Authority

Power and authority are important components of the change management process. The change agent's understanding of and adeptness in the use of power and authority may well spell the difference between a successful intervention and one that fails.

Types and Sources of Power

The following definitions of authority and power are offered:

Authority. The right to act based on the *legitimacy* of the authority figure's *position* in the organization. Authority is incumbent in the *position* not the *individual.* This assumes full control of rewards and sanctions.

Power. The individual's capacity to influence decisions *regardless* of the legitimacy of his or her *position.*

An individual could gain increased power without rising within the organization. Conversely, an individual in an authority position may, in fact, have very little power. Separation of the notions of power and authority is important in developing change interventions. Those with formal authority may have the influence needed to effect change, but then again there may be others in the organization of lesser authority that have developed power bases that afford them even greater influence over decisions, viewpoints or attitudes.

Formally based power contrasts with *charismatic* power which derives obedience because of highly personalized, almost "magical" leadership qualities. Charismatic power need not be bound by rules of any kind. A top organizational leader with high legitimate power and full control of rewards and sanctions can mobilize more support for his policies if he or she can generate charisma (Katz & Kahn, 1966, p. 318).

The majority of individuals in an organization are not in a position to perceive problems or to evaluate proposals for organizational actions in any real fashion. Hence, they are inclined to turn to a leader whose character, strength and skill give assurance the problem can be handled in the best interests of all concerned. The Army MANPRINT program was fortunate to have this type of power available to aid the early process of change intervention and management.

Persistence of organizational change based on charismatic power in a governmental bureaucracy depends greatly on such changes being institutionalized before the departure of the authority figure. If the charismatic leader is in the military command structure, his tenure is apt to be only three to four years before reassignment. It may be possible for the successor of such a leader to assume some of the aura of his predecessor as *traditional authority*. However, one might expect a general weakening of the original policies and structure and a tendency for the organization to slide backward if complete institutionalization has not occurred.

Power Distribution

Another way to view power is the manner in which it is distributed within the organization. The types of power lie on a continuum. At one end, power is *unilateral* and tends to be concentrated in the hands of a very few, high-level managers. At the other end, subordinates are essentially in control of problem definition and approach selection and wield *delegated power*. *Shared power* represents the middle ground where power is retained at the top but subordinates are allowed to participate in decision making.

Power Coalitions

Organizations are made up of individuals, but Pfeffer (1978, p. 8) suggests that it may be more realistic to think of organizations as *coalitions of interests*. Pfeffer believes that the notion of rational, objective behavior within organizations is unlikely due to different individual preferences for outcomes. It would seem that individual differences in goal-directed behavior are sufficient to create differing opinions as to what is "rational, objective behavior." Therefore, coalitions are formed to serve and protect the vested interests of individuals and groups.

Power coalitions can exist within the internal environment of the organization as well as in sectors within intermediate and external environments. As Robbins (1983, p. 173) notes, the dominant power coalition is the one that has the power to influence structural decisions. In governmental bureaucracies, coalitions may be formed around program planning and funding activities. Others may be based on control of information, expertise and any other resource essential to the organization.

Being rational beings, we would like to see organizational structures developed via rational and objective processes. Unfortunately, individual and group power can subvert such idealistic expectations. Richards (1978, p. 64) notes that since organizational structure may not follow rational grounds, the change agent (manager) must understand power and politics and be prepared to build effective coalitions which in turn can be used to derive an appropriate (desired) structure for the organization.

Power and Politics

Though managers as well as change agents may react somewhat negatively to the notion of politics or political strategy, change management most definitely involves such activities. Richards (1978, p. 67) notes that establishing goals and developing policies (along with change intervention planning) without political savvy, skills and a willingness to use them suggests that the manager (or change agent) will either fail, abdicate, or leave (be reassigned).

Politics of administration are concerned primarily with interactions between the focal organization and its external environment which would have an influence on resources important to the design, implementation, and institutionalization of organizational change. In a governmental bureaucracy, these resources include jurisdictional authority, approval of key appointments, rank within the organizational hierarchy and financial support.

ORGANIZATIONAL CHANGE PROCESSES

It is generally considered inevitable that organizations change because they are open social systems in constant interaction with their environments through input and output of material, energy and information. If the organization is an open system (otherwise it is doomed theoretically to positive entropy), it will need to respond to forces for change. Such change can be reactive or it can be planned. One of management's responsibilities is to plan for change such that it can be managed. Therefore, organizations must develop internal mechanisms which sense the need for change (boundary agents) in adequate time to allow for change to occur in a planned, orderly manner.

This section classifies the change management literature in accordance with some of the best known processes. Examples of where Army experience seemed to apply are described in the various tables provided.

Forces and Counterforces

Table 3-1 provided examples of environmental factors for the MANPRINT system. Table 3-2 lists the various forces and counterforces in the change management process including several categories of change drivers, factors which contribute to change innovation resistance, and various phases involved in change planning.

Change Drivers

Katz and Kahn (1966, p. 446) note that although organizations are usually in some state of disequilibrium, major changes are the exception rather than the rule. Sources of internal strain are not considered the most potent

Table 3-2
Forces and Counterforces in Change Management Process

Factors and Description	Army Applications
1. CHANGE DRIVERS (Kast & Rosenzweig, 1970, p. 566)	1. CHANGE DRIVERS
a. <u>Crisis or Galvanizing Event.</u> Could involve a demand from Department of Defense official, admonition from congressional subcommittee, significant noncompliance by contractor, major funding cut, etc.	a. Army crisis of soldier-machine interface culminated in 1984; concerns on effect of soldier quality on unit performance and readiness; effect of the influx of substantial amounts of technologically advanced equipment on the Army (Army Science Board, 1984; Kerwin, Blanchard, Atzinger, & Topper, 1980).
b. <u>Changed Goals or Values.</u> Value shifts, such as less tolerance for hazardous system operating conditions, which could induce a goal to design and build more "people safe" systems.	b. Army's senior leadership survey revealed manpower requirements (i.e., spaces) and personnel (i.e., faces) issues of greatest significance; Congress asked Department of Defense the cost of human resources, especially cost of recruiting soldiers from upper mental categories and to demonstrate the relationship between quality soldiers and performance.
c. <u>Technology Push.</u> Pressures to utilize complex hardware technology which may stress human capabilities (information overload) and increase the need to ensure that the human can operate and maintain the system.	c. Technology push was not a major driver for MANPRINT even though military manpower quality debate focused on issues of man-machine system failure effects on the national strategy of superior technology.
d. <u>Structural Problems.</u> Could result from overly centralized decision making, inhibited information flow, internal conflicts, inability to fill vacancies, etc.	
e. <u>Procedural Difficulties.</u> Change induced due to convoluted approval procedures, outdated budgeting guidelines, restricted information distribution policies, etc.	e. Personnel community needed better procedure for getting personnel and acquisition related information from Army materiel development community in order to plan effectively in the Army System Acquisition Review Council Decisions (General Accounting Office, 1985)
f. <u>Psychosocial Pressures.</u> Change driven by social or governmental pressures for programs such as upward mobility, substance abuse, career enhancement policies, etc.	
2. COUNTERFORCES TO CHANGE (Huse, 1975, p. 111; Robbins, 1983, p. 167)	2. COUNTERFORCES TO CHANGE
a. <u>Organizational Resistance.</u> Dynamic equilibrium of the system resists change unless sufficient legitimate force can be brought to bear to set it into a state of disequilibrium ("unfreezing").	a. The greatest resistance was from the Army materiel developers who argued personnel community would slow the acquisition process (already too slow) without contributing to the resolution of technical problems. General Maxwell R. Thurman in his new position as Vice Chief of Staff, second in command, along with other top level leaders gave personnel community the support necessary to overcome internal resistance. Resistance resurfaced in 1985 and continues in some areas.

Table 3-2 *(continued)*

Factors and Description	Army Applications
2. COUNTERFORCES TO CHANGE (continued)	2. COUNTERFORCES TO CHANGE (continued)
b. Individual Resistance. A counterforce due to perceived threat to a member's position, prestige, security or opportunity for advancement.	
c. Sunk Costs. Time, energy, funds already invested in job. The greater the investment, the greater the threat of change and the greater the resistance.	
d. Uncertainty. Threat to the individual stemming from uncertain impact or outcome from a planned change; threat of the unknown.	
e. Misunderstandings. Stems from poor communication, lack of information or simply wrong information.	e & g. The logistics community and personnel community were partners in change during the late1970s and early 1980s. In 1985, however, much of the logistics community opposed the advent of the MANPRINT effort. A "turf conflict" arose over the MANPRINT domains of manpower, personnel, and training which are also elements of Integrated Logistics Support.
f. Group Norms. Changes which go against established rules or procedures (formal or informal) on the "way the group does business" will incur massive resistance.	
g. Balance of Power. Any change which tends to threaten the balance of power between individuals or organizational units will be resisted vehemently as "turf conflicts."	
3. PLANNING FOR CHANGE (Kanter, 1983, p. 287; Kast & Rosenzweig, 1970, p. 587; Robbins, 1983, p. 273)	3. PLANNING FOR CHANGE
Phase I - Problem Definition. Collect and analyze information obtained from organizational environments that threaten or challenge the equilibrium of the organization. Include boundary agents and mid-level managers. Perform studies (a) to obtain a historical perspective of past change efforts, goals, values and cultural factors, and (b) to determine general organizational readiness (climate) for change. A positive climate for change may exist (1) if members are aware of existing pressures for change; (2) if the organization has competent boundary agents; (3) if there is a clear understanding of the problem, goals, structure, people, procedures; (4) if specific objectives being targeted are clear-cut; and (5) if results can be clearly assessed and advertised.	Phase I - MANPRINT Problem Definition. The problems associated with human factors, manpower, personnel, and training were most intensely studied during the late 1980 to mid 1984 time frame. The relatively large number of studies and analyses had a wide variety of recommended solutions; the consensus, however, was to give the personnel community a bigger role. These efforts provided a positive climate to induce change (Daws et al., 1984; General Accounting Office, 1985; Hartel & Kaplan, 1984; Kerwin et al., 1980). Problem definition culminated with Army Science Board Report Leading and Manning Army 21 (Army Science Board,1984).

Table 3-2 *(continued)*
Forces and Counterforces in Change Management Process

Factors and Description	Army Applications
3. PLANNING FOR CHANGE (continued)	3. PLANNING FOR CHANGE (continued)
Phase II - Problem Exploration. Obtain a thorough understanding of the change problems at hand. Ask such questions as (a) What is the level of intervention (individuals; groups; organizations)? (b) Who/what is anticipated to be at the root of the problem? (c) Is there a clear understanding of the problem by all concerned? (d) What are the specific objectives which are being targeted? (e) Can results be assessed clearly and advertised widely?	Phase II - MANPRINT Did This Without Further Study. The Army decided in the summer of 1984 that changes would be made. The Office of the Deputy Chief of Staff for Personnel was given staff responsibility for developing policy and to clearly define the MANPRINT responsibilities of other organizations involved in the acquisition process.
Phase III - Change Program Development. Planners reassess and confirm priorities and identify alternative paths to problem solutions. Alternative approaches are evaluated, often with advice and aid of senior management. This phase should result in development of a specific program plan. Hold meetings with units apt to be impacted by plan to obtain inputs and recommendations. Attempt to satisfy divergent views prior to program implementation.	Phase III - MANPRINT Concentrated on Long-Range Planning. Two plans were developed independently and at different levels of command within the Army. The first plan was developed in late 1984 by the Army Materiel Command (AMC). This plan was developed at headquarters level with little input from subordinate organizations. It identified and assigned responsibility, primarily within AMC, for resolution of 600+ actions. The second plan was developed by the Army Staff. It was approved December 31, 1985, by Lieutenant General Robert M. Elton, the Deputy Chief of Staff for Personnel. The plan allowed for slow adoption of the program by industry (3-5 years) as the Army got the word out about the program. Hardware successes would take 5-10 years because of the developmental time. The plan focused on the Army's efforts in six broad areas: Policy and Procedures; Education and Training; Marketing and Communication; Research and Studies; Test and Evaluation; and Resources.
Phase IV - Change Program Implementation and Conduct. Involves program implementation, preliminary evaluation and operation. Phase begins with pilot demonstration to permit early assessment and revision if required. Continual monitoring and timely corrective actions are required from this point forward.	Phase IV - MANPRINT Found Feedback Key to Senior Management Confidence. General Officer in-progress reviews were conducted twice a year to assess how the program was being implemented and to make necessary revisions. Feedback to senior officers was informally provided through meetings with MANPRINT students, industry seminars, and public speaking sessions with question and answer periods following. Both positive and negative feedback was provided to give leaders the incentive to continue their speaking efforts. The MANPRINT Team made sure that senior leaders were informed of bad news and that their support was still needed on significant issues.

causes of organizational change. Systems develop many mechanisms for handling internal conflict and, although they may change somewhat in this process, such change is usually slow and generally does not alter the basic structure of the system.

Pressures which arise from the external and intermediate environments are the critical factors in promoting significant change in organizations. This is particularly true for bureaucratic systems whether public or private institutions. Peabody and Rourke (1965, p. 817) conclude external environments, in particular, may contain the most critical of all variables influencing an organization's well being.

Counterforces to Change

Numerous forces are continually present to resist, impede and otherwise interfere with the change process. These forces must be recognized and underlying motives understood if successful counterstrategies are to be developed.

Argyris (1970, p. 17) indicates that the lack of *valid and useful* information constitutes a general counterforce to change. Valid and useful information is that which describes the environmental factors (both external and internal) which created the problem for the system. Kanter (1983, p. 75) also stresses the role of poor information flow in inhibiting organizational change. She feels that heavily segmented structures which impede information also impede change, whereas those organizations which are "open" and not dominated by tall hierarchies have better communication and are less likely to impede change.

Planning for Change

An important initial step in planning for change is to conduct an abbreviated systems analysis to insure that the problem and its organizational context are thoroughly understood and that all information necessary to developing intervention strategies are at hand. As described in detail in Table 3-2, several rather traditional programmatic phases can be employed in the change planning process.

Intervention Strategies and Sponsorship

Kanter (1983, p. 289) describes intervention as the "action vehicle." It consists of a strategy for implementing, assessing and institutionalizing a change or innovation. Actions associated with the change strategy cannot be based on concepts and abstractions, but must rest on concrete procedures, structures, policies, communications, evaluation factors, work methods and rewards. All members of the organizations involved should be fully informed of their particular roles in the new strategy.

From systems theory, this is the point at which the steady state or *equilibrium* of the system is disrupted and caused to move into a non-steady state or *disequilibrium*. Following a successful intervention, the change strategy will be institutionalized and the system moved to a new steady state of dynamic equilibrium. Should the intervention be unsuccessful, the system could revert to the previous, unsatisfactory steady state, or even evolve into a different state altogether. In such a case, any goal-directed behavior may be random at best. Table 3-3 describes various categories of interventions which may be addressed in planning for change.

According to Kanter (1983, p. 289), an individual with power, termed a "Prime Mover," is required if a new strategy, regardless of merit, is to be implemented fully. Such a force may be formed from legitimate power, charisma, or both. Also, top managers must make it clear that they sincerely *believe* in the new strategy and that they are firmly and unequivocally committed to its implementation. Through consistent support and endorsement, top managers can help coalesce potential supporters into a strong constituency supporting the change. Table 3-3 outlines the motive forces that must be involved in a successful intervention.

Top managers must not only support the change project during start-up but continue such support and involvement throughout the change effort. They would not need to be involved directly in all the details of the change plan, but they must be involved in all key steps such as defining needs, determining new structures, deciding on initial steps, and in all critical decisions.

Change Implementation

Implementing change requires completion of the planning phases described in Table 3-2, the availability of a specific plan of action for conducting, and evaluating the change intervention program and top level sponsorship (Table 3-3). It also involves appointing a change agent or manager and establishing a change intervention tracking and monitoring system.

The Change Agent

The change agent (change manager or interventionist) is the individual who will take the lead in implementing the change intervention. The change agent should be brought in at the earliest possible time and should participate throughout the various phases of the organizational change process. The change agent is often a consultant hired by the organization, although use of internal change agent's working in coordination with an external consultant is growing. The change agent may be assisted by one or more *change advocates* who are established and respected members of the client system. These individuals thoroughly understand the problem and purpose of the intervention and are committed to supporting the change. They can serve key "gatekeeper" positions, help reassure other members as to the efficacy of the change, are able to answer questions in terms comprehensible by other members, and provide an excellent source of feedback information to the change agent.

Monitoring the Change Process

Frequently in the past, change intervention programs have neglected *monitoring or tracking* the change process. Methods like observing, interviewing, and testing are needed to ascertain how well the change is holding. Further, there is a need with every change to identify second and third order benefits, as well as possible negative consequences, and to affirm that the change itself did indeed fulfill its intended purpose. Monitoring is critical to establishing whether or not institutionalization has taken place; that is, that the system has indeed been *refrozen* in a new state of *dynamic equilibrium* (Lippitt, Langseth, & Mossop, 1985, p. 103).

Change Implementation Tactics

The tactics described in Table 3-4 might be employed by the change manager in the process of installing a change intervention strategy (See Lippitt et al., 1985, pp. 99-101; Robbins, 1983, pp. 275-277). The change intervention typically would involve a plan based on a combination of the tactics in Table 3-4.

Strategies for Reducing Resistance to Change

Reducing resistance to change involves developing counteractions to the sources of resistance to change described previously. Listed in Table 3-5 are some strategies suggested by Huse (1975, pp. 113-115) which can serve that purpose. These strategies would be used in conjunction with the change implementation tactics noted in Table 3-4.

Table 3-3
Change Intervention Strategies and Sponsorship

Factors and Description	Army Applications
1. CATEGORIES OF INTERVENTIONS An intervention is a plan for implementing and institutionalizing a change innovation. This is the point in the change process at which the equilibrium of the system is disrupted. Following the intervention, the change must be institutionalized and the system returned to a new state of dynamic equilibrium (Robbins, 1983, p. 273).	1. CATEGORIES OF INTERVENTIONS
a. Structure. Distribution of power and authority. Includes changes in chain of command, degree of formalization, degree of centralization of authority, job and position structure, reward structure, supervisory span of control, etc.	a. Army made several power and authority structure changes. These included development of Army Regulation 602-2, MANPRINT in the Acquisition Process; Deputy Chief of Staff for Personnel (DCSPER) integration into the acquisition process; the creation of a Senior Executive Service position for MANPRINT within DCSPER; and the assignment of MANPRINT responsibilities for all other Army organizations.
b. Technology. Equipment, materials, processes which are incorporated in a change intervention. Could involve modifications to current equipment or installation of new state-of-the-art gear such as personal computers, electronic mail systems, teleconferencing equipment, computerized control systems, automated storage and retrieval systems, information processing and dissemination systems.	b. The Army Research Institute was provided nonsystem specific research and development funds to develop manpower, personnel and training technologies. (See Booher & Hewitt, Chapter 12.)
c. Organizational Processes. Changes to the processes used in the conduct of the organization's mission. Includes decision making structure and approval paths, patterns of communication, procedures for obtaining and transferring information, etc.	c. MANPRINT found improved supervision, not organizational change, made a difference. MANPRINT played down organizational changes. The Army's position was that significant organizational changes were unnecessary and decided instead to concentrate on improved leadership and education and training, therefore, providing MANPRINT visibility without demanding significant additional resources. This approach was accepted and generally resources were applied at the management level, not at the worker level. Only one document, the System MANPRINT Management Plan, was added to the acquisition list of reports and only one additional review, the pre-Army System Acquisition Review Council.
d. People. Includes a variety of interventions directed at changes in attitudes, values, skills, and competencies of individuals or groups. Examples are T-Groups, sensitivity training, competency training, managerial grids, job enrichment and career enhancement programs, etc.	d. Education and training programs were established to alter attitudes and develop MANPRINT practitioners throughout the Army.

Table 3-3 *(continued)*

Factors and Description	Army Applications
2. SPONSORSHIP (MOTIVE FORCE)	2. SPONSORSHIP (MOTIVE FORCE)
A "prime mover" is required if a change strategy, regardless of merit, is to be implemented fully (Kanter, 1983, p. 289).	
a. Prime Mover. An individual with power is vital to the change implementation process. A prime mover might empower a champion who validates the prime mover's commitment. Prime movers must stress the new strategy at every opportunity in speeches, reporting, staff meetings, etc.	a. MANPRINT Had a Group of Prime Movers. The strength of the MANPRINT effort has been the broadbase of support. Supporters include the Army Science Board, industry and universities, Congressional staffers, and internal Army senior officers and Senior Service Executives. Prime movers changed over time as the program progressed. No single prime mover could cause change implementation in an organization as large as the Army.
b. Top Managers. Leaders must make it clear that they sincerely believe in the new strategy and have accepted ownership of it. Only top managers can create the "vision" or policies, practices and viewpoints which allow the mission to be accomplished (Oshry & Oshry, 1977, p. 241).	b. MANPRINT Had Overwhelming Leadership Support. Many senior leaders avidly supported the program. In addition, civilian aides to the Secretary of the Army were sent one page information papers and were asked by the Secretary of the Army to address MANPRINT at any opportunity. MANPRINT also found that top level management support alone is insufficient. Dedicated mid-management support is also needed to attend to the details. The efforts on the first plan (Army Materiel Command) mentioned earlier failed largely because it neglected mid- and low-level managers.

Importance of Supporting Information

Timely and accurate information communicated in a form most amenable to the needs of specific change target groups is represented in most of the tactics and strategies noted above. As discussed earlier, valid and useful information is the most critical element in achieving success in change implementation. This means frequent, widely distributed feedback including such information as written follow-ups on meetings and generally available action plans with at least weekly if not daily bulletins on progress and accomplishments. Also included in this area are monthly newsletters, video presentations, articles in trade journals, professional bulletins, formal and informal briefings, and so forth. Such media should be designed with a specific change audience in mind with full consideration of their group and individual concerns, anxieties and goals. All levels of readership must be represented appropriately (Lippitt et al., 1985, p. 103). As described in Tables 3-4 and 3-5, much of the Army MANPRINT effort has been directed at spreading valid and useful information to as wide an audience as possible.

Processes in Institutionalizing Change

At some point, the change can be considered permanent or *institutionalized.* The processes outlined in Table 3-6 define factors which need to be present to maintain the phenomenon of institutionalization.

Setbacks and Countermeasures

Implementing change actions seldom proceeds in a straightforward, smooth manner. Implementing actions to accomplish change can take considerable time and require trying different strategies.

The six processes active during institutionalization, described in Table 3-6, are primary sources of institutional decline if they are improperly applied, compromised or failed in some manner. In addition, Lippitt et al. (1985, pp. 104-106) mentions several areas in which setbacks may be encountered: (a) a tendency for motivation to backslide; (b) deteriorating competence because of new roles, procedures, attitudes and new skills; (c) disturbed relationships due to revised organizational structure; (d) power being dispersed as members enter different units; and (e) intrinsic rewards being devalued as positions in the hierarchy are altered or promotion potential is dimmed.

Table 3-7 summarizes possible setbacks and pitfalls in implementing planned change (Schein, 1980) and offers strategies for countering them (Lippitt et al., 1985).

Assessing Results

It is essential that the change management program provide for assessing the processes and outcomes of the interventions installed. The dynamics of organizational change are complex and difficult to assess, but they can be evaluated, assuming proper planning is carried out as part of intervention strategy development. Planning requires special attention to evaluation and measurement for example.

Lippitt et al. (1985, p. 121) raises the following issues concerning the practical aspects of evaluation: (1) complexity of the change; (2) time required to gather the necessary information; (3)degree of objectivity possible; (4) relation of costs to the value of the data obtainable; and (5) nonfinancial impact involving extra support, work disruptions, number of employee hours, public relations impact, stress.

The evaluation needs to be systematic. Otherwise it will be carried out subjectively by various individuals and groups throughout the organization (Robbins, 1983, p. 20; Lippitt et al., 1985, p. 121). The outcomes of such evaluations are apt to be biased, at best producing findings of little importance and at worse actually distorting the real situation.

There are *degrees* of institutionalization. It is not an all-or-none process (Goodman & Dean, 1984, p. 229). Institutionalization can vary in terms of persistence, the number of people involved, and the degree to which it exists as a social fact. Table 3-8 lists five criteria (facets) for assessing the degree of institutionalization.

Elsewhere, Blanchard (1989) has proposed MANPRINT be defined conceptually as "Dual Behavioral Science Technologies" which comprises (a) organizational design and change management technology and (b) personnel system technology. The technologies can be assessed together as any other developmental system. Perrow (1983) has also argued effectively that human factors technology must be supported by organizational systems for long-term effectiveness. To determine overall change effectiveness requires assessing human component impact at increasingly higher hierarchial levels of system interaction. For the Army, assessment may begin at the individual soldier level, but needs to progress to the organizational unit level, to major mission level, and ultimately throughout all Army activities. The Army is currently conducting a major assessment of MANPRINT after five years of implementation for its degree of institutionalization (G. M. Hewitt, personal communication, 1989).

INTEGRATED CHANGE MANAGEMENT

In this section change theory and methodology are combined to support a system-based concept for organizational change management. An eight phase Change Management Model is proposed for utilizing the information available within the current state of the art on change.

Table 3-4
Change Implementation Tactics

Factors and Description	Army Applications
IMPLEMENTATION TACTICS The following techniques, usually in combination, might be employed by a change manager to install a change intervention. (Lippitt et al., 1985, p. 99; Robbins, 1983, p. 275).	IMPLEMENTATION TACTICS
a. Inform and Educate. Provide explanatory information to help members to understand the purpose and logic of change. Use one-on-one discussions, memos, reports, presentations, newsletters, etc.	a. An Army MANPRINT training program was established and industry personnel were encouraged to attend. The program was tiered to entry-level, mid-level and senior management. In addition to the training program, presentations were made at professional association conferences and seminars held by such diverse organizations as Association of the United States Army, Human Factors Society, Interservice Training Equipment Conference, Society of Logistic Engineers, Military Operations Research Society, as well as numerous industry briefings.
b. Involve All Employees. Solicit members opinions and ideas about problems or approaches; involve members in the decision making process wherever possible.	b. From the program's inception in 1984, a deliberate attempt was made to involve as many people as possible in the formation, development, and support of the program. Initially this was a 20 person study advisory group and later included MANPRINT points of contact. The MANPRINT Bulletin was and continues to be very popular, and a MANPRINT computer conferencing network has been established.
c. Use Feedback. Brief members frequently on plans and progress in the intervention. Provide opportunity for objections to be voiced and to be discussed. Attempt to reduce uncertainty, be reassuring, and build member commitment.	c. In-process reviews, student seminars, MANPRINT industry seminars, trade shows and meetings.
d. Facilitate and Support. Offer support in solving problems such as individual counseling, new skills training. Retain established work groups or routine, carpools, access to parking, and so forth.	d. The MANPRINT community fostered a notion that failure was assured for those who did not try. Hence, everyone left the course knowing that they were not expected to have all the answers, but they were expected to call and discuss problems and seek help. A MANPRINT point of contact list was published quarterly.
e. Negotiate. Identify something of value to the members of group in exchange for lessening resistance and making a commitment to the change. Most useful with intermediate and external environments.	

Table 3-4 *(continued)*

Factors and Description	Army Applications
IMPLEMENTATION TACTICS (continued)	IMPLEMENTATION TACTICS (continued)
f. Spread Success Quickly. Manage implementation so that initial success can be realized quickly. Promptly recognize and reward participants accordingly and advertise progress.	f. When success cannot be realized quickly, the expectations for success should be fostered, and intermediate goals should be established and tracked. The intermediate goals for MANPRINT were attendance at the course, having a point of contact, having System MANPRINT Management Plans, incorporating MANPRINT into the Request for Proposal and Source Selection.
g. Build Group Concensus. Use group norms to build group concensus and obtain group ownership of change. Key steps involve obtaining group commitment to new values, policies and procedures.	g. " MANPRINT REMEMBER THE SOLDIER" - This slogan optimized the basic tenents of leadership and was generally accepted by everyone. Although it may seem like a motherhood statement, it reflected the seriousness of MANPRINT relative to combat operations.
h. Form Coalitions. Build agreements among parties who have common goals whose combined strength and commitment will aid intervention. Parties would agree on roles to be played with an understanding of anticipated outcomes.	h. The MANPRINT coalition is reflected in each of the MANPRINT domains: manpower, personnel, training, human factors, safety, and health hazards. This coalition formed because individually these elements were disgruntled with the acquisition community. Together they saw that they could be more influential.
i. Co-opt. Manipulation to persuade the leaders of a resistance group by giving them special consideration or key roles in the change decision. Soliciting advice solely to obtain endorsement.	
j. Coerce. Use of pressure or threats to overcome resistance; use of coercive power. A risky tactic with little chance of obtaining permanence and with good chance of negative backlash.	

Table 3-5
Strategies for Reducing Resistance to Change

Factors and Description	Army Applications
STRATEGIES	STRATEGIES
The following strategies (Huse, 1975, pp. 113-115) can be used in conjunction with change implementation tactics given in Table 4.	
1. Perceived Benefits. Consider specific needs, attitudes and beliefs of individuals and groups as well as forces external to the organization. Insure that personal benefits (individual, group, division) are perceptible.	1. The perceived benefit was a better service to the customer through integration of MANPRINT domains and overall a better Army.
2. Sponsor/Prime Mover. Gain understanding of power relationships and coalitions in organization and select sponsor(s) with high prestige and influence coupled with legitimate power and charisma.	2. This was done external to the Army by having industry, university, and other prominent people meet with the Army's senior leadership and discuss the problem with them. Internally it was done at each level of command by encouraging superiors to bring in their subordinates to discuss the problem.
3. Uncertainty. Identify specific sources of member anxiety and address these directly with special meetings, counseling, tailored information, etc. Reduce the level of uncertainty concerning the positive value of the change and the chances of the desired outcome occurring.	3. The principal effort here was oriented toward the logistics community. Presentations were made at the Logistics and Acquisition Management Program sessions, Logistic Manager Conferences, and Society of Logistic Engineers meetings. In the short term, these had little impact; but in the long term, they have had a positive effect and undoubtedly contributed to the greatly improved relationship that presently exists in several areas today.
4. Tailored Information. Develop information tailored to specific groups concerning their performance in the change intervention and the predicted positive results of the installed change.	4. Every MANPRINT presentation was tailored for its audience. The success of this effort can be measured in the number of presentations that had to be turned down or scheduled months in advance. By late 1986, presentations were being scheduled three to six months in advance. In 1990, scheduling is often more than one year in advance.
5. Shared Perceptions. Invite individuals or groups to participate in gathering and interpreting data which support need for change or initial positive results of change. Group should consider such data its own.	
6. Group Prestige. Enhance the attractiveness and desirability of a group to its members with respect to quality of worklife and its capacity to satisfy the needs of its members. Try to make it a group considered on the "cutting edge" of progress and achievement.	6. Group prestige was down played to the external community, but encouraged within the MANPRINT community. MANPRINT T-shirts, posters, diplomas, and pencils were all used as well as meetings with senior officers which provided "psychic dollars."

Table 3-5 *(continued)*

Factors and Description	Army Applications
STRATEGIES (continued)	STRATEGIES (continued)
7. Role Consistency. Change orientation and training should recognize the roles of the members and be carried on in that job context. Change processes embedded in the immediate work situation tend to be more lasting.	7. Role consistency was part of the training course. Each student had to give a briefing at the end of the course on what the program was all about and how it would affect his/her organization. In the early courses, the students had to write job descriptions that included their MANPRINT responsibilities. Later these became student handouts.
8. Communication. Ensure that valid and useful information on the need for change, plans for implementation, and anticipated consequences are shared and understood by all members. Timely and accurate feedback is an important component of communication (See Table 4).	8. The *MANPRINT Bulletin* is the best example of this.
9. Goal Congruence. Gain an understanding of the goals, values and beliefs (culture) of the change target groups (and organization) and confirm that goals of the change intervention are not in conflict.	
10. Structural Compatibility. Do not present change as replacing or eliminating an existing structure without first obtaining support of change target. If climate for change is questionable, the change proposition may initially be presented as "supporting" rather than "displacing."	10. Structural compatibility received a lot of attention. Generally, MANPRINT was presented as something the Army was supposed to have been doing all along and the increased attention was needed to overcome inertia. "No new empires" was one of the public speaking themes which also focused on structural compatibility.

Table 3-6
Processes in Institutionalizing Change

Factors and Description	Army Applications
1. Socialization. Assimilating the change innovation into the beliefs, preferences, norms and values of the members of the target organization. Need to socialize new members is highly important. Belief systems must be continually supported.	1. The MANPRINT training classes, Industry Executives and MANPRINT Practitioners Seminars were principal socialization tools.
2. Commitment. Persuading members to assume "ownership" of the change innovation and to "internalize" it as a part of their own "free-choice" system of preferences. Ensure that commitment exists through all levels of the organizational system.	2. The transition of ownership of the program from Headquarters, Department of the Army (HQDA) to the field was inevitable. Timing was critical to internalization/institutionalization, therefore, HQDA has not yet relinquished "ownership." An area of exception is the training program which was transitioned to Headquarters, Training and Doctrine Command.
3. Reward Allocation. Action to reward successes in performing the new behaviors associated with the change intervention. Recognition should be immediate and public. A mix of rewards may be required over time. Rewards must be perceived as equitable by members. Strive for minimum conflict.	3. In order to establish the program within current resource constraints, rewards were not offered. However, increased morale and job satisfaction have provided considerable self rewards. Also, the MANPRINT Directorate is quick to write letters of commendation for individual achievements and publicize group success stories.
4. Diffusion. Dual strategies of maintenance and growth by transferring change innovations established in one unit to a new unit or units. Spreading institutionalized behaviors into several subsystems tends to affirm change behavior and protect against setbacks.	4. The MANPRINT Program is so new that the diffusion process is still going on, and it is difficult to assess how well that has happened.
5. Sustained Sponsorship. Stresses need to maintain sponsorship ("prime mover power") during changes in leadership or top managers. Weakened sponsorship will create a climate to return to a previous steady state. Messages from sponsors should consistently reinforce the values and beliefs of the intervention.	5. Sustained sponsorship began in the mid 1980s and has gained momentum as increasing numbers of senior Army leadership have internalized the program.
6. Sensing and Reinforcement. Variations in institutionalized behavior over time must be "sensed" and reinforced if consistent with desired change. If counterforces are detected, one or more of the above processes must be used to check possible decline. (Goodman & Dean, 1984, p. 235; Kanter, 1983, p. 300; Robbins, 1983, p. 278)	6. The plan for institutionalization of MANPRINT, developed at program initiation, is broad in scope. Highlights of the institutionalization effort include an extensive training program for both government and industry personnel from the worker to executive levels. Seminars with industry executives and MANPRINT practitioners have been a mainstay for communicating MANPRINT successes and to reinforce the program's importance. Plans are being made for incorporating MANPRINT training into other Army courses and civilian university level programs. A major success was the inclusion of MANPRINT as a separate major area in the source selection evaluation process which emphasizes to industry the need to develop MANPRINT expertise in order to win government contracts.

Table 3-7
Pitfalls and Counterstrategies

Factors and Description	Army Applications
1. PITFALLS (Schein, 1980, p. 235-238)	1. PITFALLS
a. Failure to continually scan environments to sense forces for change or weakening change interventions. Includes misreading forces for change which may be operating.	a. A potential problem still exists between MANPRINT and Integrated Logistics Support (see 2.a. below)
b. Failure to transmit relevant information to members or to action parties who can act upon it. Need to "import" information even if it is technically complex or requires revisions to policies.	b. Always a potential problem because of extensive turnover of key top civilian and military leaders. Action is to emphasize continual communication of the value of MANPRINT to individuals who have power and inclination to take action. Accomplished at Army Systems Acquisition Review Council Meetings, MANPRINT with Industry Executive Seminars, and MANPRINT Senior Training Courses.
c. Failure to sense and overcome resistance to change due to naivete, complacency or ignorance. Lack of use of tactics and strategies given in Tables 4 and 5.	
d. Failure to export and advertise results of successful changes to environments (agencies) who can remove pressure or recognize change benefits and help with institutionalization.	d. Very weak for all human resources initiatives. The MANPRINT Directorate is beginning to determine and advertise return on investment associated with new acquisitions.
e. Failure to obtain feedback by continual monitoring and tracking of change programs. Loss of motivation, failure to socialize new members, obtain ownership, and establish change in value system of members are frequent pitfalls.	
2. COUNTERSTRATEGIES (Lippitt et al., 1985, p. 106)	2. COUNTERSTRATEGIES
a. Conduct "linking" analysis to ensure that a planned change is not in conflict with ongoing standing policies or procedures.	a. The MANPRINT Directorate and Integrated Logistics Support Community are working diligently to eliminate duplication and ensure mutually supportive efforts.
b. Develop a task force made up of top officials, consultants, and in-house specialists to serve as a support system to provide guidance and control of the change program.	b. MANPRINT Enhancement Study was initiated in September 1989 to identify areas needing change and to establish long-range plans to improve the program.
c. Enhance available competence among members by selecting a "change facilitator" with expertise in a particularly troublesome area to provide specialized support.	c. Obtaining training expertise to develop and deliver programs on MANPRINT procedures.
d. Avoiding overload such as occurs when too many concurrent change projects are underway simultaneously. As a result, no one change project can gain necessary momentum. Take care to plan use of change resources carefully; avoid multiple demands.	

Table 3-8
Facets of Institutionalization

Facet	Description	Change Relevance
KNOWLEDGE OF BEHAVIOR	Degree to which people know what behaviors are required of them, e.g., if better forecasting of manpower needs are required, do they know specifically how to do better forecasts?	Necessary condition
PERFORMANCE	Degree to which actions needed to be performed actually are performed. Are more people now doing better manpower forecasts?	Necessary condition
PREFERENCE FOR THE BEHAVIOR	Degree to which people are positive about the action. Although people are doing better forecasting, do they like or dislike what they have to do?	Positive attitude by majority ultimately needed
NORMATIVE CONSENSUS	Extent to which people are aware of others performing required actions and there is consensus about the appropriateness of the action.	Institutionalization can only occur when this is in place
VALUES	Group consensus on the value of the acts required by the change. If MANPRINT is to be supported throughout an organization, consensus of its members must be that it is worth supporting as an inherent value. It is a philosophy that should be supported.	Institutionalization can only occur when this is in place
(Goodman & Dean, 1984)		

Change Theory

From the foregoing discussion, several premises can be described as reasonable postulate thought to underlay change in organizations. These, combined with system change processes described by Lewin (1951), allow a theoretical framework for our Change Management Model.

Postulates Underlying Change

Schein (1980, p. 243) makes the point that planning effective change interventions requires some kind of change theory and a clear statement of assumptions underlying that theory. The following statements, in the form of postulates, address that need:

- The change process involves not only learning something new, but unlearning something that currently exists and is well integrated into the value and belief systems of the organization.
- Change cannot occur without motivation to do so. If such motivation is absent, *inducing motivation* can be the most difficult part of the change process.
- Organizational changes (new structures, processes, procedures, people, etc.) occur only through changes in *individual key members* of the organization. That is, organizational change is *mediated* through individual change.
- Most individual change involves attitudes, values, beliefs, and aspirations. The unlearning of responses in these areas is inherently distressful and threatening and produces resistance to change.

Unfreezing, Change, Refreezing Processes

Lewin (1951) holds that change is a multistage process involving (1) *unfreezing* the status quo (disrupting the dynamic equilibrium of the system) to create the motivation for change resulting in the unlearning of behavior no longer considered consistent with organizational goals; (2) *initiating change* by which new behaviors are acquired to move the system into a new state; and (3) *refreezing* the system in a new steady state to make the change permanent (reestablishing the system's dynamic equilibrium). All stages must be completed before an institutionalized (stabilized) change can be considered to have taken place.

Implicit in this concept is the recognition that the mere introduction of change does not ensure either the elimination of the prechange condition or that the change will be permanent. Change in an organization is often followed by regression toward the old way of doing business after the forces for change have been relaxed. An example of this regression is the termination of the United States Army's Organizational Effectiveness Center and School (OECS) after ten years of successful operation (Roberts & Barko, 1986). The tendency to return to the previous state points up the importance of refreezing (*institutionalizing*) following change implementation and the need for continued monitoring over time to ensure that the change state does not weaken and fail.

Refreezing basically requires systematic replacement of temporary forces with permanent ones. With persistence, the member's own norms (values) will emerge over time to sustain the new equilibrium. There is no clear line separating unfreezing and changing in that many of the efforts made to unfreeze the status quo may, in fact, introduce change (Robbins, 1983, p. 275).

Change Management Model

In this section, an attempt is made to assimilate the various methods, techniques, principles and factors into a change management system.

Figure 3-2 illustrates eight process phases of the model. The model input is status information obtained by scanning the three relevant environments. The eight process phases are shown with other key supporting principles, factors and techniques. For a more comprehensive discussion, see Blanchard (1989).

The three primary stages of institutionalization – unfreeze, change, refreeze – are depicted in Figure 3-2 as occurring after Phase 3-Change Intervention Program, the beginning of Phase 5-Change Process, and after Phase 6-Institutionalize Change, respectively.

In its current form, the model represents a hypothesized set of relationships which interact in the production of change innovation. The level of those interactions and the differential effects of the variables on the outcome of change innovation cannot be determined at the present time.

However, even in its preliminary form, the model should prove useful to Department of Defense agencies and to private industries who recognize the need for applying change management techniques to their particular organizations and wish to take advantage of current change management technology.

System Functional Framework

A preliminary set of system functions was developed from the information base for organizational change efforts concerned with internal, intermediate and external environments.

The functions identified mirror the eight process phases shown in Figure 3-2. These functions outline a logical sequence to follow when undertaking organizational change. Most of the organizational change processes

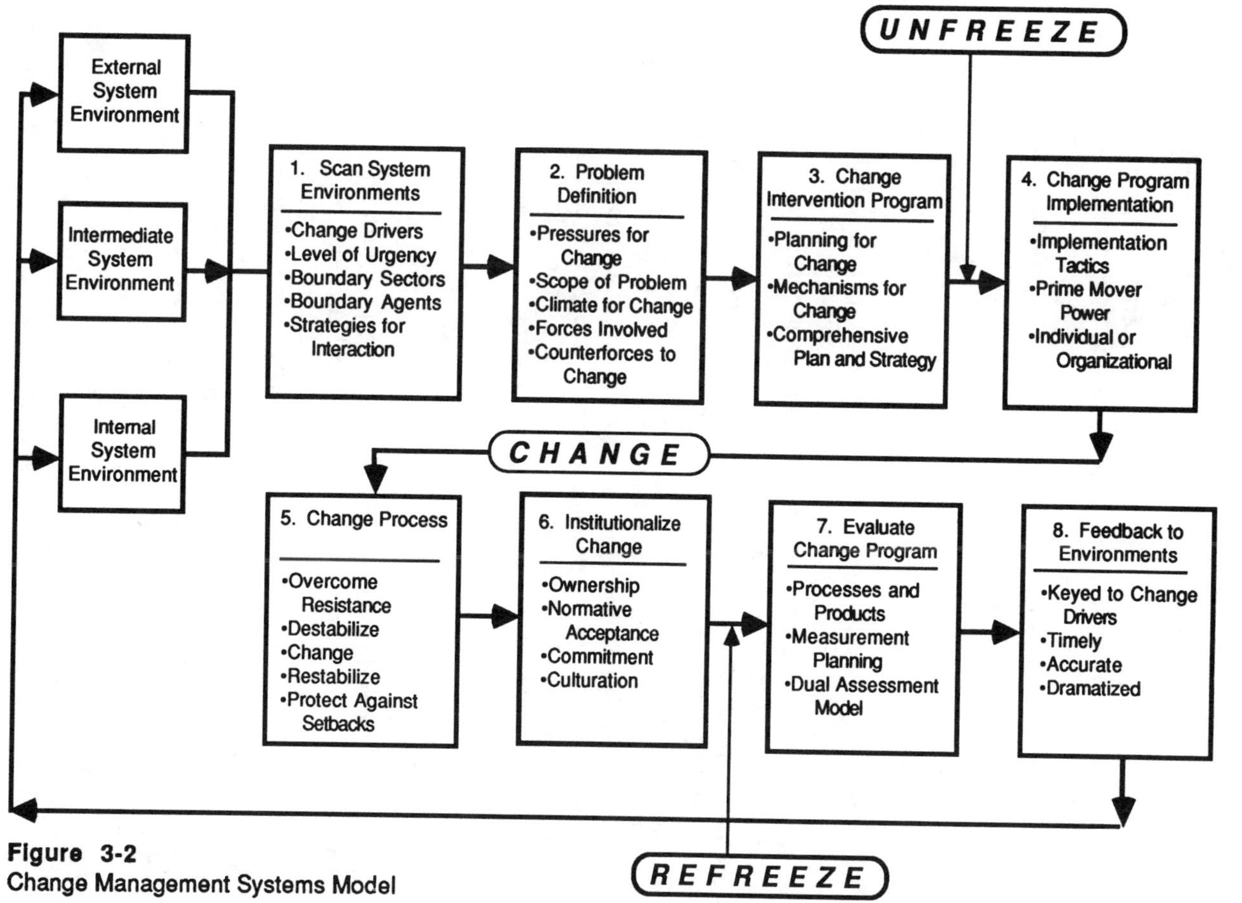

Figure 3-2
Change Management Systems Model

discussed earlier – change drivers, situational analyses, implementation tactics, countermeasures, etc. – are utilized in the model. Change drivers, for example, are important system processes during Phase 1. An important additional use of the function phases is as a framework for identifying relevant independent variables in laboratory and field studies. As indicated above, such studies are necessary to support a theoretical structure of change management. A discussion of the functions comprising the change management process follows.

Scan System Environments

This function includes the process of being aware of and sensing pressures in external, intermediate, and internal environments which may tend to force or indicate the need for change. Performing the function requires identifying the specific environmental factors involved, the complexity and urgency of the change driver(s), and specific linkages (boundary roles) to the focal organization. Scanning also involves use of strategies for managing and controlling interactions with external and intermediate environments. Included is knowledge of the dimensions of environments and the skilled use of boundary agents (gatekeepers) to aid in scanning, monitoring and interpreting developments. It should be recalled that one of the many pitfalls in achieving institutional change is failure to scan environments and to skillfully manage interactions.

Problem Definition

This is a problem definition and planning step fueled by information from environmental scanning of a crisis or "galvanizing" event. Through situational analyses, it is possible to gain a historical perspective of the sources of pressure for change, other possible encounters in the past, and the values and goals of the organizational units involved. Also important to this function is the climate for change which could be positive or negative depending upon numerous organizational and individual factors in potential target units and in the focal organization.

Many of these factors concern the culture of the target unit or organization. Therefore, as part of this function, the change manager should consider the patterns of organizational behavior and the rules and procedures followed, both formal and informal. Of critical importance is whether there is congruence between a possible new goal structure and the current structure supported by individual or group members. Included in this function is the identification of any practices, beliefs or values which are sacrosanct. The change agent must be cautious in challenging such cultural factors, at least initially.

Also required is a thorough understanding of the problem dimensions involved and a grasp of the counterforces to change that are the most pressing. Obtaining complete information on what is needed to proceed is also included; for example, conducting brief studies, holding workshops with experts, retaining technical consultants, exploring type and level of power of potential sponsors, and so forth. A final task in this function concerns strategic decision making leading to an "articulated" direction of attack.

Change Intervention Program (Strategy)

This function includes developing a program or strategy for the change intervention(s) determined by the complexity (structure, technology, processes, people) of the change. Of particular interest are mechanisms which will aid in installation and maintenance of planned changes. It involves anticipating the nature and strength of forces expected to resist or impede change and developing strategies for overcoming them, e.g., identifying change agents, forming political coalitions, engaging in negotiations, developing sponsorship with a legitimate source of power, identifying change advocates, and so forth. Although several sources of resistance to change are possible, *uncertainty* and *balance of power* are likely to be the most difficult to overcome. Resistance may be particularly stiff when the intermediate environment involves interfaces with several units or agencies who may feel threatened by perceived loss of power or influence (turf conflict).

An important facet of this function is the need to seek congruence of organizational goals with those of individual units and members. The change agent would need to identify subgoals which would fulfill individual or group needs that were, at the same time, consistent with the goals inherent in planned change.

This function also involves developing an information and education program which provides for information dissemination and feedback (briefings, bulletins, newsletters, articles, etc.). In addition, planning for test and evaluation is required at this point which involves such questions as type of data to be collected, time frame, collection devices, types of analyses, reporting, and so forth. Completion of this function should result in an action vehicle with detailed plans, steps, assignments and schedule for execution.

Change Program Implementation

This function involves the execution of the change intervention according to plan. The ability of the change agent to adapt to steps which did not go quite as planned is important to this function. It involves the notion of progressive implementation beginning with relatively "low-conflict" actions and progressing to those requiring considerable staging and perhaps some degree of demonstrable prior success. Major emphasis is on executing the communication and feedback plan depending upon whether change is organizational (external) or internal. This function involves decision making regarding initiating forces and how type of power used will influence the change process. For example, if the initiating force is unilateral power (coercive), execution will take a somewhat different path than if the change is initiated by shared or delegated power.

This function requires disrupting the equilibrium of the system (status quo) and initiating change to move the system into a new state (unfreezing). Individuals must first *unlearn* present attitudes, beliefs, or values. Present behavior must actually be *disconfirmed* by not being rewarded over a period of time. The disconfirmation must induce sufficient concern or anxiety to motivate the individual to alter his/her behavior. Valid information and feedback are important elements in accomplishing this objective. The various implementation tactics (inform and educate, negotiate, build consensus, form coalitions) can be used to effect the disconfirmation of present behavior and to provide a basis for learning the new behavior.

Change Process

This broad function is performed during the period of disequilibrium of the system in which new beliefs, attitudes and norms are established. The strategies to overcome resistant to change must be emphasized. An individual motivated to change is receptive to learning new concepts and techniques or new ways of interpreting old information. Two mechanisms are involved: (1) identification with a role model, colleague or some other person and learning to accept the other person's point of view (change agents or prime movers can play this role); and (2) selecting information from multiple sources which is relevant to one's own particular problem, which tends to provide more individualized data than a role model might provide. Schein (1980, p. 244) maintains that change is a cognitive process and is facilitated by obtaining new information and concepts. At this stage, the change manager should confirm that sufficient motivation is present before proceeding. Otherwise, new motivational forces will need to be found and the disconfirmation process in the previous function repeated.

Important to this function is monitoring the change process at appropriate sensing points to track progress and to detect signs of slipping or decline. A support program should be available to react at the first sign of weakening or backsliding. One common pitfall is to overload the program with too many concurrent changes. The support system could include use of individual experts or teams to reduce conflicts, interpret problems and recommend solutions, or to obtain needed feedback.

During this function, the change agent must be sensitive to the loss of individuals who might have been playing roles important to the change plan. Such vacancies are just as likely to be filled by individuals totally unfamiliar with the change program. Consequently, provision for socializing new members as quickly as possible and confirming their level of commitment is an important aspect of this function.

Institutionalize Change

Since there is constant pressure for the system to return to its previous state of "status-quo," institutionalization has to do with ensuring the persistence and permanence of the change and stabilizing the system in a new state of equilibrium (refreezing). This process is tracked by the monitoring system. Involved here is use of a framework comprising the six variables upon which institutionalization is based: socialization, commitment, reward allocation, diffusion, sponsorship, and reinforcement. Included in the socialization factor is the need to indoctrinate new members of the system due to turnover.

Stability of the change depends upon two mechanisms: (1) the individual must be able to test whether the new beliefs or values can be accepted and integrated comfortably, or *ownership* assumed; and (2) an opportunity should be provided for the individual to determine whether the norms of his or her social group will accept and confirm the new behavior patterns. This is an argument for *group* rather than *individual* approaches to change whenever possible.

This function also involves assessing the degree of institutionalization through use of a set of five criteria dealing with individual and group behavior. Important to this function is continued surveillance of the system until complete assurance is gained that the change is indeed refrozen and institutionalized.

Evaluate Change Program

This function involves the conduct of an ongoing evaluation program, described earlier as part of Function 2. Included here is appraising both means (processes) used in the change intervention and the ends (outcomes) which were obtained. The need for differing psychometric approaches and level of data collection should be recognized.

It should also be recognized that varying levels of criterion variables will usually be needed (varying from immediate to ultimate) as well as a multivariable approach to selecting independent variables.

The assessment function should be organized according to the approach which integrates personnel system technology and organizational systems technology (Blanchard, 1989).

Feedback to Environments

This important function involves providing feedback on the results of the change to the environmental sector and organizational unit or units that induced the change in the first place. This function deals directly with the need to export and advertise results of changes. Depending upon urgencies involved, this action should be taken in a responsive, timely manner, but with real and accurate information. Such feedback is most important when dealing with intermediate environments which are particularly important to governmental bureaucracies.

SUMMARY AND RECOMMENDATIONS

For the reasons proposed and discussed in this chapter, the concept of organizational change management is fundamental to the MANPRINT philosophy and process. Although the model proposed is preliminary and generally lacks a cohesive theoretical framework, it is believed to provide a useful concept for MANPRINT managers. For example, the model should prove useful in planning for change associated with the various MANPRINT interventions, for anticipating potential problems areas, and for designing specific actions to overcome them.

The value of the model presented here is that it represents a start in a necessary process of assimilating the body of rather disassociated literature into a systematic framework which can be studied, tested and improved over time. In the interim, it provides an overview of the process of organizational change management which should be of value to the employee or consultant who is faced with performing the role of change manager in implementing MANPRINT or similar programs.

It is important to acknowledge the need for systematic consideration of organizational change technology in conjunction with personnel systems technology. History shows where the principles, concepts and philosophy of human-machine technology are not made a permanent part of the belief systems of the institutions (both public and private) engaged in using them, the philosophy fades into disuse over time and its potential payoffs are soon lost.

Brief summaries of the roles and responsibilities recommended for various individuals in the change management arena are provided below.

Top Managers

Concern for obtaining and maintaining sponsorship for the change program is a major responsibility of top managers. Also, they must be consistent and visible in their support for the intervention on a longitudinal basis. This requires selecting technically competent change managers and backing them up to the full extent of their authority. Top managers must be sensitive to inputs concerning counterforces and be quick to support actions to intercept possible sources of decline. Above all, top managers must be knowledgeable and realistic about the fragility of organizational change and appreciate that literally years may be required until a particular change innovation can be considered fully institutionalized.

Mid-level Managers

This level of management is more likely to be involved in a change program on a day-to-day basis and to have the responsibility of implementing a change program as a change advocate or perhaps even a change manager. Mid-level managers must be knowledgeable of change management principles and techniques, in particular those strategies and tactics included in Tables 3-4 and 3-5. Along with the change manager, these managers should be sensitive to common pitfalls, be able to interpret indications of such pitfalls, and be prepared to initiate (or to directly support) counterchange reversions as noted in Table 3-7. In a real sense, mid-level managers represent the marketing and promotional force for a change. They must be prepared to deal effectively with highly resistant position levels both above and below their own.

Researchers

The research community may have the most difficult job of all since a new vision is required if the technology of organizational change and change management is to be advanced. The proliferation of techniques has expanded beyond available theory, and there is currently no basis for explaining organizational behavior in a manner that can be understood and tested empirically.

Huse suggests (1975, p. 330) the research community should seek to define a metatheory which assimilates the various techniques, theories, concepts and notions into a framework which will allow systematic test and development. In doing so, their vision must include applying the scientific method in hypothesis generation and testing (Schein, 1980). Researchers must view the area with a systems viewpoint and be prepared to exploit opportunities to design controlled studies, gather relevant data and to develop multivariate models of the organizational change processes. Such studies could provide insight as to which change methods or intervention approaches tend to work best in which situations. Unfortunately, greatest confidence in what produces lasting institutional effects needs to come from research which is carried out over long periods of time; periods much greater than usually found in current organizational management research.

REFERENCES

Argyris, C. (1970). *Intervention theory and method: A behavioral science review.* London: Addison-Wesley.

Army Science Board (1984, November). *Report on panel on leading and manning Army 21.* Washington, DC: Office of the Secretary of the Army (Research, Development, and Acquisition).

Bernstein, M. H. (1958). *The job of the federal executive.* Washington, DC: Brookings Institution.

Blanchard, R. E. (1988). *Institutionalizing MANPRINT within the organizational context of the U. S. Army: Change management methods* (Report No. 106-2). San Diego: BehaviorMetrics.

Blanchard, R. E. (1989). *Conceptual model for assessing MANPRINT's organizational effectiveness* (Report No. BM106-4). San Diego: BehaviorMetrics.

Daws, R. N., Jr., et al., (1984, June). *Reverse engineering of the STINGER air defense missile system: Human factors, manpower, personnel, and training in the weapons system acquisition process.* Alexandria, VA: Army Research Institute for the Behavioral and Social Sciences.

General Accounting Office (1985, September 27). *The Army can better integrate manpower, personnel, and training into the weapon systems acquisition process* (GAO/NSIAD-85-154). Washington, DC: General Accounting Office.

Goodman, P. S. & Dean, J. W. (1984). Creating long-term organizational change. In P. S. Goodman (Ed.), *Change in organizations.* London: Jossey-Bass.

Hall, R. H. (1977). *Organizations: structure and process* (3rd edition). Englewood Cliffs, NJ: Prentice-Hall.

Hartel, C. R., & Kaplan, J. (1984, June). *Reverse engineering of the BLACKHAWK (UH-60A) helicopter: Human factors, manpower, personnel, and training in the weapons systems acquisition process.* Alexandria, VA: Army Research Institute for the Behavioral and Social Sciences.

Huse, E. F. (1975). *Organization development and change.* New York: West Publishing Co.

Kanter, R. M. (1983). *The change masters.* New York: Simon & Schuster.

Kast, F. E., & Rosenzweig, J. E. (1970). *Organization and management.* New York: McGraw-Hill.

Katz, D., & Kahn, R. L. (1966). *The social psychology of organizations.* New York: John Wiley & Sons.

Kerwin, W. T., Blanchard, G. S., Atzinger, E. M., Topper, P. E. (1980,

August). *Man/machine interface - A growing crisis.* Alexandria, U.S. Army Materiel Development and Readiness Command.

Leibowitz, Z. B., Farren, C., & Kaye, B. L. (1986). *Designing career development systems.* San Francisco: Jossey-Bass Publishers.

Lewin, K. (1951). *Field theory in social science.* New York: Harper and Row.

Lippitt, G. L., Langseth, P., & Mossop J. (1985). *Implementing organizational change.* London: Jossey-Bass.

Oshry, B., & Oshry, P. (1977). *Power and position.* Boston: Power and Systems Training.

Peabody, R. L., & Rourke, F. E. (1965). Public bureaucracies. In J. B. March (Ed.), *Handbook of organizations.* Chicago: Rand McNally.

Perrow, C. (1983). The organizational context of human factors engineering. *Administrative Science Quarterly, 28,* 521-541.

Pfeffer, J. (1978). *Organizational design.* Arlington Heights, IL: AHM Publishing Co.

Richards, M. D. (1978). *Organizational goal structures.* St. Paul, MN: West Publishing Co.

Robbins, S. P. (1983). *Organization theory: Structure and design of organizations.* New Jersey: Prentice-Hall.

Roberts, B. J., & Barko, W. F. (1986). Organizational development in the U.S. Army: A conceptual case analysis. *PAQ,* Fall, 325-335.

Schein, E. H. (1980). *Organizational psychology.* New Jersey: Prentice-Hall.

Schein, E. H. (1983). The role of the founder in creating organizational culture. *Organizational Dynamics,* Summer, pp. 13-28.

SUGGESTIONS FOR FURTHER READING

Goodman, P. S. & Dean, J. W. (1984). Creating long-term organizational change. In P. S. Goodman (Ed.), *Change in organizations.* London: Jossey-Bass.

Katz, D., & Kahn, R. L. (1966). *The social psychology of organizations.* New York: John Wiley & Sons.

Perrow, C. (1983). The organizational context of human factors engineering. *Administrative Science Quarterly, 28,* 521-541.

Schein, E. H. (1980). *Organizational psychology.* New Jersey: Prentice-Hall.

CHAPTER 4

MANAGEMENT INTEGRATION METHODS

Elliott Mittler
Glen M. Hewitt
Charles A. Vehlow

ABSTRACT

Managing large scale product design and development requires an extraordinary amount of coordination and communication among workers and managers. Collaboration is often complicated by traditionally static hierarchical management structures which inhibit smooth product flow through development processes that cut across functional lines. New approaches to project management suggest integration principles that are useful to managers of complex systems projects. This chapter surveys the history of managerial theory, summarizes its principles relevant to MANPRINT, and describes integration methods and experiences of organizations who have recently managed MANPRINT projects.

BACKGROUND TO MODERN MANAGEMENT PRINCIPLES

Management and what managers do have been the subjects of study since early societies first employed the idea of division of work and supervisors were designated to oversee the activities of others. Consequently, there are many useful theories today which can be applied to increase the productivity of work groups and large organizations. Although no one theory has been accepted universally as providing adequate information for managers and engineers to optimize their productive efforts, much can be learned about project management from management theory. Because MANPRINT changes the way conventional organizations manage projects, it is important to know the limits and advantages of past approaches to management.

From the study of ideas, both old and new, some of the more important characteristics and approaches applicable to MANPRINT are addressed below (see Kast and Rosenzweig [1985] for a detailed discussion of management). They identify (a) the dichotomy between differentiation and integration in organizations, (b) classical organizational structure concepts, (c) human relations management concepts, and (d) modern quality control approaches.

Differentiation vs. Integration

Management theorists face a dilemma in determining what constitutes good management practice. Good management is often circumstantial; discovering pragmatic applications in which circumstances dictate what works best. In introducing their set of guiding principles, Lawrence and Lorsch (1967) concluded that complex organizations attempt to structure themselves in response to environmental demands in terms of *integration* and *differentiation*. On the one hand, through differentiation, organizations acquire a competitive advantage by developing specialists whose narrow expertise increases efficiency in accomplishing subtasks (Porter, 1985). This is most apparent from the typical functionally structured organization chart as shown in Figure 4-1. On the other hand, competitive advantage can also be gained through integration. Integration stresses unity of effort among the functions to improve the organization's ability to accomplish shared goals. Both of these concepts emerge in the history of management schools of thought.

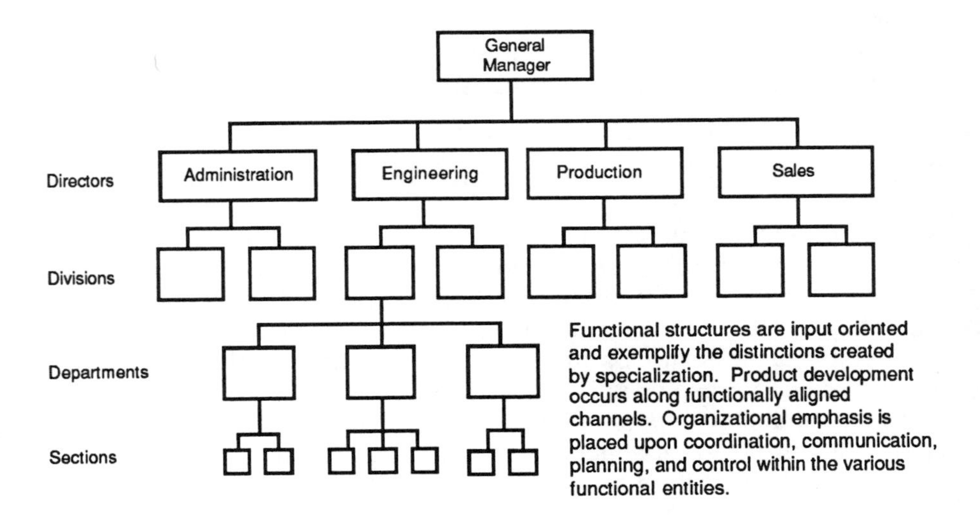

Figure 4-1
Functional Structure

Classical Approaches To Management

The fundamentals of 20th Century management thought were conceptualized by Henri Fayol in 1916 (Fayol, 1949/1916). Large corporations were just starting to develop and managers needed to be assured that large corporations could also be successful. Fayol's analysis said that all organizations, regardless of size and purpose, would function best if they followed certain laws and rules. He built upon existing theories such as those of Max Weber and Frederick Taylor. Weber's bureaucratic model postulated that complex organizations were most efficient when designed as ordered hierarchial structures, while Taylor believed scientifically designed tasks would maximize efficiency.

Fayol's fourteen principles (Table 4-1) were intended to guide managers in applying differentiation (Kast & Rosenzweig, 1985). He recognized the strength of the hierarchical pyramid where unity of direction could be established and task completion could be coordinated through the scalar chain. Achieving greater efficiency was possible because order and discipline directed employee efforts. The classical functional (differentiated) structure promoted communication among functional experts to maximize the performance of that activity.

Table 4-1
Fayol's Principles of Management
(Source: Fayol, 1949/1916)

Fayol's Principles of Management
Division of work
Authority and responsibility
Discipline
Unity of command
Unity of direction
Subordination of individual interest
Remuneration of personnel
Centralization
Scalar chain
Order
Equity
Stability
Initiative
Esprit de corps

Fayol also believed that all managers (regardless of the type of organization) perform the same basic tasks and that these tasks could be categorized into five functions – planning, organizing, coordinating, commanding, and controlling. Although the number of functions and their names may vary, most organizations accept the premise that management can be defined by a universal set of functions (Koontz, O'Donnell, & Weihrich, 1980). Fayol's fundamental ideas form the basis of the traditional management school and have become imbedded in key concepts of all organizations (Harrison, 1985) (Table 4-2).

Human Relations in Management

Most classical management theorists believe that scientific management can improve productivity. Careful job design and environmental control by

Table 4-2
Key Concepts of Organization Management
(Adapted from Harrison, 1985)

The Functional Division of Management Activities:

Management activities are departmentalized and "bureaucratized" creating a specialization of labor, a division of line and staff, a grouping of employees with similar skills, and functionally separated responsibilities, policies, and procedures.

The Hierarchical Relationship of Levels of Management:

Management develops a pyramidal structure in which superior-subordinate relationships are well defined and the flow of authority, responsibility, and communication from the highest to the lowest level imposes itself upon the decision making process.

The Principles of Organizational Management:

The traditional ideas of management (such as division of labor, span of control, and delegation of authority) play an essential role in defining the rules to which organizations must conform in order to achieve their potential efficiency and effectiveness.

industrial engineers is known to contribute to raised output levels. By studying the effects of different techniques, scientific management has aided advances in productivity. Research such as the "Hawthorne Experiments" (Mayo, 1933) also led to important conclusions about the role of human behavior on productivity (Table 4-3). The Hawthorne experiments and other studies into human motivation and interpersonal relations demonstrated the importance of human relations in management and have been integrated into the fabric of other organizational theories.

Capitalizing upon the symbiotic relationship between the individual and the organization has resulted in improved individual motivation, improved social relations, and greater employee participation and exchange in the workplace. With increased participation, all members of the organization are now considered capable of improving output. Enhanced communication opportunities in the work environment place greater reliance on individual and small group self-organization which has proven useful when organizations compete in environments containing uncertain markets, complicated interrelationships, and undetermined dependencies. For today's complex projects, dominated by rapidly changing technologies, there are few organizations without need for better human relations among managers and workers. As a result, the *Behavioral School of Management* which stresses human relations principles has grown to materially influence organizational design and development. Human relations principles improve management integration by increasing information exchange and by accommodating the dynamic nature of engineering design.

Table 4-3
The Hawthorne Experiments

Professor Elton Mayo (1933) led a Harvard group in exploring the effect of work conditions on productivity. They investigated an organization of female telephone assemblers at the Western Electric plant in Hawthorne, Illinois, in the late 1920s. First, the working conditions were improved, one element at a time, anticipating that productivity would increase. When working conditions were reversed (made worse), to their surprise, productivity still continued to rise. Several interpretations of these results include the "Hawthorne effect," a rise in productivity that occurs when people are put under management's spotlight. The most valuable conclusion emanating from the study was that productivity was not entirely related to the scientific design of jobs, but depended upon overriding psychological factors.

Quality on Time

The Second World War moved American industry toward greater employment of integration techniques. In the wartime economy with demands for accelerated deliveries, higher productivity, and improved quality of products, manufacturers were forced to develop factory capabilities and work relationships that shortened production schedules, incorporated new engineering ideas, and employed a new work force.

Under the pressures of war, industry was well-motivated to meet these challenges. Factories were run as experiments for improved productivity. Each company developed its own methods to get products out quicker while continuing to meet specifications. Advanced techniques were developed to speed up the production process, reduce costs, and improve the product (Dell'Isola, 1974). Managers discovered that the classical approach was inefficient. They began to integrate activities out of necessity. Interdepartmental cooperation and cross-functional communication blossomed.

The Ordnance Department of the United States Army facilitated the development of productivity and product quality by devising statistical testing strategies based on acceptable quality levels (Garvin, 1988). The successful use of sampling tables (replacing inspection of every part) demonstrated that statistical methods could improve productivity, that workers could carry out the tests, and that the results could be shared with other work groups to improve total product reliability. Statistical qualitative review and feedback provided a means of transferring product improvement concepts from managers to workers and established further reason why functional groups must communicate effectively.

Other advances made in the application of integrated quality management concentrated on how quality techniques could improve the efficiency of the entire organization (Juran, 1951; Crosby, 1979; and Deming, 1982). Included were methods by which to break down functional barriers and to incorporate multidisciplinary teams (like quality circles). Management integration approaches with characteristics similar to those of MANPRINT improved coordination, enhanced feedback mechanisms, and fostered increased quality control. Chapter 2 (Booher & Fender) provides a more detailed discussion of quality management approaches.

MODERN MANAGEMENT

In the early 1900s, new products were less complex, were not rushed to market, and were not dependent upon assurances of initial high quality. Functional structures worked well when time was not critical, plans and schedules were flexible, and resources ample to integrate and improve final products. Today's development, fabrication, and test environment requires earlier and more extensive integration. Attempts to solve integration problems are evident in (a) the growth of the systems movement, (b) the success of project management, (c) new developments in modern management, and (d) the challenges recognized within modern management.

The Systems Movement

Concepts enhancing integrated engineering management came in response to the need of the largest business organizations for better management techniques. Researchers from the Tavistock Institute of Human Relations in London, England, (Jaques, 1951; Trist, Higgen, Murray, & Pollock, 1963) developed sociotechnical systems theory as a way of understanding the interrelationship between human and technological resources and as a model for effective organizational design. The Tavistock researchers demonstrated at the Glacier Metal Company and in the British coal industry that productivity increased when work groups were organized as systems to attain group goals. Greater integration resulted from the synergistic effects that took place as workers learned and applied experiences and techniques across other subsystem elements.

A model for integrated organizational activities was formulated in the open systems theory (von Bertalanffy, 1950) which emphasized that organizations should be recognized as inseparable from their environments (suppliers and customers). It states that an organization's survival is dependent upon its ability to reorganize itself and to take on new functions to meet new environmental demands. To deal with change, organizations must create flexible role-assignments, avoid stereotypical procedures, and learn as work is accomplished. In this way they become "evolving self-organizing systems" (McEwan, 1971). Emery and Trist (1965) stipulated that tough competition, rapid rates of product change, and uncertain technological risk makes integration essential to successfully incorporate environmental factors in the organization. Management systems models were used to assess the optimal contribution of each functional process.

The systems concept demonstrated how integration improves the organization's handling of complex technical projects over traditional or classical approaches. There are four benefits: (1) systems thinking recognizes the importance of each person and finds ways to integrate the contribution of each, (2) management in the modern systems environment is a shared activity between different levels and different disciplines of the organization, (3) the systems approach is input, throughput, and output oriented providing a way to evaluate the contribution of system components (i.e., value added), and (4) the systems approach permits an adaptable process that can revise planning and control functions along with specific organizational relationships in order to meet changing project needs.

Project Management

When corporations developed multiple product lines based upon different technologies, product oriented divisions (see Figure 4-2) were devised (Mali, 1981). However, most companies perpetuated functional entities by creating functional structures in support of each product. But high technology firms have found it difficult to run large projects effectively from functional organizations. In 1953, in its first formal application of large-scale systems integration in a high technology field (TRW Systems Group (D), 1968), Ramo-Woolridge contracted with the Air Force to perform systems engineering and forward planning on the intercontinental ballistic missile program. Evolving from subsequent organizational refinements in this effort, the matrix organization was developed (Figure 4-3). The matrix overlaid the functional organization with the product organization. This structure reflects a project management approach which has certain advantages. These include: (a) a single point of management, (b) a team dedicated to the completion of the overall task (which can integrate the activities of the functional groups), and (c) formal recognition of the need for cross-functional communication and coordination.

Unlike the systems approach, project management is derived from the classical approach with which most managers and workers are familiar and relies on a specific organizational structure. Project managers are responsible for the overall completion of the project including schedule and budget. The functional managers plan and execute the technical aspects of

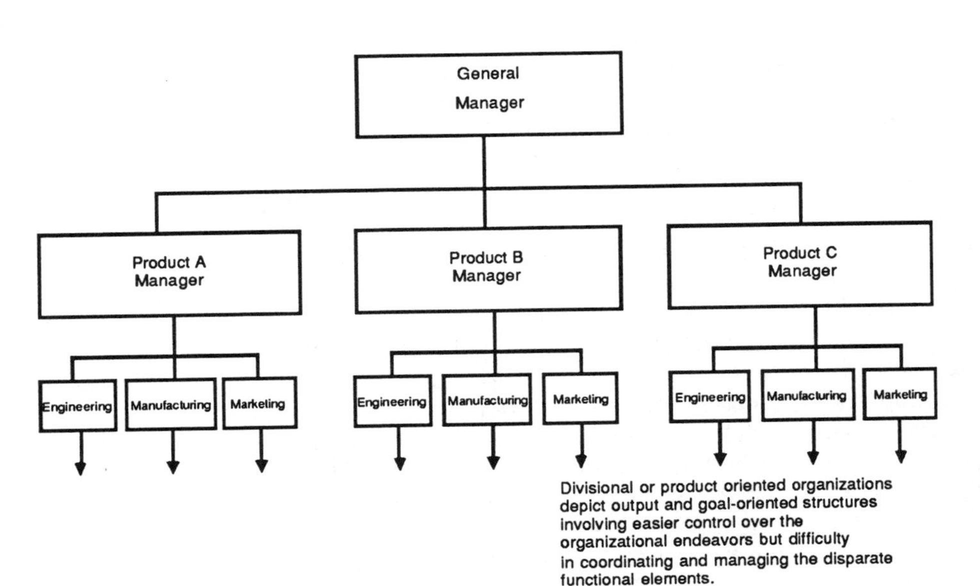

Figure 4-2
Divisional Structure

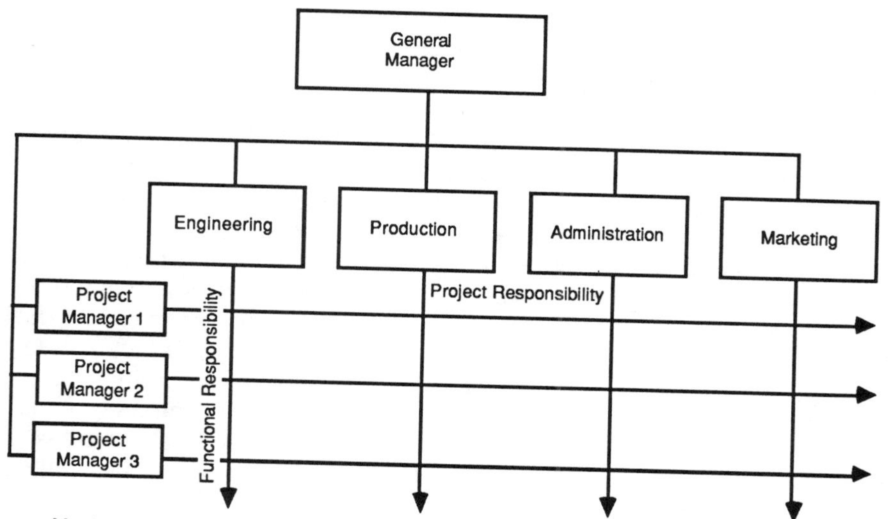

Figure 4-3
Matrix Structure

the project to attain performance criteria. The matrix organization relies upon functional control of manageable subtasks or upon the project office to integrate activities between functions. Its strengths lie in planning, control, and audit. Complex projects can be broken down into manageable parts, comparing actual to expected results. Projects can be monitored through the use of work authorizations, budgets, schedules, change notices, and variance reports. Success, however, is dependent on the accuracy of estimating techniques and standards used to establish expected outcomes of design.

Developments in Modern Management

America's excellence in management techniques for large projects' research and development has been challenged by the changing concepts of integrated production. For example, the Japanese have sought to increase their competitiveness by continually upgrading current products, through enhancements, greater variety, and lower prices (Hiromoto, 1988). To accomplish these ends, engineering, manufacturing, and marketing work together instead of apart. In industries selected for international competition such as electronics, the principles of quality and integration outlined by Deming have been adopted. Companies have begun to take the long-range view (termed "strategic intent" by Hamel & Prahalad, 1989), stress product quality, and institutionalize control systems to reach cost targets. To achieve these targets, broad functional groups have been forced to interact. Terms like "design for manufacturability" have emerged.

An early leader in the use of adaptable organizational responses to environmental demands was Texas Instruments (1982) who combined two overlapping hierarchies, one to deal with current products and one to deal with long-range innovations. The former is a traditional product-oriented profit center hierarchy responsible for short-term goals, insuring that current products remain competitive. The latter is a hierarchy which cuts across functional boundaries to link together groups which work together to foster and develop new products, thereby creating a formal means of integration. Each hierarchy has resources to meet separate goals; yet, together they support adherence to integration principles.

Challenges for Modern Management

Despite the different focus placed by various management schools (Table 4-4) (Kerzner, 1984), modern approaches to management integration all contain common characteristics. These characteristics include: (a) an emphasis upon the use of interdisciplinary teams, (b) the importance of organizational structure, (c) reliance upon the human dimension (including such factors as leadership, motivation, communication, conflict resolution, and time management), (d) the application of technical tools and managerial rules, (e) the concentration on planning, scheduling, and controlling, and (f) the need to accommodate uncertainty.

Management methods used to solve the problems associated with integrating interdisciplinary teams generally recommend organization structural solutions (functional, product, or matrix). Solutions that depend upon structure to increase interorganizational communication suffer from one or more management pitfalls (Table 4-5) (Cleland & King, 1983). Efforts to solve project management complexities through the interpersonal and technical means have resulted in rapid growth in two major fields of study: *Behavioral Science* and *Management Science*. Both fields of study provide means to resolve some management problems. *Behavioral science* identifies the importance of individual and group psychology in meeting organizational objectives. The field of *management science* provides technical means to separate, dissect, and evaluate component parts of a problem. However, each field opens a new dimension of integration complexity that must be accommodated in the engineering environment.

Methods for planning, scheduling, and controlling projects either advocate the application of control techniques (Table 4-6) (Cleland & King, 1983), or depend upon a systems approach. Control techniques offer useful methods to identify the link between activities of a project and provide insight into schedule interdependencies. However, they generally contribute little to identifying trade-offs among project disciplines, alternative designs, and variations in performance. The systems approach constructs a framework within which the problems of a complex environment are to be contained, but inadequately describes how to resolve characteristic integration problems.

Table 4-4
Management Schools of Thought
(Adapted from Kerzner, 1984)

The Classical/Traditional School:

Management places emphasis upon a universal set of concepts applied to the process of getting things done with the end item (or objective) in mind.

The Empirical School:

Managerial capabilities and effective results can be developed by studying the experiences of others.

The Behavioral School:

Management achieves its objectives through processes which emphasize interpersonal relationships and the social system that accommodates cultural change.

The Decision Theory School:

Management involves rational decision making using mathematical models and processes such as operations research and management science.

The Management Systems School:

Managerial success involves a uniquely optimized set of systems models that identifies the flow of resources (money, equipment, facilities, personnel, information, and material) necessary to obtain some objective.

Table 4-5
Management Structure Pitfalls
(Adapted from Cleland and King, 1983)

Management Structure Compatibility:

The issues in criticizing bureaucratic theory lie in the differences between "crises-centered" and "knowledge-centered" organizations. Crises-centered organizations depend upon the traditional methods of functional and hierarchical division. Knowledge-centered organizations contain free-flowing, loosely organized relationships among participants.

Fracturing by Component:

Traditional theory emphasized the functional elements of the organization (finance, marketing, engineering, production) and neglected the interfaces of business.

Directional Errors:

Emphasis upon the satisfaction of a single objective may result in ignoring the multifaceted, long-range aspects of a complex and unstable environment.

Management Principle Myths:

Reliance upon the mechanical application of basic organizational management principles may lead an organization toward disciplined mismanagement.

Individual/Organization Interdependencies:

Failure to recognize and accommodate the symbiotic relationship between individuals and the organization in which they work fails to heed the needs of the modern work force.

Project management works well when the critical components are well understood and when technical problems are foreseen. But complex projects are dominated by uncertainty and inaccurate estimates. Some emphasis has been placed on the production and operation phases, but it is in the definition and conceptual phase where the cost/effective opportunities of engineering design are realized. Great difficulty exists in influencing this early portion of project management. The inherent uncertainty of dynamic design causes adjustments to project or organizational plans and objectives. As a result, estimating, budgeting, cost

Table 4-6
Control Techniques
(Adapted from Cleland and King, 1983)

Control Techniques
Work Breakdown Structures (WBS)
Project planning (Gantt) charts
Precedence diagrams
Critical Path Method (CPM)
Program Evaluation and Review Techniques (PERT)
Network analysis and simulation
Graphical Evaluation and Review Technique (GERT)
Line of Balance (LOB)

control, and performance analysis all require an interactive, iterative, auditable process.

New management methods are also required to accommodate cultural, corporate, and technical trends (Table 4-7) (Patchin, 1983). The objectives of these new methods need to include: (a) utilizing innovative ways to force constructive communications, (b) overcoming the psychologically satisfying hierarchical structures (for both managers and workers), (c) securing consistent support for long-term goals, (d) documenting alternatives, issues, and decisions, (e) accommodating the dynamic uncertainty of design efforts, and (f) insuring early and continued plurality of opinion.

PRINCIPLES OF MODERN MANAGEMENT INTEGRATION

The history of traditional and modern management theory provides several principles which appear evident in effective management integration. Programs demonstrating successful integration methods in the modern engineering environment illustrate the importance of:

- Communication
- Physical Proximity
- Horizontal Processing
- Commitment
- Decision Documentation

- Flexibility
- Feedback Mechanisms and Measures of Effectiveness
- Design Decision Influence

Table 4-7
Trends Affecting Modern Management
(Adapted from Patchin, 1983)

Integrating Culture:

The workplace culture is changing. Management methods are being revised to recognize (a) the relationship between corporate strategies and new societal values, (b) the commonality of interests between workers and management, (c) the changing concepts of authority in hierarchical positions, (d) the emphasis on inherent individual effectiveness, and (e) an accelerating explosion of knowledge.

Corporate Exposure:

Industry is recognizing increased professional responsibility and accountability requiring greater documentation and collaboration and the need for better and early evaluation of new concepts and alternatives to decrease the costly failure rate of expensive new product development.

Life Cycles:

There is increased emphasis on shortening project life cycles to cut costs and to capture technological advantages.

Decision Making Tools:

Government and industry are developing and utilizing new technical decision making tools.

Research and Development's Unique Attributes:

Engineering and research activities are exploring new methods to handle the ambiguity of a "best" solution, the unavoidability of iteration, and unforeseen influences in a constrained resource environment.

Integrating the Process:

New methodologies are being investigated to solve the isolation of organization and operational islands.

Communication: The importance of communication is evident in all management theory and management effectiveness techniques. Contemporary innovative techniques employed to solve organizational management problems usually entail methods by which to improve internal communication (Allen, 1986). Automated information systems almost always stress objectives of enhanced communication. Nonelectronic approaches include such administrative methods as bulletins, memorandums, meetings, seminars, workshops, working groups, committees, and task forces. Other approaches include adding computer-aided design/computer-aided manufacturing (CAD/CAM) coordinating units, consolidating supervisory reporting chains, grouping units by common process, decentralizing, or downsizing (Majchrzak, 1988). Concepts of *adhocracy* informalize and decentralize the use of committees to enhance communication. Other techniques like team building, participative management, and quality circles stress human relations to increase communication and are important, but may depend upon a high degree of interpersonal skill. MANPRINT procedures stress the importance of communication and formalize the process of exchanging information among functional disciplines, organizational entities, and process documentation.

Physical Proximity: Managers often overlook the obvious but underemphasized role of collocating related activities. Studies on the effective utilization of space ascribe a relationship between communication and distance (Hall, 1966). Other studies (Van Cott & Kinkade, 1968) have shown the importance of making good information readily available to those who need it. Communication, horizontal processing, and design decision making influence can be enhanced best by physically locating those activities in close proximity. Some companies have gone to great lengths to ensure people from various functional entities are within close distance. Since MANPRINT requires trade-offs among several disciplines which cut across functional lines, strategic organizational placement is critical.

Horizontal Processing: Traditional concepts in organizational design and management have evolved from the obvious vertical relationships necessary to organize, direct, and control the essential functions of the organization. These traditional approaches lack the horizontal structure and flexibility needed to facilitate the complex interrelationships and communications of modern engineering efforts. The MANPRINT program emphasizes a management process that interconnects the research and development activities through formal documentation and assigned integration responsibilities.

Several organizations have adopted the positive aspects of human relations, systems management, project management, and integration by following what Hayes, Wheelwright, and Clark (1988) term dynamic manufacturing techniques. In doing so, companies like Chaparral Steel (Chaparral Steel, 1986; Forward, 1986) stress horizontal communication between processing functions, thereby coupling design decisions with manufacturing process decisions to improve performance and quality (Figure 4-4).

Other efforts to enhance horizontal integration of process include altering the location and structure of functional elements, breaking down organizational barriers, and facilitating interaction outside these boundaries. The complexities of modern engineering design applications demand increased attention to coordination and integration of design functions. The requirement for coordination across organization lines increases the importance of providing the proper level of authority, physical proximity among diverse disciplines, and the necessary resource support to those elements charged with the integration mission.

Commitment: For an integration effort to be properly implemented, employees at all levels of the organization must be convinced of top management's long-term commitment to MANPRINT objectives. Commitment comes in two forms: one tangible, the other intangible. Both must be present. The latter is evidenced through the attitudes and temperament of management's decisions. The former is most obvious through continuous financial support of integration endeavors. Financial resources must be controlled by an entity outside the design engineering arena. Fiscal independence and the ability to allocate resources to the integration effort must be retained without compromising the ability to influence design.

Decision Documentation: Comprehensive documentation tailored to the decision maker provides a management tool for collecting, analyzing, and prioritizing design issues. Decision documentation should include a method

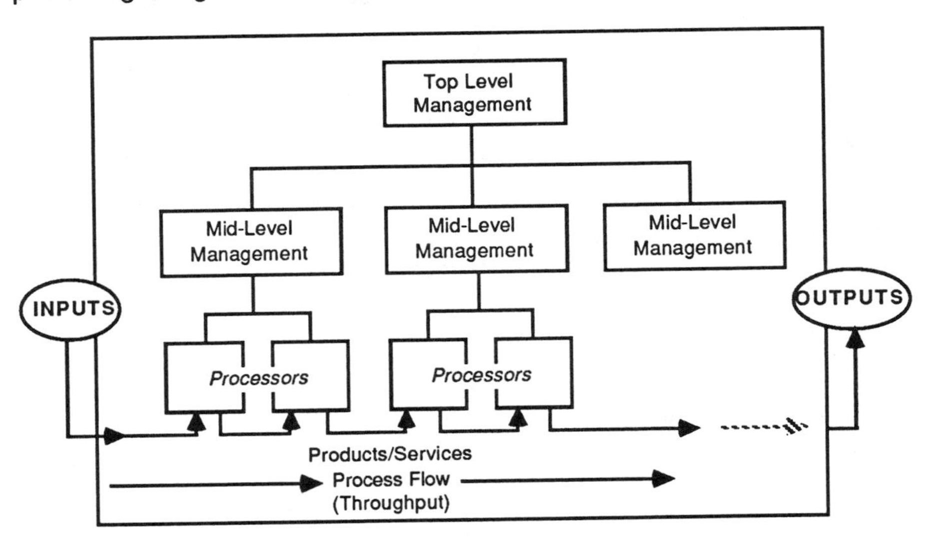

Figure 4-4
Horizontal Product Flow

to evaluate the importance of design attributes related to the human interface. The documentation should provide an audit trail of design issues, priorities for working group action, interdiscipline coordination mechanisms, and decision rationale. Such documentation should be in sufficient detail to highlight design criteria issues at the subsystem level. Comprehensive documentation will help the design team retain good ideas and prevent repeating errors that impact manpower, personnel, and training. The documentation does not replace standard data requirements of system development. It focuses upon the process of integrating design alternatives from various disciplines and involving various technologies. It offers opportunity to resolve residual problems and identify trade-offs that were made to achieve MANPRINT related goals. It provides a method to accomplish the interactive review of design decisions and provides design restart points.

Flexibility: Most management control mechanisms are geared to advise on deviations from a set of predefined variations from the plan. Few management aids readdress the planning process itself. Success in modern developmental projects depends upon integrating project components, identifying trade-offs, recycling in the interactive process, and managing change as new information and new developments arise. Modern projects require interactive, responsive, multipathed, iterative, self-organizing and adaptive ways to trade off among competing objectives and to deal with uncertainty. Integrated project management requires online, multileveled involvement of management for complex planning and control. Integration techniques must deal with both the technical and administrative activities of a project. The dynamic nature of project development and design entails capturing the opportunities for trade-offs within a constantly changing array of alternatives. The MANPRINT philosophy encourages an option oriented process that exploits opportunities and shifts its emphasis as the equipment design and development progresses. In this way, the project manager is allowed maximum flexibility to meet MANPRINT objectives through revised concepts and alternative designs.

There are three ways to make organizational structure less rigid: (1) create an integration entity as part of the organization structure, (2) break down functional barriers with methods like the use of natural work groups and work breakdown structures, and (3) dissolve organizational barriers by defining lateral process interrelationships. Creating an integration entity enhances integration by fixing the responsibility on a particular group, but it has two drawbacks: (1) it fails to distribute integration responsibility among the many key functional people in the design process, and (2) it adds one more entity with which all elements must coordinate. Breaking down functional barriers helps to share integration responsibilities, but it, too, has two deficiencies: (1) success is dependent upon harmonious relationships that usually require much interpersonal skill and effort, and (2) it may entail a structured approach that depends upon continual top-down support.

Dissolving organizational barriers through process definition entails a structured design by management, but the key benefits are two: (1) it shares the integration responsibilities among those most involved in the relevant processes, and (2) it identifies the lateral linkages among organizational elements upon which successful processing is most dependent.

Feedback Mechanisms and Measures of Effectiveness: There are two ways to evaluate effectiveness (Patchin, 1983). First, one can measure the activity level of the program. Items of measurement could include changes in attitudes among the work force; changes in the behavior of the work force (e.g., decreases in absenteeism, tardiness, turnover, grievances); accounting for the integration costs (e.g., materials, training, consultants); and counting the number of integration activities (e.g., meetings, training classes, presentations, and briefings). This method is not particularly useful since it rewards subtask accomplishment.

A second, more valid way to measure effectiveness is to measure the degree of movement toward the program objectives. This method seeks to identify the degree to which program implementation has created a new environment. Such return on investment evaluations include measures of accomplishment, impact on products, and impact upon the process itself. Modern management techniques (including those employed to meet MANPRINT objectives) must be able to demonstrate and document their utility and cost effectiveness. Well-devised assessments and criteria are necessary to achieve a successful program.

Design Decision Influence: The functions associated with the MANPRINT integrated design concept should not be confused with designated specialty engineering functions. *Specialty engineering* tends to be an audit function, reviewing the considerations made by design engineers, but generally lacks the organizational stature or authority to significantly affect change or influence design. Accommodations must be made to assist participation by the engineering staff at all levels and stages of developmental decisions. Design decision influence intended in the MANPRINT program can be achieved through organizational structure design, demonstration of management commitment, and interlocking decision mechanisms that force a plurality of opinions.

MANAGEMENT INTEGRATION APPLICATIONS

Visibility of tangible results from the application of management integration principles requires enough time for design alternatives to surface in the products manufactured. Where development cycles are long, results may not be readily apparent for years. In programs of shorter duration, or in which MANPRINT alternatives and influences have been documented, results may be easier and more quickly assessed. The following identifies programs in which the impact of MANPRINT's use of modern integration principles are already beginning to emerge. Table 4-8 provides a summary of how these principles have been applied.

The T800 Engine

The T800 engine is a 1200 horsepower turbine engine designed to power the Army's Light Helicopter Experimental (LHX) helicopter. Developed through a teaming effort won by Garrett and Allison, the T800 engine was the first aviation development program in which MANPRINT played a major role. The government demonstrated the importance of the integration role by weighting MANPRINT (including logistics support) at 25 percent of the evaluation for contract award. Industry's response included MANPRINT as an integral part of their design effort.

Reliability and maintainability standards achieved were well above those stated in the system requirements. The T800 design facilitated the training efforts by not requiring any new maintenance skills or special tools. User

Table 4-8
Industry Integration Applications

COMPANY	Communication	Physical Proximity	Horizontal Processing	Commitment	Decision Documentation	Flexibility	Feedback & Measures of Effectiveness	Design Decision Influence	SPECIAL FEATURES
Garrett-Allison Helicopter Engine	X			X	X		X	X	• No Cost to Government clause for MANPRINT • Computerized logistics support data base
BMY Improved Howitzer	X	X	X	X	X		X	X	• Unique MANPRINT data item descriptions • Emphasis on redefining requirements • Systems engineering emphasis
McDonnell Douglas Applications	X		X	X	X	X	X	X	• Enhanced by MDAT • Depersonalized decision making
Texas Instruments Advanced Antitank Weapon System	X	X	X	X				X	• Overlayed organizational structures • Alignment with quality efforts

level maintenance tools were reduced from 134 to 6. Maintenance skills were consolidated and manpower requirements were reduced. Government and industry program representatives attribute much of the success in meeting such high standards to the commitment shown for achieving maintainability goals. Top management attitude and interest provided an environment in which the integration principles of communication, commitment, decision documentation, feedback, and design decision influence flourished.

Through the use of an information system to document requirements for task analyses at all levels of maintenance, the developmental community reviewed and annotated design changes reflecting hundreds of MANPRINT issues. Circulating comments from design reviews to various discipline representatives via biweekly reports ensured the visibility and coordination of MANPRINT initiatives.

The contractor gave the areas of Integrated Logistics Support (ILS) and MANPRINT a status equal to design engineering (Figure 4-5). Their responsibilities included drawing review and signature authority enabling them to: (1) make supportability design recommendations throughout the program, (2) provide significant cost containment recommendations, (3) participate actively in the trade study process, (4) conduct early maintenance demonstrations of the T800 supportability principles, and (5) devise a comprehensive computerized logistic support data base.

While an informal organization assisted resolution of design differences,

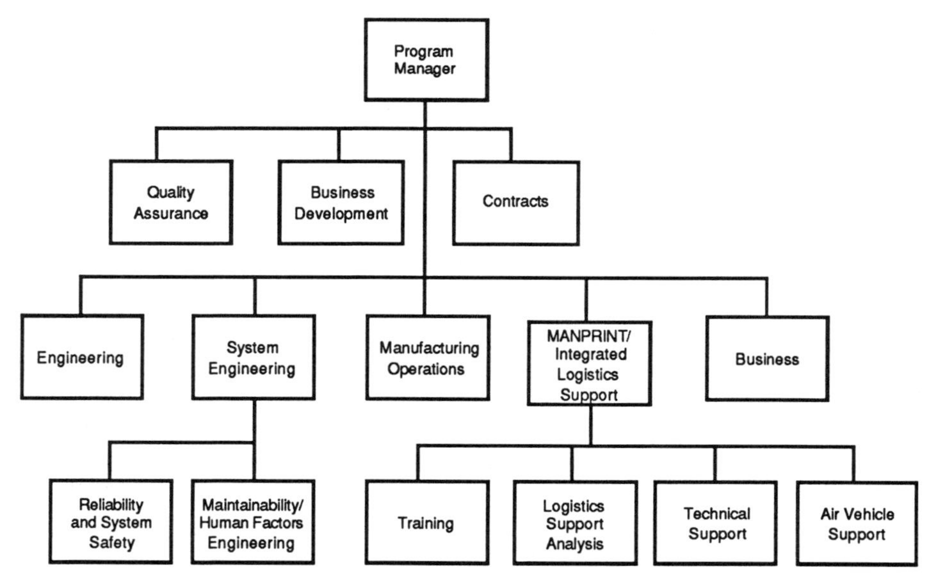

Figure 4-5
Light Helicopter Turbine Engine Company (LHTEC) Program Organization

persons responsible for the MANPRINT functions played the coordinating role. Perhaps because the program was focused solely upon maintenance (vs. operations and maintenance), increased attention to the system description was facilitated. Actual target audience populations participated in MANPRINT demonstrations which adjusted or verified design alternatives and maintenance concepts, and ensured the individual soldier was considered during the design process.

The success of the development program attests to the contribution made by strict attention to design alternatives as a result of effective communication and coordination, meticulous documentation, and from creating a strong integration function in the MANPRINT organization.

The Howitzer Improvement Program

The Howitzer Improvement Program (HIP) provides another example of how modern management integration principles have affected the design process through the MANPRINT program. The Bowen-McLaughlin-York Company (BMY) approach demonstrates the importance of effective communication, horizontal processing, decision documentation, measures of effectiveness, and design decision influence for satisfying complex design requirements.

HIP is a product improvement program to upgrade the United States Army's M109 self-propelled 155 millimeter field artillery howitzer. The improvement program involved risk technologies to increase responsiveness and firepower; enhance survivability on the battlefield; and improve reliability, availability, and maintainability. Improvements to the system include: (1) an automatic fire control system, (2) embedded maintenance diagnostics, (3) modified armament system, (4) a nuclear, biological, and chemical microclimatic conditioning system, (5) a new turret structure, (6) suspension upgrades, and a variety of other enhancements. MANPRINT requirements included reducing crew from five to four, maintaining predecessor system constraints for personnel quality and quantity within current institutional training constraints, and minimizing probability and severity of personal injury.

During HIP contract definition phase, BMY and government personnel jointly participated in an up-front effort to define MANPRINT requirements. The requirements were expanded from one paragraph in the contract Statement of Work (SOW) to three pages of objectives and constraints documented in a MANPRINT management plan. Close ties were established among the contractor, user (target audience), United States Army materiel developer, and Army support agencies. A library of reference material, analytic tools, and procedures were developed as in-house personnel resources were expanded. HIP peculiar (one-time) Data Item Descriptions (DID) were developed to properly document the manpower, personnel, and training design issues especially in referring to the operational characteristics of the system.

BMY's organizational structure contributed to the horizontal processing in the program by establishing MANPRINT as a coordinating element of their systems engineering organization. This structure (Figure 4-6) helped ensure that MANPRINT requirements were defined, allocated, validated, and achieved similar to other system performance requirements. MANPRINT engineers were responsible for developing system timelines, conducting system simulations, and validating analyses/ assumptions on system evaluators, mockups, and prototypes.

BMY attributes its MANPRINT success to:

- An organizational structure and climate which encouraged innovative approaches to MANPRINT issues.
- A cooperative government/contractor team which emphasized early definition and realistic implementation of MANPRINT design initiatives.
- A proactive systems engineering approach to measure the effectiveness of the program to meet human performance requirements/ constraints.
- Enhanced communication through physical collocation of design and MANPRINT engineers.
- Effective design decision influence facilitated by supplementing internal personnel resources with specialized MANPRINT consultants and locating them close to the center of design decision making.

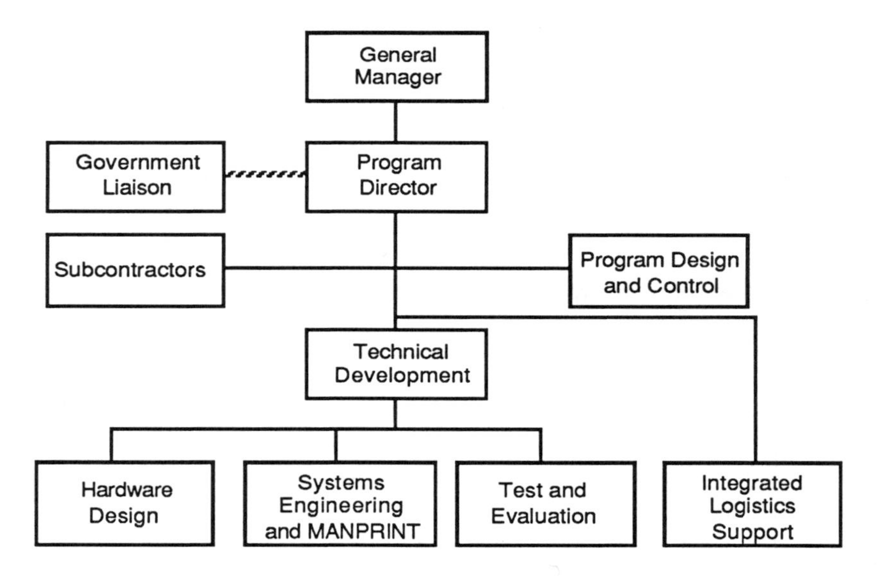

Figure 4-6
Howitzer Improvement Program (HIP) MANPRINT Functional Organization

The Advanced Antitank Weapon System – Medium Program

The Advanced Antitank Weapon System-Medium (AAWS-M) is a one-man, portable, fire and forget antitank missile. The AAWS-M program objectives include improving predecessor weapons in the areas of gunner survivability (weapon system launch signature), toxicity and blast effects, portability, and operability, and directly relate to MANPRINT concerns. The Texas Instruments' (TI) AAWS-M MANPRINT program gives evidence of the commitment, horizontal processing, and communication necessary to integrate man-machine considerations.

TI demonstrated early commitment to the MANPRINT program objectives. The suitability of the weapon for individual soldiers and marines became a continuous criterion for evaluation as successive designs were analyzed, tested, and evaluated. MANPRINT considerations were emphasized and incorporated into design concepts even as early as preproposal negotiations. The company modified funding allocations to assure committed "up-front" MANPRINT efforts. MANPRINT efforts were aligned with Total Quality Management endeavors in the company.

The TI structure for horizontal processing accommodated the integration efforts. A "MANPRINT Manager" was established to coordinate the integration functions under systems engineering for maximum design decision influence. Systems engineering was organized to include systems safety engineering, integrated logistics support, field service training, human factors engineering, maintainability, reliability, testability, product assurance, and producibility.

Communication also played a key role. Seminars among middle level managers were conducted to foster the use of multidisciplinary technical teams (including operation research and human factors specialists). A MANPRINT Program Plan was developed to incorporate all human considerations in early development work. Company training was conducted on the MANPRINT objectives of the program.

TOTAL INTEGRATION METHODOLOGY

A data base management system was developed by McDonnell Douglas to assist in the integration of MANPRINT objectives into the engineering design process. Called MANPRINT Design Analysis Technique (MDAT), it identifies system design issues and documents functional design analyses. MDAT documentation provides an audit trail and a historical record of the system design evolution. It tracks trade studies/surveys, design issues, design note analyses, design decision documents, system task analyses, life cycle costs, and MANPRINT Working Group (MWG) activities. MDAT gives analysts and decision makers online, real time interface with design activities.

Purpose of MDAT

The MDAT program is a system management tool that provides visibility to areas needing corrective action in the design process. It maximizes MANPRINT influence in design and provides an automated method of tracking system design analyses from the functional disciplines (Table 4-9). It records an analyst's summary of these analyses for evaluating the overall impact of a design and provides a variety of other capabilities (Table 4-10). The program automates the identification of system design "high drivers" by using dynamic, user-friendly, interactive data bases. Based upon these "high drivers," necessary design trade-offs are implemented to optimize the system.

Structure of MDAT

The MDAT program is designed for multiple applications (Figure 4-7) and is partitioned by system, functional discipline, and user activity. Analysts view data pertinent to their functional discipline until necessary integration occurs. The MANPRINT Manager Track provides a variety of reports to assist in the evaluation of the responsiveness of the design to MANPRINT needs. Some of the reports generated include:

Table 4-9
Functional disciplines

Functional disciplines
Manpower
Personnel
Training
Human Factors
Health Hazards
System Safety
Reliability
Maintainability
Logistics

Table 4-10
MANPRINT Design Analysis Technique (MDAT) Capabilities

Trade/Survey Tracking:
Aggregates and describes all trade studies/surveys conducted in association with new system designs.

Design Issues Tracking:
Documents MANPRINT design issues, problems, and concerns that occur at any time during the design process by all members of the design team.

Design Note Analysis Tracking:
Identifies design "high drivers" through engineering design notes and functional discipline design note analyses.

Design Decision Document Tracking:
Provides the status of documents associated with design analyses and decisions.

Task Analysis Tracking:
Compares designed tasks and task steps associated with predecessor, baseline comparison, and new system designs.

Life Cycle Cost Tracking:
Offers system specific development, operational, and support cost data information results from changes in the design configuration or changes in major subsystems.

MANPRINT Working Group Tracking:
Keeps a tally of MANPRINT Working Group activites associated with new system design development.

MANPRINT Action Tracking:
Shows MANPRINT design change options submitted in conjunction with new system designs.

Management Reports:
Summarize design note and task analysis data (organized by systems, subsystems or functional disciplines) for the status of a specific system.

Audit Trail:
Gives a historical record of functional discipline inputs maintained for the Design Note and Task Analysis areas of the MDAT program.

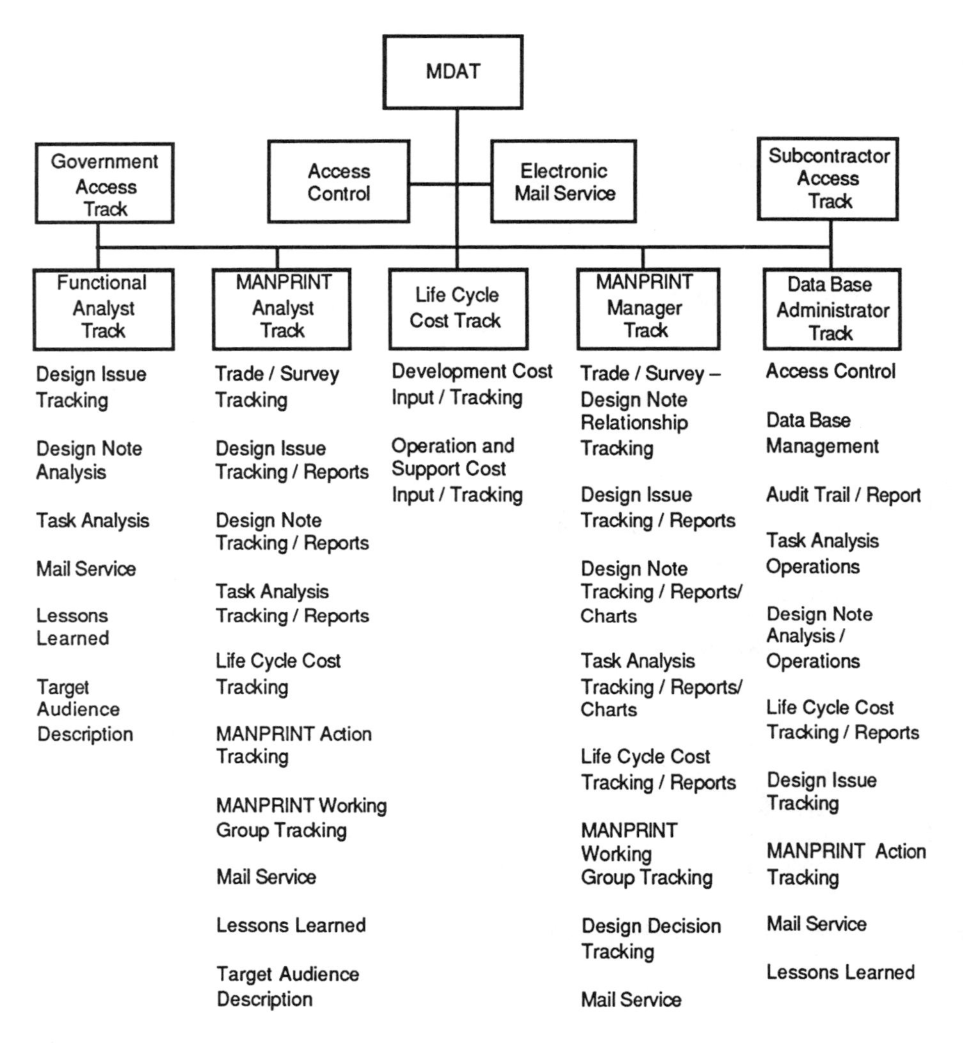

Figure 4-7
MANPRINT Design Analysis Technique (MDAT) Application

• Task Analysis Reports: These reports allow managerial personnel to input, review, and edit scoring, analyses, and comments of the various disciplinary analysts for each system and subsystem task.

• Design Note Analysis Reports: These reports generate the results of the design note review conducted by the functional analysts. A review of all functional reviews provides a total system perspective.

• System and Subsystem Analysis Reports: These reports give the manager the overall analysis status of a specific subsystem including number of notes, weighted score average, scoring range, and other sensitivities. They allow a comparison across two different systems or can be requested in a format to compare functional analyst reviews across their functional area.

Use of MDAT in the Design Process

Based upon initial customer requirements, trade studies and surveys are conducted to determine the best design approaches. This trade/survey information is entered into the MDAT system by a trade/survey tracking number, the person responsible for the information, the objective of the trade/survey, background information, a description of the trade/survey options, the design option selected, and the rationale for the selection.

These initial and design option selections are recorded as design notes for each system and subsystem and linked to the trade/survey documentation available.

To track system and subsystem design options, design note data (i.e., design note number, design note title, date of generation, cognizant project engineer, subsystem, and version) is entered into the MDAT. Finalized design notes are locked into the MDAT data base as a historical record when new revised notes are entered.

When the design note is generated, each functional discipline analyzes the note to determine the impact of the design option on their functional area. The functional analyst will determine the appropriate weighting score (Table 4-11) to assign to that design note (and a supporting rationale). Scores of 1 or 2 prompt the analyst to recommend a design change option and a description of the change necessary to reduce design impact. Notes may be changed as new information becomes available. Once a design

Table 4-11
MANPRINT Design Analysis Technique (MDAT) Weighting Criteria

Score	*Criteria*
4	No Impact
3	Minor Impact
2	Major Impact
1	Red Flag Item

note has been fully evaluated and accepted, a Design Decision Document is generated for final approval of the design. The MDAT program provides options for determining the status and date of approval of Design Decision Documents.

Upon approval of the design through the Design Decision Document process, performance task data is entered into the MDAT for operator, maintainer, and support personnel. Much the same way the design notes are weighted for subsystem impact, functional discipline analysts provide a weighting score and rationale for each task/task step. Numerous reports are available to evaluate the impact of the design upon all tasks and task steps. Changes may be made and recorded as a historical file.

At any point during the design process, any authorized user may generate a design issue. Design issues identify the subject area, the originator, date of identification, associated risks, and associated equipment, tasks, and functional areas affected. Design issues are publicized through memoranda generated at the analyst's request. Resolution to design issues are also recorded in memoranda and processed in the information system.

In this manner MDAT stores, reports, and analyzes information related to the continual proceedings of the MANPRINT Working Group, life cycle cost of the system design, as well as the data related to design task analyses.

MDAT as a Management Integration Aid

MDAT serves as a management integration aid by fostering the horizontal and vertical communication essential to an integrated design process. The MDAT information system supplements the organizational structure and the project management techniques necessary to achieve the interfaces among disparate disciplines and organizational entities. The information system also accomplishes the documentation (audit trail, restart point, concerns, rationale, and the like) that assures each discipline's opportunity to influence the design process and assures a plurality of opinion in the decision process. MDAT provides an opportunity to evaluate design alternatives in a modeling environment, thereby assessing the impact of various approaches and determining the sensitivity of operational and supportability factors to the design. It offers a continuous process for designing, rating, documenting, evaluating, and redesigning to meet MANPRINT objectives.

CONCLUSION

One of the most critical elements of a MANPRINT program remains the methodology with which the mission of "integration" is carried out. Integration defies easy explanation because it is a complex concept, often lacks definitive techniques, and applies to a variety of goals. Despite the inherent complications of integration, much can be accomplished through the application of the principles and techniques of modern managerial practices. Companies with successful integration programs demonstrate these principles and techniques. In addition to other tenets that support good organizational and project management, these firms exemplify one or more of the key MANPRINT elements (communication, physical proximity, horizontal processing, commitment, decision documentation, feedback mechanisms and effectiveness measures, flexibility, and design decision influence).

Management processes such as MANPRINT often have a difficult time getting started and, once started, experience difficulty in maintaining momentum. Their success is dependent upon the continued vigorous commitment of top management, refining the process as necessary, and clearly identifying the value added to the system being developed.

REFERENCES

Allen, T. J. (1986). Organizational structure, information technology, and R&D productivity (Vol. EM-33, No. 4). *IEEE Transactions on Engineering Management,* pp. 212-217.

Chaparral Steel (Abridged), Case #9-687-045 (1986). Cambridge, MA: HBS Case Services, Harvard Business School.

Cleland, D. I., & King, W. R. (1983). *Systems analysis and project management.* New York: McGraw-Hill Book Company.

Crosby, P. B. (1979). *Quality is free.* New York: McGraw-Hill Book Company.

Dell'Isola, A. J. (1974). *Value engineering in the construction industry.* New York: Construction Publishing Company, Inc.

Deming, W. E. (1982). *Quality, productivity, and competitive position.* Cambridge, MA: Massachusetts Institute of Technology, Center for Advanced Engineering Study.

Emery, F. E., & Trist, E. L. (1965). The causal texture of organizational environments. *Human Relations*, pp. 21-32.

Fayol, Henri (1949). *General and industrial management* (C. Storrs, Trans.). London: Sir Isaac Pitman & Sons. (Original work published 1916).

Forward, G. E. (1986, May-June). Wide-open management at Chaparral Steel (interview by A. M. Kantrow). *Harvard Business Review, 64*, 3, pp. 96-102.

Garvin, D. A. (1988). *Managing quality: The strategic and competitive edge.* New York: The Free Press.

Hall, E. T. (1966). *The hidden dimension.* Garden City, NY: Doubleday and Company, Inc.

Hamel, G., & Prahalad, C. K. (1989, May-June). Strategic intent. *Harvard Business Review, 67*, 3, pp. 63-76.

Harrison, F. L. (1985). *Advanced project management* (Second Edition). New York: John Wiley and Sons.

Hayes, R. H., Wheelwright, S. C., & Clark, K. B. (1988). *Dynamic manufacturing: Creating the learning organization*. New York: The Free Press.

Hiromoto, T. (1988, July-August). Another hidden edge – Japanese management accounting. *Harvard Business Review, 66*, 4, pp. 22-26.

Jaques, E. (1951). *The changing culture of a factory*. London: Tavistock Publications Limited.

Juran, J. M. (Ed.) (1951). *Quality control handbook*. New York: McGraw-Hill.

Kast, F. E., & Rosenzweig, J. E. (1985). *Organization and management* (Fourth Edition). New York: McGraw-Hill Book Company.

Kerzner, H. (1984). *Project management*. New York: Van Nostrand Reinhold.

Koontz, H., O'Donnell, C., & Weihrich, H. (1980). *Management* (Seventh Edition). New York: McGraw-Hill Book Company.

Lawrence, P. R., & Lorsch, J. W. (1967). *Organization and environment: Managing differentiation and integration*. Boston: Harvard University.

Majchrzak, A. (1988). *The human side of factory automation*. San Francisco, CA: Jossey-Bass Publishers.

Mali, P. (Ed.) (1981). *Management handbook: Operating guidelines, techniques and practices*. New York: John Wiley and Sons.

Mayo, E. (1933). *The human problems of an industrial civilization*. New York: The Macmillan Company.

McEwan, J. D. (1971). The cybernetics of self-organizing systems. In C. G. Benello & D. F. Roussopoulos (Eds.), *The case for participatory democracy* (pp. 179-194). New York: Viking Compass.

Patchin, R. I. (1983). *The management and maintenance of quality circles*. Homewood, IL: Dow Jones-Irwin.

Porter, M. E. (1985). *Competitive advantage: Creating and sustaining superior performance*. New York: The Free Press.

Texas Instruments, Inc. – Educational products. Case #9-683-001 (1982). Cambridge, MA: HBS Case Services, Harvard Business School.

Trist, E. L., Higgen, G. W., Murray, H., & Pollock, A.B. (1963). *Organizational choice: Capabilities of groups at the coal face under changing conditions, the loss, rediscovery, & transformation of a work tradition*. London: Tavistock Publications Limited.

TRW Systems Group (D). Case #9-413-066. (1968). Cambridge, MA: HBS Case Services, Harvard Business School.

Van Cott, H., & Kinkade, R. G. (1968). Human simulation applied to the functional design of information system. *Human Factors, Vol. 10, No. 3*, pp. 211-216.

von Bertalanffy, L. (1950, January 13). The theory of open systems in physics and biology. *Science*, pp. 23-29.

SUGGESTIONS FOR FURTHER READING

Cleland, D. I., & King, W. R. (1983). *Systems analysis and project management.* New York: McGraw-Hill.

Karger, D. W., & Murdick, R. G. (1980). *Managing engineering and research.* New York: Industrial Press, Inc.

Kast, F. E., & Rosenzweig, J. E. (1985). *Organization and management* (Fourth Edition). New York: McGraw-Hill Book Company.

Kerzner, H. (1979). *Project management.* New York: Van Nostrand Reinhold.

CHAPTER 5

MANPRINT IN A SYSTEMS ENGINEERING ORGANIZATION

John B. Shafer

ABSTRACT

The intent of this chapter is to show how a large, complex company institutionalized an organizational and management process which integrates people, technology and organizations. This was accomplished primarily through major changes within its systems engineering practices and through education and training of its systems engineers. More specifically, it discusses (1) the guiding principles and practices for the company's current systems engineer, (2) the tools that enhance the systems engineer's capability to implement the practices, and (3) the courses used to train systems engineers.

BACKGROUND LEADING TO INSTITUTIONAL CHANGE

Systems integration in a large competitive business requires long-term investments in comprehensive integrated design processes. Successful applications focus on balanced designs which effectively integrate people concerns into equipment and software design. People concerns should be given equal if not greater weight at the beginning of the development cycle not only to ensure effective short-term solutions but also a competitive future position for the company.

In the past, this approach to systems integration was not fully appreciated since customers were eager to use new technology and any leading technology breakthrough would provide a comfortable competitive market. In this environment, hardware and software engineering managers tasked with developing new systems tended to focus on their areas of expertise and place little emphasis on improving the user interface. Technical management did not have an adequate process for continuously producing new high quality systems on a predictable and competitive schedule.

In the late 1970s, it was recognized, however, that the key to continuous improvements in system design lay in improvements to the systems engineering process. At the direction of top management of IBM Systems Integration Division, a systems engineering board was formed comprising representative systems engineering managers from the Division's sites spread across the country. Recognizing the importance of a central focus with strong technical input, the board recommended the formation of a focal group at Division Headquarters. Staffed with its best system engineering talent, the focal group was formed and immediately instituted a Division-wide Systems Engineering Excellence Program Plan. There were three major elements to the plan (Figure 5-1) under the headings of practices, tools, and training. To ensure middle management would evolve a better systems engineering process, executive level managers were to use Program Management Reviews (PMR) and System Assurance Reviews (SAR) to monitor risk, progress and quality of the process. Finally, it was recognized that the systems engineering process was incomplete without the full consideration of the system user. The integration of the user into the system engineers practices, tools, and training became the impetus for a major institutional change in the Systems Integration Division at IBM.

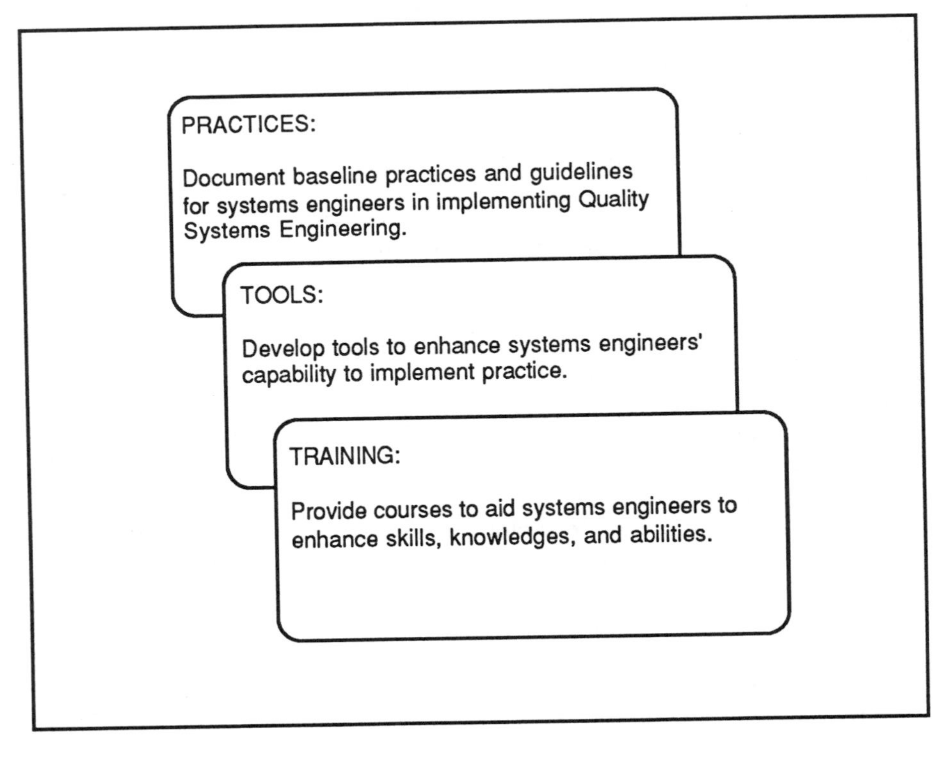

Figure 5-1
Systems Engineering Excellence Program Plan

Many of the principles of organizational change outlined by Blanchard and Blackwood (Chapter 3) will be seen to apply to the company change. The effects on product quality from MANPRINT as discussed by Booher and Fender (Chapter 2) are directly relevant. Because this book is very much concerned with quality management and the effectiveness of institutional change in organizations, the editor, Dr. Booher, asked me a number of questions about the experience we had at IBM. Questions and answers are presented in Table 5-1.

The following sections of this chapter show more specifically how MANPRINT principles were applied to its systems engineering process. The chapter also describes training programs and standards used in bringing about institutional change and shows methods used to increase continuous process improvement.

BASELINE PRACTICES

A primary goal of the systems engineering board was to document baseline practices (IBM, 1987) to guide the systems engineer through the systems development process (Figure 5-2). The foundational standard (or "practice") specifies *quality systems engineering.* The systems engineer is required to apply the practice document but may, depending upon the project requirements, apply other documents (called "bulletins") as appropriate. The bulletins of Figure 5-2 include software and hardware interfaces for the systems engineer but only indirectly spell out user interface requirements. This is provided in a special bulletin developed by human factors specialists.

User-Computer Interface Bulletin

A Human Computer Interface Requirements Specification (IBM, 1988) (hereafter referred to as the User-Computer Interface Bulletin) contains specific details arranged in two levels for application to different stages of product development. The first level, for example, presents requirements best written in the front-end of a program whereas the second level presents the details more relevant to the later stages of development.

The user interface requirements comprise the following areas: user specification; function and task elaboration; accessing the computer; input devices; output devices; major interface considerations; human-computer interface configuration; and documentation.

When the User-Computer Interface Bulletin has been presented to human factors engineers, one of the questions commonly asked has been: "What is the relationship of the User-Computer Interface Bulletin to other human factors guidelines and standards such as *Guidelines for Designing*

Table 5-1
Questions and Answers About the Institutionalization of MANPRINT in an Industrial Organization

Questions	Answers
1. The MANPRINT approach states that someone or several people who understand what MANPRINT means to systems integration need to be present in top level decision positions in order to provide the necessary voice, authority and support to institutionalize change in the organization. Who at IBM provided this top level support for the change? What specific actions did they take to bring about the change? Also, who in a mid-management position became the change agent (or agents) to carry out top-level policy?	1. The president of the Systems Integration Division decided to address the change, convening a task force of systems engineering managers to determine what to do. A permanent systems engineering staff with a "middle-manager" was created at headquarters to become the "change agent."
2. EDUCATION AND TRAINING. Both the MANPRINT and Total Quality Management (TQM) approaches emphasize education and training for all people in the process. What does IBM see as the relationship between its TQM training and your MANPRINT training?	2. The training courses that pertain to the MANPRINT domains are subactivities of the Systems Engineering Excellence Plan which is a TQM plan.
3. USER REQUIREMENTS AND SPECIFICATION. MANPRINT makes user requirements a part of system specifications and test plans to ensure the designer understands that "the system" includes the user and that system performance effectiveness measures are made with a real user. How does IBM assure that the human factors requirements are always included as an institutional feature of the system engineering organization since human factors does not appear to be an inherent part of either the Practice or the supporting Bulletins (which themselves are only guides)?	3. The MANPRINT domains are addressed in various subsections of the Corporate Bulletins C-B 0-2507-022 (operational need) and C-B 0-4010-002 (software requirements specification). The implementation of the practices and bulletins are left to the project systems engineering management.
4. INTERDISCIPLINARY. How do the MANPRINT disciplines, (e.g., human factors engineering, safety, health hazards, manpower, personnel, training) play as equal partners in IBM system engineering trade-off decisions? Also, how do the various people-oriented disciplines provide any integrated front to those who make people/hardware trade-off decisions?	4. The lead systems engineer has the responsibility to ensure that the MANPRINT disciplines play as equal partners and are appropriately integrated into people/hardware trade-off decisions.
5. SCAN SYSTEM ENVIRONMENTS. Blanchard and Blackwood (Chapter 4) point out this is the first step of change management in which there are numerous factors in the various environments (external, intermediate, and internal to the organization) which tend to force or indicate the need for change. What were some of these environmental factors that drove IBM to change and what is now in place at IBM to continue scanning, monitoring, and interpreting environmental developments so that change is truly institutionalized and will not slip back to the "old ways"?	5. IBM's desire for continued growth was the stimulus for change. Growth meant expanding into new business areas and changing the systems engineering process to address the expanded business areas. IBM marketing groups continually monitor IBM's success in the expanding areas. Executive management with support from Division headquarters staff perform Project Management Reviews (PMR) and Systems Assurance Reviews (SAR) to ensure the quality of each project's systems development process.
6. PROBLEM DEFINITION PLANNING. What were some of the past practices, beliefs or values that could be considered IBM cultural values that had to be challenged with the MANPRINT change? How did IBM's change agent(s) deal with this?	6. It has long been assumed that if the hardware and software requirements of a system were satisfied the system would be a success. The production of the Human Factors in Systems Engineering (HFISE) course and Human Computer Interface Requirements Specification (HCIRS) are meant to deal with that cultural attitude and to include management of the development of the user interface as well.

Table 5-1 (*Continued*)

Questions	Answers
7. CHANGE INTERVENTION STRATEGY. What was the specific strategy for change intervention? How was resistance to change by those identified in Question 6 handled? How was uncertainty of the "new way" and turf conflicts handled? Were goals identified which would fulfill individual and group needs that at the same time were consistent with the planned change goals? How was information disseminated and feedback (briefings, bulletins, newletters, etc.) accomplished? What plans for test and evaluation to demonstrate the worth of the change were made?	7. The specific change intervention strategy was for top management to direct the implementation of the Systems Engineering Principles and Practices through a document called a Management Instruction. Since resistance to change came from projects with existing contracts which did not require the system engineering process, the Management Instruction only applied to future systems engineering contracts. Besides the Management Instruction, information about the Systems Engineering Excellence Plan was disseminated through staff meetings and the systems engineering courses. To demonstrate the worth of the program, certain projects such as Combat Talon II and Army Special Operations Aircraft (SOA) were selected for implementation. No plans for test and evaluation of the Systems Engineering Excellence Plan were made.
8. CHANGE PROGRAM IMPLEMENTATION. How well was IBM able to carry out the plan? Give one or two interesting examples of unfreezing. Was a project selected and failure-of-old-way or success-of-new-way documented to convince others? How did certain hold out attitudes become unlearned? How did they learn new behaviors?	8. The Systems Engineering Excellence Plan is still evolving with an increasing number of contracts applying the principles and practices each year. In terms of "unfreezing" there are some dramatic career limiting situations where the systems engineering manager failed to take user concerns into account. In one situation the customer refused to accept the system and top management directed the user interface to be redesigned. Such situations are used as negative examples in the Systems Engineering Principles and Practices (SEPP) course and have a strong influence on career oriented systems engineering managers.
9. CHANGE PROCESS. How did change managers confirm that people were motivated to adopt the new process? What new concepts and information were provided? How was (or is) progress monitored? What backup is there in case of weakening or backsliding of the program?	9. Feedback from the systems engineering courses convinced the "change manager" that people were motivated to adapt. Progress is monitored through Project Management Reviews and Systems Assurance Reviews. If the Systems Engineering Excellence Plan were to weaken, a new program would replace it.
10. INSTITUTIONALIZE CHANGE. How well does IBM carry out a monitoring system? How are the variables mentioned by Blanchard and Blackwood (socialization, commitment, reward allocation, diffusion, sponsorship) used?	10. The monitoring process is rather informal using feedback through courses, Project Management Reviews and System Assurance Reviews to the headquarters staff. The variables for institutionalization are operating but they are not used as a set of criteria to assess change.
11. EVALUATION CHANGE PROGRAM. Does IBM have an ongoing assessment program as envisioned by Blanchard and Blackwood for MANPRINT? I doubt it, but maybe it could tie to the corporate overall evaluation program for systems engineering or a Total Quality Management Program.	11. IBM continually assesses the health of its business including the efficiency of the systems engineering process. The measures of health deal with the costs and profitability of the systems engineering process. The effectiveness of the systems engineering courses are monitored on a class-by-class basis and are upgraded as required.
12. FEEDBACK TO ENVIRONMENTS. How has IBM advertised results of implementing MANPRINT (human factors) both outside and within IBM?	12. The systems engineering courses (SEPP and HFISE) have been described in the IBM publication "FSD Technical Directions" which reaches 10,000 people outside of IBM. The HCIRS has been presented at National Electricians Association Conference and will be a Behavior and Information Technology journal article.

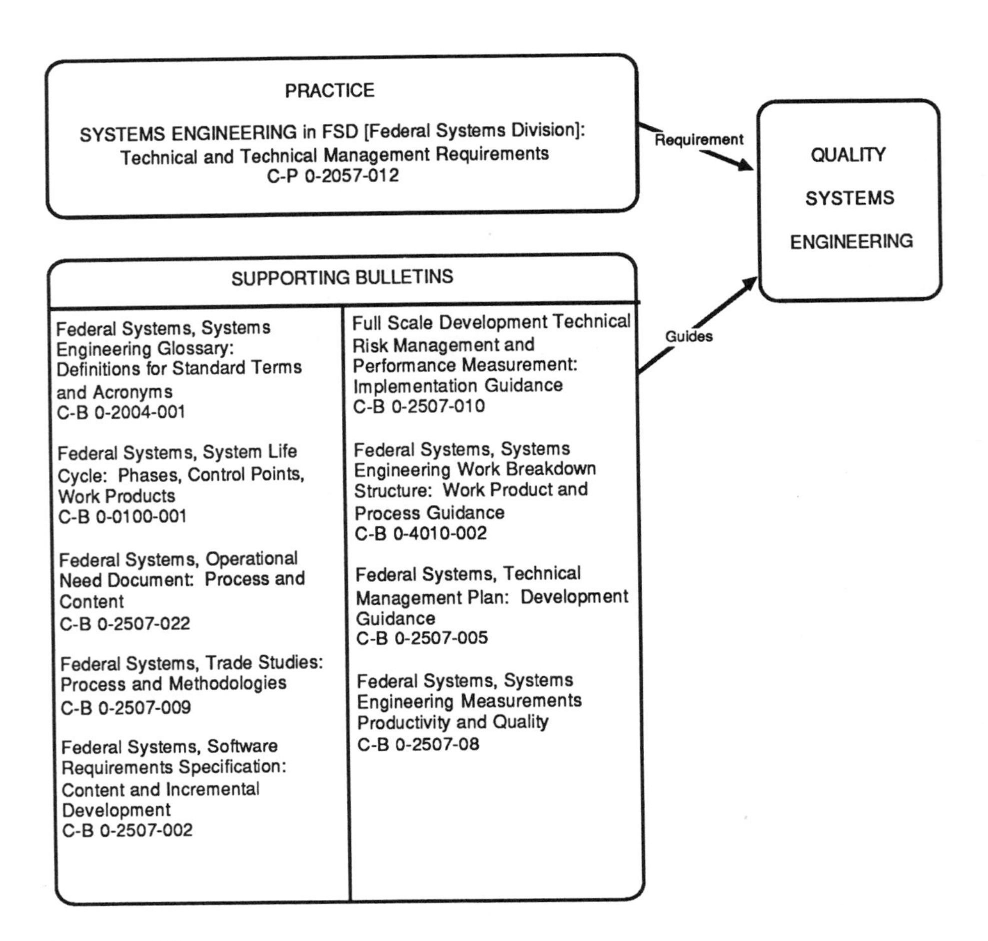

Figure 5-2
Systems Engineering Practice and Bulletins

User Interface Software (MITRE Corporation Guidelines) (Smith & Mosier, 1986), *Military Standard 1472D* (MIL-STD-1472D, 1989), and the *American National Standard for Human Factors Engineering of Visual Display Terminal Workstations* (ANSI/HFS 100-1988)."

MITRE Corporation Guidelines: The MITRE guidelines provide general guidance for human computer interface design, but requires greater designer interpretation. For instance, the MITRE guidelines would ask the engineer, if he were specifying the form of dialogue, to consider a number of different dialogue forms but would leave the alternatives choice to the designer. The IBM bulletin must specify exactly the form of dialogue so that the programmer may code to that format.

MIL-STD-1472D: This military standard also provides general guidelines and useful information for the design of the human-computer interface. Like the MITRE guidelines, it can be useful when combined with human factors analysis in developing project specific requirements. But they are only guidelines, and it is the human factors engineer who must select the specific design point.

ANSI/HFS 100-1988: The ANSI/HFS technical standard specifies requirements for visual display terminals, associated furniture and the office environment where the use of a visual display terminal is required. Applications considered are text processing, data entry, and data inquiry. While some visual display parameters, keyboard designs and measurement techniques are covered, the greatest emphasis in the standard is on the anthropometrics of workstation furniture (workstation surfaces, seats and footrests). Many topics (i.e., dialogues, menus, languages, response time, the development process, etc.) related to visual display workstations are not covered by this standard but are in the IBM User-Computer Interface Bulletin.

User-Computer Interface Process

The process for generating a user-computer interface specification is outlined in Figure 5-3 as a series of eight steps, broadly categorized to answer four basic questions:

Who is the user?
What is the user's job?
How does it work?
Does it work?

These basic MANPRINT questions are also described in several of the earlier chapters of the book. Shields, Johnson, and Riviello (Chapter 10), for example, use very similar questions in the acquisition decision process and Booher and Fender (Chapter 2) describe the "voice of the user" in quality management terms.

Who is the User?

Determining who is the user has been broken down into two steps (Steps 1-2 of Figure 5-3):

Step 1. Generate an Operational Concept: The generation of an operational concept follows a statement of an operational need. Once it has been determined that a system is needed, then the concept for deploying

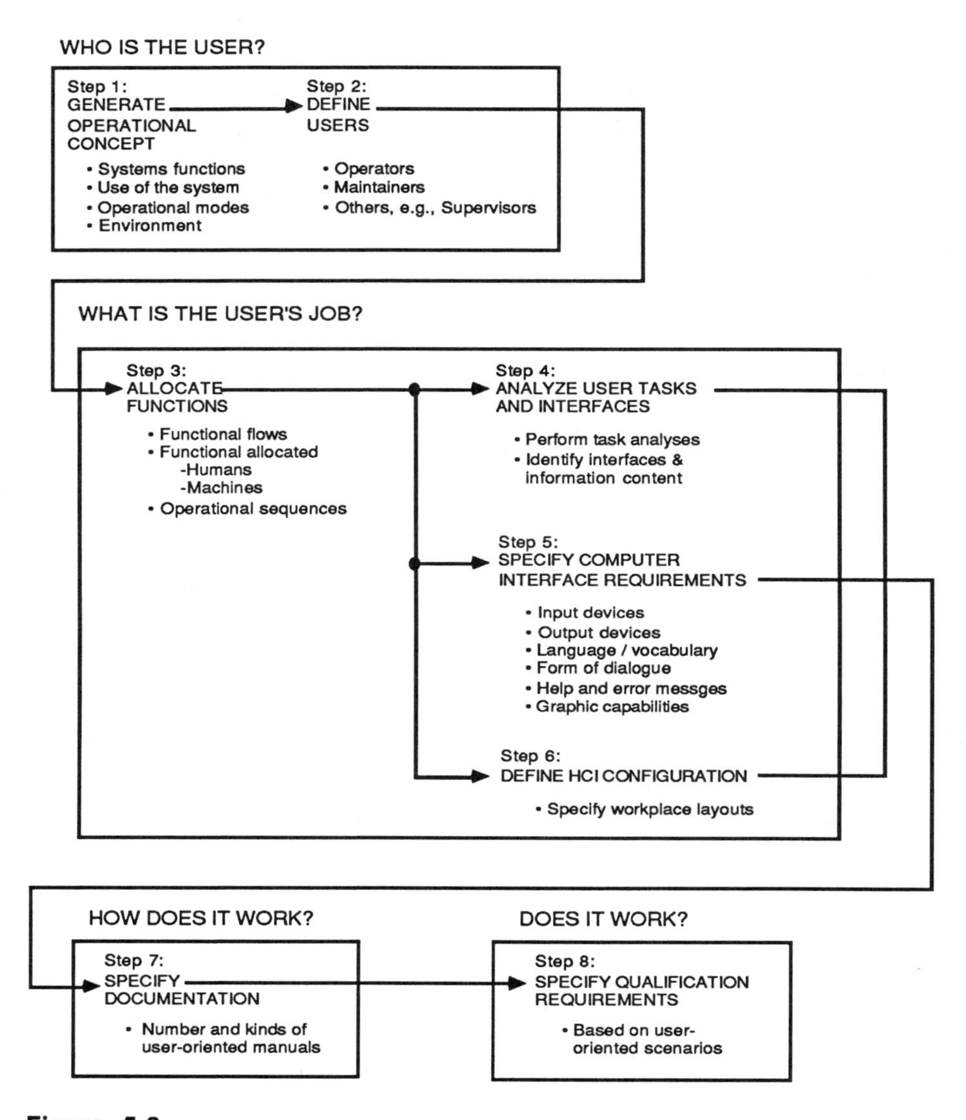

Figure 5-3
Human Computer Interface (HCI) Requirements

and operating that system must be defined. Usually the operational concept is a top level description of how the system functions will meet the operational need. This will usually take the form of a scenario against some form of mission timeline. There may also be functional flow diagrams which decompose the system to a level where the functions can be allocated. It will often indicate how users are expected to use the system in various operational modes and any special environmental conditions that will constrain the operation of the system. The key factor in generating an operational concept is to be sure that it is user oriented and generated in sufficient detail to permit further analysis.

Step 2. Define Users: Although operators are frequently thought of as users of the system, they are not the only users. Maintainers certainly are among the population of users, and they represent a significant impact on system design. In addition, there will frequently be a system supervisor (as in command control systems) with various levels of supervision. Typically, there will also be an operations manager to ensure system availability. For complex systems, it is not unusual that each one of these user types (operators, maintainers, managers, etc.) will require a significant different human-computer interface.

What Is the User's Job?

IBM has divided the task of determining user jobs into four steps (Steps 3-6 of Figure 5-3):

Step 3. Allocate Functions: Allocation of functions is a procedure for assigning each system function to hardware, software, the user, or some combination of them. Function allocation is an iterative process that trades off the state of the art in equipment and software against the capabilities and limitations of the users. Once the functions have been allocated, then the systems engineer has to define the baseline for crew task definitions, crew skills and information needs, estimates of staffing and training requirements, workload assessments, and crew station configuration concepts.

Step 4. Analyze User Tasks and Interfaces: True integration of a system takes place through task analysis. A particular user's task cannot be defined without an understanding of how the equipment and software work together to permit the user to perform that task. The task analysis (see e.g., Van Cott & Kinkade, 1972, Figure 1-4) must be performed from a user orientation, and it will define action requirements, information requirements, environmental considerations, working conditions, training requirements, and staffing requirements.

These types of information are essential for each crew station and each user interface. The task analysis may take various forms. One of the most popular forms is the operational sequence diagram (OSD) (see e.g., Van Cott & Kinkade, 1972, Figure 1-7) which shows the interaction and integration of various elements of a system over the course of a number of events. The OSD is particularly useful to those systems engineers who may wish to simulate the real world situation with some form of rapid prototyping.

Step 5. Specify Computer Interface Requirements: Once the interaction and information of the task has been defined, then the systems engineer must specify the devices which will permit that task to be performed. This step is the central focus of the IBM bulletin. Within this step, the systems engineer must specify input devices, output devices, language/vocabulary, form of dialogue, help and error messages, and graphic capabilities.

Step 6. Define Human-Computer Interface (HCI) Configuration: The task and the devices used to perform that task are usually done at a workstation. The workstation configuration must also be defined. The human factors engineer will prepare diagrams showing the placements, locations, dimensions of input devices, output devices, workspaces and auxiliary equipment such as communication equipment. A number of human factor techniques may be used to determine the configurations such as link analysis, anthropometric analysis, reach and vision envelopes and mockups. As can be seen from the diagram in Figure 5-3, Steps 4, 5, and 6 may very well be done in parallel and be traded off against each other. It may be that the workspace will not hold all the devices that are required to perform the task. In that case then the task may be reallocated. It may also be that the information which is to be presented may not fit on a particular display device. It may be better presented through voice or some other medium. As the design progresses, these kinds of trade-offs will gradually iterate and converge into a design.

How Does It Work?

Step 7. Specific Documentation: There will be some form of documentation for each type of user, and there very well may be many forms of documentation for each user. Since the task analysis was user oriented, manuals and documentation that are based on the task analysis will also have a user orientation. For instance, a device oriented training document might say, "Here is the display format and this symbol shall be used to designate a target" leaving the user to discover how to operate the system. Whereas, a user orientation might say "Your task is to designate a target and these are the following steps." The advantage of considering documentation at this point, whether it be hard copy or electronic, is that a number of downstream processes will be enhanced. The later training and the logistics of handling documentation will be much simpler.

Does It Work?

Step 8. Specify Qualification Requirements: The specification of a user interface must be done in such a way that it can be tested. Each statement that specifies an interface should be capable of being reduced to operationally defined terms that will permit quantitative evaluation. The human factors engineer can, for example, specify the task to be performed by a defined user population within a range of actions in some amount of time within some error limit. User-oriented operationally defined terms permit accurate usability tests (e.g., rapid prototyping) to be performed throughout the development cycle.

User Interface and the Development Process

There are different approaches to inserting the design of the user interface into the development process. One would be to generate the documentation and rationale, using the process that was just described, and then insert that data into the Software Requirements Specification. When this is done, there is a risk that the user interface will be described from a software orientation. In fact many years of experience with that approach has shown the risk to be very real and one of the factors giving impetus to MANPRINT. A more direct effort is suggested wherein the human factors engineer is tasked to provide design documentation and rationale at designated control points in the development process. One approach to the timing of those tasks and responsibilities is shown in Figure 5-4.

System Development Control Point	User-Computer Interface and Related Work Products	Responsibilities: Lead/Support
	Operational Concept	SE/HFE
SRR _____		
	User Interface Requirements	HFE/SE
	User Interface Baseline Design	HFE/SE/SWE
	User Interface Requirements and Design Rationale	HFE/SE/SWE
	User Interface Prototype Implementation (Baseline Design)	SE/SWE
	User Interface Prototype Evaluation	HFE/SE
SDR _____	User Interface Requirements Decomposition and Allocation	SE/SWE
	Software Requirements Specification	SE/SWE
SSR _____	Software Top Level Design Document	SWE/SE/HFE
	Level 2 Software Design	SWE/SE/HFE
PDR _____	User Interface/Software Requirement Updates	SE/SWE/HFE
	User Interface Prototype Updates	SE/SWE/HFE

LEGEND

SRR - System Requirements Review
SDR - System Design Review
SSR - Software Specification Review
PDR - Preliminary Design Review

SE - Systems Engineering
HFE - Human Factors Engineering
SWE - Software Engineering

Figure 5-4
User-Computer Interface Timing and Responsibilities

This chart points out that the human-computer interface development process must be product oriented to permit the assignment of responsibilities and management commitment. Clearly it shows the major design effort must be completed before the System Design Review and is a combined effort of systems engineering, human factors, and software engineering. The software architecture of a system must be based on the user's tasks which in-turn are derived from the operational need and operational concept. Practically all of the human computer interface design will be completed before the Preliminary Design Review. If the user-computer interface development process is not well executed before this point, there is a high risk of a poor user interface with little chance to recover.

THE TOOLS

One of the advantages of assigning the construction of the user-computer interface documentation to human factors is that the selection of tools and methods to generate the data required becomes the responsibility of the human factors engineer. The methods available to the human factors engineer to solve the design challenge have evolved over many generations of practical applications. The informality of that evolution led the National Academy of Sciences to formally document these methods for future research and education (National Academy of Sciences, 1983). The flow diagram in Figure 5-5 is a summary of that effort and indicates there is a relationship among the methods. The flow of methods is universal to problem-solving and is very similar to the systems engineering development process. Some of the methods have their origin in other disciplines but have been adapted to become tools and techniques for human factors. Not all types of analyses are shown on the chart. Accident investigations, anthropometrics analyses, and human error analyses are others that may be used in the development process to address MANPRINT issues. Not all these methods are used in every development project, and there are typically feedback loops, not shown in the figure, that require the human factors professional to go back into the sequence to deal with unexpected findings. For example, a workload assessment might disclose that the projected workload for an operator or maintainer might be more than one person can handle. This might then require the human factors professional, together with the systems engineer, to reexamine functional allocation decisions made earlier giving emphasis to automating some functions that had not been previously automated, thus relieving the operator of some of the workload.

The application of these methods is iterative since they are applied with various degrees of precision throughout the system development cycle. Initially, during concept exploration (see Shields, Johnson, & Riviello, Chapter 10), the methods may be applied roughly by relying on expert

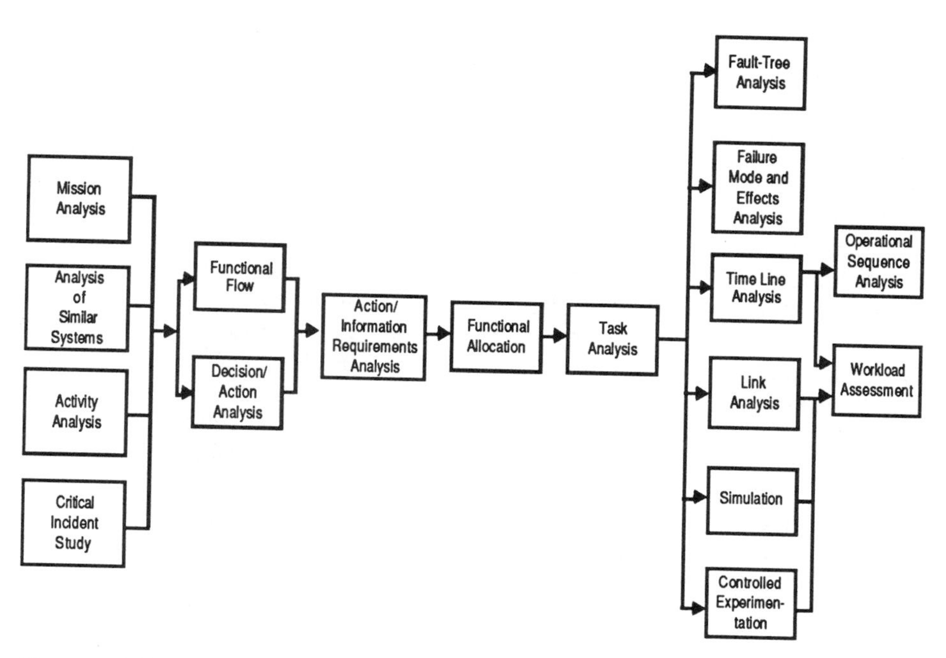

Figure 5-5
Sequence of Methods

opinion or previous data, to decide whether specific system concepts are viable from an operator or maintainer point of view. At this stage, the products of the human factors specialist are inputs to the first draft of the system specification and to the System Requirements Review (SRR).

During concept demonstration and validation, as attention moves from overall system definition to a consideration of particular items of equipment, human factors methods are applied with somewhat greater precision to provide inputs to the final draft of the system specification, the System Design Review (SDR), and the first drafts of the lower-level specifications, that deal with configuration items.

During the initial stages of full-scale engineering development, human factors methods are applied with still greater precision for the final drafts of the specifications, the Preliminary Design Review (PDR), Critical Design Review (CDR), Test Readiness Review (TRR), and the operational evaluations.

When the system is in use by the customer, the methods, especially Activity Analysis and Critical Incident Studies, are useful primarily for discovering unanticipated difficulties that may have appeared during use. These provide inputs to engineering changes.

Methods vary in complexity. Some are simple, involving no more than paper-and-pencil analyses, others are more complex and may require mockups, simulators, or elaborate experimental laboratories. One thing common to them all, however, is that they take time. Maximum effectiveness of the systems engineering-human factors team occurs when the partnership is formed early in a program. Not only does that permit human factors to prevent gross errors in user versus system task allocation before such errors become expensive to correct, but it also allows adequate lead time for simulation and the application of other methods.

Data that are gathered through the application of various methods are gathered for the purpose of producing a design specification that will lead to the production of the system. The design specification must reflect a particular operator-machine interface which may be modeled in various ways. The model IBM uses shows that the operator may be conceived as a system element and that one model of an operator parallels the model of machine functions (Figure 5-6). That ideal is further elaborated in Figure 5-7 which shows that the stimuli or inputs to the human operator are displays (dials, gages, CRT's, buzzers, annunciator systems, others). These displays have to be sensed, the information has to be processed and human decisions have to be affected through controls (cursors, aircraft controls, cranks,

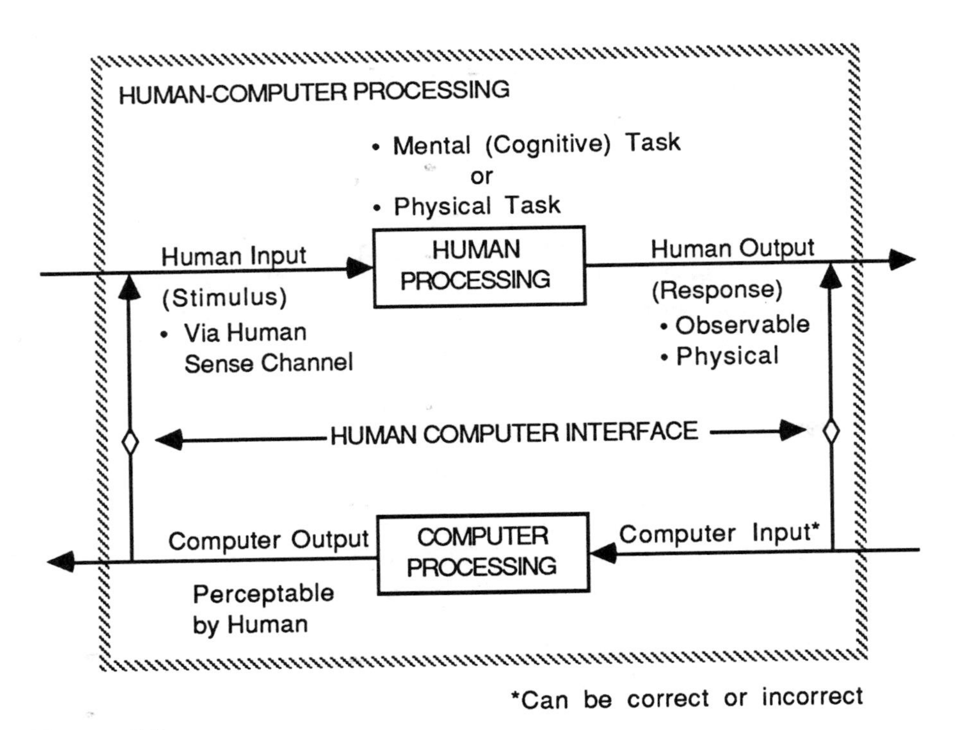

Figure 5-6
Parallelism Between the Machine and Operator Needs

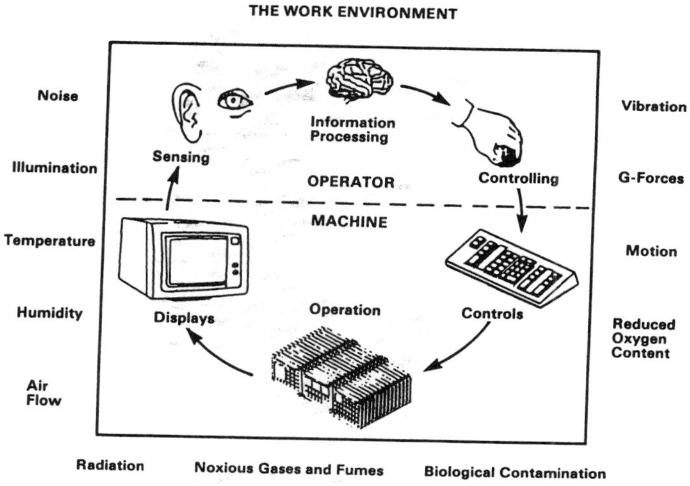

Figure 5-7
Operator Interfaces with Machines and the Work Environment

levers, switches, keyboards, and pedals). Control actions change the behavior of the machine, which in turn changes the displays so that the cycle is complete. The operator-machine combination does not work in a vacuum, it works in an environment of some kind and the border of the illustration shows the environmental factors, some commonplace, others more exotic, that influence human and system performance.

Although the model is a simple one, it is one way of viewing many human-machine interactions such as driving an automobile, operating a small computer, flying an aircraft, or using a power lawn mower. Moreover, it is a useful way of structuring information about people and applying that information to the design of the operator-machine interface.

THE TRAINING

It is rare indeed to find a systems engineer with the natural talent to both technically manage and at the same time comprehensively integrate all the factors to be implemented in the systems development process. But given the right aptitudes and inclination, it is possible through training and years of practice to "technically grow" a considerable number of highly competent engineers. That technical growth process for IBM systems engineers has been enhanced by a special three-day course called "Systems Engineering Principles and Practices" or simply SEPP (Budurka, 1984).

Creating the SEPP Course

With the increased intensity for competition, the need for people skilled in systems engineering was increased. The systems engineering board was keenly aware that practicing engineers were searching for courses and proposing training programs. Courses found at universities were limited to theoretical or generic approaches thus prompting the creation of the SEPP course. It was determined that the SEPP course should:

- Emphasize those principles and practices that remain constant across customer applications.
- Be practical, founded on sound technical principles, and focused on the production of technical and management work products.
- Concentrate on what systems engineers need to do to maximize the added value of engineering specialties (i.e., MANPRINT domains).
- Prepare a structure for the evolution and adaptation of standards that would emerge in future years.

The focal group coordinated the work with the help of many experts from across the division. The course developers tested and validated their material through a long succession of reviews and dry runs with extensive constructive critiques. In a matter of months a practical and focused SEPP Course became a reality.

SEPP Course Structure

The SEPP Course (IBM, 1989) is structured to provide the systems engineer with a top level awareness of the systems engineering elements. There are a series of follow-on courses on special subjects (i.e., software, human factors, etc.) which carry particular elements to greater depth. Since the first introduction of the five-day SEPP Course in 1982, the course has undergone a series of revisions and streamlining so that today the course is three days long. The SEPP Course (Figure 5-8) is presently focused on three major areas:

- The technical work the systems engineer has to do through a prescribed process resulting in a definitized set of work products.
- The engineering specialties including the MANPRINT domains that the systems engineer has to integrate.
- An introduction to the technical management of the systems engineering process.

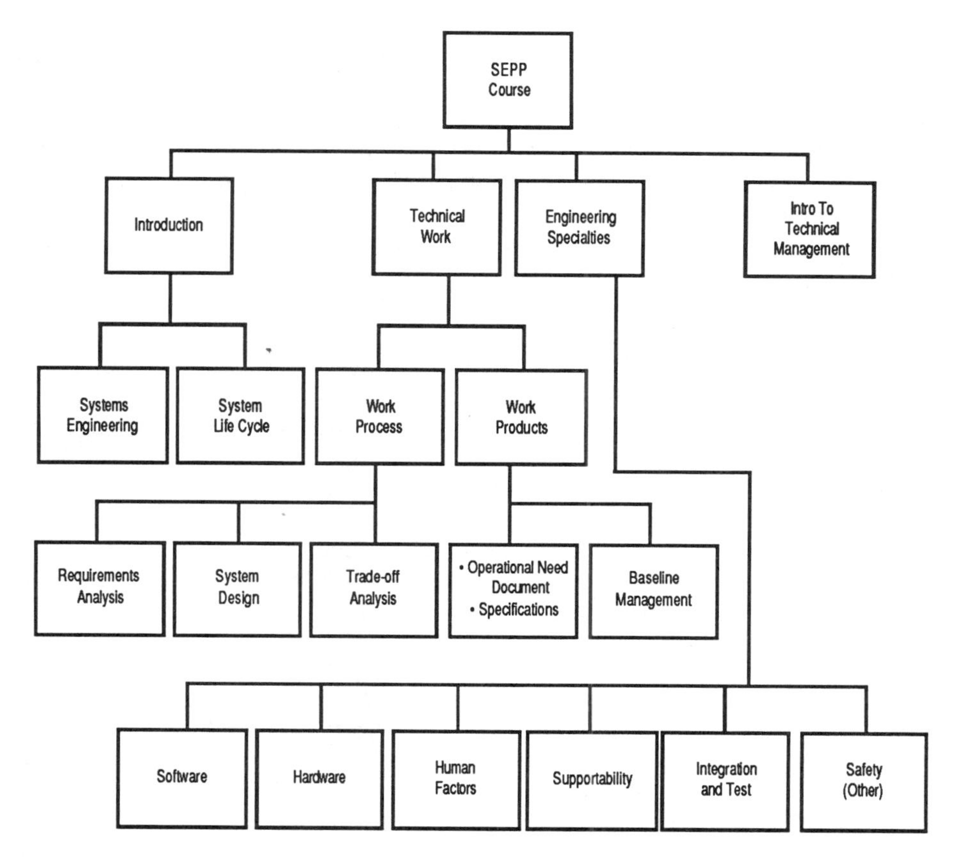

Figure 5-8
Systems Engineering Principles and Practices (SEPP) Course Structure

SEPP Progress

The SEPP Course continues to evolve and change reflecting the dynamic needs of the business, but the basic tenants of the system engineering process remain constant. Students are exposed to these basic tenants and the breadth it encompasses. Their performance on the job after taking SEPP indicates enthusiastic support for practicing a thorough and comprehensive development process.

Human Factors in Systems Engineering Course

Providing courses for the training of systems engineers extends beyond the SEPP. A number of spin-off courses have been developed to give the systems engineer more in-depth training in special subjects. One of those was the very successful two-day Human Factors in Systems Engineering (HFISE) course (Chapanis & Shafer, 1986). A basic tenant of the course is that system performance depends upon human performance, and it is the systems engineer's responsibility to ensure that human factors expertise is applied to maximize system efficiency. Systems engineers usually do not naturally take a user orientation when developing a system. They find it difficult to interpret the "soft" human factors inputs and integrate those inputs into their designs. The two-day course provides a structure that clarifies what human factors can be and demonstrates the necessity of early human factors contributions to the systems development process.

Although there are a great many short courses on human factors taught in universities and in professional development seminars, they did not meet our needs. Our review indicated that the academic approach often concentrates on measuring, developing, and evaluating human factors data, but fails to show how those data are used in systems development. We sought to prepare a course that spoke directly to systems engineers, showed them how human factors fit into the systems development cycle, how human factors professionals went about their work, and what work products could be expected of the human factors professional. The result is a course that is oriented towards systems engineers and the systems engineering process. Two additional features of the course that have contributed greatly to its popularity are: (1) a number of student participation exercises to illustrate principles, and (2) actual examples of systems and systems components that have benefited from human factors in their design.

Organization of Human Factors in Systems Engineering Course

The two-day course consists of a sequence of seven teaching modules, as shown in Figure 5-9. Each module, which lasts from three quarters of an

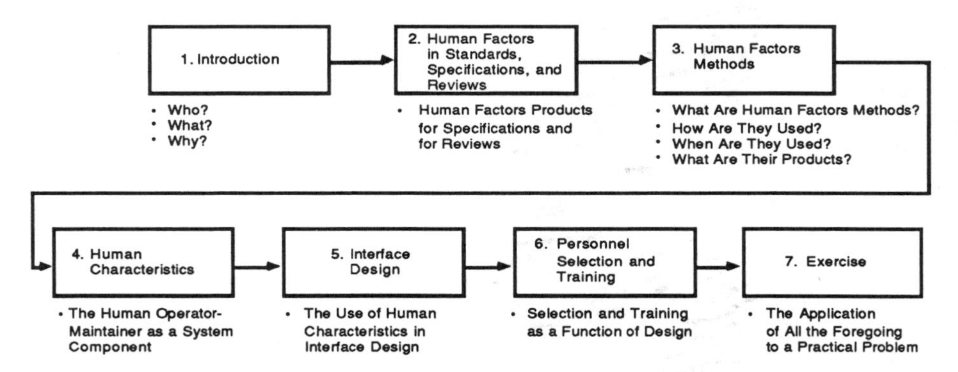

Figure 5-9
Human Factors Teaching Modules

hour to three hours, is devoted to a specific topic. The seven modules are: (1) introduction, (2) human factors in standards, specifications and reviews, (3) human factors methods, (4) human characteristics, (5) interface design, (6) personnel selection and training, and (7) exercise.

As the figure suggests, and as later sections will point out, the course subject matter flows from one topic to the next.

Module 1: Introduction

It is the systems engineer's responsibility to provide the soldier with the system he needs to get the job done. When the first warrior selected a club, he did not select one that was too light or one that was too heavy, but one that was an extension of his own fighting style. With that challenge the course begins. Credibility and the necessity for a user orientation are developed through a series of dramatic presentations of a disaster, government documents, articles, and advertisements. Students participate in two stimulating exercises that subtly but firmly draw them into a human factors viewpoint.

Module 2: Human Factors in Standards Specifications and Reviews

Most government contracts to develop systems come with an overwhelming structure of standards, specifications, and reviews. The structure is there to reduce the risk of failing to achieve user requirements and to ensure a good business relationship. Some commercial customers contracting for large scale systems integration fail to impress such a structure on the systems development. This module points out that tempting as it may be for the systems engineers to "wing-it" without a structure it is a dangerous business risk. Whether it is a government or IBM structure, good business and system development practice require appropriate standards, specifications, and reviews be applied for a successful project.

As a government contract is developed, there are literally hundreds of requirements the systems engineer may encounter. The module focuses on the principle specifications and standards that are most commonly involved and that contain human factors requirements. These (Figure 5-10) are grouped into two major categories: (1) those the systems engineer will most likely be required to meet, and (2) those that are primarily for the human factors professional. As the figure shows, there is a kind of continuity, or flow, to these various documents because each cites others which ultimately lead directly to those concerned with human factors. The module then discusses several of these documents in greater detail and points out similarities and differences among them. The module concludes with a

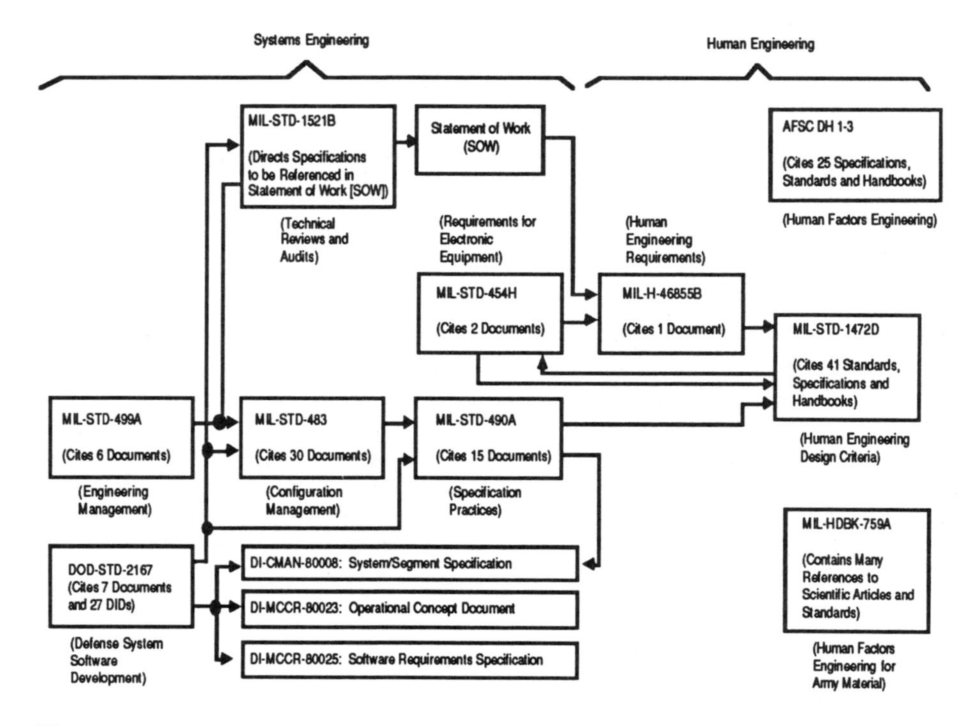

Figure 5-10
Relationships Among Systems Engineering and Human Engineering Specifications, Standards and Handbooks

discussion of the human factors inputs to some of the reviews, e.g., the System Requirements Review (SRR), System Design Review (SDR), Preliminary Design Review (PDR), Critical Design Review (CDR), and Test Readiness Review (TRR), that the systems engineer and human factors engineer must prepare for.

Module 3: Human Factors Methods

Having shown in the previous module what human factors inputs and products the systems engineer needs, module 3 discusses the methods by which those inputs and products are generated.

Each of the methods in the sequence shown earlier in Figure 5-5 is discussed in a standard format: a succinct description of the method, a statement of the inputs required to use the method, the procedure followed in using the method, and the products that the method yields. A feature that has met with universal student approval is the use of a number of genuine examples showing how several of these methods have been used to solve human factors problems in system development. The examples come both from within and from without IBM.

Module 4: Human Characteristics

Human factors is concerned with the discovery and application of information about human abilities, limitations, and characteristics to design. In other words, human characteristics are the raw materials with which the human factors professional has to work. Or, to put it into the context of the course, the human factors methods discussed in the previous module are applied to human characteristics to generate the human factors systems engineering inputs. The information required by the systems engineer to comply with and to prepare the various specifications and reviews that were discussed in the second module.

Individual differences are generally pervasive and troublesome to systems designers. For instance, engineers can have trouble designing systems to be used by a population of users which has the full range of human abilities. To dramatize this point, the lecture illustrates the range of individual differences in sensory, motor and mental abilities, in anthropometric dimensions, and in physiological functions. In that context the lecture explodes the concept of "the average man," and emphasizes that design must always accommodate a range of people. The final section of the module shows some of the long-term changes in human characteristics and in societal demographics (i.e., fewer personnel) that impact system design.

Module 5: Interface Design

In this module the methods discussed in module 3 are applied to the human characteristics of module 4 to produce interface guidelines and recommendations for design. A special section is devoted to human factors issues in the design of the user-computer interface. The concluding section of the module shows by an actual example how the design of an IBM system started out as a simple drawing, identifying only major components by name, and how in six stages the design became successively more refined and detailed as human factors and other requirements were taken into consideration to arrive at the final production version. That example illustrates the iterative nature of complex system design, its progressive elaboration, and the various human factors and trade-off considerations that influence that elaboration.

Module 6: Personnel Selection and Training

A new system is of little use if operators and maintainers have not been selected and trained to operate it. This module points out that human resources are scarce and expensive and that the design of a system impacts life cycle costs directly by the kinds and numbers of people who must be selected to operate it and by the training they must undergo to do so. Figure 5-11 provides the focal point for this discussion. The design of a system directly affects equipment and software operability. In general, the more complex the system, the more stringent are personnel subsystem requirements, both in terms of the kinds and numbers of people who must be selected and the amount of training they must undergo. For example, system design affects crew size. Simpler systems can often be operated by fewer personnel. The design of a system also impacts the kinds of skills the operators and maintainers must have. The more complex the system, the greater the skills that will be required.

Meanwhile, from the population at large, applicants are screened and selected to serve as operators and maintainers. Those selected have certain basic skills. The difference between the skills required and those available must be accommodated through training. But training requires trainers, curricula, methods, media, and devices. The greater the skill, the longer and more complex training must be. So, as the figure illustrates, skill is a function of the operability and maintainability of the system, that is, its design. Training costs, in turn, are a function of that skill. This, in part, is a thrust of the Army's MANPRINT initiative. Although engineers often tend to ignore these selection and training costs, engineers can make the difference between systems that are easily operable and those that are not.

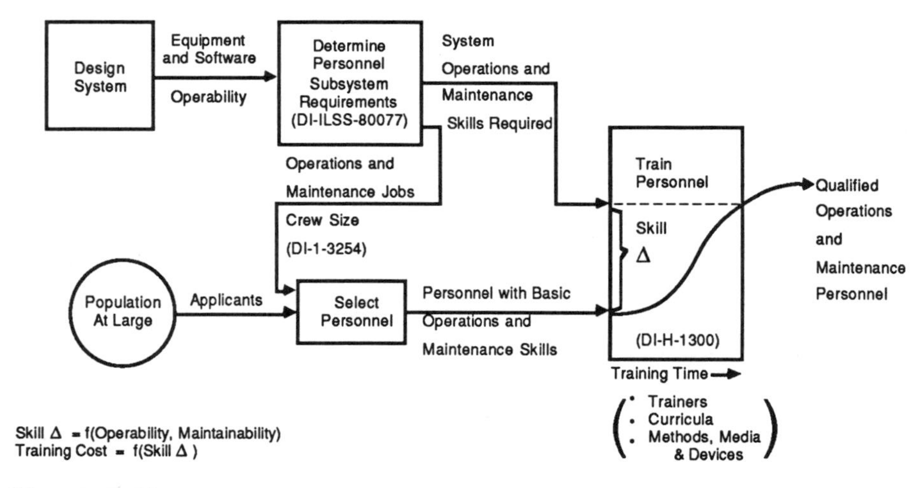

Figure 5-11
System Design Impacts Training

Module 7: Summary

The summary reviews, module by module, what the course has tried to convey to students, and it closes with the thought that machines do not function alone. Moreover, systems engineers cannot assume that operators and maintainers can do functions that are arbitrarily assigned to them. Those functions have to be verified and validated throughout the systems development cycle. The major purpose of human factors is to predict in advance how a system (the combination of hardware, software, and personnel) will perform after the system has been built and put into operation. If major operational problems turn up at that point, someone – the systems engineers, the human factor professionals, or both – did not do their jobs well. IBM's aim is to never permit such failures to occur.

TWO EXAMPLES

Military

The business need to upgrade systems engineering combined with the ground swell pressure from the courses and standards, has resulted in implementing systems engineering principles and practices in many of the emerging programs. One of those programs was Army Special Operations Aircraft (SOA). The Army SOA Request for Proposal (RFP) also had a unique feature – it requested a MANPRINT volume. The intent of the Army SOA program was to provide the soldier with an advanced high-technology helo cockpit to perform special operations missions. The systems engineering challenge was to provide the soldier with consistent cockpit functions and operating procedures across two diverse helicopter cockpits, the MH-60K and MH-47E.

The first action was to translate the Army's operational need into system specific operational concepts. This meant creating a more detailed mission scenario for each of the stated operational needs: i.e., Rendezvous and Refuel, Communications, Attack, Scramble Take-off, and Emergency Carrier Landing. The situation was further compounded by restrictions placed upon the competitive situation. Competing companies were not permitted to use active Special Operations Force (SOF) personnel as subject matter experts. The top-level mission scenario (Figure 5-12) was constructed using in-house expertise knowing there was a risk that current tactics may not be reflected. For that reason the subsequent scenarios were kept generic and structured so that tactical details could be added at some future time. From these scenarios various types of requirements were derived which related to MANPRINT concerns.

The human factors requirement of pilot oriented controls and displays become immediately apparent:

1. Support "first-tour" Warrant and Commissioned Army Pilots in the accomplishment of mission objectives.
2. Provide all information to the pilots necessary to perform low-level flight in visual and instrument meterological conditions.
3. Provide all information to the pilots necessary for the conduct of flight operations in controlled and uncontrolled airspace in the United States and foreign countries.
4. Provide support to the pilots in the performance of normal and emergency procedures.
5. Provide the cues and information necessary for the pilot to maintain a complete picture of the aircraft's situation.
6. Provide a clear and natural interface between the aviation-oriented operator and the avionic system for the control of devices, the entry of data and access to information.
7. Provide the information and tools necessary for the pilots to determine current aircraft and avionic state and operational readiness.
8. Provide logical, unambiguous formats that support the performance of specific tasks.
9. Reduce pilot tasking through the use of automation:
    a. Automate time-consuming and clerical/housekeeping functions.
    b. Reconfigure the system automatically when equipment failures occur to return required function.
    c. Where possible, automate the accomplishment of equipment-required reconfiguration that results from an operator requested task.
    d. Alert the operator only when a failure has occurred, a system reconfiguration takes place, or when a requested action cannot be accomplished.
    e. Provide the pilots with the means to "time-shift" effort by performing setup steps prior to the need to perform the task.
10. Provide controls to support operator tailoring of avionic function and display to specific mission requirements.
11. Support qualified crew chiefs and maintenance personnel in diagnosis, isolation and repair of casualties to installed avionic systems by providing failure information and suggested repair actions.

The front-end generation of soldier-oriented requirements then led to the creation of cockpit controls and displays which were compatible with the mission objectives. Subsequent meetings of joint working groups involving pilots, airframe contractors and the integration contractor helped refine the control/display design as system development evolved while still maintaining the soldier orientation.

Maintenance and logistic concerns flowed from these early requirement statements. Normally these concerns would be stated early on but more important in this case was the fact they were approached with a soldier orientation. System safety and health hazards were drawn into the normal

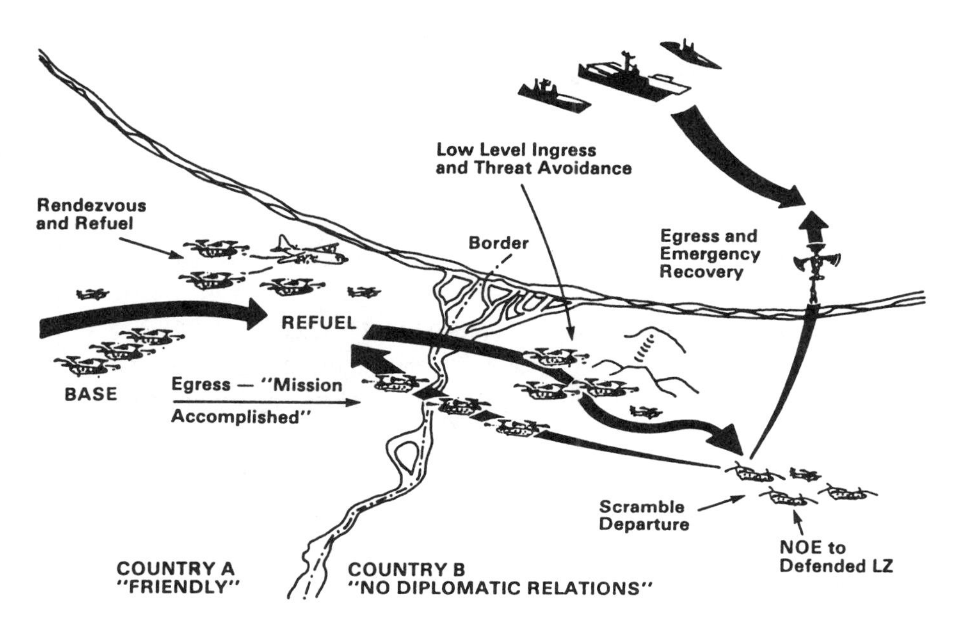

Figure 5-12
Mission Scenario

flow of activities thus saving a lot of unnecessary downstream contention in the development process. Training concerns were very much a part of the early systems engineering discussion resulting in a software architecture with the potential for embedded training and mission rehearsal. Putting the soldier up front in the development process produced a soldier oriented weapon system.

Commercial

Sometimes IBM has the opportunity to develop systems for customers that are not as structured in their procurement practices as the military. Suppose for the moment you the reader have an opportunity to develop an automatic teller for a bank. It is not likely the bank would have a structured systems procurement process, and it is very likely you would have to impress some formal structure on the development process to establish and maintain a good business relationship. In the process of impressing a tailored set of standards specifications and reviews on to the automatic teller project the User-Computer Interface Bulletin could be applied.

It might be helpful, therefore, to explore briefly what some sections of an User- Computer Interface Bulletin might look like for a top-level response to a proposal for an automatic teller. The first requirement would be to scope the system, as shown in Table 5-2.

Table 5-2
Sample Computer Interface Bulletin-Section 1.0

1.0 Scope

1.1 Identification

This System Level User-Computer Interface Bulletin establishes the requirements for the interfaces between the customers, service personnel, maintenance personnel, and supervisors of the Bank and its Automated Teller System.

1.2 Purpose

This interface provides bank customers with the ability to conduct selected banking transactions at any time of the day without having contact with bank tellers to do so.

1.3 Introduction

The requirements in this document will be employed as pass-fail criteria in evaluating the usability by bank customers of the automatic teller machine system provided and developed by the HCI Group. They are derived by the HCI Group on the basis of human factors analysis of Bank specific needs and operational concept.

The analyses and particular methods used to generate a user-oriented operational concept will then lead the human factors specialist to document the specific human interface requirements for the Automatic Teller. That would be Section 3.0 of the User-Computer Interface Bulletin (Table 5-3).

Although greater depth is beyond the scope of this chapter, an examination of the questions that must be answered shows a people-oriented tone has been set. The automatic teller must be user-oriented to reflect the Bank's concern for its customers. The user-computer interface documentation permits the software developers to focus their task on what they do best – program software. The Bank gets a system that is efficient to operate and the customer has an automatic teller that is easy to use.

Table 5-3
Sample Computer Interface Bulletin-Section 3.0

3.0 Interface Requirements

3.1 User Specification

The Human-Computer Interface (HCI) shall be designed to be usable [1] by bank customers with the following characteristics:

- Have a checking and/or savings account with the bank.
- Will carry out transactions with either: standing on crutches, seated in a wheelchair, seated in a car.
- Will have visual acuity no more than 20/100.
- Will range in height from 58 to 75 inches.
- Will cover a wide range of social-economic status.
- May have hearing disabilities.

3.2 Function and Task Elaboration

The HCI shall be designed to allow a bank customer to accomplish the following transactions:

- Withdraw cash from either the customer's checking or savings account.
- Deposit cash or checks into either the customer's checking account or savings account.
- Transfer funds from the customer's checking to savings account or savings to checking account.
- Determine the balance in the customer's checking or savings account.

3.3 Accessing the Computer

To be determined.

3.4 Input Devices

Input devices shall be provided to allow bank customers having the characteristics identified in "User Specification" on page 1 of this specification to:

- Enter the customer's account number (8 digits) and personal identification number (6 digits).
- Select one of the transactions specified in "Function and Task Elaboration" on page 2 of this specification.
- Enter the amount of money involved in the transaction in U.S. dollars using the numerical format $xxx.xx.
- Confirm or correct any entry selected.

---

[1] Usable is defined as: The system shall be considered usable if 19 of 20 persons having the characteristics defined in 3.1 can carry out the functions defined in 3.2 within X units of time, with no more than Y errors, after Z units of instruction (or after T trials).

Table 5-3 *(continued)*
Sample Computer Interface Bulletin - Section 3.0

3.5 Output Devices

Output Devices shall be provided to allow bank customers having the characteristics identified in "User Specification" in 3.1 of this specification to:

- Display verification of all actions taken by the customer.
- Inform the customer of any customer errors in carrying out a transaction including the customer actions required to correct them.
- Inform the customer of any system malfunctions which prevent carrying out or completing the transaction.
- Provide the customer with a hard copy record of the transaction.

3.6 Major Interface Considerations

3.6.1 Language/Vocabulary

All text provided on input and output devices shall be in the English language with vocabulary chosen to accommodate an education level not to exceed the eighth grade.

3.6.2 Form of Dialogue

Dialogues shall be designed to lead the customer through all customer required actions in performing a transaction and allow the customer a maximum of x-units of time to perform the action before being prompted.

3.6.3 Cursors

If a cursor is used, it shall automatically be positioned at the point of data entry and shall be moved to the next position as each character/number is entered.

3.7 Human-Computer Interface Configuration

To be supplied by Ergonomics Design Inc.

3.8 Documentation

Documentation provided to banking customers shall be limited to:

- Instructions on how to become eligible to use the system.
- A 3x5 inch pocket card, printed on one side, describing the transactions provided by the system and an explanation for its use.
- A hard copy record of the transaction indicating time, location, amount, kind of transaction, and before and after balances.
- A monthly report of all transactions as part of the monthly bank statement for the account.

Table 5-3
Sample Computer Interface Bulletin - Section 3.0

4.0 Qualification Requirements

4.1 General Qualification Requirements

"Response Time" in Paragraph 3.6.4: Compliance with this requirement shall be verified by having a test person perform each of the transactions on a functional mockup and measuring elapsed times.

"Error Messages" in Paragraph 3.6.6: Compliance with this requirement shall be verified by having a test person make each of the kinds of errors on a functional mockup and observing the resulting error messages.

"Documentation" in Paragraph 3.8: Compliance with this requirement shall be verified by inspection of the final drafts of the documentation.

SUMMARY AND CONCLUSIONS

By developing a standard set of principles and practices and producing a series of courses to educate systems engineers, IBM has changed the way it approaches systems integration. Top management determined through business contacts and futures research that sustaining growth in an increasingly competitive world market required an upgraded systems engineering capability. Through a Total Quality Management Plan of Excellence, the domains of MANPRINT were given the opportunity for high level visibility in enhancing systems engineering design and development processes. Table 5-1 provides a series of questions and answers about the institutionalization of MANPRINT in an industrial setting raised by the Blanchard and Blackwood Change Management Model (Chapter 3). From this it can be seen that tremendous changes have taken place in how IBM has moved from primarily a hardware/software orientation to one in which user concerns are strongly represented in top level management systems and equipment decisions. Obvious strengths have been (1) a management instruction applied to all future systems engineering contracts which directs implementation of the new principles and practices; (2) availability of user computer interface documentation as guides to quality systems engineering; (3) the clear relationships of TQM, MANPRINT, and systems engineering in its education and training programs; and (4) the visible support of management in redesign of user interface if a systems engineering manager has failed to take user concerns into account.

There is still room for growth, however, since the process is continually improving. The current system for monitoring the institutionalization of change is an informal one and the specific contributions of MANPRINT in terms of cost-effectiveness parameters are still difficult to assess and are largely unknown. There is much yet to learn and apply from the MANPRINT approach, but programs which ignore the user are now the exception rather than the rule.

ACKNOWLEDGMENTS

Grateful acknowledgment is given to Bill Budurka, who provided me with many creative insights into the Systems Engineering process and its relationship to human factors. It is Bill's efforts that are reflected in the Systems Engineering Principles and Practices course. The Human Factors in Systems Engineering course was the combined effort of Al Chapanis and the author, with Bill Budurka's constructive critique. The Human Computer Interface Requirements Specification was the collaborative effort of Al Chapanis and Bill Budurka with stimulation and inputs by the author.

REFERENCES

ANSI/HFS 100-1988 (1988). *American national standard for human factors engineering of visual display terminal workstations*. Santa Monica, CA: Human Factors Society.

Budurka, W. J. (1984). Developing strong systems engineering skills. *FSD Technical Directions, 10* (4), 41-50.

Chapanis, A. and Shafer, J. B. (1986). Factoring Humans into FSD Systems. *FSD Technical Directions, 12* (1), 15-22.

IBM Corporation, Systems Integration Division (1987). *Systems Engineering in FSD* (C-P 0-2057-012). Bethesda, MD: IBM.

IBM Corporation, Systems Integration Division (1988). *Federal Systems, Human Computer Interface Requirements Specification* (C-B 0-2507-011). Bethesda, MD: IBM.

IBM Corporation, Systems Integration Division (1989). *Systems Engineering Principles and Practices*. Bethesda, MD: IBM.

MIL-STD-1472D (1989, March 14). *Military standard: Human engineering design criteria for military systems, equipment and facilities*. Washington, DC: Department of Defense.

National Academy of Sciences, Committee on Human Factors (1983). *Research needs for human factors*. Washington, DC: National Academy Press.

Smith, S. L., & Mosier, J. N. (1986). *Guidelines for designing user interface software* (Technical Report EDS-TR-86-278). Hanscom Air Force Base, MA: United States Air Force Electronic Systems Division.

Van Cott, H. P., & Kinkade, R. G. (1972). *Human engineering guide to equipment design* (2nd Edition). Joint Army-Navy-Air Force Steering Committee. Washington, DC: U.S. Government Printing Office.

PART 

USER-CENTERED DESIGN ADVANCES

Organizations committed to a user-centered design and manufacturing process for their products adhere to four fundamental integration *rules*. These are illustrated in Figure 1 for Part II. *First*, the systems engineering and support organizations are so well integrated that they are nearly indistinguishable so far as the timeline for product design and development is concerned. *Second*, system requirements are truly integrated so that product requirements and user requirements are specified in total system performance language. *Third*, the product is actually tested in operational environments with people who use and fix the product. *Fourth*, tools and techniques are used in the process which facilitate an integrated design. This part presents four concepts for user-centered design, each of which is relevant to one or more of the integration rules.

In Chapter 6, *Conceptual System Design and the Human Role*, Price describes many of the extensive problems that occur from inadequate consideration of the human role in the conceptual stages of system design. As an aid to the systems managers and design engineers of future systems, he then provides critical information needed by them to avoid continuation of these problems. In particular, he emphasizes the need to (1) better understand the economic payoff from investing in human factors technology in conceptual design stages; (2) recognize the principles for defining the human role in systems; (3) appreciate special issues associated with the human role and advanced technology; (4) exercise intelligent decision making in the allocation of functions to man and machine; and (5) understand user acceptance (or rejection) of new technology.

Chapter 7, *Computer-Aided Ergonomic Design Tools,* by McDaniel and Hofmann describes recent developments in computer-aided design (CAD) tools which includes three-dimensional motion envelopes useful to workplace designers. The CAD models allow the system designer to perform the functions of an expert ergonomist and also make it easier to evaluate how well different individuals can operate or repair the product.

In Chapter 8, *Designing for Human Error,* Rouse presents the concept of designing error tolerant man-machine interfaces. This concept deviates significantly from the conventional human factors approach in systems design which concentrates primarily on reducing or eliminating human error. The significance of the error tolerant approach is to reinstitute the primary reasons to include people in engineering systems, i.e., to utilize human ability as adaptor and innovator, particularly in pattern recognition, making associative "leaps" and persevering in ambiguous situations.

Chapter 9, *Workload Assessment and Prediction,* by Hart and Wickens discusses the concept of workload and its importance to systems performance. The usefulness and limitations of qualitative and quantitative methodologies for defining integrated human capabilities in various operational scenarios receive particular emphasis. Issues relevant to implementing and interpreting workload measures are discussed and a brief description of models useful in predicting workload is provided.

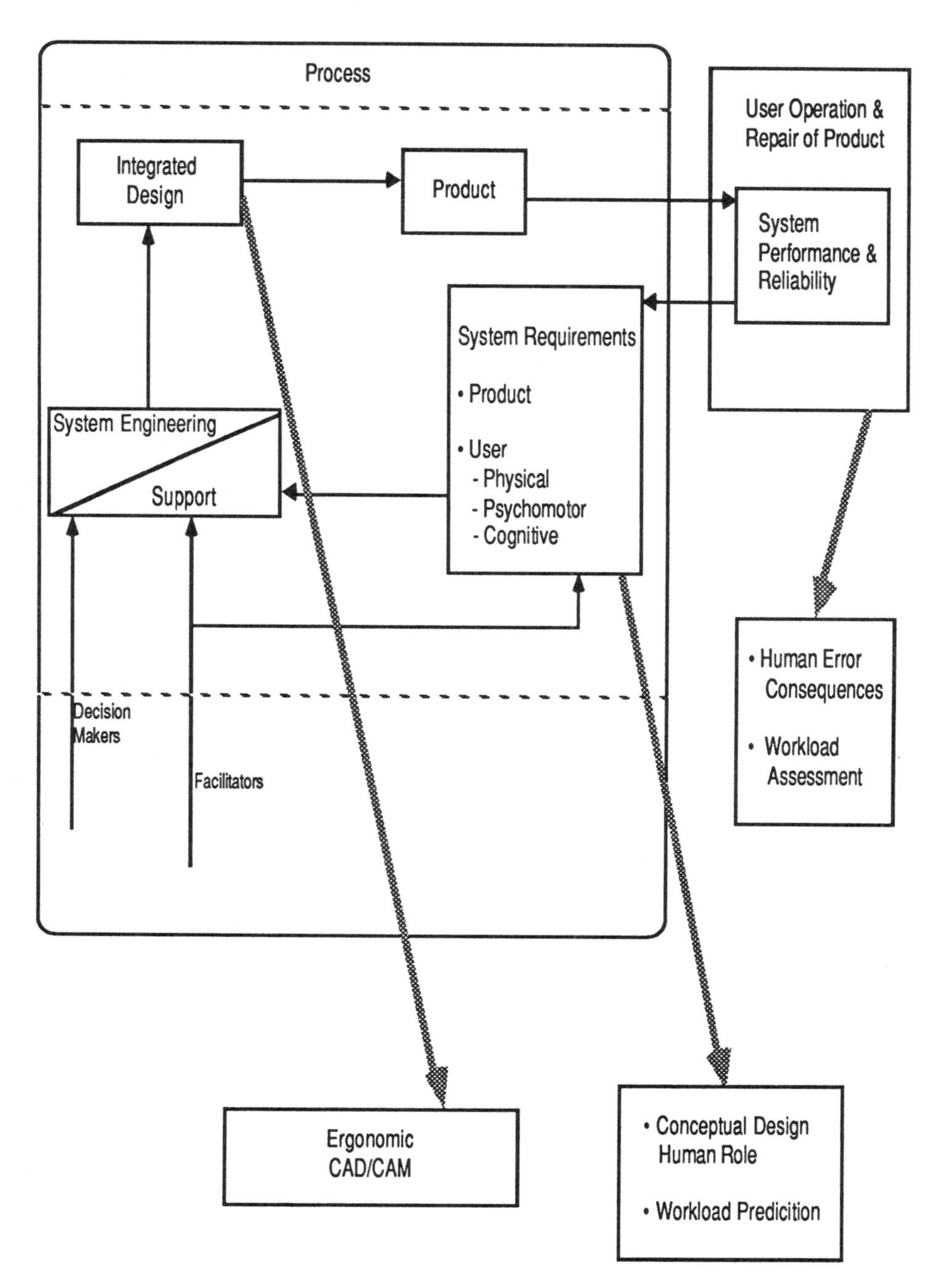

Part II, Figure 1
User-Centered Design Advances

CHAPTER **6**

CONCEPTUAL SYSTEM DESIGN AND THE HUMAN ROLE

Harold E. Price

ABSTRACT

There is no such thing as an unmanned system. Every military, industrial, commercial or social system has people involved in its operation or maintenance. Sometimes they are remote from the system, but they are, nevertheless, in control. If these systems are to perform effectively, efficiently and safely, the human role must be a deliberate design effort during concept design while critical choices can still be made. Otherwise, the human role decision will be made by default. Later in system development, when degrees of freedom are limited, we will be forced to make those human design choices to operate and maintain the system as it has evolved; or else, we will be forced to make retrofits and changes to the system which can be costly. Human factors technology and the human role must be deliberate design efforts during conceptual system design – this is the theme of this chapter.

BACKGROUND

The earliest systems devised by mankind as an aid to survival made use of the full range of human capabilities. People developed and implemented plans, initiated and stopped action, provided muscle-power, manually controlled processes, monitored results, and solved problems. These simple systems merely augmented or extended human capabilities.

The industrial revolution and mass production led to the distinction between mental and physical work. Different human roles evolved for planning, production or operations, and maintenance. Within these groups, further division of work took place and was highly structured and specialized. This specialization of work was accompanied by an elaborate hierarchical structure in which nobody would get more than his share of thinking and decision making according to Margulies and Zemanek (1982). Human choice was what remained after higher levels had their choice. As time-passed, many functions previously performed by people were gradually allocated to machinery. Sensors magnified, extended, or replaced human vision and hearing. Controls replaced human manipulation. Displays were added to integrate sensor data.

Beginning with the end of the Second World War, the technological revolution and the dawn of automation began. Over the next 30 years or so, advances in technology with respect to new materials, new sources of energy, automated methods of production, and new organizational theory changed the human role to even less of a hands-on activity in most jobs. From about 1975 until the present, high technology based on the computer chip has dominated the human role in systems.

With each increment in technology, the role and task of man as a system element changed, and he was gradually removed to the periphery of closed-loop control. Nonetheless, man's role in these new high tech systems was still considered essential which according to Margulies and Zemanek (1982) led to the adoption of a new human factors technology – ergonomics. Because of the importance of man to make sure that the new equipment was being adopted, man has to be treated, maintained, and looked after like any other part of the system. Thus ergonomics became recognized as necessary for user oriented design, interface friendliness, reduction of stress, adjustable tables and chairs, noise reduction, and other things to enhance human performance. Research and case studies in ergonomics or human engineering have shown that wherever these were implemented, certain job or task difficulties were reduced and, no doubt, improvement achieved.

The human role in contemporary systems will be driven by many forces, but policy makers and designers should recognize and respond to the most forceful drivers that must be accounted for in new or modernized systems. Three that will influence the human role in these systems are:

The changing characteristics of the population;
The ever increasing advances in technology;
The consequences of human error and system malfunction.

The Changing Characteristics of Our Population

It has been known for some time that the characteristics of the United States population will be changing significantly in the remainder of the 20th Century and into the beginning of the 21st Century. One of these characteristics that has deeply concerned both military planners and civilian employers is the declining youth population. As illustrated by Figure 6-1, the available young people for the military and civilian job market will decline steeply in the next several years. Similar demographic troughs are predicted for Britain and western Europe. The population and the work force will grow more slowly than at anytime since the 1930s.

When combined with the other important demographics for the next decade, Johnston and Packer (1987) conclude ". . . the new workers entering the workforce between now and the year 2000 will be much different from those who people it today. Non-whites, women, and immigrants will make up more than five-sixths of the net additions to the workforce between now and the year 2000, though they make up only about half of it today."[1]

The military has already begun to notice a drop in the number of high school graduates they have attracted and a drop in mental categories I through IIIA, the top three mental categories. In addition to the demographics noted above, the basic literacy and skill levels of the population must also be considered. Chisman (1989) reports 75 percent of the people who will constitute the American work force in the year 2000 are

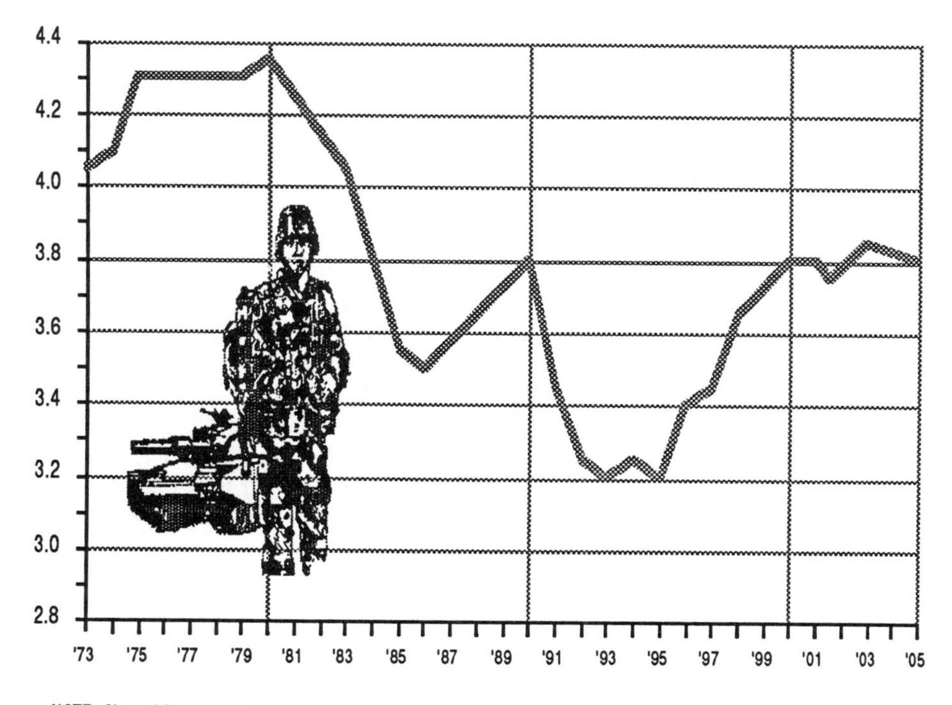

Figure 6-1
Estimated Population in Millions of U.S. 18-Year-Olds, 1973 to 2005
(Source: Bureau of the Census)

adults today and out of school. Experts differ about exactly how many adult Americans are struggling with basic skills problems, but most estimate the number is at least 20-30 million, and by many measures, it is far higher. According to Chisman (1989), it is a conservative estimate to say that 20 million-plus adults have serious problems with basic skills.

The Ever Increasing Advances in Technology

The second driver is the proliferation of technologies. As stated in a recent Office of Technology Assessment report on maintaining the defense technology base (U.S. Congress, 1989), "As we approach the 21st Century, much has changed. The model of U.S. technology leading the world, with defense technology leading the United States, still retains some validity, but it is a diminishingly accurate image of reality . . . At the same time, the U.S. military has been plagued with complex systems that do not work as expected, work only after expensive fixes, or simply don't work."

Technology will advance in many areas including information storage and processing, automation, communications, advanced materials, biotechnology, superconductivity, and others. Undoubtedly, those technologies based on emerging electronics will have the most effect on the human role in systems. In particular, what may be called information technology and automation technology will have the biggest impact. Information technology based on revolutionary computer developments and digital electronics will directly affect the way people are presented information and are able to communicate back to the system.

Zuboff (1988) concludes that computer-based technologies are not neutral; they embody essential characteristics that are bound to alter the nature of work within our factories and offices. She notes that information technology can be viewed as either the basis for automating jobs or the basis for harmonizing labor and management. In discussing this duality of information technology, she states, "In this way, information technology supersedes the traditional logic of automation. The word that I have coined to describe this unique capacity is *informate*. Activities, events, and objects are translated into and made visible by information when a technology *informates* as well as *automates*."[2]

The Consequences of Human Error in Complex Systems

The third driver affecting the human role is the consequences of human error in complex systems. Most human error in complex systems is induced. Human error is directly related to system malfunctioning and breakdown. The consequences of human error in tightly coupled, highly interactive systems can be devastating with respect to lack of operational performance, loss of economic viability, and human suffering or death. When the intended purpose of a system is not achieved, whether it is a military, industrial, commercial, or social system, there is some kind of loss. In complex systems, this loss can be acute; yet, we often make the operation and maintenance of the system the responsibility of humans, but the authority we give them is not optimal in terms of their role or their interfaces with the system. This leads to error, which leads to malfunction, which leads to system breakdown.

In view of the foregoing, it would appear obvious that human factors technology will play an ever increasing role in the design of future systems. But in order for this to occur routinely as part of systems integration efforts, systems managers and design engineers will need to (1) better understand the economic payoff from investing in human factors technology in conceptual design stages; (2) recognize the principles for defining the human role in systems; (3) appreciate special issues associated with the human role and advanced technology; (4) exercise intelligent decision making in the allocation of functions to man and machine in conceptual design; and (5) understand user acceptance (or rejection) of new technology. The remainder of the chapter concentrates on these five areas.

BENEFITS OF HUMAN TECHNOLOGY INVESTMENT

Perhaps the reason human factors technology has so seldom been applied early or integrated routinely into the system design process is the lack of recognition of the unique benefits to be had from its investments. There are, however, several very important payoffs for early investment in human factors.

Prevent Disasters and Accidents

The costs associated with lost lives and injuries from technology which has inadequately considered the human role cannot be overstated. More importantly from a systems design viewpoint, it is becoming more widely recognized that technology produced problems will not be eliminated by still more technology.

System designers must recognize the importance of the human factor and the role humans play in systems. This case has been rather eloquently stated by Goodman (1987):

> We know that more lives hinge on fewer "things": on nuclear missiles and plants, on chemicals and computers. It may be easier to think on "systems" that can be perfected than on people who aren't perfectible.

> The disaster at Chernobyl, the near-disaster at Three Mile Island, each had its human factor, and yet most of the original attention focused on the buildings, the systems. The Challenger explosion initially was billed as a technological disaster. It was a while before the inquiry shifted from the state of the O-rings to the state of the decision makers.
>
> At Bhopal, India, where some 17,000 people died, and at Basel, Switzerland, where the Rhine River was poisoned, we heard first of chemical leaks and spills and impersonal safety "procedures." We heard only secondarily of workers who might not have sounded alarms or even known enough not to hose chemicals. Even in the recent low-tech Amtrak disaster, the attention was first on the state of signals and only then on the signal readers.
>
> . . . Most of us choose to think of the human role in our sophisticated technological society as a minor part of the equation. We accept a walk-on part in the modern world and give the machines, the systems, the lead. Again and again, in the wake of a catastrophe, we look for solutions that will correct "it" rather than "us." . . . But no machine is more trustworthy than the humans who made it and operate it. So we are stuck. Stuck here in the high-tech, high-risk world with our own low-tech species, like it or not. No mechanical system can ever be more perfect than the sum of its very human factors.[3]

Perhaps the single most important concept for policy makers and system designers to better appreciate is that of *human error*. Resolving human error issues in the context of current technology should be the primary thrust for managing future disasters and accidents. Three important points about human error during concept design are:

- Human error is a primary causal factor in most system breakdowns;
- Most human error is induced by some system characteristic;
- Human factors technology can anticipate and thereby avoid human error through system design.

Human Error Is Primary Causal Factor

The problem of human error, its causes and contributions to system malfunction and accidents and its consequences, became widely recognized with the Three Mile Island-II nuclear power plant accident. In the ten years since then, there have been many other catastrophic accidents in which human error has been a primary causal factor. Notable among these are Chernobyl, Tennerife, Bhopal, Challenger, KAL 007, and the shooting down of the Iranian passenger aircraft. Investigations of all of these accidents showed that human error and human factors deficiencies were primary causal factors. In fact, human error, as reviewed by many researchers and investigators (for example, see Sanders & Shaw, 1988), have concluded that human error is a contributing factor in 60 to 80 percent of all complex system breakdowns or accidents.

Induced Human Error

Many human factors researchers, including the present writer, believe much, perhaps even most, human error or degraded human performance is "induced" by poor design, procedures, training, organizations, or other system characteristics (see, e.g., Conway, Muckler, & Peay, 1982; Hawkins, 1987; Perrow, 1984).

This also means that most human error inducing design can be avoided by better human factors design decisions from the very beginning of system design to the very end. Perrow (1984) reviews many contexts where human error was an induced and causal issue. Concerning marine accidents he states, "The problem, it seems to me, lies in the type of system that exists. I will call it 'error-inducing' system; the configuration of its many components induces errors and defeats attempt at error reduction . . . In an error-inducing system, the tendency to attribute blame to operator error is particularly prominent. Such studies as there are all report operator error as the cause of 80% or more of the marine accidents."[4] Speaking of systems in general, he further notes that, ". . . every system we will examine places 'operator error' high on its list of causal factors – generally, about 60 to 80% of accidents are attributed to this factor. But if, as we shall see time and time again, the operator is confronted by unexpected and unusually mysterious interactions among failures, saying that he or she should have zigged instead of zagged is possible only after the fact."[5]

Human Factors Technology Can Avoid Human Error

If most human error is induced, then it, along with degraded human performance, can be avoided by better human factors design decisions during system design.

Conway, Muckler, and Peay (1982), in a review of human errors, injuries, and accidents, suggest a clear separation of induced human errors from those truly the fault of the operator or maintainer. They conclude that a set of tools, procedures, and criteria is required for separating out operator errors from errors created by poor hardware and/or system design, inadequate training, and poor organization.

Wiener (1988), elaborates further, "If human errors can be induced by design, they can also be prevented at the design stage, but in order to do so, the behavioral consequences of possible design decisions must be taken into account."[6] Moreover, he notes, "A variety of techniques for the study, prediction, and prevention of human error at the design stage now exist. These must be applied not only to the design of hardware, but all supporting materials, procedures, language, graphics, and operating details."[7] Wiener concludes there is a front end cost associated with human factors technology in the conceptual stages, but it is negligible compared to the everyday operating costs induced by inadequate design:

> There is an "iron law" that should never be ignored. To consider human factors properly at the design and certification stage is costly, but the cost is paid only once. If the operator must compensate for incorrect design in his training program, the price must be paid every day. And what is worse, we can never be sure that when the chips are down, the correct response will be made.[8]

Improve System Performance

In addition to the unnecessary costs associated with obvious breakdowns in the machine and human interface (as described above), there is an even greater cost associated with everyday degradation in overall system performance. Because of inadequate consideration of the human role during conceptual design, systems frequently do not perform as expected.

The IEEE Spectrum (1987) describes a number of typical problems which have resulted from ignoring human factors in early system design:

> Leaving the human factors until too late may result in a whole variety of problems, which often come to light only when the prototype is built and someone tries to make it work. When the M-1 tank was prototyped, the seat gave many test drivers back and neck strain that required medical attention. And equipment maintenance may be close to impossible if designers to not carefully consider what the human maintenance team has to do. Each time the Army AHIP helicopters electronic fuel control was removed, all the nuts and bolts fell to the ground. Such poor design can escalate equipment maintenance costs, and once the hardware is out, changing the design takes up large amounts of money and time. . .[9]

Similar problems were reported with respect to the Redeye and Stinger missiles, automobiles with radios and trip computers that require drivers to take their eyes off the road in order to use them, and difficulties with monitoring equipment and controls in nuclear power plants, chemical plants, operating rooms, air traffic, and spacecraft systems.

The case for incorporating human factors technology early and continually in system design is usually made by reference to "horror stories" that document the consequences of human error or degraded human performance on system performance (or lack thereof). In truth, the best measure of human factors effectiveness is the lack of problems. This has been substantiated by Lane (1987) who states:

> Good HFE (human factors engineering) design is marked by an *absence of problems* in the use of a system by humans and its effects are thus "invisible" in the final operational system. Its contributions become an integral part of each component or subsystem, and *cannot be readily isolated* from overall system functioning or "credit" to the HFE inputs.

Reduce Manpower, Personnel, Training Costs

The realization that most of a system's acquisition costs are determined during the concept exploration phase of system development has been known for over 20 years (U. S. General Accounting Office, 1985; Graine, 1988). The U. S. General Accounting Office (1985) concludes "Identifying and analyzing a system's MPT [manpower, personnel, and training] needs are necessary during each phase, but they are particularly critical during the concept exploration phase because this phase has the greatest effect on the system's life cycle costs. Many studies of life cycle and weapon systems supportability show that about 70% of a system's life cycle costs are determined by decisions made prior to Milestone 1. After the concept exploration phase, as development proceeds and the design becomes more set, changes to ensure that trained personnel can operate and maintain the system are more difficult and costly" (see Figure 6-2). It is also important to recognize that over 50 percent of the entire Department of Defense budget is for personnel costs. According to Graine (1988), people and associated training requirements combined cost close to 60 percent of the life cycle costs of a weapon system.

If a broader and longer-term view is taken to the costs associated with the introduction of new technology, the implications are enormous. Too often in the past, military programs have been drive by the short-term view. Pirie (1987) states, "Within the current military procurement system, incentives are heavily weighted to meeting short-term cost and performance goals at the expense of the longer term. Thus, one finds situations in which equipment is left out of an aircraft to keep procurement costs down, with the result that downstream operations and maintenance costs are multiplied far beyond the initial savings."[10] Pirie goes on further to report that the Packard

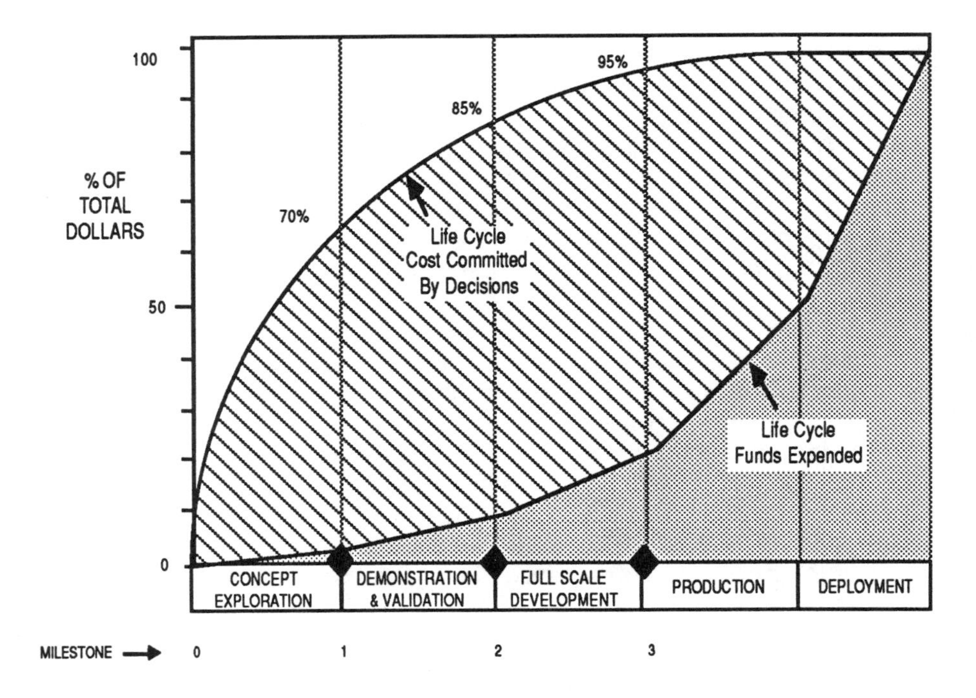

Figure 6-2
Schedule of Commitment Decisions and Life Cycle Costs
(Modified from Graine, 1988)

Commission said in effect that the Department of Defense should place a much greater emphasis on using technology to reduce costs – both directly by reducing unit acquisition costs and indirectly by improving the reliability, operability, and maintainability of military equipment.

But how can technology actually reduce MPT costs? According to Binkin (1986), the answer is to simplify the complexity of systems at the early stages and thus reduce the qualifications and numbers of people required to operate and maintain these systems later on:

> The number and skills of personnel needed by the Armed Forces depend on many factors, such as what tasks specific units are expected to do (workload); how they are organized (combat-to-support ratio); and guiding personnel policies (how people are assigned and used). The influence of technology comes into play when calculating the number and qualifications of specialists and technicians needed to operate and maintain military equipment. As new systems and advanced technologies are introduced, the effect on the military workforce will depend largely on the degree of equipment complexity.[11]

Pirie (1987), however, sees this as more of an "institutional" problem than a technological one. "It should require fewer man-hours, less highly skilled personnel, less preliminary technical training and the like to assure that systems work properly and reliably. This does not require dramatic shifts or increases in military technologies so much as shifts in institutional priorities and patterns of behavior."[12]

Human Factors – A High Leveraged Investment

The cost of human factors technology and determining the human role during concept design is a high leveraged investment. Systems that are designed from the beginning to integrate human capabilities, limitations, and expectations into design not only reduce the potential for human error, but are easier to learn and thus reduce the ultimate investment in training. Also, human factors engineering is a one-time investment – it becomes a permanent part of the system/equipment. Conversely, investments in personnel, manpower, and training are recurring costs. Thus, human factors integrated into system design from the very beginning is one of the surest ways available to the systems engineer to avoid later costs.

DEFINING THE HUMAN ROLE

Perhaps the most widely held perception of system managers, designers, and the public at large regarding human errors associated with accidents and poor system performance is that they are produced only by operations and repair people. Perrow (1984) has shown this clearly is a misperception. In introducing the notions of "interactive complexity" and "tight coupling" – system characteristics that will inevitably produce an accident and in reviewing numerous accident scenarios in military, commercial, industrial and social systems – he concludes, "At each turn, even in the best of the industries, we found rampant attribution of operator error to the neglect of errors by the Great Designers and the Centralized Managers . . . These systems are human constructions, whether designed by engineers and corporate presidents, or the result of unplanned, unwitting, crescive, slowly evolving human attempts to cope . . . Better training alone will not solve the problem, or more gadgets, or promises that it won't happen again."[13]

Can People Be Eliminated?

Many system designers view humans as unreliable and inefficient and they should, therefore, be eliminated from the system. In the 1980s, the promise of artificial intelligence and automation fueled this viewpoint. It is natural for system designers to exploit affordable technology to its limits. However, it is unrealistic to think that machine functioning will ever entirely replace human functioning. This issue has been addressed most recently in aircraft cockpit automation (Wiener, 1985, 1988, 1989a, 1989b; Lauber, 1989). According to Wiener (1985):

> Designers responded to "pilot error" and the increasing cockpit workload by attempting to remove the error at its source, that is, to replace human functioning with device functioning; in their view, to automate human error out of the system. But there were two flaws in this reasoning. (1) The devices themselves had to be operated and monitored by the very humans whose caprice they were designed to avoid; human error was not eliminated but relocated. (2) The devices themselves had the potential for generating errors that could result in accidents. Overall, the movement toward cockpit automation has undoubtedly enhanced safety, but new problems have been created that are only now being appreciated.[14]

Increased Cognitive Workload

Automation is almost always introduced with the expectation of reducing human error and workload. However, what frequently happens is that the potential for error is simply relocated and workload is shifted from a psychomotor to a cognitive base (Bainbridge, 1982; Moray & Huey, 1988; Nobel, 1984). Typical examples of unforeseen consequences from automation are listed in Table 6-1.

More often than not, automation does not replace people in systems; rather, it places the person in a different, more demanding role. The role of man in automated systems has been characterized by Bainbridge (1982) and others as consisting of two general categories of tasks. He may be expected to monitor that the automatic system is operating correctly, and if it is not, he may be expected to intervene or seek more experienced help. To take over and stabilize the process requires perceptual-motors control skills, and to diagnose the fault for shut down or recovery requires cognitive skills.

Pivotal People

The importance of people in a technological society is further reflected in the concept of *pivotal people*. Pfeiffer (1989) emphasizes the irreplaceability of pivotal people in stressful environments like aircraft carrier flight operations, air traffic control, and power utility grid control. Also, Lubove (1987), in describing certain people vital to success, concludes that

Table 6-1
Unforeseen Consequences of Automation

Type of Automation	New Problems
Decrease pilot workload with aircraft cockpit automation.	1) Automation devices have to be operated and monitored; 2) Automation devices themselves generate error. (Weiner, 1985)
Shift from manual to supervisory control in steel product process.	Productivity fell because of failure to support new supervisory control demands. (Moray & Huey, 1988)
Shift from tile annuniciator alarm to computer-based alarm in power plant control room.	Forced to return to older technology because strategies to meet cognitive demands of fault management that were implicit in the older were lost in the newer technology. (Moray & Huey, 1988)
Shifts from paper-based procedures to computerized procedures.	Failed due to disorientation problems arising from inadequate consideration of technological demands for human problem solving. (Moray & Huey, 1988)

even in an age of impersonal silicon chips and robotic arms, certain people with particular skills mean success or failure for their employers. These workers are pivotal people whose knowledge often reflects experience, not education. Some typical examples from Lubove are: (1) A blast furnace engineer who must ensure that the raw materials – iron ore, limestone, and coal – ultimately combine in a huge furnace at 3700°. The decisions and control of this process cannot be automated; (2) A pattern cutter in a women's clothing factory is pivotal. Machines can't be programmed to sense the hang and balance of a dress or to ensure that a blouse won't be too revealing; (3) A brewmaster can make or break a brewery.

History has shown us over and over again that in complex systems, no matter how automated, the human is the last vote in deciding a critical issue and the human is the last line of defense in case of system breakdown. If we expect people to function efficiently, effectively and safely, the person's role must be an integrated decision and not a default decision.

Decisions in Determining the Human Role

The principal human factors products of each phase of the military acquisition process have been described in some detail (Price, Fiorello, Lowry, Smith, & Kidd, 1980). In their report they define the principal product of the concept exploration phase (see Figure 6-2) as a "Role of Man Statement." The basis for developing this product and the decisions required are discussed below.

The Basis for Determining the Human Role

In military systems, human factors should begin by analyzing mission scenarios that are expected to be encountered in combat. Analysis is conducted to identify the critical role humans will play to succeed with any particular mission. Will man be an operator, maintainer, sensor, manager, analyzer, decision maker, information manager, backup to equipment, or some mix of the above. A very important decision is whether man will be local or remote from the mission equipment. To define that role, all functions that are needed to achieve the mission objectives must be specified first. To identify all functions, the operational and environmental conditions under which the system is to operate must be determined. For example, will the system operate in temperature extremes, during day and night, in unusually rugged conditions, or for unusually sustained periods of time, etc.? In addition, analyses of existing similar systems (if any) should help identify operational and environmental conditions as well as other positive and negative aspects. For example, what was the role of man in the predecessor or similar system(s)? What man functions and man-machine functions have been successful and unsuccessful? All of the information can be used in performing a functional analysis.

Performing trade-off studies with the major factors (e.g., logistics, maintenance, costs, advantages and disadvantages of using man in alternative roles) should result in cost-effective system configurations, given system constraints. Such human factors analyses also lower the probability of major changes in design downstream to accommodate the idiosyncracies of man.

"Role of Man Statement" Decisions

The "Role of Man Statement" should be developed as a part of the Mission Element Needs Statement at the conclusion of the concept exploration phase (see Price, Fiorello, Lowry, Smith, & Kidd, 1980). Decisions inherent in developing the "Role of Man Statement" are annotated in Table 6-2.

Table 6-2
"Role of Man Statement" Decisions

ASSUMPTIONS
• A separate "role of man" analysis will be provided for each alternative system concept selected. • Human factors specialists will develop "role of man" concepts and interact with mission analysis team in development of Mission Element Needs Statement (MENS). • "Role of man" components are listed according to probable order of presentation in MENS (not according to their development sequence).
DECISIONS
1. List effects envisioned for overall system as a result of "role of man" devised for each alternative system concept as configured (e.g., operability, maintainability, mission effectiveness). 2. List effects envisioned for man's role/human issues as a consequence of each alternative system concept as proposed (e.g., safety, habitability, user acceptance). 3. Determine location of man in system (local or remote) to perform designated role. 4. Specify advantages accorded man's role for each alternative concept (e.g., facilitate operation of system, allowance for contingencies). 5. Specify disadvantages accorded man's role for each alternative concept (e.g., manpower reserves consumption, level of training requirements). 6. Determine required human performance, behaviors, capabilities, and performance limits (e.g., sensing, processing, information storage, decision making, responding) identified for each functional category. 7. Determine personnel constraints impacting man's role for each alternative system concept such as the following: a. maximum and minimum numbers of personnel who can be used in the system; b. types of personnel (e.g., skill level and aptitude) available for system assignment; c. anthropometry of identified personnel population (existing and projected); d. user acceptance problems projected and their effects; e. effects of system and mission as configured on personnel vulnerability (e.g., environmental hazards); f. communications requirements and limits (system and other personnel).

Table 6-2 *(Continued)*
"Role of Man Statement" Decisions

DECISIONS

8. Determine implications envisioned for each alternative system concept upon requirements for:
   a. training (e.g., level of training, trainability, training support and facilities, training devices);
   b. manpower (e.g., manpower levels, performance availability);
   c. life support;
   d. "-ilities" support (e.g., logistics, reliability, maintainability);
   e. social/organizational impact (e.g., MX basing).
9. Select contributions to functional analysis in Mission Analysis Phase:
   a. identification of threat;
   b. need demonstration: new system or modification to current system;
   c. requirement;
   d. mission;
   e. system objective definition (and required input/output);
   f. mission segment;
   g. scenario(s);
   h. functional categories;
   i. functional flow and operational event sequences;
   j. system specification
      (1) manual
      (2) hardwired
      (3) automated: Facilitate system functioning
      Override (bypass) system malfunctioning
      Control system graceful degradation
      Permit system to operate
10. List human factors characteristics that will facilitate successful system development and mission success for each alternative concept (design, development, testing, production, deployment, and operation):
    a. advancement in state-of-the-art human factors technology;
    b. currently available human factors technology.
11. List impacts upon cost and system effectiveness for each alternative concept in association with human factors inputs:
    a. research and development, training, personnel, manpower;
    b. mission success, vulnerability, survivability.
12. Prepare Human Factors Research and Development Program Plan tailored to each alternative concept for balance of system life cycle.

While material in this section was taken entirely from a report dealing with military system development, in principle the same notions apply to industrial, commercial, or business systems. That is, the human role in terms of responsibilities, authority and expected decisions should be clearly defined, and the method for measuring successful performance clearly stated during concept designs.

FUTURE TECHNOLOGY IMPACT

The Defense Authorization Act, fiscal year 1989, stipulated that a Critical Technologies Plan be submitted annually to the Senate and House Committees on Armed Services. The first such plan which was submitted in early 1989 listed 22 critical technologies (see Table 6-3). Human factors and

Table 6-3
Critical Technologies
(Source: Lake, 1989[15])

	Critical Technologies	Funding ($ Million)
1.	Microelectronics Circuits/Fabrication	200
2.	Preparation of GaAs and Other Compound Semi-Conductors	100
3.	Software Producibility	70
4.	Parallel Computer Architectures	80
5.	Machine Intelligence/Robotics	70
6.	Simulation and Modeling	115
7.	Integrated Optics	25
8.	Fiber Optics	20
9.	Sensitive Radars	170
10.	Passive Sensors	170
11.	Automatic Target Recognition	75
12.	Phased Arrays	80
13.	Data Fusion	115
14.	Signature Control	N/A
15.	Computational Fluid Dynamics	30
16.	Air-Breathing Propulsion	300
17.	High-Power Microwaves	50
18.	Pulsed Power	65
19.	Hypervelocity Projectiles	100
20.	High-Temperature/High Strength/Lightweight Composite Materials	110
21.	Superconductivity	100
22.	Biotechnology Materials/Processing	100

other MANPRINT domains are relevant to many, if not all of them. No matter what advances are made in microelectronics, materials, chemistry and physics, these advances must be integrated into a design with humans in order to produce a system. Several of the technologies listed directly imply human factors technology. For example, Machine Intelligence/Robotics has as an objective the incorporation of human "intelligence" and actions into mechanical devices; Data Fusion has as an objective the machine integration and/or interpretation of data and its presentation in convenient form to the human operator.

Technological Trends

It appears that the most critical technological trends affecting human factors and the effective integration of humans into systems will be founded in microelectronics/computers. Where is this technological trend going? In general, computer hardware capabilities are doubling every 18 months. Johnston and Packer (1987) provide a projection for the next decade:

> By the year 2000, microcomputers will be as powerful as today's mainframes. Today's memory chips that can store 256,000 or a million bits of information will have given way to chips with 10-30 times more storage capability. Today's microprocessors that can process one or two million instructions per second will have been succeeded by chips that are an order of magnitude more capable. Desktop storage will not be measured in today's megabytes (million words) but in gigabytes (billion words) and terabytes (trillion words). Machines that can analyze many different types of information at once, and enormously sophisticated software, will have finally resolved traditional debates over whether computers can ever really think like people. Artificial intelligence will be real.[16]

The technologies that will affect the human role in systems the most are *information and automation*. Information technology will foster profound changes in areas such as communications, flat panel displays, head-up displays including holography, voice interactive techniques, data link, and artificial intelligence/expert systems. Automation technology will likewise foster significant progress in areas such as robotics, teleoperations, distributed control, and digital control systems including "control-by-wire." While a thorough review of technologies and the implications for human factors will not be attempted, some brief notions of what is being contemplated by the military, commercial aviation, and business sectors will be presented to emphasize critical changes in the human role which will occur.

Military Technology

The military will make drastic changes because of the modernization programs. As reported by Binkin (1986), the Army, with a growing emphasis on offensive minded "deep strike doctrine," is developing concepts that will require integration of C^3 (Command, Control, and Communication), active and passive surveillance, real-time tactical information dissemination, targeting, and weapons delivery employing a standoff wide area, or point suppression guided by precise navigation or active target designators. Other technological advances will extend tactical intelligence capabilities by enhancing night and poor weather conditions vision, detection and tracking enemy formations, locating the enemy's electronic emissions and artillery batteries, listening to his communications, and all the while blinding the enemy's intelligence sensors. All of these systems are expected to be more fully automated and informated.

The Air Force, which has always been in the forefront of electro-technology, likewise anticipates dramatic effects on future application of air power. Most of these advances will be in avionics, "smart" and "brilliant" weapons, and, of course, the electronic cockpit.

The Navy will experience some fundamental changes in its ability to install potent, long-range weapon systems in smaller spaces, to construct smaller and perhaps less detectable ships, to have distributed control systems on-board for running a "smart" ship, and integrating tactical information over large ocean-sized areas.

Finally, the Marine Corps, not known for a high technology love affair, will have all electronic weapon stations on its amphibious assault vehicles and a family of "smart" munitions. Binkin (1986) sums up as follows:

> In sum, given the sheer magnitude of the modernization effort launched in the early 1980s, and the reasonable prospects for breakthroughs in several exotic technologies, the next generation of U.S. weapons is sure to represent a major improvement in military capability. But whether the armed forces can achieve the full performance designed into their systems in an open question, whose answer depends largely on the extent to which military personnel will be up to the task of operating and maintaining the new weaponry.[17]

Commercial Technology

In commercial aviation, new digital avionics technology is discussed by Sexton (1988). He notes that the crewsystem designer of today has the opportunity to create new functions because of the tremendous capabilities of the on-board computers to analyze, sort, integrate, and route information from a variety of sensors and subsystems. Sexton also discusses the impact of computers, head-down displays, head-up displays, fly-by-wire, control technology, data link, and lighting. Some of these technologies, e.g., fly-by-wire, have been questioned with respect to the pilot's role and safety (Beatson,1989). Sexton summarizes his discussion of technology by noting that it is tempting for designers to include a new technology into a design just because it is new and it seems like a "neat" idea. That temptation must be strictly controlled, however, through good design practices. Mission requirements must drive the design and the determination of which technologies to apply.

Changes in business or commercial systems are evident everywhere we look. It is likely that by the turn of the century, a digital telecommunications network will connect most businesses and many homes with fiber-optic links of enormous capacity. Through terminals in the homes, such things as working from home, shopping, banking, tailored news broadcasts, electronic mail delivery, and selected audiovisual entertainment will be possible. Johnston and Packer (1987) talk of "silicon secretaries that can take dictation and edit letters, reservation clerks that understand speech in any language, or robotics that can load a truck or pick strawberries . . . "[18] These are but a tiny fraction of the technological changes that will affect the human role and the interaction with humans in systems in the not too distant future.

Computer Error

In view of the central role microelectronics/computers play in the technological future, it is important to recognize that new and more insidious blunder – the computer error –is infiltrating our newest and most sophisticated systems. Computer errors are, of course, human errors because the source of error (in most, but not all cases) is a designer or programmer.

Computer errors can be expected to increase because of the extensive programming required for modern sophisticated systems. For example, according to Lake (1989):

> In reality, the magnitude of the numbers of lines of code multiply with each new system or program. The C-5 uses 25,000 lines of code, the C-17, 750,000, and the Aegis cruiser, over 2 million. As the lines of code go up linearly, the error elimination, debugging and testing effort required to complete the programming increases exponentially. Failures of software programs far and away outnumber the successes. Even in the latter cases the cost for debugging, updating and maintenance is a built-in program lifetime funding requirement.[19]

As with other kinds of system errors, computer errors do not always show up until the system is in operation and a specific set of circumstances prevails. The only way to be sure a program has been properly designed is to exhaustively test it. Jacky (1989) states that flaws can be overlooked with "even the most rigorous of trials," because there are a limitless number of situations a computer program can be exposed to. It is routine to let the market discover and correct errors left in software "over time." Jacky (1989) explains that programmers expect themselves "to introduce inadvertently, about 50 errors in every thousand lines of code." Although testing weeds out most of these programmer errors, down to "two or three errors in a thousand lines," a good program (e.g., 50,000 code lines) "may contain more than 100 errors."

Computer errors can be categorized as:

- Dedicated Program Execution: These are errors in the sense that computers equivocally pursue their instructions, even though the system that they control may be heading for a catastrophe (economic or safety-related).
- Software Bugs: These are programming errors that do not usually show up until some unique situation is presented to the computer.
- Truth in Computers: These are errors of perception in that we are often willing to accept the results if it is done by a computer.

Dedicated Program Execution Errors. The dedication with which computers pursue their programmed instructions is at once awesome and scary. Could they be programmed incorrectly or could data errors lead to some insidious result? Many incidents prompt us to say yes.

Who can forget the tragedy of the destruction of Korean Airlines Flight 007? It appears that the crew made a data entry error to the inertial navigation system, causing the aircraft to fly off course over Russian territory where it was shot down. Like most accidents, this misfortune had multiple causes. Nonetheless, the computer has some culpability as it resolutely guided the aircraft into disaster. Software problems have also invaded military systems, despite elaborate precautions. Jacky (1989) describes the firing of a missile from a wing-mounted launcher when a computer-controller failed to release its grip as "creating . . . the world's largest pinwheel, when the aircraft went violently out of control."

Wiener (1985) reminds us that unanticipated computer errors are now appearing in new automated aircraft.

> In the last two years, several accidents and incidents, rightly or not, were laid to automation. A DC-9-80 (MD-80) lost both engines due to fuel starvation because the center-tank pumps were not turned on after takeoff (Aviation Week and Space Technology (AW&ST), 1983). Digital fuel gauges were blamed

> . . . Soon after, another 767 during descent, had to have both engines shut down and restarted to bring them out of idle power. An over-efficient computer was seen as the villain (Beck 1983; Miami Herald 1983).[20]

Software Bugs: Latent Error. As computer programs get more massive and complicated, software bugs will become impossible to fully detect during program testing and validation, and these defects will show up after "operations" have begun. Jacky (1989) vividly describes some of the incidents from software bugs in medical equipment. A computer-controlled radiation therapy machine, Therac-25, manufactured by Atomic Energy of Canada, led to the death of two patients from an overdose of radiation. It was a software error, involving the operation of a switch, that killed them.

A *Wall Street Journal* article describes a variety of software bug incidents. The article noted that tiny software bugs can fell mighty machines – often with disastrous consequences. Software defects over the past five years have killed sailors, maimed patients, shaken corporations, caused the telephone system to collapse, and threatened to cause the government-securities market to collapse. As a costly financial example, the Bank of New York in 1985 had a computer error which blocked the bank from delivering a substantial amount of government securities to customers and accepting payment. As a result of this "computer error," the bank had to borrow $23.6 billion from the Federal Reserve Bank in New York to cover the shortfall overnight and pay $5 million interest on the loan.

Truth in Computers: Perceptual Errors. Computers may induce a kind of overconfidence that can lead to accepting questionable or even wrong results at face value. Petroski (1985) discusses this phenomenon.

> Some structural failures have been attributed to the use and misuse of the computer, and not only by recent graduates, and there is a real concern that its growing power and use will lead to other failures . . . Thus, while the computer can be an almost indispensable partner in the design process, it can also be a source of overconfidence on the part of its human bosses . . . And as more complex structures are designed because it is believed that the computer can do what man cannot, then there is indeed an increased likelihood that structure will fail, for the further we stray from experience, the less likely we are to think of all the right questions.[21]

Computer models and simulations can be excellent design aids, but the result from these analyses should also be questioned and perhaps calculated or approximated some other way before being relied upon. Just because it was done on a computer does not make it right or righteous.

The Impact of Technology on Manpower, Personnel, and Training

As indicated earlier in this chapter, human factors applied during concept design can make the role of humans and the human interfaces more effective, efficient, and safer and, therefore, can reduce the burden on manpower, personnel, and training to achieve the same objectives. In modern systems, the implications for manpower, personnel, and training depend on complexity which seems to go hand-in- hand with technology in military systems. Technology can make complexity real or transparent.

Complexity has no standard definition. It may be equated to job difficulty, time for training to a certain standard, the number of components or parts in a system, the amount of documentation, or the number of personnel errors. Whatever the definition, complexity generally carries a negative connotation with respect to manpower, personnel, and training.

Military Systems

"Three important issues are likely to be at the top of the U.S. military's manpower agenda in the late 1980s" according to Pirie (1987). One of these "is the emergence of opportunities and technology to create military equipment that will be highly effective when operated and maintained by people of moderate skills and aptitude."[22] Discussing military manpower management, Pirie further states:

> Beyond mere questions of intelligence management, however, lies a more fundamental issue, the question of whether technology is not aggressively adapted to making weapons systems easier and cheaper to produce, to maintain, and to operate? Military reformers of the modern Luddite school equate high tech with high cost and great complexity. Their answer is to emphasize low tech. Others are perhaps too optimistic about the promise of technology and about the capacity of the present system to adapt technology properly to military needs.[23]

The implications for manpower personnel and training in military systems is discussed at some length by Binkin (1986) who predicts:

> In the future as in the past, advances in technology will play a prominent role in shaping the work force. Much will depend on the design characteristics of the military hardware, and specifically its complexity, reliability, and maintainability. If the technologists are correct, the next generation of hardware,

> albeit more complex than the current one, will also be more reliable and easier to maintain, and thus the services will not need as many highly qualified people to keep it in working condition. If the historical experience with high-performance systems is any guide, however, these promises warrant a healthy measure of skepticism. The weight of the evidence is that both new and replacement weapon systems will demand ever-more-skillful operators and maintainers, especially if the capabilities of new systems are to be fully exploited. Thus prudent planners should anticipate that the services' requirements for bright, technologically literate individuals are unlikely to diminish in the years ahead, and it is more than likely, given the present course, that the need for such people will grow commensurate with the complexity of the systems being fielded.[24]

Industrial and Commercial Systems

Complexity in new technology systems is not limited to the military, but is also seen in aviation systems, business systems, and manufacturing plants among others. Concerns for the impact on manpower, personnel, and training are the same.

In commercial aviation, for example, cockpit automation has had a considerable and controversial effect on the crew roles. The number of crew members required (and certified) to fly modern jets has been reduced and there is no evidence that this has impacted safety. The controversy seems to revolve around issues of workload and maintenance of proficiency. Wiener (1989b) speaks of this controversy as follows:

> There appears to be ample evidence to support both positions; however, as usual, the truth undoubtedly lies somewhere between the extremes. On the positive side, the new digitally based equipment is extremely reliable, generally works "as advertised" and offers opportunities to reduce flight time and costs, navigate more precisely (laterally and vertically), operate power plants more efficiently, and augment highly imperfect human monitoring ability with a variety of warning and alerting systems. On the negative side, the digital systems seem to invite new forms of human error in their operation, often leading to gross blunders rather than the relatively minor errors which characterize traditional systems. Furthermore, the equipment does not appear to live up to its expectations in reducing crew workload or increasing time available for extra-cockpit scanning (Curry, 1985; Wiener, 1985), since while the manual tasks may be declining, monitoring and mental workload have increased.[25]

From an MPT perspective, the questions seem to be – has new technology reduced the qualifications and training for aircrews? The answer, I think, is no. Designers will never anticipate everything required in cockpit automation, or in the event of an emergency, the crew will have to take over and fly the aircraft with possibly less direct control than older technology aircraft. Their skills will, therefore, have to be as good or better than before the new technology cockpits. It is also interesting to note that in 1989, several commercial aviation emergencies were handled heroically by pilots in their late 50s, about to be forced to resign. This has opened up another controversy – should pilots (or, for that matter, key personnel in other occupations) be forced to retire at some arbitrary age, or do we need their years of experience and intelligence for those critical times when technology simply won't do the job?

The impact of technology on industrial workers' skills is treated at some length by Zuboff (1988). She explores the changes in centralized control of manufacturing processes and their need for more mental skills as opposed to "body" skills. Patterson (1988) discusses the costs of complexity with several managers and consultants in the manufacturing industry. His article discusses how technology can lead to simplicity or complexity. He quotes Mr. John Hagel, a principal with McKenzie & Company, New York, who talks about design complexity. "You can never anticipate the range of complexity involved when you begin a project. And, further along, the progress becomes so rapid that the temptation to build in complexity is very great because you don't 'unanticipated' to 'unnoticed' to 'unavoidable'. What we are just beginning to realize is how much it costs."[26] Susan B. Bassin, a principal with King Casey, Inc., observes, "It takes maturity to make things simple, to see the value behind simplicity."[27] She also notes that a critical point is when the project team is in concept generation, weighing features and benefits, and someone argues for adding a feature because it's doable. "It's forcing the familiar," notes Ms. Bassin, "that establishes a high comfort level, but it's just another nail in the new-product coffin."[28]

Fastest Growing Jobs

Johnston and Packer (1987) present evidence that the fastest-growing jobs in the next decade will be in the professional, technical, and sales fields which require the highest education and skill levels. When jobs are ranked according to skills, only 27 percent of all new jobs fall into the lowest two skill categories which carries 40 percent of current jobs. By contrast, 41 percent of new jobs are in the three highest skill groups, compared to only 24 percent of current jobs. Johnston and Packer further conclude:

> As the economies of developed nations move further into the post-industrial era, human capital plays an ever-more-important

> role in their progress. As the society becomes more complex, the amount of education and knowledge needed to make a productive contribution to the economy becomes greater. A century ago, a high school education was thought to be superfluous for factory workers and a college degree was the mark of an academic or a lawyer. Between now and the year 2000, for the first time in history, a majority of all new jobs will require postsecondary education.[29]

Technology vs. User Driven Designs

There is a growing temptation to incorporate some new technological pizzazz in a design just because it can be done rather than because it is necessary. In other words, designs may be driven by technological feasibility rather than the needs of the users who must operate and maintain the products of these designs. Nowhere is this more evident than in the military. Urban (1988) describes this mind-set as follows:

> The Department of Defense considers new technology to be the cure for just about any ill that exists in the operating forces. Yet, the hasty introduction of technology over the years has made the job of operating and maintaining those new systems an overwhelming burden. New technology of the next decade will not benefit the operating forces to the fullest extent unless the procurement establishment implements it with human capabilities in mind.
>
> Frequently, the disenchantment with new systems is not with the new technology but with the human interface with that technology. Interface design has been neglected in the headlong rush to implement new hardware capabilities. Engineers have to realize that implementing new technology is more than bolting on or cabling up new systems that just make the operator's job harder to do.[30]

This same concern has been voiced by Lind (1985) who indicates that if technology would be used to meet combat-based requirements rather than technologists' dreams, then the design of advanced-technology components would be simple and robust for the battlefield, not just the laboratory.

The penchant for technology is not limited to the miliary. The concern for technology-drive designs with computer system displays was covered very well by Schwalm and Samet (1989):

> Over the last few years, breakthroughs in computer hardware and software technology have made it cost-effective to present

information to users in a variety of new and sophisticated ways. With the continued development and acceptability of such innovations as desktop and windowing environments, videodisc-based and digital map systems, enhanced graphics and animation, and sophisticated data base and query systems, it is now possible to create and implement a variety of so-called hypermedia for supporting cognitively demanding tasks. For example, technological advancements now permit integrated display, user manipulation, and analysis of a wide range of media including text, maps, imagery, symbology, photographs, drawings, voice and sound effects, and video sequences.

Furthermore, the advent of automation has seen the operator's role in systems change from that of an initiator of task activities to more of a monitor and intervenor in systems in which the computer can attain greater degrees of control – the "mixed initiative" systems.

Overwhelmed by the complexity of hypermedia capabilities for accessing, processing, and presenting data, users are falling prey to a high-technology ailment that promises to plague them well into the 1990s. It is an ailment brought about by chronic difficulty in information processing and decision making that ensues from nonstructured exposure to too many high-technology tools and capabilities. It is an ailment that results in the typical user's inability to effectively employ those tools and capabilities to manage and utilize information. It is "feature shock."[31]

ALLOCATION OF FUNCTIONS

In most complex systems, there is a close sharing of tasks between humans and machines. How these tasks are apportioned is determined by the design of the system – the physical equipment and software, and the human organization; that apportionment is called the "allocation of functions between humans and machines." It is one of the most fundamental of system design decisions and is made during the early design phases, but a systematic approach to function allocation has been an elusive goal of human factors specialists for almost forty years. "Most often it is a decision that the designers make unconsciously rather than by deliberation, and it is a decision that establishes the framework within which job or task analysis, and the requirements for personnel selection, training, procedures development, and design of the human-machine interface are developed"[32] (Price, 1985).

Requirements, Methods, and Guidelines

The military as early as the 1950s and more has recognized the requirement for a systematic approach to the allocation of functions. A current specification, MIL-H-46855B, (Department of Defense, 1984) states under General Requirements – "Starting with a mission analysis developed from a baseline scenario, the functions that must be performed by the system in achieving its mission objectives shall be identified and described. These functions shall be analyzed to determine the best allocation to personnel, equipment, software, or combinations thereof." In the same specification under Detail Requirements, the following statement is provided.

> Defining and Allocating System Functions – The functions that must be performed by the system in achieving its objective(s) within specified mission environments shall be analyzed. Human engineering principles and criteria shall be applied to specify personnel-equipment/software performance requirements for system operation, maintenance and control functions and to allocate system functions to (1) automatic operation/maintenance, (2) manual operation/maintenance, or (3) some combination thereof. Function allocation is an iterative process achieving the level of detail appropriate for the level of system definition.

MIL-H-46855 originated in the 1960s. More recently, the need for a systematic allocation of functions has been recognized by industry (Institute of Electrical and Electronic Engineers, 1988):

> Function allocation refers to the conscious design decisions which determine the extent to which a given job, task, function, or responsibility is to be automated or assigned to human performance. Such decisions should be based upon aspects such as relative capabilities and limitations of humans versus machines in terms of reliability, speed, accuracy, strength and flexibility of response, cost, and the importance of successful and timely task or function accomplishment to successful and safe operations. A wide variety of allocations is possible, ranging from totally automated functions with personnel merely overseeing and monitoring machine performance, through totally human dominated manual tasks. At the finest level of refinement, function allocation also includes determining specific roles and responsibilities of various personnel operating as a team to accomplish the function.[33]

There are requirements, therefore, in both military and civilian system design specifications for the systematic consideration of the allocation of functions.

Methods and Principles for Functions Allocation

There is no standard method, however, for the allocation of functions. Perhaps the first formal treatment of allocation of functions appeared in a landmark article by Paul M. Fitts (1951) of the Ohio State University Aviation Psychology Program. He attempted to characterize qualitatively those functions performed better by machines than by humans, and those performed better by humans than machines. He did this in the form of a list of human and machine capabilities, and similar lists have generally been referred to as 'Fitts lists.' For years, this paper was regarded as the definitive work on allocation of functions, although Fitts said that using the criteria in his list as the sole determinant of allocation of functions was to lose sight of the basic nature of a system containing humans and machines. Considerable progress has been made in the ensuing years and while standardization has not been reached, there are some critical commonalities in most approaches. As a minimum, there is general agreement that there are some functions that only humans can do or only machines can do; and, the majority of the functions are shared between humans and machines. For further discussion of methods of allocating functions, the reader is referred to Clegg, Ravden, Corbett, and Johnson (1989); Kantowitz and Sorkin (1987); and Pulliam and Price (1985). The approach prescribed by Pulliam and Price (1985) is illustrated in Figure 6-3.

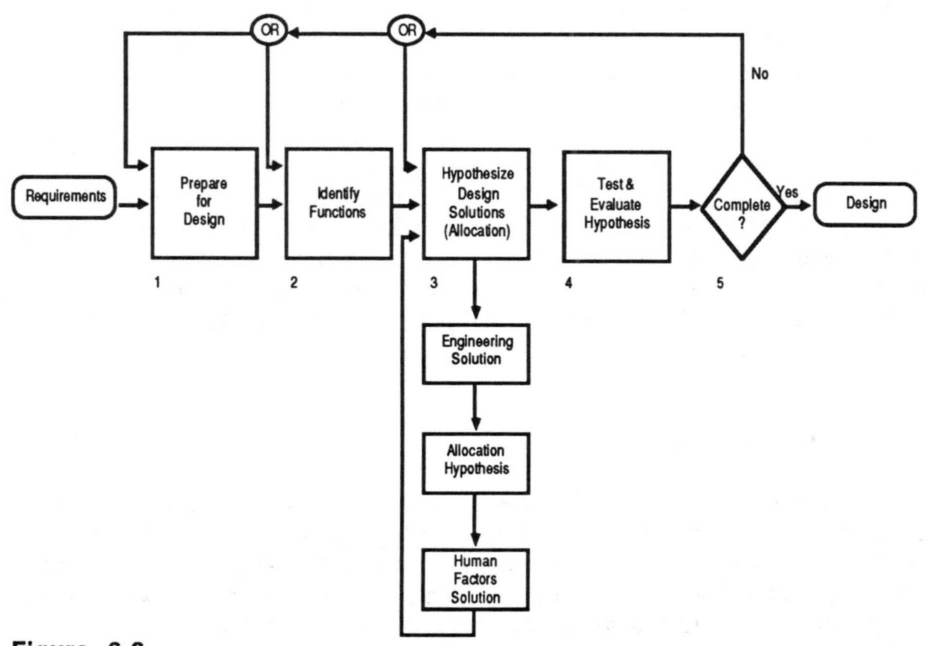

Figure 6-3
The Design Cycle and the Allocation of Functions Process

This approach is characterized by the theory of design – hypothesis and test. That is, the allocation method is embedded in system design activities in which the designer hypothesizes a way to achieve some function and then subjects that hypothesis to empirical or analytical tests. The process is iterated until a satisfactory allocation is achieved.

Inherent in this method is the notion that some allocations will be predetermined by the overall engineering concept or the nature of the tasks. There are several principles which apply to human-machine allocation. These are:

- Mandatory Allocation
- Shared Allocations
- Keeping Man-in-the-Loop
- People as Monitors
- Automated Systems Unique Requirements

Mandatory Allocations

Some functions or tasks *must* be performed by machines and some *must* be performed by humans. This mandatory allocation may be because of the fundamental nature of the system or the demands of the task performance. In the first case, for example, strategic military explosive weapons can be bought to bear on enemy targets with the use of ballistic missiles or manned bombers. In either case, certain mandatory allocations are implied. In the more general case, automation may be mandatory because of such factors as working conditions (hostile to humans), regulation, or task demands beyond human capability because of response time, perceptual requirements or complexity. On the other hand, human performance may be mandatory because of some labor or policy agreement, or because the alternatives cannot be specified in advance and thus cannot be preprogrammed – requiring human decision-making and control. In most cases, tasks will be shared rather than performed by humans exclusively or by machines exclusively.

Shared Allocations

A logical error of many allocation methods is an assumption that the choice between human and machine is a zero-sum problem. As noted by Price (1985), "It is assumed that if a human is a poor controller, then a machine must necessarily be a good one. Once stated, we recognized that this is clearly not true: there are tasks, such as high-speed estimation of risk, that neither a human nor a machine does well, and other tasks that both can do superbly."[34] Thus, most allocations will have an acceptable solution across a range of automation and people, and the final design may be based on utilitarian (is a human available in any event?) and cost considerations. Finally, it is now technologically feasible to have adaptive control systems that can be modified (manually or automatically) to redistribute the allocation of responsibilities to humans and machines. It is a real challenge to system engineers and human factors specialists to identify optimum shared allocation solutions, particularly in situations where the ultimate control and responsibility rests with human operators.

Keeping Man-in-the-Loop

Automated systems, at least for the remainder of this century, will continue to be human-machine systems that will require the human to monitor machine functions and to decide when to override them. After all, are we ready to let automated weapon systems roam the battlefield and fire at will? Are we ready to let aircraft take off and land without a "pilot in command?" Are we ready to let computers run our nuclear power plants or chemical plants and make the ultimate decision concerning the public health and safety? If we are not, we must keep man-in-the-loop (MIL). Or, as a recent Food and Drug Administration draft policy stated (Peterson, 1988), the concept of "competent human intervention" sets the dividing line between what is and is not regulated in computer-controlled medical devices. We can prepare for competent human intervention by keeping man-in-the-loop.

The extent to which humans are kept in-the-loop for purposes of maintaining necessary skills, knowledge, and attitudes needed to function effectively was recently addressed by the National Research Council (NRC) (Moray & Huey, 1988) as a critical behavioral science issue related to questions of automation. This issue is not the level of automation, but the relation between the human and the machine roles in controlling and managing a complex machine process. According to the NRC report, the result of increased automation in the past was to move the human role farther away from direct contact with the controlled process. Thus, the human becomes a supervisor and manager of partially autonomous machine resources (e.g., Sheridan and Hennessy, 1984). The implication is that, with increases in the level of automation, the human is moved "out-of-the-loop." Recent research on the effects of automation (e.g., Wiener, 1985) suggest that this architecture may be poor for human-machine control of complex processes; and there is a need for new architectures in which the level of machine involvement is high (a highly automated system) but in which the human plays a more continuously active role in the control and management of the process or is more "in-the-loop."

Effectively keeping man-in-the-loop requires an understanding of people as monitors and their unique information needs for competent intervention.

People as Monitors

A person's effectiveness as a supervisory monitor/controller depends both on design considerations to enhance monitoring performance and on specific information to satisfy the mental needs to be ready for intervention. Many researchers and designers have acknowledged the possibility that automatic systems may not provide the necessary information for the operator or supervisor to remain in-the-loop mentally. A person can be an effective supervisory monitor only if cognitive support is provided at the control station. Cognitive support refers to the human need for information to be ready for actions or decisions that may be required. Could displays or other sources of information be provided to furnish all the information needed? Can a work sequence be established so that the human will maintain an adequate mental model of the system and its conditions by being actively involved in controlling or approving the key system changes? Could the human operator be given sufficient activity to ensure alertness?

Sharit (1985) provides a perspective about supervisory control of manufacturing systems as expressed by this excerpt:

> Although the potential exists for an active supervisory role, the job design philosophy assumed by most manufacturers regarding supervisory control . . . implicitly negates human decision-making functions. Most of the human's actions result by default – that is, when something goes wrong. This makes the human's involvement more passive, and in the process, violates various human factors principles of job design. A coherent role would demand a significant portion of the human's attention and abilities by requiring the properties or displayed information to be studies instead of being only scanned for problems. By not being actively involved in system control, the human is less likely to be efficient in reacting to critical system events. The presence of such events could, in fact, go entirely unnoticed. The human's supervisory role essentially reduces to monitoring, a task human may perform poorly over prolonged periods of time.[35]

We must remember that humans are inherently poor monitors and we must provide for their unique information needs if we are to keep man-in-the-loop.

Automated Systems Unique Requirements

One way to keep the human actively involved and to maintain their "mental model" is to provide him or her with information concerning the "intent" of the automated system. That is, given the current decisions made or about to be made by the automated system, what will the situation look like in the future. Essentially, the system should not only identify a potential problem but suggest alternative solutions and show the implications of the best, next best, etc., solution. The operator could agree or disagree with the system's plan based on the projected situation. Another option is to allow the operator to test hypotheses and enter alternate plans into the system. For each plan the system would display the projected outcome. Operators have certain expectations and are continually forming solutions to the current situation. They need to know what the computer is doing and why. Also, they may want to compare the results of their strategies with those recommended by the computer.

There is a long history of research on the use of quickened and predictive displays to assist pilots, controllers, and nuclear power plant operators in making decisions about the best course of action to pursue. Predictive displays use a fast time model of the system to predict the future excursion of the system and display the excursion to the operator on a scope. In the past, predictive displays have been used to show the operator the implications of his or her actions – an if-then process. In this capacity, they have provided an excellent training tool.

Predictive displays can play an important role in keeping the operator's mind actively engaged in the control process. The displays could be designed to show what the automated system's "intent" and what the implications are. The operator could then concur or disagree with the system. If the operator instituted a change, the results of the change would be displayed.

The problems of factory workers relating to centralized automated control are even more unique. This area is dealt with at great length by Zuboff (1988). She talks about workers who, instead of using their bodies as instruments, now depend on a "data interface" represented by computer terminals monitored from central control rooms. Zuboff goes on to note that workers were at first overwhelmed with the feeling that they could no longer see or touch their work as if it had been made both invisible and intangible by the computer. This dramatically indicates the plight of keeping the factory worker in-the-loop as plants are upgraded to centralized digital control systems.

The technology is just around the corner to allow people to effectively communicate with computers. In many cases, it is just a question of properly displaying information (visually or aurally) already available in the system. Thus, it is up to designers to ensure that operators (or maintainers) and computers each know what the other is doing. The person must understand the objectives of automatic control and the allocation of responsibility to human operation and machine logic. In the same way, the automatic logic must be informed about the actions and objectives of the operator. Otherwise, the human and hardware/software systems may work at cross purposes.

USER ACCEPTANCE

The introduction of automation can result in significant changes in the jobs being performed by people. In some instances automated devices are built as aids; in other instances they are designed to replace human operators. Successful implementation relies heavily on user acceptance and understanding and on the compatibility of the automation device features with the task requirements in the operational environment. The first requirement for anyone hoping to introduce new automation in the workplace is to better understand why people resist change.

Resistance to Change

A recent conference hosted by the National Academy of Sciences (Vare, 1989) discussed how United States businesses can extend their skill at developing technologies to actually utilizing the innovations in their own workplaces. "Engineers and behavioral researchers at the meeting said employees now resist technological change for diverse reasons . . . new technologies can have unintended human consequences." The conferees further agreed that technology alone is unlikely to help a company unless attention also is paid to the people who use it. Staff involvement is essential and the conferees cautioned against engineering departments making technology decisions without the input of the people who will use it.

The key, according to one conferee, "Is to look past technological wizardry to whether innovations will really improve the workplace. Change is likely to be rejected if it is poorly understood, imposed from above, or perceived as risky or threatening."

There is a long history of resistance to change resulting from the introduction of automation. According to Rogers and Shoemaker (1971) there are subjective, psychological risks associated with all innovations, particularly those that involve changes in responsibility. With advances in technology, automation is beginning to assist and in some cases take over intellectual functions such as expert judgment and decision making. Automated systems which embody these characteristics are viewed by many as threats and are rejected on the basis of fear. Others resist automation because they feel that the devices are unreliable and do not match with either the operational requirements of the job or the ways that the user perceives the job and the operating environment. There are several instances reported by Mackie and Wylie (1985) and Price, Maisano and Van Cott (1982) in which equipment was either misused, underused, or even sabotaged by operators due to different perceptions of users and designers about the nature of the job. Another major contributor to resistance to automation is the requirement for the individual in charge to take responsibility (be held accountable) for the output of the automated system. Without proper involvement of the user in the design and development, process there is a clear tendency not to trust these systems.

Principles for User Acceptance

Price, Smith, and Behan (1964), in a study of automated landing systems, have provided a number of acceptance principles which are still applicable today (see Table 6-4). Ayral and Conley (1987) provide suggestions for gaining operator acceptance of advanced controls. Suggestions include (1)holding kickoff meetings with each of the operating shifts; (2) conducting plant tests with operators involved for their opinions; (3) conducting control

Table 6-4
Acceptance Principles from a Study of Commercial All-Weather Landing Systems (Source: Price, Smith, & Behan, 1964)

ROLE EXPECTANCY
1. People are generally accepting of system roles which give them an opportunity to exercise and, therefore, maintain skills which they feel are important to maintaining their position in the occupational and social status system in which they are immersed.
2. People are generally accepting of system roles which permit them to vary their procedures and the manner of accomplishing their tasks, on their own initiative. Roles that fail to permit people to vary their procedures on their own are generally labeled mechanical.
3. People are more accepting of roles which permit them to learn.
AUTOMATION ACCEPTANCE
1. The more system experience people have with this experience, including exposure to automated equipment, the more accepting they are of the automated equipment and the more they will use it in the prescribed manner.
2. Those with more status, responsibility, and authority tend to be more accepting of and make more use of automated equipment than others.
3. Where failure of the performance of its function by automated equipment can endanger the life of people, they are less likely to accept and use it despite prescribed procedures.
4. There is generally high acceptance, within the limits of the above three principles, of the automation of servo tasks, particularly those which must be performed over long periods of time.
5. There is generally rather low acceptance of automation of decision making functions.

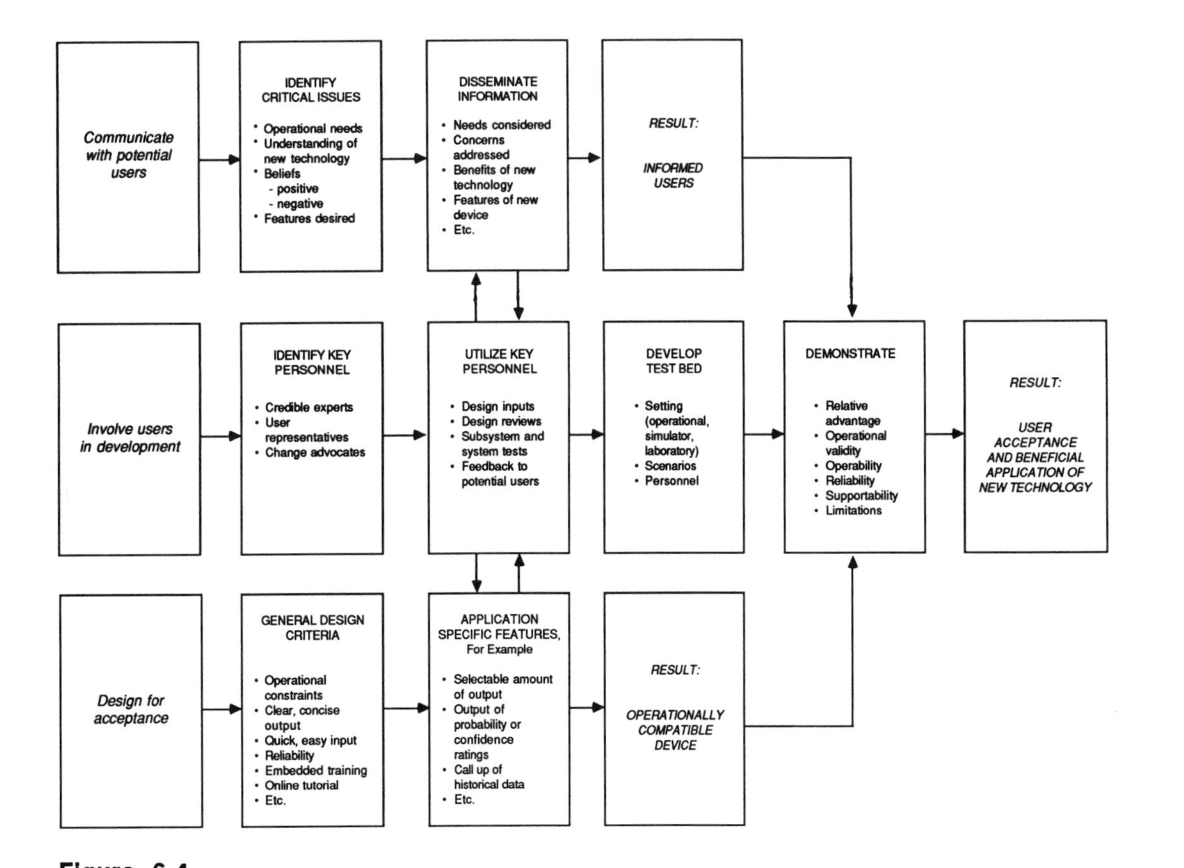

Figure 6-4
A Model for Increasing User Acceptance and Technology Transfer (From Mackie and Wylie, 1985)

simulations and demonstrating how the new control system works to operations supervisors and operators themselves; (4) during control commissioning, talking to operators about potential unit problems encountered and making sure the control design addresses all of these problems; and (5) implementing the new design gradually and achieving a reputation for success.

Model for User Acceptance and New Technology Introduction

In a more integrated approach, Mackie and Wylie (1985) present a model for successful transfer and application of new technology (see Figure 6-4) in the workplace. The three key steps in this process are: communicate with potential users; involve users in the development process; and design for acceptance.

Communication should be initiated early in the design process. The purposes of the communication are (1) to inform potential users about the technology; (2) to gather information from users about operational issues; and (3) to provide users with feedback on how the planned technology will address operational requirements.

Actively involving users in the development includes working closely with users to (1) describe task and users' perceptions of task procedures; (2) set up a test bed; and (3) demonstrate the validity, reliability, supportability, and limitations of the new system.

There are several steps involved in designing for acceptance. The output of these steps should be an innovation which is compatible with operational requirements. That is, the innovation should be designed with operational and training needs in mind. The three general design criteria of greatest importance are that (1) the design incorporate both operational needs and constraints; (2) the technology is reliable, fast, and easy to use; and (3) training support is provided on-line.

In general, user acceptance will be enhanced if the users are brought into the design process early and participate in all stages of the system development. This requires working with the users to obtain operational and task information, providing the users with briefings and demonstrations, and making modifications on the basis of users' feedback.

It is important to remember that the same users who turn on automated systems can also turn them off. Therefore, it is the users who are the real judges of automation.

CONCLUSION

People's role in future systems will become even more demanding, despite the promises of automation. While new and advanced technology will undoubtedly reduce some human workload (both physically and mentally) and machines will relieve many people from some tedious, repetitive, unhealthy and unsafe tasks, the human will always be required to monitor and have authority to start, stop, and intervene in system operations. There are those who say that getting rid of people gets rid of human error – but when we replace people with computers, we get computer error. Who will agree to let autonomous military, industrial, and civil systems make decisions and implement them without some human oversight – particularly when the consequences of malfunction or accident in many high technology systems can lead to economical, environmental, or human disasters?

Technological discovery and implementation in the future has incredible potential, particularly in the information and automation areas. Yet we must not let the designs of systems be driven by technological feasibility instead of user needs. Automation and the allocation of functions, when done with deliberation during concept design, can result in the appropriate responsibility and authority for people; and, when integrated with manpower, personnel, and training requirements, will provide high performance at the lowest cost of ownership. History has taught us that, unless we plan for the optional human role during concept design, we may be in for costly retrofits or redesign when we discover later that humans cannot or will not use the system as designed. But spending our resources in an attempt to replace rather than integrate humans in authoritative roles will be a costly lesson. As Donald Latham, Undersecretary of Defense, said at a Harvard University Seminar in 1985 (Pew and Jarvis, 1986):

> . . . there has been a lot of writing and speculation in the papers vis-a-vis Star Wars and other issues, speculating that somehow we will automate this process and take people out of the loop. I think it is exactly the opposite . . .
>
> When you're handling hundreds of megabits of information and you must reconstruct it into images on the ground with processors, you need computers. But in the end, you still want to have humans who can do things much better than machines can today, even with all our sophisticated artificial intelligence techniques, and probably always will. The human mind has the ability to do things that we don't begin to understand. It correlates facts and looks at things . . .
>
> We'll always, always keep people there to look at these scopes, and do that final assessment and correlation function. Even in SDI, where things could happen very rapidly, you're just simply not going to turn the system on "automatic." No way.

There is no such thing as an unmanned system.

NOTES

[1]Johnston, W. B., & Packer, A. H. (1987). *Workforce 2000: Work and workers for the twenty-first century.* Indianapolis, IN: Hudson Institute.

[2]Zuboff, S. (1988). *In the Age of the Smart Machine.* New York: Basic Books, Inc.

[3]ELLEN GOODMAN, © 1987, The Boston Globe Newspaper Company/Washington Post Writers Group. Reprinted with permission.

[4]Perrow, C. (1984). *Normal accidents: Living with high-risk technologies.* New York: Basic Books, Inc.

[5]Ibid.

[6]Reprinted with permission © 1988 from *Human Error Avoidance Techniques Conference Proceedings*, P-204.

[7]Ibid.

[8]Ibid.

[9]*IEEE Spectrum* (1987, June). Too much, too soon: Information overload. New York: Institute of Electrical and Electronic Engineers, pp. 51-55. © 1987 IEEE.

[10]Pirie, R. B., Jr. (1987). Manpower. In J. Kruzel (Ed.), *American Defense Annual* (pp 155-169). New York: Lexington Books.

[11]Binkin, M. (1986). *Military technology and defense manpower.* Washington, DC: The Brookings Institute.

[12]Pirie, R. B., Jr. (1987). Manpower. In J. Kruzel (Ed.), *American Defense Annual* (pp 155-169). New York: Lexington Books.

[13]Perrow, C. (1984). *Normal accidents: Living with high-risk technologies.* New York: Basic Books, Inc.

[14]From Human Factors, Vol. 27, No. 1, 1985. Copyright 1985 by the Human Factors Society, Inc., and reprinted with permission.

[15]Reprinted with permission from the July 1989 issue of *Defense Science* Magazine.

[16]Johnston, W. B., & Packer, A. H. (1987). *Workforce 2000: Work and workers for the twenty-first century.* Indianapolis, IN: Hudson Institute.

[17]Binkin, M. (1986). *Military technology and defense manpower.* Washington, DC: The Brookings Institute.

[18]Johnston, W. B., & Packer, A. H. (1987). *Workforce 2000: Work and workers for the twenty-first century.* Indianapolis, IN: Hudson Institute.

[19]Reprinted with permission from the July 1989 issue of *Defense Science* Magazine.

[20]From Human Factors, Vol. 27, No. 1, 1985. Copyright 1985 by the Human Factors Society, Inc., and reprinted with permission.

[21]*To engineer is human: The role of failure in successful design*, Henry Petroski, © 1985, St. Martin's Press, Inc., New York.

[22]Pirie, R. B., Jr. (1987). Manpower. In J. Kruzel (Ed.), *American Defense Annual* (pp 155-169). New York: Lexington Books.

[23]Ibid.

[24] Binkin, M. (1986). *Military technology and defense manpower.* Washington, DC: The Brookings Institute.

[25] Wiener, E. L. (1989). *Cockpit automation.* In E. L. Wiener & D. C. Nagel (Eds.), *Human factors in aviation.* San Diego: Academic Press.

[26] Reprinted with permission from *Industry Week*, June 6, 1988. Copyright, Penton Publishing, Inc., Cleveland, Ohio.

[27] Ibid.

[28] Ibid.

[29] Johnston, W. B., & Packer, A. H. (1987). *Workforce 2000: Work and workers for the twenty-first century.* Indianapolis, IN: Hudson Institute.

[30] Reprinted from Proceedings with permission; copyright © (1988) U.S. Naval Institute.

[31] From the *Human Factors Society Bulletin*, June 1989. Copyright 1989 by the Human Factors Society, Inc., and reprinted with permission.

[32] From *Human Factors*, Vol. 27, No. 1. Copyright 1985 by the Human Factors Society, Inc., and reprinted with permission.

[33] Reprinted from IEEE Std 1023-1988, IEEE Guide for the Application of Human Factors Engineering to Systems, Equipment, and Facilities of Nuclear Power Generating Stations, copyright © 1989 by The Institute of Electrical and Electronics Engineers, Inc., with the permission of the IEEE Standards Department.

[34] From *Human Factors*, Vol. 27, No. 1. Copyright 1985 by the Human Factors Society, Inc., and reprinted with permission.

[35] Ibid.

REFERENCES

Ayral, T. E., & Conley, R. C. (1987). Gaining operator acceptance of advanced controls. *Hydrocarbon Processing, Vol. 66* (6), pp. 42-43.

Bainbridge, L. (1982). Ironies of automation. In G. Johannsen & J. E. Rijnsdorp (Eds.), *Analysis, design, and evaluation of man-machine systems, Proceedings of the IFAC/IFIP/FFORS/IEA Conference* (pp. 129-135). New York: Pergamon Press.

Beatson, J. (1989, April 2). Is America ready to "fly by wire"? *The Washington Post.*

Binkin, M. (1986). *Military technology and defense manpower.* Washington, DC: The Brookings Institute.

Chisman, F. P. (1989). *Jump start: The federal role in adult literacy.* Sponsored by the Southport Institute for Policy Analysis.

Clegg, C., Ravden, S., Corbett, M., & Johnson, G. (1989). Allocating functions in computer integrated manufacturing: A review and new method. *Behavior and Information Technology, Vol. 8, No. 3*, pp. 175-190.

Conway, E., Muckler, F., & Peay, J. (1982). *Human error, injuries and accidents–A review and analysis.* Paper presented at the 26th Annual Meeting of the Human Factors Society, Seattle, WA.

Davis, B. (1987). Costly bugs: As complexity rises tiny flaws in software pose a growing threat. *Wall Street Journal.*

Department of Defense (1984). *Human engineering requirements for military systems, equipment, and facilities (MIL-H-46855).*

Fitts, P. M. (Ed.). (1951). *Human engineering for an effective air-navigation and traffic control system.* Ohio State University Research Foundation.

Goodman, E. (1987, January 13). What is the "Human Factor"? *Washington Post.*

Graine, G. N. (1988). The engineering syndrome vs. the manpower, personnel and training dilemma. *Naval Engineers Journal*, pp. 54-59.

Hawkins, F. H. (1987). *Human factors in flight.* Gower Publishing Company, Vermont.

IEEE Spectrum (1987, June). Too much, too soon: Information overload. New York: Institute of Electrical and Electronic Engineers, pp. 51-55.

Institute of Electrical and Electronic Engineers (1988). *IEEE STD-1023, Guide for the application of human factors engineering to systems, equipment, and facilities of nuclear power generating stations.* New York: Institute of Electrical and Electronic Engineers.

Jacky, J. (1989, September/October). Programmed for disaster: software errors that imperil lives. *The Sciences*, pp. 22-27.

Johnston, W. B., & Packer, A. H. (1987). *Workforce 2000: Work and workers for the twenty-first century.* Indianapolis, IN: Hudson Institute.

Kantowitz, B. H., & Sorkin, R. D. (1987). Allocation of Functions. In G. Salvendy (Ed.), *Handbook of Human Factors* (pp. 355-369). New York: John Wiley and Sons.

Lake, J. S. (1989, July). Critical Technology. *Defense Science*, pp. 17-22.

Lane, N. E. (1987). *Evaluating the cost effectiveness of human factors engineering* (Institute for Defense Analysis Contract MDA 903 '84 C 0031). Orlando, FL: Essex Corporation.

Lauber, J. (1989, February). New philosophy need for cockpit automation. *Avionics*, pp. 8-9.

Lind, W. S. (1985). A doubtful revolution. *Issues in Science and Technology.*

Lubove, S. H. (1987). In the computer age, certain workers are still vital to success. *Wall Street Journal.*

Mackie, R. R., & Wylie, C. D. (1985). *User considerations in the acceptance and use of AI decision aids.* Goleta, CA: Essex Corporation.

Margulies, F., & Zemanek, H. (1982). Man's role in man-machine systems. In G. Johannsen & R. E. Rijnsdorp (Eds.), *Analysis, design, and evaluation of man-machine systems, Proceedings of the IFAC/IFIP/FFORS/IEA Conference.* New York: Pergamon Press.

Moray, N., & Huey, B. (1988). *Human factors research and nuclear safety.* Washington, DC: National Academy Press.

Nobel, D. F. (1984). *Forces of production: A social history of industrial automation.* New York: Alfred A. Knopf.

Patterson, W. P. (1988, June 6). The costs of complexity. *Industry Week*, pp. 63-68.

Perrow, C. (1984). *Normal accidents: Living with high-risk technologies.* New York: Basic Books, Inc.

Peterson, I. (1988, March 12). A digital matter of life and death. *Science News, Vol. 33*, pp. 170-171.

Petroski, H. (1985). *To engineer is human: The role of failure in successful design.* New York: St. Martin's Press.

Pew, R. W., & Jarvis, M. (1986). *An Air Force advanced engineering program for SDI BM/C3 human factors* (BBM Report Number 6382). Cambridge, MA: BBM Laboratories, Inc.

Pfeiffer, J. (1989). The secret of life at the limits: Cogs become big wheels. *Smithsonian*, 27(4), pp. 38-48.

Pirie, R. B., Jr. (1987). Manpower. In J. Kruzel (Ed.), *American Defense Annual* (pp 155-169). New York: Lexington Books.

Price, H. E. (1985). The allocation of functions in systems. *Human Factors, 27(1).*

Price, H. E., Fiorello, M., Lowry, I. C., Smith, G. M., & Kidd, I. S. (1980). *The contribution of human factors in military system development: Methodological considerations* (Technical Report - 476). Alexandria, VA: U.S. Army Research Institute for the Behavioral and Social Sciences.

Price, H. E., Maisano, R. E., & Van Cott, H. P. (1982). *The allocation of functions in man-machine system: A perspective and literature review* (NUREG-CR-2623). Oak Ridge National Laboratories.

Price, H. E., Smith, E. E., & Behan, R. A. (1964). *Utilization of acceptance data in a descriptive model for determining man's role in a system* (NASA Contractor Report No. CR-95). Washington, DC: National Aeronautics and Space Administration.

Pulliam, R., & Price, H. E. (1985, February). *Automation and the allocation of functions between human and automatic control: General method* (AFAMRL-JTR-85-017). Wright Patterson Air Force Base, OH: Aerospace Medical Division, Air Force Aerospace Medical Research Laboratory.

Rogers, E. M., & Shoemaker, F. F. (1971). *Communication of innovations.* New York: Free Press.

Sanders, M., & Shaw, B. (1988). *Research to determine the contribution of system factors in the occurrence of underground injury accidents* (Bureau of Mines Contract J0348042). Westlake, CA: Essex Corporation.

Schwalm, H. D., & Samet, M. G. (1989). Hypermedia: Are we in for "future shock"? *Human Factors Bulletin, Vol. 32* (6).

Sexton, G. A. (1988). Cockpit-crew systems design and integration. In E.L. Wiener & D.C. Nagel. (Eds.), *Human Factors in Aviation.* San Diego, CA: Academic Press.

Sharit, J. (1985). Supervisory control of a flexible manufacturing system.

Human Factors, 27, (1), pp. 47-59.
Urban, C. D. (1988, October). The human factor. *Proceedings of the U. S. Naval Institute*, pp. 58-64.
U. S. General Accounting Office (1985, September 27). *The Army can better integrate manpower, personnel, and training into the weapon systems acquisition process* (Report GAO/NSIAD-85-154).
U. S. Congress, Office of Technology Assessment (1989). *Holding the edge: Maintaining the defense technology base* (OTA-ISC-420). Washington, DC.
Vare, E. A. (1989). Worker fears can deter adoption of new technology. *News Report, National Research Council, Volume XXXIX 5*, pp. 10-12.
Wiener, E. L. (1985). Beyond the sterile cockpit. *Human Factors, 27*, (1), pp. 75-90.
Wiener, E .L. (1988). Management of human error by design. *Human Error Avoidance Techniques Conference Proceedings*. Society of Automotive Engineers, Inc.
Wiener, E. L. (1989a). *Human factors of advanced technology ("glass cockpit") transport aircraft* (NASA Contractor Report 177528). Washington, DC: National Aeronautics and Space Administration.
Wiener, E. L. (1989b). *Cockpit automation.* In E. L. Wiener & D. C. Nagel (Eds.), *Human factors in aviation.* San Diego: Academic Press.
Zuboff, S. (1988). *In the Age of the Smart Machine*. New York: Basic Books, Inc.

CHAPTER **7**

COMPUTER-AIDED ERGONOMIC DESIGN TOOLS

Joe W. McDaniel
Mark A. Hofmann

ABSTRACT

Maintaining the competitive edge on the battlefield or in the private sector requires increasing productivity at a reasonable cost. In the process of designing, engineering, and manufacturing equipment/software, progress is being made toward improving efficiency of the design process through the application of computer-aided design, engineering, and manufacturing (CAD/CAE/CAM).

The availability of these new techniques provides a new environment for implementing the MANPRINT philosophy, focusing research, and improving the technology transfer of knowledge and its applications in the field of human performance. Modern CAD systems allow conceptual designs to be used as *electronic mockups*. Also, new ergonomic design tools provide a quick and accurate means of evaluating the interaction between a maintenance technician and a system design. Without such tools, these types of evaluations were rarely cost effective and often of insufficient quality. The state of the art for these and other ergonomic design techniques is presented along with current limitations and the need for a fully interactive dynamic ergonomic model.

INTRODUCTION

By changing the very nature of the engineer's work, computers are revolutionizing the design, engineering, and production process. Low-cost computers and user-friendly software are changing the way engineers work. Desktop and mini-computers are now commonly found throughout industry and government. Designers and engineers often use computer graphics to sketch a basic configuration, computer analysis tools to perform calculations formerly computed with a slide rule, computer engineering aids to solve equations and make trade-offs, simulation tools for processes once done manually by stress or aerodynamics departments, three dimensional (3-D) computer-aided drafting software to avoid blueprints, and computer-aided manufacturing processes to machine and inspect the parts.

Demands for improved quality, reliability, and performance have placed additional requirements upon the design process itself. Early in the design of a new system, decisions are made which commit the project to an approach with implicit limitations in quality, reliability, and performance. The process still depends on the creativity of the designer or engineer using the traditional *try and test* method. However, drafting lofts with hundreds of detail-generating draftsmen are now replaced by tens of computer graphics stations. People are still in the loop, but they are fewer, more costly, and use a work medium which integrates the entire process of design, engineering, and manufacturing. The end result is time compression for the entire process and, therefore, enhanced productivity.

In addition to increasing productivity of the design, engineering, and manufacturing process, the CAD/CAE/CAM revolution also provides a unique opportunity to enhance utility of the resulting equipment when in the hands of the end user. Considering characteristics of the user (operator and maintainer) in the design yields equipment better designed to accommodate the user (i.e., equipment designed to capitalize on user's known capabilities and augment user's limitations). Traditionally, the human factors engineer, the user's advocate in the design process, applies established human factors engineering (HFE) standards to design the system/user interface. This interface design focuses on increasing the end performance of the equipment and decreasing the costs associated with manning and training. Design refinements and trade-off estimates are often made through the use of mockups and/or simulations depending on the complexity of the system. Today's challenge is to ensure that human factors are integral to the CAD/CAE techniques so that the resulting equipment is designed to enhance utility and performance while lowering the overall cost.

Integrating the User Into CAD

The human factors community is becoming more sensitive to the need for developing comprehensive, interactive, dynamic ergonomic models and applying them through popular industrial CAD systems. Over 100 human-models having various capabilities have been developed and are now being used by industry early in the design process. Also, efforts to synthesize human performance data required to develop comprehensive ergonomic models are progressing, as evidenced by recent publications like the *Handbook of Perception and Human Performance* (Boff, Kaufman, & Thomas, 1986) and the *Engineering Data Compendium: Human Perception and Performance* (Boff & Lincoln, 1988). It is encouraging to note that innovative work and expressions of enthusiasm for ergonomic model development are now coming from academia, the private sector, and the government, and include mathematicians, engineers, physicians, psychologists, computer scientists, anthropologists, and physicists.

The potential application of such ergonomic models is broad and includes not only equipment design but also vulnerability assessments, operability assessments, training requirements and safety assessments.

Key Ergonomic Model Attributes

Ideally, a comprehensive, dynamic, interactive, ergonomic model would encompass the range of human anthropometry as well as physical and cognitive capabilities and limitations. It would be able to interact with and be appropriately affected by new interface technology, environmental conditions, task demands, and social or crew interactions. Such a model would subsequently permit one to assess subsystem and system performance as a function of changes in human performance parameters, procedures, and/or changes in design configurations. Such a model would possess an architecture which would not only permit its use in popular CAD systems but would also be easily modifiable to accept new information coming from the human performance research.

Although we are working toward the future, we must remember that many usable *ergonomics* models already exist, several of which will be described later in this chapter. Much data concerning human characteristics and how these characteristics are affected by the environment of variables are available but not yet modeled. For example, if one knows the effects of whole body vibration on one's visual acuity, one should certainly be able to model what one would see from an acuity perspective across various potential vibration design alternatives.

MODERN ERGONOMIC CAD TOOLS

Traditional anthropometric and biomechanical accommodation analyses require a system prototype or mockup which has a mature design as a prerequisite and which is difficult to change. Even a hardware mockup has a significant cost. Designs on a CAD system, on the other hand, have little or no additional cost because modern designs are already created on CAD systems.

Customarily, mockup evaluations have tested only four or five subjects, too few to represent the user population. Because of variation in proportionality among body dimensions, there is a significant possibility of erroneous conclusions. Considering the world community, adult human stature varies from 2 to 8 feet, and adult body weight ranges from approximately 20 to 1000 pounds. Fortunately for the designer, these extreme individuals are so rare they need not be considered in general purpose designs. The great range of body sizes is not found in the military populations because the military has admission standards that screen out the rare individuals. Typically, military populations have statures ranging from 4.5 to 6.5 feet and weights from just below 100 pounds to slightly over 250 pounds (Department of Defense, 1980). But even these ranges are still too large for some systems to accommodate. More stringent body size requirements are necessary for aircraft pilots, and many of the physically demanding jobs now have minimum strength requirements as well.

Accommodation analysis with human subjects placed in an actual mockup is risky because of the lack of proportionality among people. Two people with exactly the same stature will have different length arms, legs, etc. Typical design specifications may require accommodation of a size range from a 5th percentile (small) female to a 95th percentile (large) male. There are, however, no 5th percentile women and there are no 95th percentile men. The term *percentile* refers to a statistical property of a single measure. Fifth percentile refers to that measure which 5 percent of the people are at or below and 95 percent are above. People near the extremes of a distribution in one dimension are usually nearer the average in other dimensions. So, if one selects a person with a 5th percentile stature (short), that person's arms (and therefore arm-reach characteristics) are likely to be much longer than a 5th percentile arm reach and can even be much longer than average. If one evaluates reach to controls with such a subject, one will likely overestimate the arm reach and arrive at an erroneous conclusion.

A major obstacle to anthropometric accommodation has been finding a sample of subjects which represents the exact measures on one or more dimensions which are pertinent to the design. The major advantage of using computer models of the human body to evaluate accommodation is that all dimensions can be adjusted to the required values. Thus, the computer-generated subjects chosen to evaluate the design are as good as one's ability to define the needed values.

Another reason anthropometric evaluation has been difficult is because the body-size measures are unidimensional segment lengths and circumferences, not the 3-D motion envelopes required by workplace designers. Figure 7-1 shows some of the more common anthropometric dimensions. Each is the measure of an important body segment or characteristic, but the measures are not attached to each other nor to any common coordinate system. For example, the measure *shoulder-elbow length* and *elbow-wrist* are not measured to the same point on the elbow, nor is either *elbow landmark* at the center of rotation of the elbow joint. Even if a design engineer were given these measures, he or she would not be able to assess arm reach capability.

Anthropometric measures are made on the surface of the body, while the engineer usually needs internal dimensions associated with the skeletal link system, that is, from one joint center to another. The science of relating

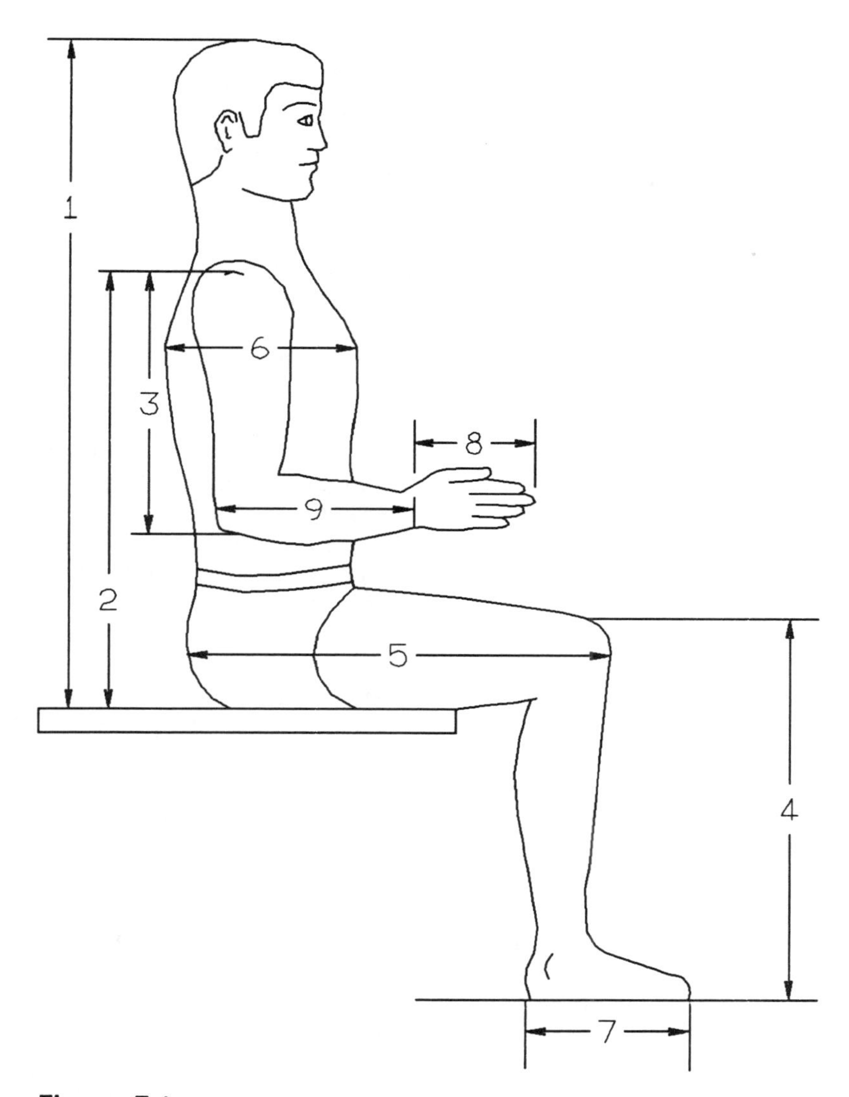

Figure 7-1
Anthropometric measures are isolated dimensions on body segments. They are not related to each other, nor to any base coordinate system. These are external measures and do not describe the skeletal links which describe the functionality of the body.

1. Sitting height; 2. Acromion height/sitting; 3. Shoulder-elbow length; 4. Knee height/sitting; 5. Buttock-knee length; 6. Chest depth; 7. Foot length; 8. Hand length; 9. Elbow-wrist length.

external and internal body dimensions is too complicated for a designer to apply routinely, but it is not too difficult to incorporate into a computer model and calibrate against empirical task analyses. Development of task-related data is usually very difficult because it is also task-specific. While computer models are based on more general data, the designer must not apply them to unusual designs which fall outside the region of the model's data bases. For this reason alone, there will continue to be a need for some physical accommodation analyses with human subjects.

Although considerable data are found in published studies, the data were gathered for specific applications and can rarely be generalized to other applications. However, all of the task-related data bases in the three-dimensional CAD models COMBIMAN (COMputerized BIomechanical MAN-model) and CREW CHIEF were gathered especially for these models. The modelers have compiled and developed the most accurate representations of the link system and task representations available. The engineer can now use the model without having to develop the model.

COMBIMAN and CREW CHIEF have realistic size, strength, mobility, and automated task analysis capabilities for simulating problems with operability and maintainability early in the design. These models function as an electronic mockup, simulating the human operator or maintainer throughout the entire range of 1st through 99th percentile body dimensions.

The military aircraft cockpit is a particularly difficult design problem because the need to save weight and volume require the cockpit to have the minimum acceptable size and adjustment. When designing a complex workplace such as an aircraft cockpit, several dimensions must be accommodated simultaneously. To be specific, pilots must be able to see out the window and to reach the hand controls and the foot controls simultaneously. This is referred to as *multivariate accommodation*. While the concept of multivariate accommodation is not new (see Moroney & Smith, 1972), the procedures for accomplishing it have been costly and laborious. Computer models can facilitate multivariate accommodation by allowing a designer to represent several specific individuals lying at the limits of a multidimensional distribution of critical body dimensions. Each individual in this set represents the worst case of several critical dimensions. For example, one model may have the shortest arms and legs appropriate for a person with a tall sitting height. Another model in the set may have short sitting height as well as short arms and legs, etc. By defining a population with combinations of realistic extremes, a multivariate accommodation can be verified if all the members of the set are accommodated. CAD models are the only practical way to represent the unusual dimensions needed for multivariate analyses. To achieve a similar quality of evaluation with human subjects, a large number would have to be sampled to find the critical ones. Computer models can do this by searching existing surveys for the required values, then dimensioning the model with those values.

These CAD models are actually expert systems which allow the system designer to perform the functions of an expert ergonomist. These models also make it easy to evaluate the accommodation of different population subgroups. Other 3-D models provide complex animation with multiple human figures to aid in task visualization. A third model, Virtual Mockup, allows designers or evaluators to create an animated task sequence showing the complex interactions between a group of workers and their workplace.

In the sections that follow, three CAD tools for ergonomic evaluation will be briefly described. Then, through a series of actual case studies, the major features of each tool will be discussed in the context of the complex issues of anthropometric accommodation.

Description of COMBIMAN

COMBIMAN was developed by the Armstrong Aerospace Medical Research Laboratory (AAMRL) as a computer-aided design model of an aircraft pilot (Korna & McDaniel, 1985). COMBIMAN is a three-dimensional interactive computer-graphics model of an aircraft pilot used to evaluate the physical accommodation of the pilot in existing or conceptual crew station designs. COMBIMAN is an expert system which performs four types of analyses: fit analysis, visibility analysis, reach analysis, and strength for operating controls with the arms and legs. The user of COMBIMAN does not have to be an expert in ergonomics or anthropometric accommodation because that expertise is automated in the software. The user defines and directs a task to be performed, just as one would a human subject in a mockup review. The program generates accurate body sizes, proportions, strength, clothing restraints, harness restraints, and vision obscuration.

COMBIMAN has been distributed to the major aerospace industries since 1978. It has been used by both the U.S. Army and the U.S. Air Force to evaluate design changes, saving the costs associated with hardware mockups and prototypes. COMBIMAN is the predecessor of CREW CHIEF (a model of a maintenance technician), and much of the COMBIMAN technology was the basis of the CREW CHIEF program. The link system, joint system, reach algorithms, vision analysis, strength algorithms, and other features originated in COMBIMAN.

COMBIMAN provides an alternative to the traditional crew station evaluation. Because it uses a computer graphics generated crew station, it is possible to analyze many conceptual designs and variations of designs early in the program. COMBIMAN has two methods for dimensioning the human model. First, one can specify one or two critical dimensions, then the model will generate the other dimensions using regression equations. Second, where multivariate analysis is desired, one may read-in dimensions for a series of human models. Postures can be manipulated automatically by task-driven commands or manually by positioning individual joints. Every link dimension and every joint angle of the human model may be set to any desired value.

Description of CREW CHIEF

CREW CHIEF, a computer graphics model of a maintenance technician, was developed by the Air Force's Armstrong Aerospace Medical Research Laboratory (AAMRL) and the Air Force Human Resources Laboratory (AFHRL) (Korna, Krauskopf, Haddox, Hardyal, Jones, Polzinetti, & McDaniel, 1988; McDaniel & Askren, 1985). This model is interfaced to several existing commercial CAD systems employed by aerospace manufacturers and may be used to evaluate the maintainability of aircraft and other complex systems. CREW CHIEF is an expert system which lets the designer perform the functions of an expert ergonomist. The CREW CHIEF model allows the designer to simulate a maintenance activity on the computer-generated image of the design and to determine if the required maintenance activities are feasible. As with COMBIMAN, the user of CREW CHIEF need only define the task to be performed, just as one would define a task for a human subject in a mockup review.

Approximately 35 percent of the lifetime cost of a military system is spent for maintenance. Excessive repair time is caused by failure of the system design to adequately consider maintenance. The maintenance technician will spend hours making a repair which could have been completed in minutes if accessibility had been adequate. Ultimately, development engineering costs and acquisition time as well as life cycle costs and maintenance time will be reduced.

CREW CHIEF generates ten sizes of human models (five male and five female) with the encumbrance of four types of clothing, personal protective equipment, and mobility limitations for twelve postures. Physical access for reaching into confined areas (with hands, tools, and objects), automated obstacle avoidance, visual accessibility, and strength analysis (for using hand tools and manual materials, handling of objects) are provided so that designers, human factors engineers, and maintainability engineers may use CREW CHIEF as an electronic mockup. Analyses, which would have required weeks or months to perform by conventional methods, can be performed in minutes. Version 1 of the CREW CHIEF system of programs became available for use in May 1988.

Description of Virtual Mockup

The Virtual Mockup program is being developed by the Army as a tool to create animated 3-D models with complex animation and multiple human figures to aid in task visualization. Virtual Mockup allows designers or evaluators to create an animated task sequence showing the complex interactions between a crew and a crew-operated system. The Virtual Mockup program incorporates multiple programs including Ptech (developed by Associative Design Technology, Westborough, MA) and JACK (developed by the Computer Sciences Department of the University of Pennsylvania [Badler, O'Rourke, & Kaufman, 1980] and used by the National Aeronautics and Space Administration (NASA) in TEMPUS [Brown, 1982; Woolford & Lewis, 1981]). Automation Research Systems, Ltd. (ARS) is integrating and applying these tools to form the Virtual Mockup system.

The Ptech program provides a framework for defining a complex task sequence (script) in natural-language flowchart format. The flowchart boxes, each representing a task segment, are linked to tailored script sequences which drive the animation of the task described by the boxes. Ptech allows the program user to enter salient knowledge (actions, movements, situations, etc.) and the user's concept of the work setting and the work segments. The block diagram can have alternative paths at each node, and when all are defined, the user has only to select the path through the diagram to assemble the entire task sequence. Each different path represents a variation of the task sequence scenario.

The JACK part of Virtual Mockup animates the sequence defined by Ptech. It provides visualization of the activity sequence in 3-D color graphics using a Silicon Graphics IRIS display system. Compared to a video tape of an actual task, a JACK animation can support much more extensive analysis because all the digital data used to create the 3-D models are accessible. Every frame has time and 3-D geometry representing not only the workplace, but also the posture and motion of each of the multiple human figures in the complex task sequence. Interactions among works can be inspected at any scale and resolution and can be replayed and reanalyzed as required. Because it is three-dimensional, the animation can be viewed from any angle, and surfaces can be fully rendered or suppressed for *stick man* illustrations. By partitioning the screen, the animation can be viewed from multiple viewpoints simultaneously on the single display screen.

Virtual Mockup may be applied to create task analyses, operator manuals, maintenance manuals, and training documents. It is being linked to a Measure of Effectiveness (MOE) model developed by ARS and Communications Technology Application for the Army Human Engineering Laboratory (HEL). The MOE model is a personal computer program linked into the IRIS workstation through an AT&T UNIX operating system window. The activity sequence generates and passes parameters to the MOE model which computes and displays performance summaries associated with the action.

Often, operator dynamics and interactions are not well understood during concept development of a complex workplace. Some engineers may have equations describing the dynamics of a new system, but have no way of visualizing them. Virtual Mockup provides a viewing mechanism. Additionally, a time history (timeline) of the activities is produced to aid in task analysis or even cognitive demand analysis. A preliminary hazard evaluation is often possible from the information provided by the animation.

Virtual Mockup is best used in situations in which real prototypes are expensive. At present, Virtual Mockup does not predict action, but rather describes predefined activities. The system supports interactive modifications in system design and rapid visualization. This is a means of integrating information that now is spread over many people, steps, and time. Cutaway images and simultaneous views make more information available. Constraints can be wide-ranging, including operator constraints (size, kinematics, senses, and coordination) and material constraints (dimensions, response time, and dynamics).

SAMPLE APPLICATIONS

Rather than provide a detailed description of these tools, actual case studies of real design problems and the difficult issues of anthropometric accommodation will be presented to illustrate how the CAD tools assist the designer.

COMBIMAN Applications

Case Study 1

Problem. Can raising the entire pilot crew station overcome limitations in over-the-nose vision? When testing a prototype helicopter, it was discovered that when landing, the helicopter maintained a significant nose-up attitude. To compensate for this, pilots would approach the landing pad at an angle to see out the side window since the raised nose obscured forward visibility. The helicopter has a tandem seating arrangement with the pilot sitting behind the copilot-gunner. The copilot-gunner had to be placed in front to use a direct view optical sight. The pilot's vision is limited by the copilot-gunner's helmet which is approximately 6.5 feet forward of the pilot's eye position and the nose of the helicopter which is about 12 feet forward. There were two proposed solutions to the pilot's vision problems. The less expensive of the two involved raising the entire pilot crew station about 6 inches to improve forward vision. The more expensive option involved redesigning the tail of the aircraft to permit a level landing attitude.

Solution. The COMBIMAN program has a function which plots the crew station from the pilot's viewpoint. Military Standard 850 (MIL-STD-850) (Department of Defense, 1970) requires such visibility plots to ensure adequate vision over the nose and around the window frames. While

MIL-STD-850 requires only plots from the arbitrary *Design Eye Position*, COMBIMAN vision analysis additionally considers realistic combinations of body size and seat adjustment, as well as realistic offset of the eye position due to head motion and obscuration overlays which represent the masking of helmets, respirators, night vision goggles, and other head gear. COMBIMAN's vision plots can also show objects outside the aircraft, such as other aircraft, runways, landing pads, tanker aircraft, etc.

To evaluate the proposed alternative, a graphics model of the pilot's and copilot-gunner's crew stations were entered into the COMBIMAN model (Figure 7-2). To simulate the landing pad, several pads at different altitudes and attitudes relative to the crew stations were also entered. Notice that for the purpose of this type of analysis, the graphics representation is quite simple, containing only the relevant features such as instrument panels, controls, and windows. Once the crew station model is created and entered into the program, the analysis effort typically requires less than two hours. In performing the analysis, visibility plots were made using different sized crew members and different seat adjustments. After the standard configuration was evaluated, the pilot's crew station was raised 6 inches and a second set of visibility plots was generated. The resulting plots clearly showed that the landing pad was still obscured by the nose and that raising the pilot's crew

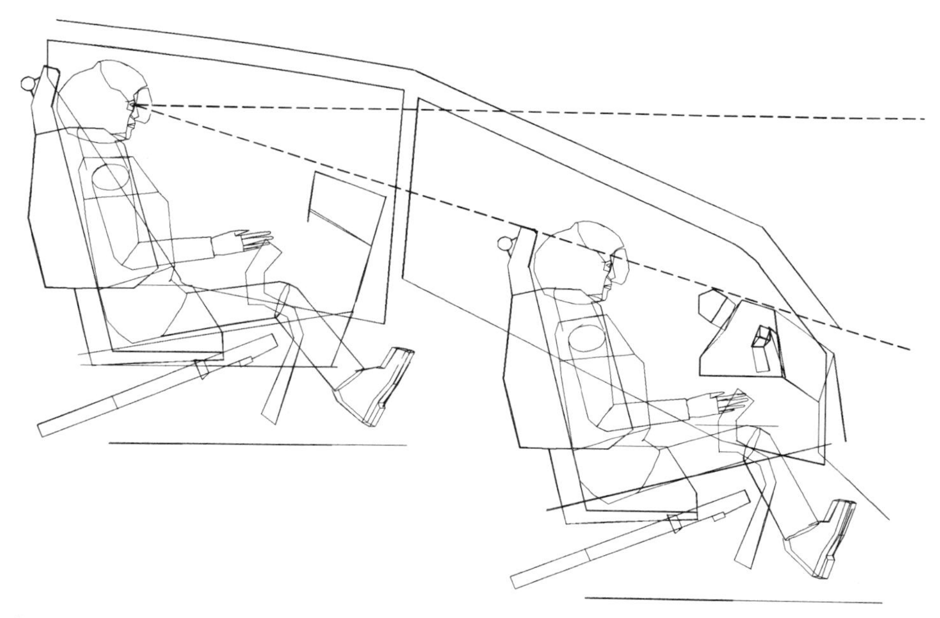

Figure 7-2
Electronic mockup of helicopter crew station represented in the COMBIMAN model. Although projected on a 2-D Display, the model itself is 3-D and can be viewed from any angle.

station was not the answer. The cost savings of using COMBIMAN to evaluate the problem, rather than modifying the prototype helicopter as the contractor had proposed, was quite significant.

Case Study 2

Problem. How does lack of seat adjustment affect reach to controls? Aircraft ejection seats adjust vertically, but usually not horizontally so the pilot's knees or feet will not strike the instrument panel or canopy frame when ejecting. This would happen if a long-legged pilot adjusted the seat too far forward. Since the result of a knee strike is catastrophic, ejection seats adjust only vertically. The *ejection clearance line* is based on the large pilot's buttock-knee length. However, when flying with the shoulder harness locked, as often happens during maneuvers, the small pilot's arms may not be long enough to reach controls on the forward instrument panel.

Solution. The seat and human-model in COMBIMAN can be adjusted without limit in three dimensions. The Reach Analysis Model is based on data gathered with three types of controls, six types of clothing, three types of harnessing, and seven reaching planes. The data collection required over two years of full-time effort, providing the most comprehensive description of arm reach available anywhere.

The COMBIMAN program evaluates arm reach with three different harness configurations: (1) lap belt only, representative of some non-pilot crew members, (2) shoulder harness on but not locked, representing the non-critical situation, and (3) the shoulder harness on and locked, representing the situation where maneuvering or inverted flight is anticipated. The harness is usually locked when practicing for combat. When the shoulder harness is locked, arm reach envelopes are limited to arm and shoulder movement.

COMBIMAN has three different hand postures for the types of controls found in aircraft and other vehicles: grip center for reaching to hand grips, functional grip for turning knobs, and finger tip for pushing buttons. Each type of hand posture has a different reach envelope.

One of the most important features of the COMBIMAN model is its ability to generate accurate body sizes and proportions representing the characteristics of a population. First, there are six anthropometric populations built into COMBIMAN: Male Air Force Pilots, Female Air Force Pilots, Air Force Women, Male Army Pilots, Army Women, and Male Navy Pilots. The user can also add additional populations. The program user need only select the population and dimensions critical to the task and COMBIMAN will create realistic body size, proportion, and clothing characteristics. For example, the user may select arm length as the critical variable for evaluating reach to hand controls or select sitting height for analysis of head clearance.

The encumbrance of clothing also affects the reach envelope. To allow the designer to accurately evaluate the effects of clothing, COMBIMAN has six different types: summer-weight flying suit; summer suit with survival vest; winter-weight suit; winter suit with survival vest; winter suit with survival vest, jacket, and flotation device; and the chemical defense protective clothing.

In this example, the summer-weight flight suit and 50th percentile arm length are also selected. Figure 7-3A shows a reach to a forward instrument panel with the shoulder harness unlocked allowing upper body movement. Figure 7-3B shows the same pilot failing to reach the forward instrument panel with the shoulder harness locked. For aircraft with ejection seats, this illustrates why critical controls which must be operated when the shoulder harness is locked cannot be placed on the forward instrument panel. The preferred location for such controls is on the stick and throttle (or cyclic and collective for a helicopter). The result is that these handgrips are being overloaded with controls, some having as many as 18 functions on a single grip.

Case Study 3

Problem. Can the pilot reach controls located on the side consoles if the

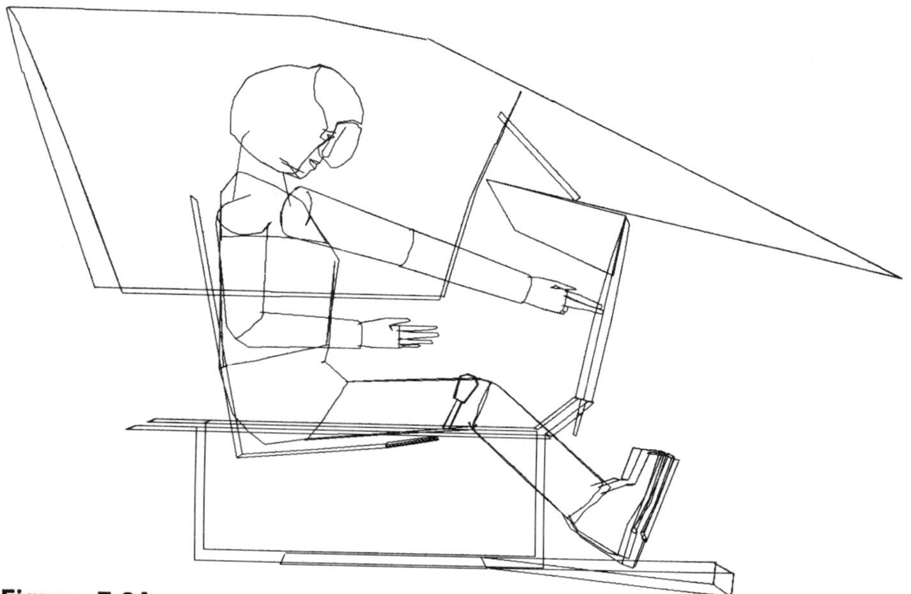

Figure 7-3A
"REACH: SUCCESSFUL." Reach analysis with 50th percentile arm length and shoulder harness unlocked. In this condition, the pilot can lean forward with the upper body.

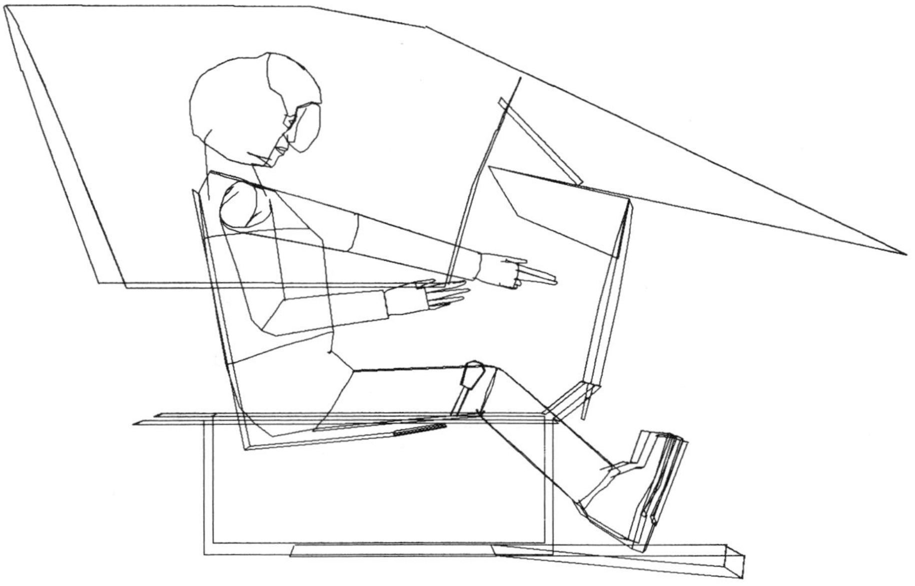

Figure 7-3B
"MISS DISTANCE: 4.83 INCHES." Reach analysis with shoulder harness locked, allowing only arm and shoulder movement, but not upper body leaning. The picture and the data provide details of the nature and magnitude of the problem.

seat is adjusted upward to achieve adequate outside vision? As in the previous example, body size, seat position, and vision requirements provide conflicting design requirements. A large pilot may have to adjust the seat down to achieve head clearance beneath the canopy. For the large pilot, reach and vision are seldom problems; however, the fit may be cramped in extreme cases. For the small pilot, everything is a problem. If the small pilot adjusts the seat up to get adequate vision through the windows, he or she is moving farther from the controls and side consoles.

Solution. COMBIMAN can compute the intersection of the reach envelope and any surface of the crew station. This is called the reach envelope analysis and it considers body size, the type of clothing worn by the pilot (more layers and bulk reduce the mobility of the joints), and the type of harness (lap belt, shoulder locked, and shoulder unlocked). Figure 7-4A shows the result of a reach envelope analysis to right side console of a helicopter for a male pilot with a 50th percentile arm length and the seat adjusted to the full down position. By contrast, Figure 7-4B shows the same conditions for the seat adjusted to the full up position. The difference in the part of the side console that can be reached without leaning is very dramatic.

Other types of reach analysis COMBIMAN can perform involve the different types of flight clothing. The type of clothing chosen for the analysis not only affects the appearance of the model, but it also automatically changes the joint mobility limits for realistic reach analysis. If the analysis described above was repeated with a more restrictive clothing type, the reach curve would have been closer to the body. This capability is very important because crew system designers usually do not have access to different types of pilots' clothing, especially all the sizes of each. So, even if the designer had a mockup, he or she would still not be able to evaluate the effects of restrictive clothing worn by the population of users of the crew system.

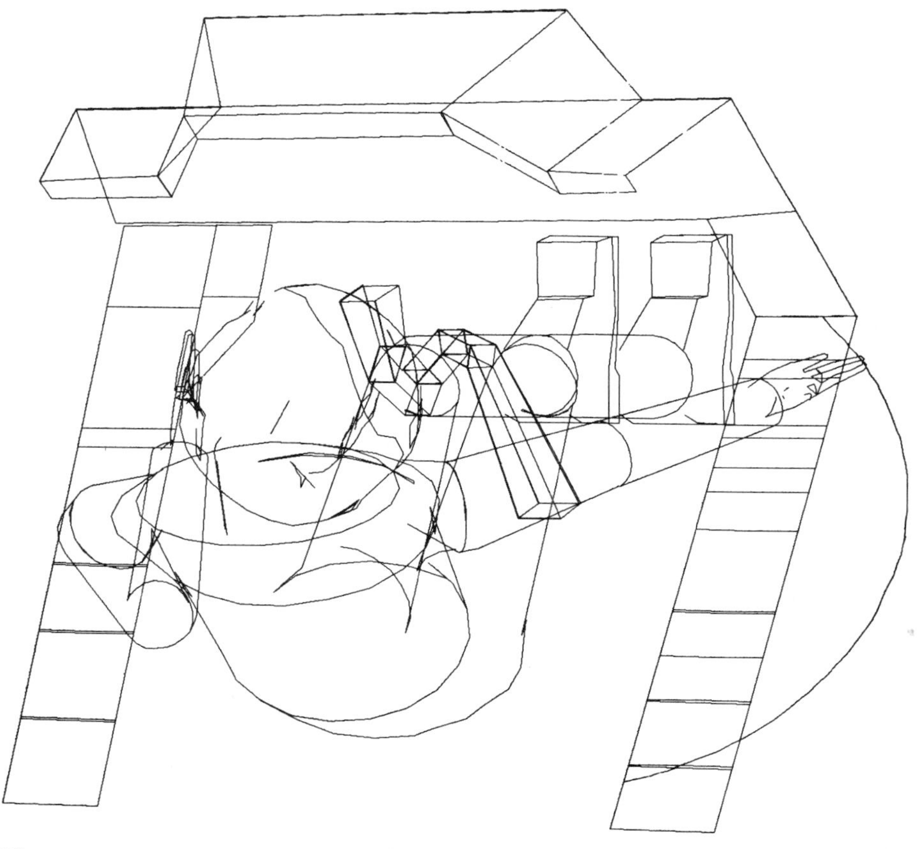

Figure 7-4A
The result of a COMBIMAN reach envelope analysis to right side console of a helicopter for a male pilot with a 50th percentile arm length and the seat adjusted to the full down position. The reach curve superimposed on the side console defines the area that can be reached with a functional grip.

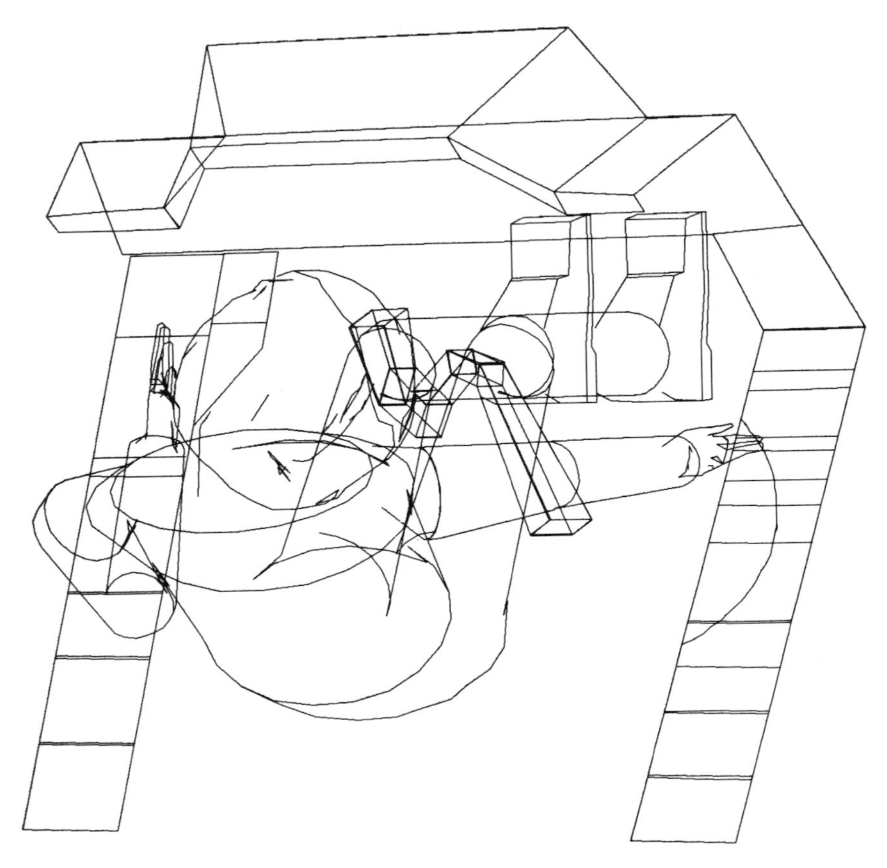

Figure 7-4B
The same conditions for the seat adjusted to the full up position. The reach curve superimposed on the side console defines the area that can be reached with a functional grip. Comparing the two reach envelope plots, the difference in the part of the side console that can be reached without leaning is very dramatic.

Case Study 4

Problem. How much force can a pilot exert on an ejection handle? Now that women are flying trainer aircraft equipped with ejection seats, the question of actuation resistance must be reconsidered. The fact that women, as a group, are less strong than men, as a group, raised an important design question: How much resistance should ejection handles have? The objective is to make the control resistance so great that the ejection controls cannot be activated inadvertently, but low enough so that all pilots can operate it. When the seat was originally designed, there were no female military pilots, so only the male strength characteristics were considered.

Solution. The COMBIMAN program has models to predict the strength a seated operator can exert on eight types of controls: stick-type aileron and elevator, wheel-type aileron and elevator, general-type hand grip (any location in the forward hemisphere reach envelope), side-mounted ejection handles, center-mounted ejection handles, and rudder pedals. The controls can be operated by left, right, or both hands, and the pedals with left or right foot. The strength models are based on measurements of over 1,000 subjects. The strength analysis models provide 1st, 5th, 50th, 95th, and 99th percentile forces for the task defined.

The strength available for operating a control depends on the type of control, location of control, orientation of the control, and the direction of force applied. These control characteristics are important because different joint angles and combinations of muscles are involved in exerting forces in different directions and locations. The orientation of the handgrip may produce awkward postures which can reduce the available strength.

In the case of the ejection handles, the side-mounted handles are located in line with the shoulders and the direction of force is toward the shoulders. Pilots can exert a larger force on side-mounted handles than on a center-mounted handle, which is not in line with the shoulders and requires a force at an angle to the line between the hand and shoulder.

The results of the strength analysis for the center pull ejection handle are shown in Figure 7-5. The analysis simulates left hand operation of a center-mounted ejection handle by a female pilot. The results show that 1st

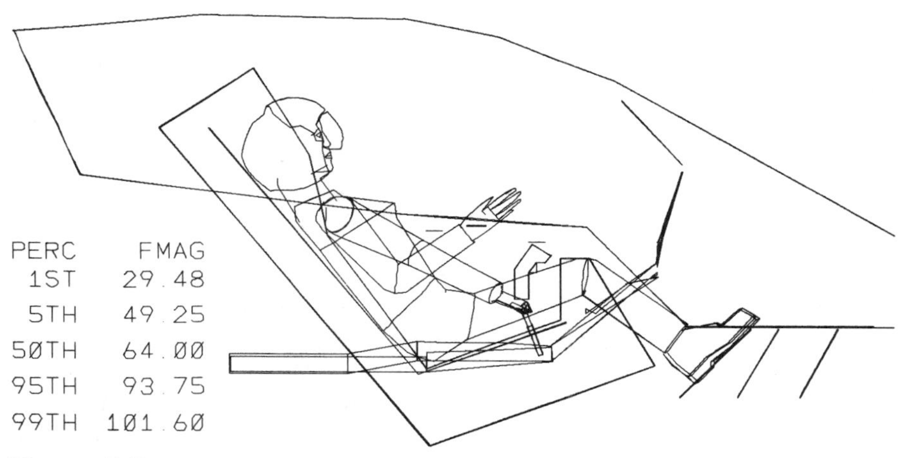

Figure 7-5
The results of the COMBIMAN strength analysis for the center pull ejection handle with the left hand only. The strength analysis models provide 1st, 5th, 50th, 95th, and 99th percentile forces. The center-mounted handle is not in line with the shoulders and requires a force at an angle to the line between the hand and shoulder. The results show that 1st percentile female one-hand strength values were 29.48 pounds, less than the 49-pound resistance of the controls.

percentile female strength value is 29.48 pounds, less than the 49-pound resistance of the controls. The 5th percentile force is 49.25 pounds, right at the value of the control resistance. This indicates that almost 5 percent of the women would not be able to operate the center-mounted handle with one hand.

Currently, only the F-16 fighter plane is equipped with the center-mounted ejection control. Since women are currently prohibited from flying combat aircraft, the situation does not now constitute a deficiency. The seat specification requires one-hand operation, however, because of the possibility of incapacitation of one hand and also because the effective forces may be reduced further by acceleration or buffeting of the aircraft. Trainer and other ejection seat-equipped aircraft have the side-mounted handles.

If the center-mounted handle were analyzed with both hands simultaneously, the 1st percentile values would have been 62.35 pounds, well above the control resistance. The two-hand force of 62 pounds is more than double the one-hand force of 29 pounds because the symmetry of the two hands pulling outward balance each other, causing the sum of the two forces to be aligned with the direction of control movement.

CREW CHIEF Applications

Case Study 5

Problem. What weight can a worker carry through an access tunnel with a ceiling height of 30 inches? There are many passageways inside large systems that require workers to work in other than a standing posture.

Solution. CREW CHIEF Task Analysis evaluates strength and accessibility for eight categories of manual materials handling activities: carry, lift, hold, push, pull, turn, grasp, and reach. In all of these tasks, the program considers the posture, as well as the location and orientation of the work.

For example, in a carry task, a low ceiling may force the CREW CHIEF human-model to bend, stoop, or even crawl while carrying the load. The CREW CHIEF program contains 12 initial postures. The designer chooses the one that is appropriate for the workplace. When the task analysis simulation is run, CREW CHIEF automatically adjusts the posture to accomplish the task. This reposturing is totally automatic, causing the human-model to reach around obstacles in the workplace if necessary, and showing graphically whether or not the human-model can perform the reach part of the task. The posturing automatically keeps the body's joint angles within realistic limits, adjusted for the additional restriction of the type of clothing chosen for the analysis. CREW CHIEF has four types of clothing: fatigues, cold weather, arctic, and chemical defense.

The 12 initial postures are standing, sitting, stooping, squatting, kneeling on one knee, kneeling on two knees, lying prone, lying supine, lying on the side, crawling, climbing, and walking. Strength analyses are available for all appropriate combinations of tasks and postures. For example, carrying strength is evaluated in the walking, crawling, and stooping postures. Strength for torquing with wrenches is appropriate for all postures except walking, crawling, and climbing. More than 100,000 strength measures were collected to develop the strength models in CREW CHIEF.

For this example, illustrated in Figure 7-6, a 5th percentile female model is chosen because the small female represents the most limiting condition. (Remember, in CREW CHIEF, a 5th percentile female model has only stature, arm reach, and weight parameters set at 5th percentile population values; the other dimensions are regressed from these three.) If the population of workers includes both men and women, Military Standard

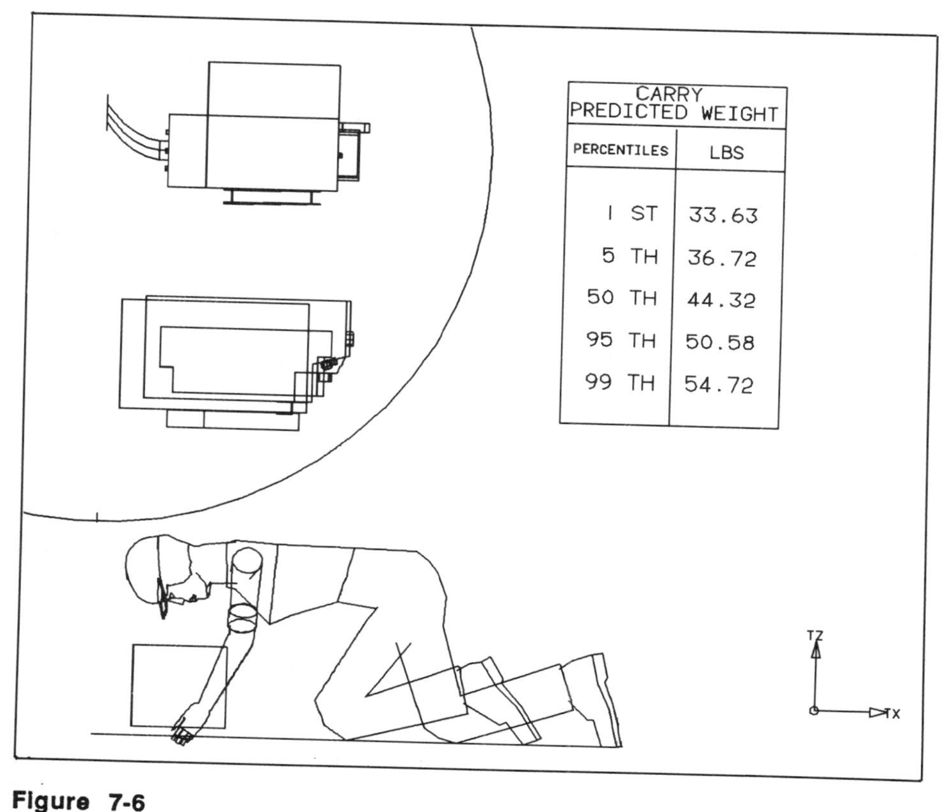

Figure 7-6
After the designer sets up the task to be performed, CREW CHIEF automatically executes the task and displays the available strength for the 1st (weak), 5th, 50th, 95th, and 99th (strong) percentile worker.

1472 (MIL-STD-1472) (Department of Defense, 1989) requires accommodation of a 5th percentile female strength as a minimum. The box carried is medium size, approximately 10 inches high and deep and about 18 inches wide. The 30-inch passage ceiling is denoted by the bottom of an aircraft fuselage. To perform the analysis, the designer identifies the location of the work (in this case, the box to be carried), the task (CARRY), and the location of the worker. The CREW CHIEF program will assemble the human-model automatically, posture it, and place it in the workplace in the location specified. The available strength for those conditions is then computed and displayed as shown in Figure 7-6. Note the 5th percentile strength value is 36.72 pounds. If the object weighs this amount or less, the design criterion is satisfied.

Case Study 6

Problem. What is the minimum height of an aircraft fuel tank which will admit the full range of cleaning personnel? A recently discovered design-induced maintainability problem involves the in-wing fuel tanks of a new aircraft. Combat aircraft have self-sealing fuel tanks for survivability. After several years of aging, the gummy sealant hardens and must be replaced. After the fuel is flushed from the tank, maintenance personnel, wearing protective clothing and breathing apparatus, must enter the tank and remove the old sealant. It was found that larger-than-average fuel maintenance personnel get stuck in wing tanks. When the aircraft was designed, the maintainability and human factors people thought to test the access opening to make sure a large person could get through. But, evidently, equal consideration was not given to the height of the tank itself.

Solution: In the case cited, it is too late to redesign or retrofit. The limited accessibility requires small workers. What was originally a design problem now becomes a personnel and recruiting problem. The career field for a major system must now be populated with small men and women.

Had an analysis model like CREW CHIEF been available, this problem could have been prevented, for with CREW CHIEF the designer can select from a range of body sizes of male and female technicians and may choose a 1st, 5th, 50th, 95th, or 99th percentile model for male and female maintenance technicians. In CREW CHIEF, the human-model sizes are much simpler than in COMBIMAN where body proportions are critical. In maintenance, proportions have less significance and were made simpler to allow the more complicated tasks to be evaluated. The CREW CHIEF human-models are based on stature, arm length, and weight, with all other dimensions regressed from these three. It would also be a poor design if it considered only gross body fit in the case of this fuel tank and not the mobility required for the cleaning task.

Figure 7-7 shows the fuel tank problem represented with the CREW

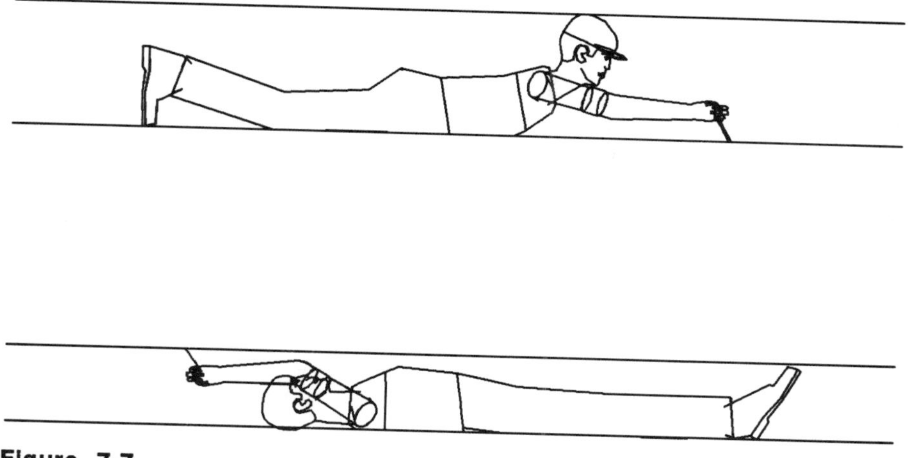

Figure 7-7
The wing fuel tank problem represented with the CREW CHIEF man-model. Here the prone and supine postures for a 99th percentile male are pictured.

CHIEF human-model. Here, the prone and supine postures for a 99th percentile male human-model are pictured. The prone posture shows the working posture.

Case Study 7

Problem. What is the minimum number of tools needed to maintain a helicopter's jet engine? Maintenance on the flight line is usually done outdoors in all kinds of weather. Maintainers in this environment have little sympathy for aircraft designs which make their jobs more difficult. While guidelines have recommended no common tools and minimum fasteners for better maintainability for many years, they have not been rigorously followed. A notable exception to this is the Army's Light Helicopter Experimental (LHX) program (Dornheim, 1989). The T800 engine, to be produced by the Light Helicopter Turbine Engine Company, requires only a literal handful of tools to maintain on the flight line. The engine has only two sizes of bolt heads and can be maintained with a ratchet, extension, two sockets, and two wrenches. Having few tools lightens the load maintainers must carry to and from the helicopter and speeds up maintenance because fewer tool changes are necessary to get the job done. This degree of design for maintainability does not just happen. It requires commitment to the MANPRINT goals and work to verify that these goals are met.

Solution. CREW CHIEF can be used to verify that a particular tool can fit into a tight clearance, that there is sufficient range of movement of the tool to remove or replace a fastener, and that the maintainer wearing even arctic or chemical defense clothing has the accessibility to perform the operation. There are 105 types of tools available in the CREW CHIEF tool box. Users can add other tools if desired. Figure 7-8 shows a female CREW CHIEF human-model using a ratchet wrench to remove a bolt from a jet engine. Because the bolt head is in a recessed area, the maintainability engineer must determine if a standard ratchet/socket combination will do the job. In practice, the CREW CHIEF human-model, not only shows that the standard tool is adequate, but also displays the range of torque strength available to

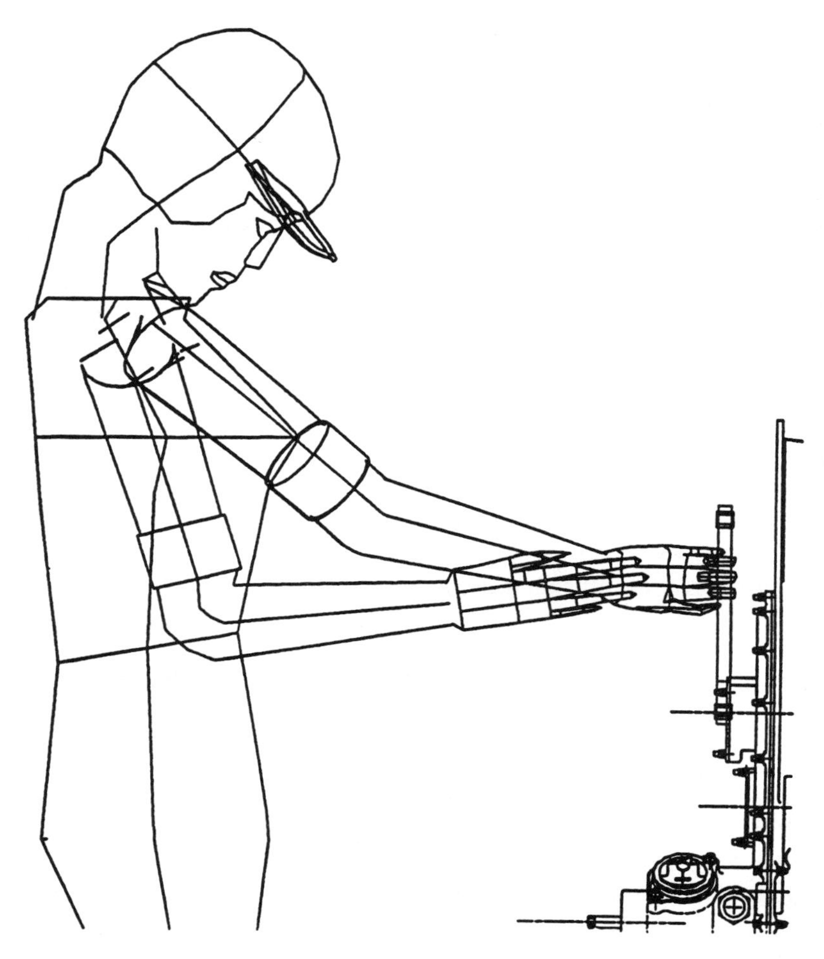

Figure 7-8
Female CREW CHIEF man-model uses a ratchet wrench to remove a recessed bolt from a jet engine, demonstrating adequacy of standard ratchet/socket. The table in the figure shows the torque strength available to perform the task.

do the task. In cases in which an unrelated component interferes with the movement of the tool handle, the tool envelope analysis function of CREW CHIEF shows the free range of movement of the tool. In Figure 7-9, the obstructing handles on a box limit the rotation of a ratchet wrench. When an extension rod is added between the ratchet and the socket, the rotation is not obstructed. This tool envelope function works for all of the tools in CREW CHIEF's tool box. The designer could have selected chemical defense clothing or arctic clothing for analysis if hand clearance had been limited.

There are several advantages to using CREW CHIEF instead of real human subjects for this type of analysis. First, CREW CHIEF can be used on the design itself, even before a hardware mockup is built. Second, there is no need to perform a difficult search for small and large subjects. Third,

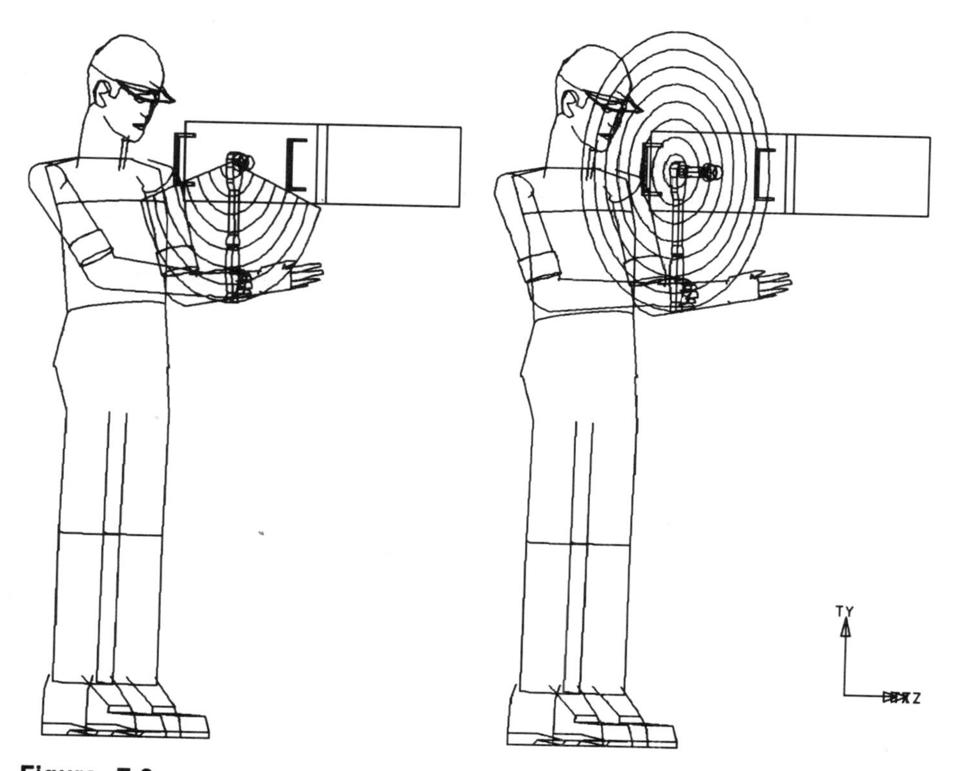

Figure 7-9
The tool envelope analysis of CREW CHIEF (left) shows ratchet wrench interferes with handles on a box. When an extension rod is added between the ratchet and the socket, the rotation is not obstructed.

there is no need to procure the protective clothing. It is expected that analytical models like CREW CHIEF will reduce the need for maintainability demonstrations on prototype equipment. This is particularly important for aircraft where maintainability demonstrations are concurrent with flight testing. Flight test schedulers are reluctant to allow maintainability demonstrations in which damage to a one-of-a-kind component could delay the flight test until a replacement part is custom made.

Virtual Mockup Applications

Case Study 8

Problem. Why is the Dragon anti-tank missile difficult to fire accurately? The Dragon (see Figure 7-10A) is a shoulder-fired Tube-launched, Optically tracked, Wire- guided (TOW) missile. The missile is in a disposable tube with a bipod support on the front of the tube. The operator sits on the ground and holds the rear of the launch tube on his right shoulder, providing the third point of support, and aims the weapon by moving the aft-end of the tube by posture adjustment. To change the aim to the left, the operator shifts his upper body to the right. To raise the aim, the operator must crouch downward to lower the rear of the tube. During the missile flight, the operator must continue to aim the sight at the target. The optical sight senses the deviation between the missile's position and the line-of-sight to the target and sends steering corrections to the missile through wires that trail out of the aft end of the missile. In other words, the missile follows the aim of the sight. The operator wraps his right arm under, around, and over the tube to the trigger. When the missile is fired, the noise far exceeds the maximum allowable for impact noise, and the operator is briefly enveloped in the exhaust flash and smoke.

Solution. The Dragon firing scenario was programmed into the Virtual Mockup program (see Figure 7-10B). Initially, the program user selected to view the scenario from the Dragon operator's viewpoint, that is, through the optical sight. When the simulated launch was initiated, several of the problem conditions were demonstrated and recorded.

Most of the 28-pound weight of the Dragon system disappears when the missile fires, causing the operator to recoil upward, sometimes steering the missile into the ground. With the weight and inertia of the launcher changing so drastically, the feel of tracking the target prior to and after firing are completely different. To counter this change, the operator should pull the tube down against his shoulder with about 150 pounds of force to minimize the recoil effects. If the dynamics of this recoil are programmed into the Virtual Mockup scenario, the effect on aim is clearly visible in the through-the-sight-view of the scenario.

Figure 7-10A
Soldier positioned to fire Dragon missile. The soldier's body is the third leg of support for the launcher, so the soldier tracks a moving tank by shifting his torso.

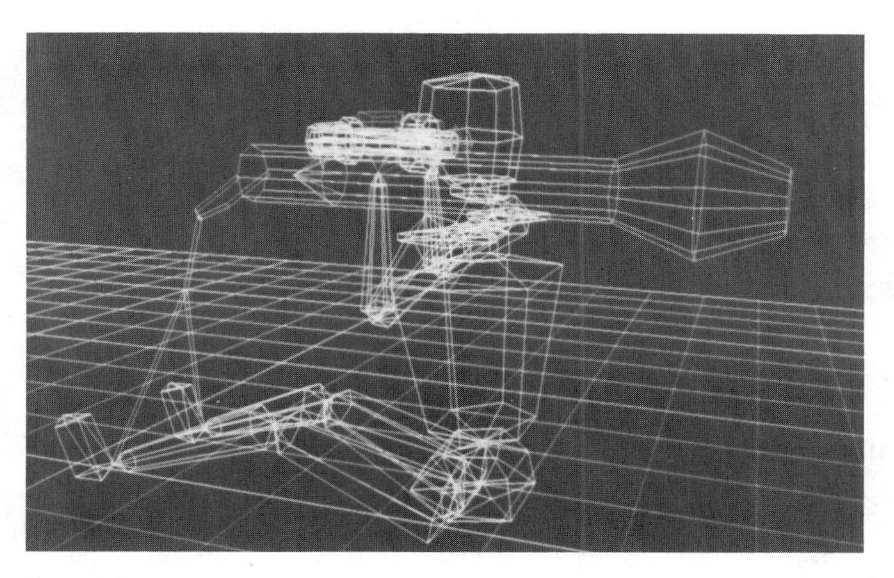

Figure 7-10B
Virtual mockup of operator firing the Dragon missile. Although not apparent in this plot, the image is 3-D and in color. Because the image is 3-D digital information, it can be viewed from any viewpoint. The 3-D model can be passed from a CAD system via the International Graphic Exchange System (IGES) protocol. Once defined, the segments of the model can be repostured and animated.

In the first seconds after firing, the missile can appear larger in the sight than does the target. The missile exhaust is bright and attention-getting, whereas the target is camouflaged and difficult to see against the background. Sometimes the Dragon operator begins tracking the missile instead of the target, causing the missile to fly off course. If the Virtual Mockup tracking equations are programmed to track the missile, the effect on aim can be made visible in the through-the-sight-view of the scenario (Figure 7-11).

Breathing while the missile is in flight causes the aim to deviate. Operators are supposed to hold their breath (12 to 25 seconds) during final aiming, firing, and the missile flight, but this may not be easy if the soldier was running just prior to firing. When the dynamics of breathing are programmed into the Virtual Mockup scenario, the effect on aim is clearly visible in the through-the-sight-view of the scenario; oversteering is bad for two reasons. One is that it may exhaust the missile's limited rocket fuel before reaching the target. The second is that it may inadvertently command a miss. An inadvertent 1-degree movement of the launch tube causes the aim to deviate off the target by 9 meters at a typical firing range of 500 meters. The Virtual Mockup through-the-sight-view of the scenario clearly shows deviations of several degrees due to breathing. The simulation can also compute the difference between the actual flight path in the simulation and the ideal straight-line path.

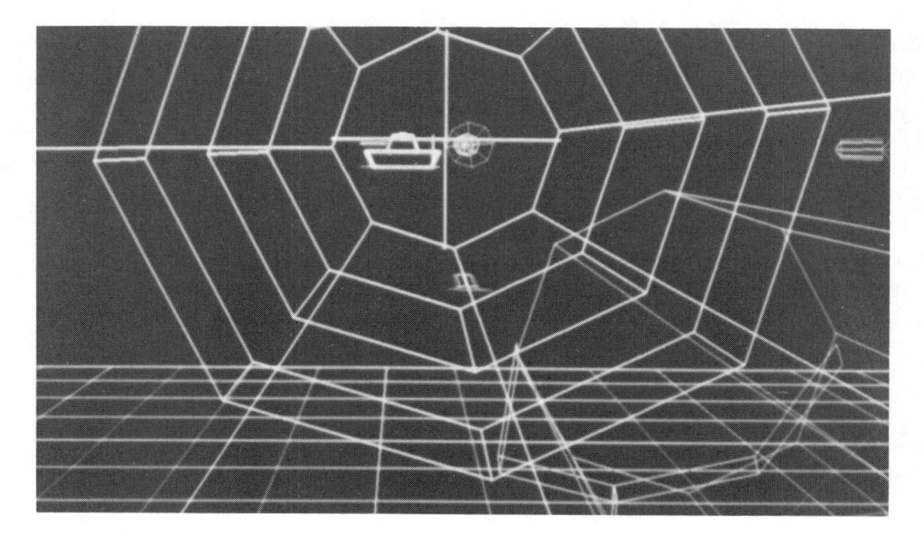

Figure 7-11
Virtual mockup "through-the-sight" of Dragon operator's view of target (tank center left) and missile (center right) at 2 seconds after launch. Sometimes the operator begins tracking the missile, causing the missile to fly off course, missing the target.

In summary, the Virtual Mockup can help the designer visualize and understand some of the very complex interactions of using a system such as the Dragon missile. The dynamic models (breathing, oversteering, etc.) which describe each of these interactions typically do not exist at this early stage of Virtual Mockup development. Each model must be custom made and linked into the program. Later, if one of the models is useful in another case, the program user can select it from a library of tasks, rather than create it from scratch as must now be done. In developing an animated scenario using Virtual Mockup, all the information is accumulated necessary for task analysis and operator and maintenance manuals. These scenarios, therefore, provide a double utility for a developmental system. It is also likely that these animated models could serve as training aids, another important aspect of MANPRINT.

OTHER MODELS

While the three models presented above are useful for ergonomic accommodation, there are many other computer models to help designers (National Research Council, 1988). The Crewstation Assessment of Reach (CAR IV) is a 3-D model developed by the Navy and uses a Monte Carlo technique (a technique for integrating a function by sampling random variants) to synthesize a population from means, standard deviations, and correlations. CAR IV then has each of the synthetic human-models reach to each of the controls in a crew station, finally computing the percent of the synthetic population able to reach each of the controls (Harris & Iavecchia, 1984). The CAR IV model was derived from the Computerized Accommodated Percentage Evaluator (2-D CAPE) model developed by Bittner (1976).

Another human modeling program popular in England and Europe is the System for Aiding Man Machine Interaction Evaluation (SAMMIE) developed by Bonney and Case (1976). SAMMIE is a 3-D animated model with the human-models defined by using somatotypes, as well as user-entered dimensions. SAMMIE can analyze both reach and vision. SAMMIE was developed to aid in evaluation of ground vehicles, and part of the vision analysis capability shows the vehicle operator's view with a rear-view mirror.

While the above-mentioned programs help the designer, others are aimed at the safety engineer and life support equipment. The Articulated Total Body (ATB) model (Leetch & Bowman, 1983) models the human body in crash and impact. It is a dynamic mechanical model of the involuntary trajectory of the body and its segments responding to impact accelerations, such as crashes and ejection from aircraft. The model contains not only accurate representations of body segment masses and moments of inertia, but also models of the passive displacement/resistance characteristics of joints. This model has been used to design harness and restraining systems for aircraft and automobiles. In evaluating ejection from aircraft, it can compute and display the trajectory of knees and feet to determine if they will strike the instrument panel.

SAFEWORK is a 3-D model which analyzes the reactive forces and torques on the major joints and spine of the human body while applying a force while doing work (Fortin, Gilbert, Beuter, Laurent, Schiettekatte, Carrier, & DeChamplain, 1988). This type of model predicts the injury potential of a workplace. Chaffin, Herrin, Keyserling, and Garg (1977) describe a similar 2-D modeling program. Both of these programs can run on personal computers.

Even anthropometric data are available on a dial-up computer system at the Center for Anthropometric Research Data (CARD) (see Robinson, Robinette, & Zehner, 1988). The data base contains hundreds of measurements from numerous surveys taken all over the world and also includes multivariate models of body proportions. As new data become available, they are added to the data base.

TRENDS FOR THE FUTURE

Future computer models may be able to depict all the details of the human body in three dimensions. Already, new technology uses a laser scanner to gather data on head and face shapes (Beecher, 1986; Zehner, 1986). In the near term, such data may be useful for head gear design, such as helmets and protective masks. High density face data are currently being used for face contour comparisons, landmark spatial distribution analyses, and subregion delineations. The volume scanned by current laser mappers is too small to scan an entire body at one time, but work is in progress to develop a system which will.

Models must not only use new technology, but should also anticipate the design problems of the future. Planned enhancements for CREW CHIEF include a model of space suits and space tools. In addition to the currently used *shuttle suit* which operates at 4.3 psi pressure, NASA is designing a high pressure suit which will operate at 8.3 psi. There are several advantages to the higher pressure suit. First, the high pressure suit will require a much shorter oxygen prebreathe period to expel nitrogen from the blood. Second, the high pressure suit will have quick-disconnect interchangeable limb components and size adjusters, so fewer suits will be needed in orbit. (The older shuttle suit had sewn adjustment segments, and sizes could not be changed in orbit.)

Development of Predetermined Time Systems (PTS) for the factory environment began in 1911 and is a mature technology. In the factory, however, jobs are designed for efficient operation with equipment designed to be free of accessibility problems. In aircraft maintenance, limited accessibility is the rule, rather than the exception. The Predetermined Time

Systems do a very poor job of predicting mean-time-to-repair on the flight line. Driven by procurement policies requiring manufacturers to warrant their products, designers need better tools for determining time to repair. The future CREW CHIEF system of programs will have a task-time estimating capability for remove/replace tasks involving very limited accessibility, as is common in aircraft maintenance. CREW CHIEF will be rehosted on several other CAD systems to make it available to a greater number of designers. A personal computer version of CREW CHIEF is also being considered.

The CREW CHIEF development program also has by-products supporting equipment design standards and providing research data bases describing the ergonomics attributes of the maintenance technician.

Finally, every effort should be made toward the goal of considering human characteristics early in the design process in the most efficient manner. One approach to reach this goal, which is compatible with the new CAD, is the development and application of a comprehensive, integrated, interactive dynamic human-model. In fact, such an approach is probably mandatory if a competitive edge is to be achieved and/or continued. It is also important to note that in addition to the changes in the design process precipitated by the CAD techniques, the human factors engineer represents less than 0.2 percent of all practicing engineers and, therefore, requires all the productivity enhancement capability he/she can get to achieve system user accommodation in the design process.

Adequate and timely progress in achieving an interactive dynamic ergonomic model for use in the CAD process and for defining needed research requires an overall strategy/plan. The development of this strategy and plan itself will require the efforts of many disciplines across academia, industry, and the government. The Department of Defense (DoD) will continue to play a leading role because of its resource base, diversity of potential application, and the payoff in system operability and performance from applying such a model to the design of equipment the government buys. Also, the DoD generally recognizes the potential usefulness that such a model would have when incorporated into gaming models to aid decisions about the types and amounts of equipment it procures. The DoD also has the ability to promote the application of an interactive dynamic ergonomic model by giving operator and maintainer accommodation appropriate consideration during contract evaluation. At the same time, if the DoD avoids including proprietary elements in the software tools it develops, the tools can be placed in the public domain and made available to the private sector at nominal cost.

REFERENCES

Badler, N. I., O'Rourke, J., & Kaufman, B. (1980, July 3). Special problems in human movement simulation. *Computer Graphic, Proceedings, Special Interest Group on Computer Graphics Association for Computing*

Machinery (SIGGRAPH 80), Volume 14, 189-197.

Beecher, R. M. (1986). Computer Graphics and Diagnostics. *Proceedings of the Human Factors Society 30th Annual Meeting*, 211-215.

Bittner, A. C. (1976). Computerized accommodated percentage evaluation (CAPE): Review and Prospectus. *Proceedings of the 6th Congress of the International Ergonomics Association.* College Park, MD: University of Maryland.

Boff, K. R., Kaufman, L., & Thomas J. P. (1986). *Handbook of perception and human performance* (Vols. I-III). New York: John Wiley & Sons, Inc.

Boff, K. R., & Lincoln, J. E. (1988). *Engineering data compendium: human perception and performance* (Vols. I-III). Wright-Patterson Air Force Base, OH: Harry G. Armstrong Aerospace Medical Research Laboratory.

Bonney, M. C., & Case, K. (1976). The development of SAMMIE for computer aided work place and work task design. *Proceedings of the 6th Congress of the International Ergonomics Association.* College Park, MD: University of Maryland.

Brown, J. W. (1982). Using computer graphics to enhance astronaut and systems safety. *Space Safety and Rescue Symposium of the 33rd International Astronautical Congress*. Science and Technology Series, American Astronautical Society, Vol. 58, 31-42.

Chaffin, D. B., Herrin, G. D., Keyserling, W. M., & Garg, A. (1977). A method for evaluating the biomechanical stresses from manual materials handling jobs. *American Industrial Hygiene Association Journal, 38*, 662-675.

Department of Defense (1970). *Aircrew Station Vision Requirements for Military Aircraft* (MIL-STD-850).

Department of Defense (1980). *Anthropometry of U. S. Military Personnel, Military Handbook* (DOD-HDBK-743).

Department of Defense (1989). *Human Engineering Design Criteria for Military Systems, Equipment and Facilities* (MIL-STD-1472D).

Dornheim, Michael A. (1989, March 6). McDonnell Douglas/Bell Team introduces its LHX concept. *Aviation Week & Space Technology*, pp. 56-59.

Fortin, C., Gilbert, R., Beuter, A., Laurent, F., Schiettekatte, J., Carrier, R., & DeChamplain B. (1988). *SAFEWORK: Micro computer-aided workstation design and analysis, new advances, directions, and future.* Montreal, Quebec: Genicom Consultants, Inc.

Harris, R. M., & Iavecchia, H. T. (1984). *Crewstation assessment of reach revision IV (CAR IV) user's guide* (Technical Report TR1800.10A). Willow Grove, PA: Analytics, Inc.

Korna, M., & McDaniel, J. W. (1985, May). *User's guide for COMBIMAN programs (Computerized Biomechanical Man-Model), Version 7* (Technical Report AFAMRL-TR-88-034). Wright-Patterson Air Force Base, OH: Armstrong Aerospace Medical Research Laboratory.

Korna, M., Krauskopf, P., Haddox, D., Hardyal, S., Jones, M., Polzinetti, J., & McDaniel, J. W. (1988, May). *User's guide for CREW CHIEF: A computer graphics simulation of an aircraft maintenance technician, Version 1 - CD20* (Technical Report AAMRL-TR-88-034). Wright-Patterson Air Force Base, OH: Armstrong Aerospace Medical Research Laboratory.

Leetch, B. D., & Bowman, W. L. (1983). *Articulated Total Body (ATB) "View" program software report, Part II, user's guide, Vol. 2* (AFAMRL-TR-81-111). Wright-Patterson Air Force Base, OH: Armstrong Aerospace Medical Research Laboratory.

McDaniel, J. W., & Askren, W. B. (1985, December). *Computer-aided design models to support ergonomics* (Technical Report AAMRL-TR-85-075). Wright-Patterson Air Force Base, OH: Armstrong Aerospace Medical Research Laboratory.

Moroney, W. F., & Smith, M. J. (1972). *Empirical reduction of potential user population as the result of imposed multivariate anthropometric limits.* Pensacola, FL: Naval Aerospace Medical Research Laboratory.

National Research Council (1988). *Ergonomic models of anthropometry, human biomechanics, and operator-equipment interfaced, proceedings of a workshop.* Washington, DC: National Academy Press.

Robinson, J. C., Robinette, K. M., & Zehner, G. F. (1988, January). *User's guide to accessing the anthropometric data base at the center for anthropometric research data (CARD)* (Technical Report AAMRL-TR-88-012). Wright-Patterson Air Force Base, OH: Armstrong Aerospace Medical Research Laboratory.

Woolford, B. J., & Lewis, J. L. (1981, April). Applications of digital image acquisition in anthropometry. *Proceedings of the SPIE Technical Symposium East, Vol. 283, 3-D Machine Perception.* Washington, DC.

Zehner, G. F. (1986). Three dimensional summarization of face shape. *Proceeding of the Human Factors Society 30th Annual Meeting,* 206-210.

CHAPTER **8**

DESIGNING FOR HUMAN ERROR: CONCEPTS FOR ERROR TOLERANT SYSTEMS

William B. Rouse

ABSTRACT

Human error is frequently judged to be a primary contributor to high-consequence accidents in complex systems. This chapter explores this issue and argues that total elimination of human error is a futile pursuit. Instead, systems should be designed so that they are error tolerant in the sense that errors can occur without leading to unacceptable consequences. The idea of error tolerance is described in terms of its empirical basis and an evolving conceptual architecture for error tolerant interfaces.

INTRODUCTION

Occasional high-consequence accidents appear to be part of the normal pattern of technologically-oriented societies (Perrow, 1984). A sampling of such accidents include:

- Flixborough (1974 in the United Kingdom)
- Tenerife (1977 in the Canary Islands)
- Three Mile Island (1979 in the United States)
- Bhopal (1984 in India)
- Chernobyl (1986 in the Soviet Union)

This small sample illustrates that such accidents are not limited to particular industries – these five examples include aviation, nuclear power, and process control. Further, this small sample shows that high-consequence accidents are not limited to particular types of country or government.

Analysis of these types of accident, as well as many other less-dramatic incidents, quite often lead to the conclusion that human error was a substantial contributor to the initiation or evolution of the accident. Regardless of type of industry, it is common to see estimates of 60-90 percent of major accidents having human error as an underlying cause. It appears that there is a widespread and significant problem.

Inadequate Solutions

The approach that most organizations have adopted for addressing this problem is to "declare war" on human error. One version of this approach emphasizes inhibiting human performance from deviating from prescribed paths. This leads to installation of interlocks, use of rigid procedures, and when feasible use of automation.

A second and more recent version of this approach focuses on management practices. This approach is premised on the notion that unacceptable human performance is primarily due to lack of management support and sensitivity. With appropriate training, guidance, and leadership, it is argued that human error will disappear.

These two strategies of improved equipment design and management practices can be of great value. However, these changes are not adequate for dealing with human error. It is seldom possible to completely prescribe human behavior and eliminate all possibilities for deviations. When it is possible, it is likely that automation is feasible and, therefore, human performance need not be an element of the system.

Beyond the issue of feasibility, there is the issue of desirability. Do we really want humans to be precluded from behaving in unexpected ways? Or, do we want them to follow standard operating procedures when appropriate, but flexibly innovate and adapt when unusual situations arise? Flexible innovation and adaptation are often primary reasons for including humans in systems – when all else fails, we expect humans to muddle through and save the day! How can we enjoy the "benefits" of human innovation and adaptation without having to pay the "costs" of human error?

The Ultimate Solution

This apparent dilemma is resolvable if design and management philosophies are changed. Why are errors undesirable? Errors in themselves are not particularly troublesome. The real problem is *consequences.* If undesirable consequences could be avoided, human errors could become, to a great extent, a non-problem.

This shift in emphasis from errors to consequences leads to a strategy of designing systems to be *error tolerant* (Rasmussen, 1983; Rouse, 1983). Such a system tolerates the occurrence of errors by avoiding their consequences. Error tolerance can be achieved in 3 complementary ways:

- Feedback about current consequences
- Feedback about future consequences
- Intelligent error monitoring

These methods are described in detail later in this chapter.

It is important to address at the outset how it is possible for a system to identify human errors and provide appropriate feedback to enable humans to correct and/or compensate for their errors. Intuitively, it would seem that a system sufficiently sophisticated to perform these functions would also be capable of replacing the humans it was monitoring. However, error tolerance can often be achieved much more easily than automation.

The key to this possibility is the fact that a system need not know what a human *should be doing* in order to determine whether or not what the human *is doing* is consistent. A system can assess the internal consistency of a stream of actions relative to each other and, with a bit more difficulty, the external consistency of the actions relative to current goals. A system can do this in complete ignorance of the "best" set of actions. Later discussion illustrates how this can be accomplished.

Beyond the issue of feasibility and practicality, another important issue is the relative value of alternative ways of achieving error tolerance. The above discussion implies that a system can be sufficiently "smart" to help humans avoid errors and/or their consequences – in other words, humans can be aided in dealing with errors. Alternative approaches to error tolerance can involve selection and training of personnel or equipment and job design.

A model-based evaluation of these five approaches (i.e., selection, training, equipment design, job design, and aiding) has been conducted. The analysis focused on assessing the impact of alternative allocations of system development resources on the frequency of errors that are uncorrected or evolve without compensation. It was found that a disproportionate share of the resources should be allocated to aiding (Rouse, 1985). In retrospect, the reason for this is very simple.

In contrast to aiding, the other four approaches focus on "upstream" aspects of the problem. The concern is with potential errors that *might* occur – typically, there are many potential errors. Error tolerance via aiding involves on-line feedback and compensation for errors that *have* occurred. The difference between directly dealing with errors that have occurred vs. indirectly dealing with errors that might occur is the reason why aiding usually should be allocated more resources than the other approaches.

It is important to note, however, that this conclusion should not be interpreted categorically. Error tolerance via aiding is not a panacea for dealing with all human performance issues. Error tolerance should be balanced with error reduction (Rouse, 1985). Further, dealing with human error should also be balanced with performance enhancement.

Recent model-based evaluations have illustrated potential trade-offs between training and aiding as means for enhancing human performance

(Rouse, 1987). This effort considered task complexity and frequencies, personnel aptitudes and abilities, and other variables. The results of this study led to the conclusion that a mixture of training and aiding may be the best overall approach to performance enhancement. This may also be the case for error reduction and tolerance.

Overview

This introductory discussion has summarized the nature of the human error problem and outlined the concept of error tolerant systems. The remainder of this chapter provides considerable more detail concerning the empirical basis for the concept, the evolution of the notion of error tolerance, and the salient issues involved with adopting the concept. The discussion of these topics occurs in the context of operations and maintenance of aircraft, supertankers, and nuclear power plants. This chapter, therefore, summarizes both the extensive basis and broad applicability of error tolerant systems.

INITIAL STUDIES

The feedback that an error tolerant system provides to humans should not be a simple ERROR message – we all have experienced the frustration of getting this type of message from computers. Providing a more useful message requires that the system not only identify anomalous actions, but also explain the actions. Explanations need not be directly presented to the humans involved, but they provide an important basis for synthesizing cogent feedback.

Two types of explanation are needed. One type is *task-oriented* in the sense that an anomaly is explained in the context of the task being performed (e.g., a particular step was omitted during execution of a specific operating procedure). The other type of explanation is *psychologically-oriented* and provides a behavioral diagnosis of the anomaly.

Norman (1981) and Reason and Mycielska (1982) have devised a particularly useful behavioral dichotomy. It allows one to distinguish between the humans' intentions and their execution of actions. If the intention is appropriate for the situation, but the execution is incorrect, the error is termed a *slip*. Types of slip include attentional capture, misperceptions, and losing track of one's place (Reason & Mycielska, 1982).

In contrast to slips, if the intention is inappropriate, even though the execution may be correct with respect to this intention, the error is termed a *mistake*. Reason (1983) has suggested types of mistake to include oversimplifications (bounded rationality), appearances of uncalled schema (imperfect rationality), overreliance on familiar cues and well-worn solutions (reluctant rationality), and irrational acts. In other words, mistakes can result from human information processing limitations, processing errors, or unwillingness to invest the effort necessary to formulate intentions more carefully.

The distinction between slips and mistakes has practical implications for the feedback provided by error tolerant systems. Errors such as unintentional closing of the wrong valve or pushing the wrong button are likely to be immediately corrected if a human receives feedback that these events have occurred. On the other hand, doggedly correct execution of an inappropriate emergency recovery strategy may be difficult to remediate without providing a human with a substantive explanation of why this intention is wrong.

Data Collection

The goal of the studies described below was to devise a systematic scheme for characterizing task-oriented and psychologically-oriented explanations. The emphasis in these studies was manual analysis of extensive data collected via observation of professional operations and maintenance personnel in complex systems.

Supertanker Control Rooms

A first study (van Eekhout & Rouse, 1981) involved seven crews of professional engineering officers who were being trained using a high-fidelity supertanker engine control room simulator. Troubleshooting exercises were observed. Measurement methods included verbal protocols, computer logs of all discrete events, interviews, questionnaires, and observer ratings.

The resulting data were analyzed for human errors by two independent analysts. Errors were independently classified using a variation of Rasmussen's scheme (Rasmussen, Pederson, Mancini, Griffon, & Gagnolet, 1981). Contributing factors were also assessed, including lack of knowledge of system functioning and automatic controller functioning, human factors design inadequacies, and simulator fidelity inadequacies.

It was found that errors associated with inappropriate identification of failures were highly correlated with a lack of knowledge of the functioning of the basic system as well as the automatic controllers within the system. These errors tended to be mistakes (rather than slips) in that inappropriate intentions were formed based on inadequate or incorrect knowledge.

Errors related to execution of procedures were highly correlated with inadequacies of the layout of the control panel and simulator fidelity inadequacies. These errors tended to be slips due to, for example, control knobs that were easily confused or labels that were difficult to read.

The types of slip that occurred were anticipated, but the nature of the mistakes provided a new insight. Errors in diagnosing failures of automatic controllers were often associated with a lack of understanding of failure modes of the controllers. In these situations, the cues presented by the automation were misleading. Crew members became confused and/or reached incorrect conclusions upon which they proceeded to act. Succinctly, basic misunderstandings led to mistakes. This phenomenon was encountered throughout the series of studies.

Aircraft Power Plant Maintenance

This effort included two studies involving 58 trainees in a Federal Aviation Administration certificate program in aircraft power plant maintenance (Johnson & Rouse, 1982). These studies involved comparing three methods for training troubleshooting. Two of the methods included computer-based power plant simulations. The third method used instructional television and a fairly traditional lecture and demonstration format.

The data collected in these experiments were analyzed in a similar manner to the supertanker study. An important difference, however, was the error classification scheme. The aircraft maintenance tasks of interest were sufficiently different from supertanker operations to require modifications of the error categories. In particular, categories were added to provide a finer-grained view of troubleshooting errors. This tailoring of the classification scheme to the domain of interest was found to be very important.

The results of analyzing the data from the first of the two experiments showed traditional video-based instruction to be superior. For the most part, this was due to procedural errors by the mechanics who were trained with the computer-based methods. In particular, it was found that mechanics trained with the low to moderate fidelity computer simulations knew *what* troubleshooting tests to make, but not *how* to make them.

This conclusion led to combining the two computer-based methods and adding material on how to make tests. The second experiment in this effort compared this hybrid computer-based method to the traditional instruction. Results indicated that the previous types of error no longer occurred, and the two methods of instruction yielded similar performance.

This effort showed how fine-grained analysis of performance can lead to insights and improvements that might not be evident from global performance measures. In particular, lack of understanding of test procedures led to execution errors. These errors are probably best categorized as mistakes. This insight helped to modify the training to provide the requisite understanding.

Aircraft Operations

This study involved four two-person crews flying a high-fidelity, twin-engine aircraft simulator. Each crew flew several full-mission, commuter airline scenarios. The purpose of the study was to evaluate a computer-based system for retrieving, displaying, and assisting in executing aircraft procedures.

The data collected were analyzed using methods similar to those for the previous studies. As before, the one exception was the classification scheme which was modified to focus on procedure execution, with several categories added in this area. This tailoring of the classification scheme to the domain and goals of the study was essential.

The results of the error analysis indicated that errors of omission (i.e., leaving out procedural steps) were virtually eliminated. In this manner, the interface was, to an extent, slip tolerant. However, errors of commission (i.e., adding unnecessary steps) were relatively unaffected by the computer-based display. The reason was simple – the computer had no way of discriminating incorrect actions from irrelevant actions. This important issue is returned to in a later section.

Resulting Methodology

The experiences and results associated with the above studies led to a general methodology for analysis and classification of human error (Rouse & Rouse, 1983). Central to this methodology is the classification scheme shown in Tables 8-1 and 8-2. The 6 general categories and 31 specific categories shown in these tables are sufficient to encompass all of the operations and maintenance tasks studied. It is important to emphasize, however, that any particular study is likely to require a much-reduced subset of this overall scheme.

The general methodology for analysis and classification of errors involves five steps:

- Data collection
- Identification of errors
- Identification of causes and contributing factors
- Classification of errors
- Statistical analysis

While the limited length of this chapter does not allow description of these steps in detail, a few points are important and should be noted.

First, it is important to study those humans who actually perform (or will perform) the jobs and tasks of interest. It is quite difficult to simulate the level of expertise, attitudes toward errors, and overall value structures of professional personnel using "generic" human subjects (e.g., college students). Too many investigators have learned this lesson the hard way.

Table 8-1
Human Error Classification Scheme

GENERAL CATEGORY	SPECIFIC CATEGORY
1. Observation of System State	a. excessive b. misinterpreted c. incorrect d. incomplete e. inappropriate f. lack
2. Choice of Hypothesis	a. inconsistent with observations b. consistent but very unlikely c. consistent but very costly d. functionally irrelevant
3. Testing of Hypothesis	a. incomplete b. false acceptance of wrong hypothesis c. false rejection of correct hypothesis d. lack
4. Choice of Goal	a. incomplete b. incorrect c. unnecessary d. lack
5. Choice of Procedure	a. incomplete b. incorrect c. unnecessary d. lack
6. Execution of Procedure	a. step omitted b. step repeated c. step added d. steps out of sequence e. inappropriate timing f. incorrect discrete position g. incorrect continuous range h. incomplete i. unrelated inappropriate action

Table 8-2
Definitions of Specific Human Error Categories

SPECIFIC CATEGORY	BRIEF DESCRIPTION
1a. excessive	improper rechecking of correct readings of appropriate state variables
1b. misinterpreted	erroneous interpretation of correct readings of appropriate state variables
1c. incorrect	incorrect readings of appropriate state variables
1d. incomplete	failure to observe sufficient number of appropriate state variables
1e. inappropriate	observations of inappropriate state variables
1f. lack	failure to observe any state variables
2a. inconsistent	could not cause particular values of state variables observed
2b. unlikely	could cause values observed but much more likely causes should be considered first
2c. costly	could cause values observed but very costly (in time or money) place to start
2d. irrelevant	does not functionally relate to state variables observed
3a. incomplete	stopped before reaching a conclusion
3b. acceptance	reached wrong conclusion
3c. rejection	considered and discarded correct conclusion
3d. lack	hypothesis not tested
4a. incomplete	insufficient specification of goal
4b. incorrect	choice of counter-productive goal
4c. unnecessary	choice of non-productive goal
4d. lack	goal not chosen
5a. incomplete	choice would not fully achieve goal
5b. incorrect	choice would achieve incorrect goal
5c. unnecessary	choice unnecessary for achieving goal
5d. lack	procedure not chosen
6a. omitted	required step omitted
6b. repeated	unnecessary repetition of required step
6c. added	unnecessary step added
6d. sequence	required steps executed in wrong order
6e. timing	step executed too early or too late
6f. discrete	discrete control in wrong position
6g. continuous	continuous control in unacceptable range
6h. incomplete	stopped before procedure complete
6i. unrelated	unrelated inappropriate step executed

Second, identification and classification is best performed by multiple independent, subject matter experts. Typically, 95 percent initial agreement is found among experts, with the 5 percent disagreements usually resolved in an objective manner. Disagreements are often due to technical issues that can be resolved by resorting to technical manuals or other documentation.

Finally, statistical analysis can be straightforward, often only involving paired t-tests and correlation methods. Univariate and multivariate analysis of variance are also frequently appropriate. Statistical analysis is usually aided by having chosen the appropriate subset of categories in Tables 8-1 and 8-2. Having many categories which inherently produce zero or minimal entries can statistically obscure causes that only affect one or two categories.

The analysis and classification methodology was used to reanalyze the data from the aforementioned study of procedure displays (Rouse & Rouse, 1983). This extensive analysis indicated that errors of omission (i.e., omitted procedural steps) were often associated with possible confusion, distraction, and/or communication problems. In contrast, errors of commission did not have any obvious causes and contributing factors. This result served to emphasize the earlier conclusion about the importance of understanding the basis for extra actions that are not necessarily incorrect.

A reduced version of the methodology was also used to evaluate the impact of computer-based training for maintenance and operation of auxiliary diesel generators in nuclear power plants (Maddox, Johnson, & Frey, 1986). It was not possible to discriminate computer-based training from traditional instruction using solely overall performance measures. However, an error analysis led to the conclusion that traditional instruction resulted in 500 percent more consequential errors than computer-based instruction. This substantial effect was found to remain even 20 weeks after training had been completed.

ERROR TOLERANT INTERFACES

Error tolerance is a concept that can be realized in a variety of ways. In this section, the notion of an interface architecture that monitors human actions for consistency and provides appropriate feedback is emphasized. An alternative and complementary approach is embodied in Rasmussen and Vicente's (1987) guidelines for interface design. They suggest that error tolerance can be achieved by:

- Providing visible performance boundaries for initiating error recovery.
- Supporting the understanding necessary for recovery.

- Providing cues that define which actions should be taken.
- Supplying tools that allow testing of hypotheses.
- Supporting peripheral monitoring of ancillary routines.
- Giving symbolic significance to action cues.
- Supporting externalization of humans' mental models.
- Supporting perceptual and conceptual data integration.
- Supporting strategies rather than procedures.
- Supporting memory of isolated items.

This comprehensive set of suggestions deals with many of the problems that appear to underlie the occurrence of human errors. To the extent that these guidelines can be operationalized in any particular domain, they are quite likely to both reduce the likelihood of errors and provide a degree of error tolerance.

The primary limitation of these guidelines is that they leave much of the burden of error tolerance on the humans in the system. Put simply, these guidelines imply that people still have to watch themselves, catch their own errors, and compensate appropriately. A system designed according to these guidelines may lessen this burden, but it is possible to provide people with much more support.

Initial Concept

The first effort in developing an error tolerant interface involved the aircraft procedure displays discussed earlier (Rouse, Rouse, & Hammer, 1982). This system displayed requested operating procedures and monitored their execution. By monitoring aircraft controls, buttons, switches, etc., as well as the dynamic state of the aircraft, it was usually possible to determine which procedural steps had been completed. Based on this information, the display indicated which steps had been completed and, by implication, which steps remained to be done. As noted earlier, this type of monitoring and feedback resulted in almost completely eliminating errors of omission.

This system did not, however, explicitly identify errors. To provide this capability, Hammer (1984) developed a method for the computer to monitor the data stream and detect anomalies. His method involves representing aircraft procedures as "scripts" which describe typical courses of events in procedure execution. Elements of scripts are matched to elements of the data stream using if-then rules.

An evaluation of Hammer's method using the data from the procedures display experiment (Rouse et al., 1982) showed that automated error identification is quite feasible. This development suggested more comprehensive notions of error tolerance were needed.

General Architecture

Error identification is only the "front end" of error tolerance. As shown in Figure 8-1, error identification leads to error classification which, in turn, leads to error remediation. These three functions must operate in a coordinated manner to provide an error tolerant system.

Figure 8-1 illustrates the components and functional relationships of a system for intelligent monitoring, identification, and remediation of errors (Rouse & Morris, 1987). The boxes labeled physical interface, at the left and right of this figure, denote the controls and displays whereby humans interact with the system of interest. Physical interface is not discussed in this section.

There are three models indicated in Figure 8-1: the world model, system model, and human model. The purpose of these models is to estimate the current and predicted "state" of the world, system, and human. Thus, they can be used to ask "what is" questions (i.e., estimated current state) and "what if" questions (i.e., predicted state).

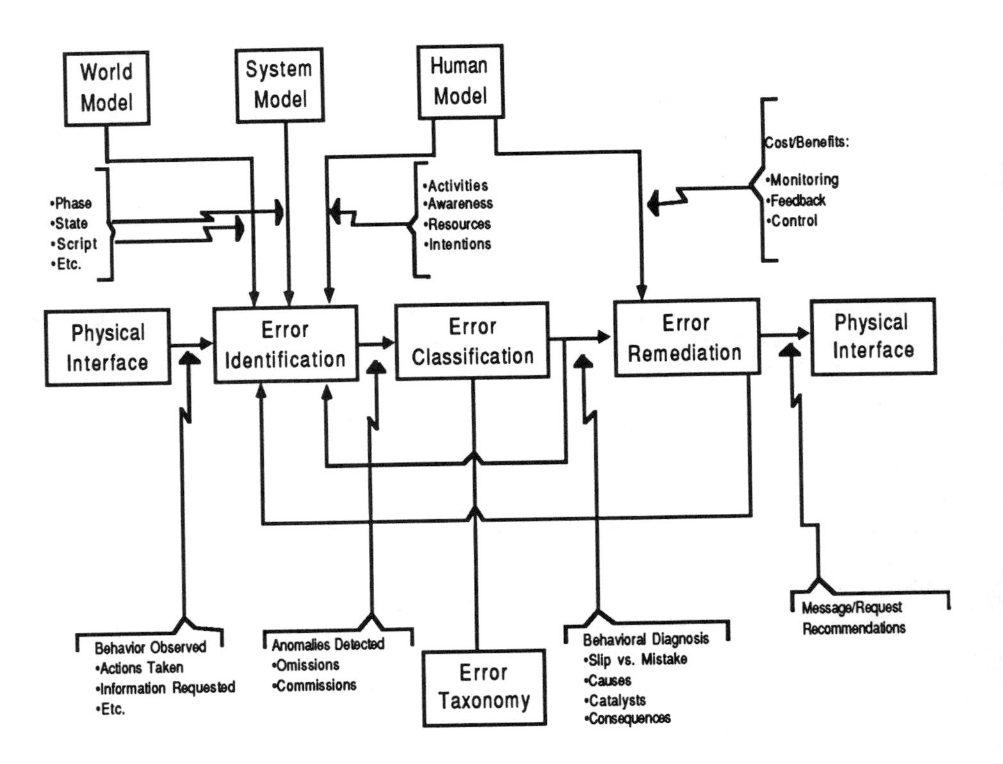

Figure 8-1
Architecture for Error Tolerant Interface

While the specific definition of "state" differs for world, system, and human, the general concept of state is that a set of variables can be defined such that knowledge of the *current* values of these variables, as well as knowledge of any external inputs, can be used in conjunction with an appropriate model to predict *future* states. The state of the world may include information on operational demands, weather, etc. The state of the system includes its dynamic state, the modes and failure status of subsystems, and information on current and upcoming operational phases, applicable procedures, etc. The state of the human includes his or her activities, awareness, resources, and intentions.

Geddes (1985, 1986, 1989) has developed a model that provides estimates of activities, awareness, and intentions. By interpreting the actions taken by humans, as well as the information that they request, relative to a context-specific goal/plan hierarchy, it is possible to provide on-line inferences of a human's intentions and, partially as a by-product, inferences concerning a human's activities and awareness. Estimates of human information processing resource requirements can be based on Wickens' (1984) multiple resource theory, with various heuristic refinements and extensions for the types of task and domain of interest.

Error Identification

Error identification involves correlating the histories of a human's behavior and the system's response with operational procedures, scripts, etc., to detect anomalies between expected and observed behavior. Hammer's method was used with elaborations to take advantage of the information provided by Geddes' model. This approach works quite well in highly-structured situations such as found in aviation and process control.

Fortunately, less-structured tasks also tend to have much looser time constraints. This allows time for error identification to be more of an interactive process between human and system. For example, the human may be able to tell the computer of his or her intentions, rather than having the computer infer them.

Error Classification

Once an anomaly has been detected and identified as an error, it is necessary to classify it if other than a global ERROR message is to be provided. As discussed earlier, the slips vs. mistakes dichotomy is an important element of explaining an error and can strongly influence how an error is remediated.

Classification of errors in terms of causes is not necessary to help humans reverse or avoid consequences. Nevertheless, identification of causes can be useful for determining how an error might be dealt with so that it is less likely to reoccur. For example, misunderstandings that cause errors may be subsequently remediated with instruction, perhaps via embedded training. It is important to emphasize, however, the consequences of the present error should be dealt with prior to being concerned with avoiding future errors.

Catalysts are factors that aggravate error-likely situations. Good examples are distractions and excessive work load. Eliminating the aggravating effects of catalysts can be a key aspect in dealing with an error situation.

Knowledge of likely consequences can be essential to effective remediation. For errors of omission, the projection of consequences can be straightforward if design models and knowledge bases are available. Errors of commission require more sophisticated models and levels of reasoning because they can involve actions that designers never anticipated.

Error Remediation

Once an error is identified and classified, there are three general types of remediation as shown in Figure 8-2. One possibility is to continue *monitoring*, perhaps looking for particular events or consequences that might trigger more active remediation. This type of remediation reflects a more active level of reasoning than associated with passive monitoring of action sequences. More specifically, remediation at this level involves active exploration of alternative explanations and courses of action.

The next level of remediation is *feedback* which involves providing messages regarding the identification and classification of an error, and perhaps advice on appropriate compensatory actions. At a minimum, this type of feedback includes traditional alerting and warning systems. However, when appropriate and possible, feedback is much more intelligent involving, for example, synthesis of multiple warnings into an integrated explanation and recommendation.

The highest level of remediation is *control* whereby the propagation of consequences (in terms of evolving undesirable states) is actively prevented and, in some situations, compensatory actions automatically initiated. Obviously, at this level it is crucial that humans not be inappropriately preempted from acting. This possibility can be avoided by using "error advisories" that inform humans of the system's evolving conclusions and, thereby, allow the humans to preempt higher levels of remediation.

Application in an Intelligent Interface

The above architecture for error tolerance has been developed as a function called the Error Monitor within a broader concept called an *intelligent*

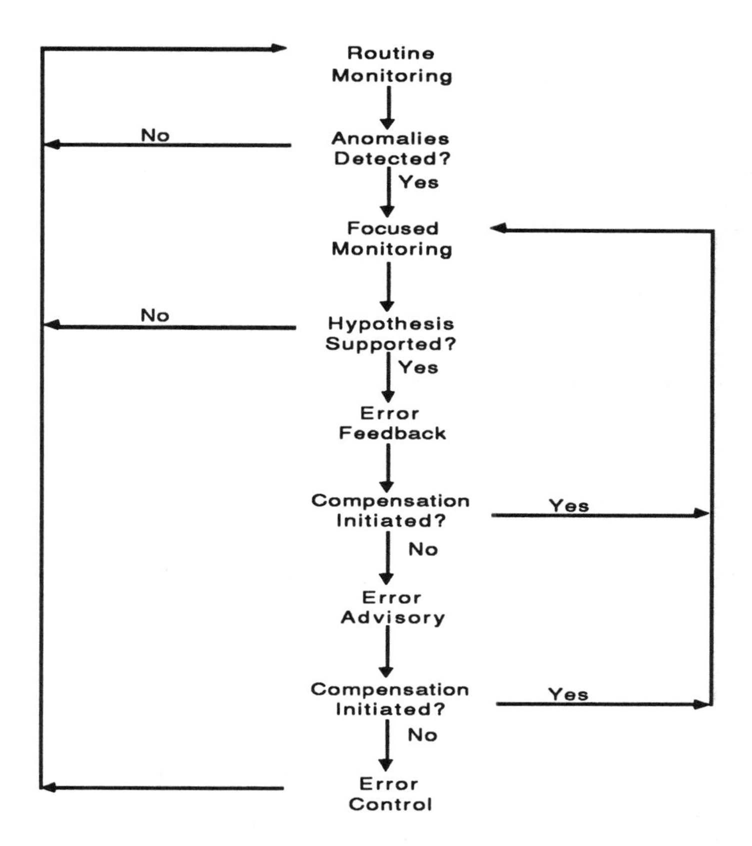

Figure 8-2
Adaptive Error Remediation

interface (Rouse, Geddes, & Curry, 1988). This interface was developed for aircraft operations, but it is broadly applicable to complex engineering systems. While space does not allow description of this comprehensive concept, it is useful to consider how error tolerance fits within this general performance enhancement framework.

A primary issue concerns how error remediation should be integrated with other types of support and, in general, with other types of ongoing activities. The intelligent interface deals with this issue in two ways. First, remediation *recommendations* from the Error Monitor are not necessarily communicated to the human. Instead, a subsystem called Adaptive Aiding considers these recommendations as only one (or a few) of the many things requesting the human's attention. As a result, error control rather than error feedback might be implemented because the human is too busy to assimilate the recommended error feedback.

The second aspect of how the intelligent interface deals with remediation recommendations involves a subsystem called the Interface Manager. This subsystem is responsible for managing and integrating all information which will be displayed to the human. This broader purview may, for example, cause the Interface Manager to streamline or reformat error remediation messages.

Application of the error tolerance concept within the overall intelligent interface led to a variety of enhancements of the error identification methods. Probably most important was the development of a set of independent "critics." Each critic inspects each anomaly with a specific explanation in mind. For example, one critic only looks for right control/wrong position or wrong control/right position explanations. If one of these explanations "fits" the anomaly, the Error Monitor communicates this explanation to Adaptive Aiding.

ISSUES AND CONCLUSIONS

Success of the error tolerance concept depends on resolving many issues. In this section, three particularly salient issues are discussed.

What Matters and What Doesn't

The first issue, and perhaps the most technically difficult, is knowing what matters and what doesn't. More specifically, how do you decide an action is anomalous or apparently inconsistent, and how do you decide if the anomaly is an error?

This issue can be resolved in part by developing a means for the computer to have *expectations* about what actions and information requests are likely. If actual behavior does not agree with expectations, then a candidate error has been detected. At this point, the computer has to reason in terms of *consequences*. If the consequences are unacceptable, an error has occurred and remediation is needed.

The technical challenge, therefore, is developing means of forming expectations and projecting consequences. For highly structured jobs and tasks, such as aircraft piloting and process control, this has been found quite feasible, although it tends to be rather laborious. For less structured tasks, increased human-system dialogue can help to form expectations and project consequences. However, humans tend to react rather negatively if the "overhead" of this interaction becomes too great.

Human Adaptation to Error Tolerance

A second important issue involves changes in the ways that humans interact with the system. These changes may be due to learning or fatigue. Another possibility recently studied involves human adaptation to system characteristics (Morris & Rouse, 1988). This includes the possibility of humans adapting to the intelligent support system. Unfortunately, there is little, if any, data available to provide insights into how the nature of human error is affected by the presence of an error tolerant interface.

Design Philosophy

The third and final important issue concerns the necessary changes of design philosophy if error tolerant interfaces are to be viable. It requires that the emphasis from counting errors and assigning blame be shifted to compensating for consequences. Errors are seldom bad and often good if they promote learning and innovation. Consequences can be bad, and providing help to humans is needed so that they can avoid or reverse potentially bad consequences.

This change of design philosophy can radically alter allocations of design resources. The result should be a decrease in the number of high-consequence accidents. From a broader perspective, error tolerance should provide an important means for helping people to cope in general with increasing technological complexity.

REFERENCES

Geddes, N. D. (1985). Intent inferencing using scripts and plans. *Proceedings of the First Annual Aerospace Applications of Artificial Intelligence Conference.*

Geddes, N. D. (1986). The use of individual differences in inferring human operator intentions. *Proceedings of the Second Annual Aerospace Applications of Artificial Intelligence Conference.*

Geddes, N. D. (1989). *Understanding intentions of human operators in complex systems.* Doctoral dissertation. Georgia Institute of Technology.

Hammer, J. M. (1984). An intelligent flight management aid for procedure execution. *IEEE Transactions on Systems, Man, and Cybernetics, SMC-14*, 885-888.

Johnson, W. B., & Rouse, W. B. (1982). Analysis and classification of human errors in troubleshooting live aircraft power plants. *IEEE Transactions on Systems, Man, and Cybernetics, SMC-12*, 389- 393.

Maddox, M. E., Johnson, W. B., & Frey, P. R. (1986). *Diagnostic training for nuclear plant personnel. Volume 2: Implementation and evaluation* (Technical Report NP-3829, Vol. 2). Palo Alto, CA: Electric Power Research Institute.

Morris, N. M., & Rouse, W. B. (1988). *Human operator response to error-likely situations in complex engineering systems* (Technical Report 177484). Moffett Field, CA: National Aeronautics and Space Administration.

Norman, D. A. (1981). Categorization of action slips. *Psychological Review, 88*, 1-15.

Perrow, C. (1984). *Normal accidents: Living with high-risk technologies.* New York: Basic Books.

Rasmussen, J. (1983). Design for error tolerance. *Proceedings of the Winter Meeting of the American Nuclear Society.*

Rasmussen, J., Pederson, O. M., Mancini, G., Griffon, M., & Gagnolet, P. (1981). *Classification system for reporting events involving human malfunctions* (Technical Report RISO-M-2240). Roskilde, Denmark: Riso National Laboratory.

Rasmussen, J., & Vicente, K. J. (1987). Cognitive control of human activities and errors: Implications for ecological interface design. *Proceedings of the Fourth International Conference on Event Perception and Action.*

Reason, J. (1983). On the nature of mistakes. In N. Moray & J.W. Senders (Eds.), *Reprints of NATO Conference on Human Error.*

Reason, J., & Mycielska, K. (1982). Absent minded: *The psychology of mental lapses and everyday errors.* Englewood Cliffs, NJ: Prentice-Hall.

Rouse, S. H., Rouse, W. B., & Hammer, J. M. (1982). Design and evaluation of an onboard computer-based information system for aircraft. *IEEE Transactions on Systems, Man, and Cybernetics, SMC-12*, 451- 463.

Rouse, W. B. (1983). Elements of human error. In N. Moray and J.W. Senders (Eds.), *Preprints of NATO Conference on Human Error.*

Rouse, W. B. (1985). Optimal allocation of system development resources to reduce and/or tolerate human error. *IEEE Transactions on Systems, Man, and Cybernetics, SMC-15*, 620-630.

Rouse, W. B. (1987). Model-based evaluation of an integrated support system concept. *Large-Scale Systems, 13*, 33-42.

Rouse, W. B., Geddes, N. D., & Curry, R. E. (1988). An architecture for intelligent interfaces: Outline of an approach to supporting operators of complex systems. *Human-Computer Interaction, 3*, 87-122.

Rouse, W. B., & Morris, N. M. (1987). Conceptual design of a human error tolerant interface for complex engineering systems. *Automatica, 23*, 231-235.

Rouse, W. B., & Rouse, S. H. (1983). Analysis and classification of human error. *IEEE Transactions on Systems, Man, and Cybernetics, SMC-13*, 539-549.

van Eekhout, J. M., & Rouse, W. B. (1981). Human errors in detection, diagnosis, and compensation for failures in the engine control room of a supertanker. *IEEE Transactions on Systems, Man, and Cybernetics, SMC-11*, 813-816.

Wickens, C. D. (1984). Processing resources in attention. In R. Parasuraman & R. Davies (Eds.), *Varieties of attention* (pp. 63-101). New York: Academic.

SUGGESTIONS FOR FURTHER READING

Rasmussen, J., Duncan, K., & Leplat, J. (Eds.). (1987). *New technology and human error.* Chichester, UK: Wiley.

Reason, J. (1989). *Causes of human error.* Cambridge, UK: Cambridge University Press.

Reason, J., & Mycielska, K. (1982). *Absent minded: The psychology of mental lapses and everyday errors.* Englewood Cliffs, NJ: Prentice-Hall.

Rouse, W. B., & Morris, N. M. (1987). Conceptual design of a human error tolerant interface for complex engineering systems. *Automatica, 23,* 231-235.

Rouse, W. B., & Rouse, S. H. (1983). Analysis and classification of human error. *IEEE Transactions on Systems, Man, and Cybernetics, SMC-13,* 539-549.

CHAPTER 9

WORKLOAD ASSESSMENT AND PREDICTION

Sandra G. Hart
Christopher D. Wickens

ABSTRACT

The concept of workload and its relationship to performance is introduced in this chapter. Four categories of workload measurement techniques (ratings, primary and secondary task measures, and physiological indices) are reviewed, examples of each category are described, and their strengths and weaknesses are summarized. The importance of carefully formulating the question which a measure is to address is emphasized, and it is argued that the question should guide the selection of measures. Issues relevant to implementing and interpreting workload measures are discussed and some of the reasons that different measures provide apparently conflicting information about the same situation (i.e., dissociation) are addressed. Finally, the chapter concludes with a brief description of models that can be used to predict workload.

WHAT IS WORKLOAD?

If people could accomplish all of the requirements imposed on them quickly, accurately, reliably and with little effort using available resources, the concept of workload would have minimal practical importance. However, they often cannot; in some cases task demands simply exceed operators' capabilities while in others apparently *human* limitations reflect poorly designed controls or displays, inappropriate or inadequate automation, or insufficient training. Such decrements in the performance of an individual operator, which may occur if workload is either too high or too low, can result in a reduction in overall system effectiveness. Although human adaptability and creativity are essential to the effective functioning of complex systems, human capabilities and limitations also represent a limiting factor in overall system performance. For this reason, operator workload is an important factor that must be considered in evaluating the adequacy and feasibility of operational requirements, system designs, and training procedures.

Workload is a general term used to describe the cost of accomplishing task requirements for the human element of man-machine systems. This "cost" may be reflected in the depletion of attentional, cognitive, or response resources, inability to accomplish additional activities, emotional stress, fatigue, or performance decrements. Workload measures are generally obtained to evaluate the effects of different systems (or operating conditions) on any human operator or to quantify the effects of individual differences in abilities or training of specific operators working with a given system. The fact that workload varies as a function of differences between systems as well as between operators highlights the locus of the workload concept at the interface between a particular operator and a specific system. Given the number of factors that might influence workload (and which are, in turn, affected by variations in workload), specific definitions disagree about the source(s) and consequence(s) of workload and the techniques recommended for its measurement. In a single sentence, we define workload as the effort invested by the human operator into task performance; workload arises from the interaction between a particular task and the performer, as represented in Figure 9-1.

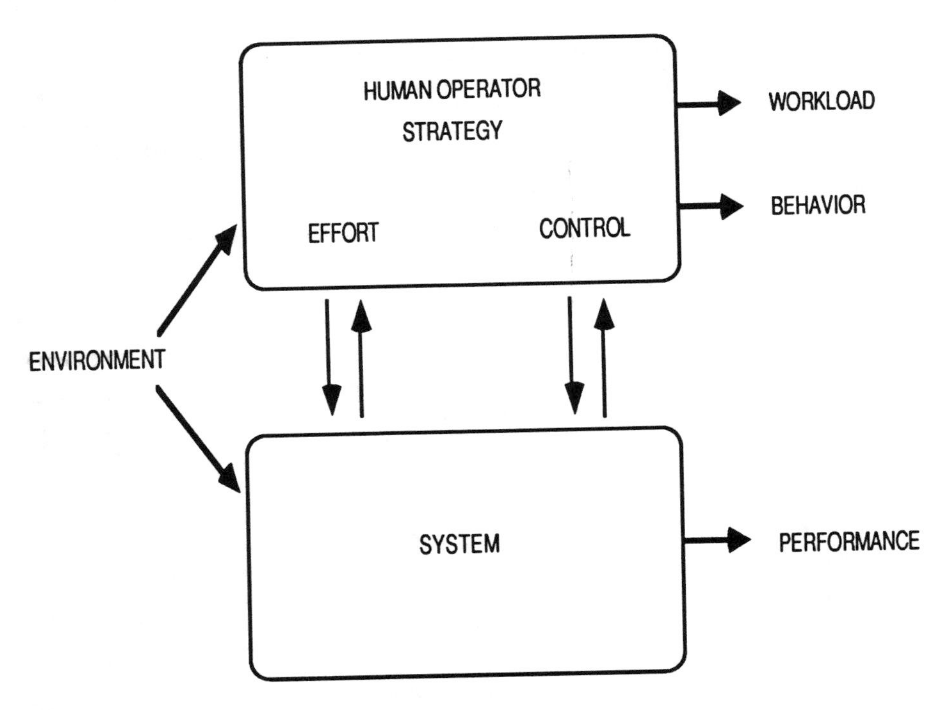

Figure 9-1
Conceptual Framework Relating Operator Performance and Workload

Although workload and performance are clearly related, the nature of this relationship is not straightforward; measures of operator and system performance and operator workload may be influenced by similar as well as different factors. In fact, operators may trade off workload and performance against each other. As task demands vary over time, operators may increase or decrease their effort (to maintain a consistent level of performance), maintain a constant level of effort (thereby allowing performance decrements to occur), defer or shed less important tasks when task demands are too high (to maintain performance on critical tasks), or complete some tasks ahead of schedule during low-workload periods (to maintain performance during high-workload periods in the future). Thus, the relationship between workload and performance from moment to moment or averaged across intervals of time depends on the strategies operators adopt and the degree to which they are able (or willing) to exert additional effort to achieve better performance, as well as on task requirements, the design of controls and displays, and environmental factors.

Formally, the relationship between workload, effort, performance, and strategies may be characterized by performance resource functions (PRF), examples of which are shown in Figure 9-2. The PRF is a hypothetical relationship that reflects the level of performance that can be obtained by a specific system, given the resources (or effort) that a particular operator invests in the performance of a specific task in a given environment (Norman & Bobrow, 1975). For example, in Figure 9-2a, performance improvements are positively, but not linearly, related to an increase in invested resources. When the PRF is steep, revealing perfect performance with few resources invested (curve B in Figure 9-2b) the system-operator interface is well designed. When it is shallow (curve A of Figure 9-2b), the system may have problems. The following hypothetical examples illustrate a number of possible relationships between workload and performance (as characterized by the PRF) to provide a framework for understanding how workload is measured and why these measurements are important.

(1) Figure 9-2b depicts the PRFs for two systems which allow operators to specify the coordinates of a point on a map. One requires a digital readout of x-y coordinates (System A). The other allows positioning of a light pen directly on the map (System B). For the light pen system (B), the operator only invests enough resources (20 percent, the vertical line) to attain maximum performance. Further investment will not improve perform-ance. Hence, lower workload and higher performance are found with System B than System A because near perfect performance can be achieved with System B by a minimal investment of resources (i.e., low effort).

(2) Figure 9-2c depicts the PRFs for helicopter flight control under day and night conditions. Equivalent performance (as measured by flight path deviations) can be achieved under both conditions when full resources are invested in flying. But, the same high level of performance can be achieved during the day with significantly less effort (as measured by an increase in

control activity) than can be achieved at night. Hence, workload is generally higher at night.

(3) Figure 9-2d depicts the PRFs obtained with two navigation displays: System A displays ownship's position and System B displays this position plus an accurate prediction of future position based on current trends. The conventional system (A) yields poorer performance, but lower workload than the enhanced system (B), the opposite relationship to that obtained in Example 1 (Figure 9-2b); the more precise information conveyed by the predictor display gives pilots more opportunity to exercise precise flight path control and encourages them to plan ahead and project the impact of their momentary control inputs into the future. Thus, they invest more resources (thereby experiencing higher workload) to exercise that control (thereby producing better performance).

(4) Figure 9-2e depicts PRFs for digital data entry obtained with either voice recognition or keyboard devices. Due to delays associated with the computer algorithm for voice recognition, performance with this system is poorer than with the keyboard. But, the use of voice, a more natural output channel than keyboard entry, requires the investment of fewer resources. Hence, the effort associated with using this system is lower.

(5) Figure 9-2f depicts PRFs that represent the improvement in pilots' skills that occur during simulator training. Such data might be used to determine when a pilot should transition to the actual aircraft. On day 4 (curve A) the pilot's routine performance is good, but he would be incapable of handling an in-flight emergency without sacrificing flight control. On day 6 (curve B) flight performance has not improved further, but the pilot now has adequate resources available to handle the emergency. As in Figure 9-2b, flight performance is equivalent, although the degree of effort that must be invested to achieve that level of performance is different.

(6) Figure 9-2g depicts the performance of two potential air traffic controllers. After extensive training, Operator A has more total resources available to allocate to the task, as predicted from basic tests of intelligence (Wickens & Weingartner, 1985). Thus, Operator A will be able to outperform Operator B in both single and dual task performance when full resources are invested, as well as when some resources are diverted to a concurrent task.

The previous examples all represent situations actually encountered in the field of system design, operator training, and assessment. They illustrate the interrelationships among the concepts of workload, performance, and effort and suggest why system performance is not a concept that can be considered in isolation from operator workload. In the following section, how workload can be assessed in operational environments and the role of operator strategies in that assessment will be considered. Then, some methodological problems related to dissociations among measures will be considered. Finally, the issue of workload prediction will be addressed.

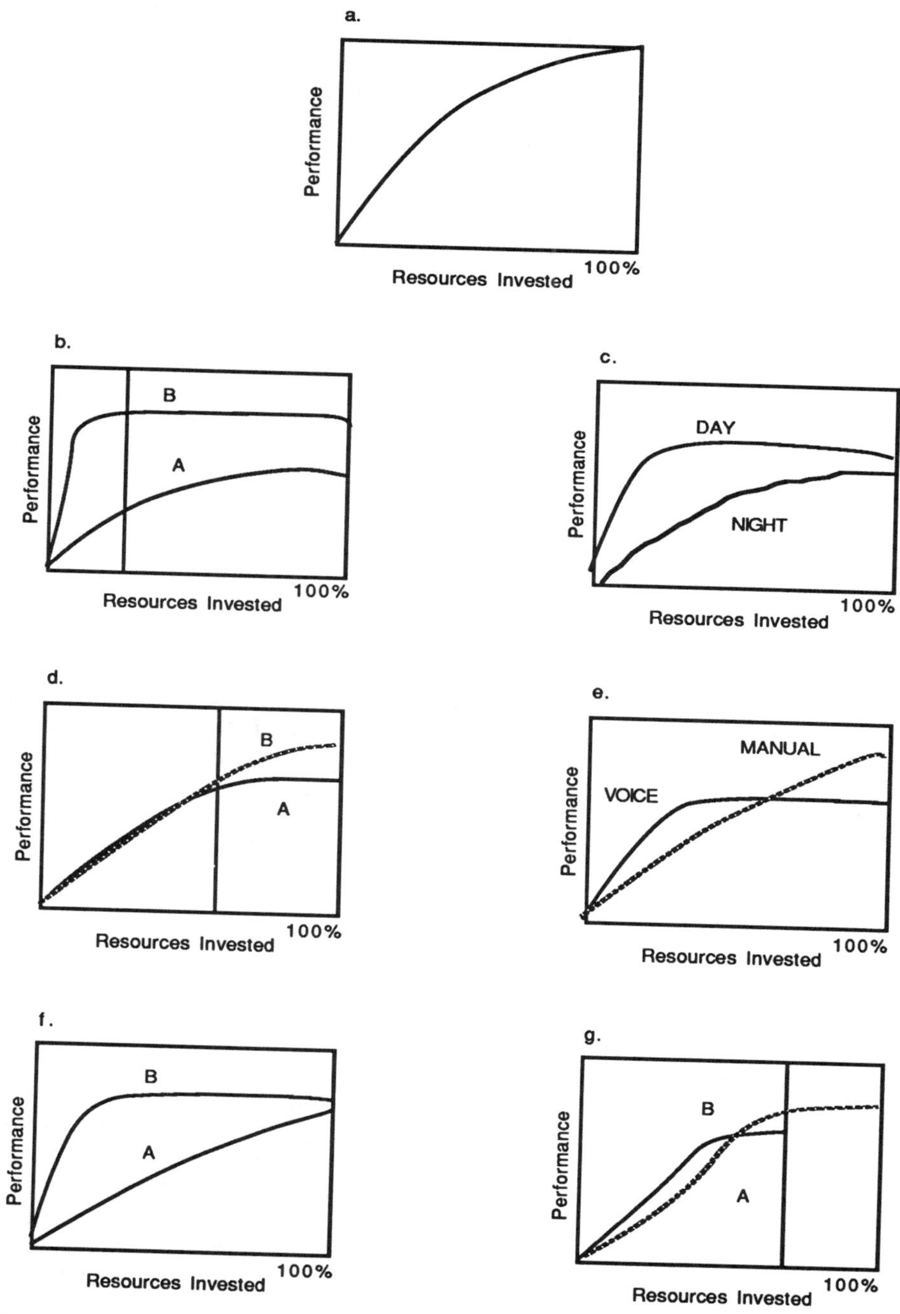

Figure 9-2
Different Examples of the Performance-Resource Function

HOW IS WORKLOAD ASSESSED?

Workload is clearly a complex phenomenon – a constantly changing, multidimensional target for analysis and assessment. The range of questions that might be asked and practical constraints further add to the complexity. However, a number of valid and practical measures have been developed that can quantify different aspects of the workload experienced by operators performing a wide variety of activities in very different environments. The measures, which will be described below, are distinguished by their objectivity, sensitivity, diagnosticity, and practicality. Since the type and quality of information provided by available measures differs, it is always a good practice to use multiple measures to develop a complete workload profile and to obtain converging evidence from different sources.

Numerous documents exist that provide detailed descriptions of available measures, the situations in which they have proven to be useful, and information about how to implement them. A limited list might include: Gopher and Donchin (1986), Hancock and Meshkati (1988), Hart (1986), Lysaght et al. (1989), Moray (1979), Moray (1988), O'Donnell and Eggemeier (1986), Roscoe (1987), Stassen, Johannsen, and Moray (1988).

Techniques used to assess operator workload generally fall into four categories: (1) ratings provided by operators or observers, (2) measures of primary task performance, (3) measures of performance on additional, "secondary" tasks introduced for the purpose of measuring residual attention or capacity, and (4) measures of covert behavior (e.g., changes in heart rate, eye blinks, eye movements, or electrical activity of the brain) which are generally referred to as "physiological" measures. Table 9-1 summarizes the categories and types of measures reviewed below.

Rating Scales

Ratings provided by task performers, observers, or experts are the most widely used measures of workload, and they often serve as the criteria against which other measures are judged. They provide an integrated summary of workload, from the perspective of the operator or an observer, and are the most direct method for evaluating the human cost of task performance. However, formally quantifying workload experiences using a structured rating scale is not a natural or commonplace activity, even though workload is experienced as a natural consequence of many daily activities. Thus, ratings might be qualitatively different than informal, spontaneous evaluations; different scales address only a subset of the factors that an individual might consider relevant and may not provide an appropriate range of alternative responses.

Table 9-1
Overview of Measures Described

Category	Examples
Rating Scales	Unidimensional Ratings Pilot Objective/Subjective Workload Assessment Technique (POSWAT) Hierarchical Ratings Cooper-Harper Handling Qualities Rating (HQR) Modified Cooper Harper (MCH) Bedford Scale Multidimensional Ratings Subjective Workload Assessment (SWAT) NASA-Task Load Index (NASA-TLX)
Primary Task Measures	Reaction Time Capture Time Tracking Error Control Variability Errors Communications
Secondary Task Measures	Reaction Time Monitoring Time Estimation Mental Arithmetic Memory Search Tracking Embedded Tasks
Physiological Measures	Heart Rate Pupil Diameter Measures of Eye Function Event-Related Potentials (ERP)

Rating scales generally consist of an ordered sequence of response categories. Labels on the scales define the correspondence between stimuli (workload experiences) and responses (rated levels). However, there is no direct relationship between values on any workload scale and specific, measurable phenomena in the physical world; "true" zero and the upper limit are undefined and intervals may not be psychologically equal. Thus, most scales provide ordinal information and index relative differences, rather than absolute levels.

Ratings may be obtained while the task is being performed, during intervals between task segments, or upon its conclusion. The former may interfere with task performance, while the latter may result in the loss of important information, unless supplemented by memory aids. The intervals of time evaluated range from several minutes to many hours, may include relatively homogeneous or diverse activities, and represent arbitrary intervals of time or meaningful units of activities. There is an obvious trade-off between the sensitivity and precision that may be achieved by obtaining ratings frequently and the possibility of task interference and rater "burn-out." Finally, since ratings are necessarily based on memory, they reflect only a subset of the information that was available during task performance (Ericsson & Simon, 1980). However, ratings obtained immediately after a task or after a considerable delay are highly correlated and their absolute values are similar.

Since workload is not completely defined by objective task demands, the experiences of different individuals faced with identical task requirements may be quite different. In addition, people may consider different variables when providing a rating (because their personal definitions of workload vary) and express different subjective biases (which may not be related directly to workload). This results in the primary drawback of subjective ratings: relatively high between-rater variability.

Although ratings are generally obtained from the person actually performing a task, observers, who can mentally project themselves into the situation experienced by the operator, can provide useful information about workload. Observer ratings may serve as the basis for operator debriefings and provide additional information that a busy operator might miss or forget to report. However, observer ratings are limited to the assessment of observable actions, task requirements, system performance, and environmental factors; it is difficult for observers, no matter how familiar they are with the task and environment, to infer the mental effort, stress, and psychological consequences of performing a task. This limits the information observer ratings can provide.

Several rating scales have been developed that have provided valuable information in a wide variety of applications. These scales may be grouped into three categories: (1) unidimensional; (2) hierarchical (Modified Cooper Harper [MCH], Bedford); and (3) multidimensional (National Aeronautics and Space Administration-Task Load Index [NASA-TLX], Subjective Workload Assessment Technique [SWAT]).

Unidimensional Ratings

Global ratings are easy to obtain (a single number can be given verbally during performance of almost any task) and provide a convenient summary value. However, unidimensional ratings provide no diagnostic information and between-rater variability is generally high (Hart & Staveland, 1988; Byers, Bittner, & Hill, 1989) – different raters base their evaluations on different aspects of the situation. There is no standard format for unidimensional ratings, although most require raters to provide numeric values (using scales that range from 1-7, 1-10, or 1-100), descriptive labels (e.g., very low, low, moderate, etc.), or magnitude estimates marked on scales presented on paper or a computer screen that are later converted to numeric values. A portable system, the Pilot Objective/Subjective Workload Assessment Technique (POSWAT) has been developed with which global workload ratings can be obtained and recorded in-flight (Mallery & Maresh, 1987) using an array of 10 labeled buttons.

Hierarchical Ratings

The Cooper-Harper Handling Qualities Rating (HQR) Scale (Cooper & Harper, 1969) was one of the earliest rating scales developed and is still used widely. Although it was developed to obtain subjective evaluations of aircraft handling qualities, subsequent applications have shown that it is also sensitive to many of the same factors that also influence workload. Raters make a series of decisions, each of which discriminates between two or three alternatives. Each decision leads the rater to another choice or to a final numeric rating ranging from 1-10. Raters may read the scale each time they provide a verbal or written rating or do so from memory. Several modified versions of the scale have been developed that retain the decision tree format and 10-point scale, but substitute terms that address workload more directly.

The Modified Cooper-Harper (MCH) Scale (Wierwille, Casali, Connor, & Rahimi, 1986) has been tested in the laboratory, simulated flight, and Army field tests of ground-based systems. Although it has been found to be sensitive to workload variations in simulated flight, it has proven to be less useful in other environments.

The Bedford Scale (Roscoe & Ellis, in press) was explicitly developed for use in flight. The wording of the scale is focused on spare capacity. It has been used widely in England and Europe to evaluate pilot workload in military and civilian aircraft and in simulation and flight research in the United States, but has received limited use in non-aviation environments.

The advantage of these and other decision tree formats is that they separate the evaluation process into a series of explicit decisions. These scales are easily implemented and scored and can be used without creating unacceptable interference in even demanding operational environments. However, they do not provide diagnostic information and have not received the extensive evaluation and application that other scales have received.

Multidimensional Ratings

Two of the most commonly used rating scales, SWAT and NASA-TLX, involve a procedure whereby ratings on several subscales are combined to derive a summary score. These scales are based on the assumption that people can evaluate component factors more reliably than they can the global concept. One of their greatest strengths is that they provide diagnostic information about the specific source of a workload problem, as well as a global summary.

SWAT (Reid, 1985; Reid & Colle, 1988; Reid & Nygren, 1988) consists of three subscales: time load, mental effort load, and psychological stress load. Each dimension is represented by a three-point rating scale (1=low, 2=medium, and 3=high). The 27 possible combinations of three levels of each of three scales are presented on individual cards for subjects to rank from lowest to highest workload prior to an experiment. These rankings are used to create a 100-point, unidimensional scale that has interval properties. Each combination of ratings on the three subscales has a unique position on this scale which is assigned a numerical global workload value. The sorting process, although time-consuming, is valuable as it provides an interval scale and considers individual differences in workload definition. The limited number of dimensions and interval scale values make SWAT attractive for use in operational settings; however, this limited range of allowable ratings (e.g., low, medium, and high) also reduces the sensitivity of the scale. The primary drawbacks are relatively high between-rater variability and low sensitivity in situations where overall workload levels are generally low (see, for example, Reid, 1985). SWAT has been used successfully to evaluate workload in the laboratory, simulation and flight research, and various ground-based operations. Thus, there is accumulated evidence that SWAT provides valid results and can be implemented in most environments.

NASA-TLX provides an estimate of overall workload based upon a weighted average of six subscale ratings: mental demand, physical demand, temporal demand, own performance, effort, and frustration (Hart & Staveland, 1988). Subscale ratings, which range from 1-100 in 5-point increments, are given verbally or by selecting a position along a scale presented on a rating form or computer screen. In addition, raters quantify the relative importance of each factor in creating the workload they experienced. These values, which range from 0 to 5, are used to weight the magnitude ratings when computing the overall workload score. Diagnostic information is provided by variations in subscale ratings as well as the weight given to each factor. NASA-TLX has been used successfully in environments ranging from the laboratory to military and civilian helicopter, general aviation, transport, and military jet simulators and aircraft, and in ground-based systems. Between-rater variability is consistently lower than typical of other rating scales, significant correlations with other measures of workload and performance are generally found, and subtle as well as gross workload differences have been discriminated throughout the workload range.

Summary: Rating Scales

A 10-point unidimensional scale, SWAT, NASA-TLX, and MCH were compared in a series of field tests conducted for two Army air defense systems: (1) the Line of Sight Forward (Heavy) (LOS-F-H) Forward Area Air Defense System (FAADS) (Hill, Zaklad, Bittner, Byers, & Christ, 1988) and the Pedestal Mounted Stinger (Byers, 1989) and for the Aquila Remotely Piloted Vehicle (Byers, Bittner, Hill, Zaklad, & Christ, 1988). Ratings were obtained, for at least some of the scales, either: (1) prospectively – raters familiar with the basic system were asked to evaluate the *potential* impact of a hardware modification, different tactical situations, or a reorganization of crew functioning (Hill, Byers, Zaklad, Bittner, & Christ, 1989); (2) during or soon after the field tests; or (3) retrospectively – many months after initial field experience, operators and subject matter experts rated the "generic" workload associated with mission conditions, task segments, and operator position (Bittner, Byers, Hill, Zaklad, & Christ, 1989). Generally, high correlations were found among the four scales, suggesting that they all reflect the same "overall workload factor." Many of the scales were able to distinguish meaningful workload differences among mission segments, variations in task demands, and crew position. However, NASA-TLX ratings were more closely related to objective measures of performance (good performance was generally associated with low workload), they had the lowest between-rater variability and the highest overall workload factor validity across tests, and the best user acceptance (see, for example, Byers et al., 1988). Although NASA-TLX was considered to provide the most accurate description of workload and was recommended for use in field evaluations of Army systems, the unidimensional ratings, which consistently out-performed SWAT and MCH, were also recommended for use in screening gross levels of workload, in preparation for a more diagnostic evaluation of workload problem areas (Hill et al., 1988). Although the unidimensional scale did not provide the diagnostic information given by the NASA-TLX subscales, it took considerably less time and effort to use than did NASA-TLX.

Rating scales are the most practical and generally applicable measure of workload. They are easy to implement and score, appropriate in almost any environment, acceptable to most operators, and have a certain amount of face validity. The rank ordering of ratings is quite stable across raters, although the absolute values of the ratings exhibit relatively high variability. In addition, subscale ratings can provide diagnostic information. However, ratings can represent no more that the rater's memory of what was experienced, integrated across time. Thus, most ratings are insensitive to momentary variations in workload and they are subject to rater biases. Although improvement in the psychometric properties of these scales is warranted, their practical utility and the wealth of information they provide outweigh their drawbacks.

Primary Task Measures

Although improving performance may not be the only motivation for measuring workload, measures of performance are an essential component of any workload analysis; it is difficult to interpret workload measures without knowing the level of performance the operator was able to achieve. In addition, some performance measures can provide objective answers to workload questions. Since ensuring adequate performance is the motivation behind most applied workload analyses, it seems reasonable to consider performance measures first when selecting a workload metric. However, as many of the examples in Figure 9-2 illustrate, these measures reflect what the man-machine system was able to accomplish, rather than the cost of doing so for the human operators, and they are also influenced by factors other than workload (e.g., system response time). In addition, accurate measures of performance may not be available in field evaluations. For highly automated systems, system performance depends on the operator's inputs at a very high level; moment to moment variations depend entirely on the performance of the automated system, while operator inputs occur infrequently.

Types of Primary Task Measures

There are three classes of performance measures: (1) Accuracy (number of correct responses, control error compared to a target value), (2) Speed (response time measured in seconds or fractions of seconds), and (3) Number (how many tasks or task elements were completed correctly within an interval of time). Some measures summarize the effectiveness of the operator's activities, while others also provide information about the fine-grained structure of the operator's control strategies. The former reflect the combined output of the operator's behavior and system output, while the latter measure operator effort (and, thus, workload) more directly.

Limitations of Primary Task Measures

For the purpose of workload assessment, the usual assumption is that decrements in performance indicate higher workload; more errors, longer response times, higher control error, and fewer tasks completed are taken as evidence of increased workload. However, the actual relationships between workload and performance are more complex. O'Donnell and Eggemeier (1986) suggested that: (1) for relatively easy tasks, consistent performance is maintained over a range of difficulty levels (although workload increases), (2) for moderately difficult tasks, performance deteriorates linearly as task demands increase (and workload increases), and

(3) for very difficult tasks, operators may ignore some tasks and maintain a consistent level of effort on others, even in the face of increasing demands, so that performance deteriorates, but workload does not increase further. In general, measures of performance appear to correlate with task demands and workload for moderately difficult tasks only. For very easy or very difficult tasks, measures of workload and performance dissociate.

Numerous models of attention and performance have been proposed to explain the relationships found between task demands, workload, and performance (e.g., O'Donnell & Eggemeier, 1986; Vidulich, 1988; Wickens & Yeh, 1988), as suggested by the examples in the beginning of the chapter. The basic assumption is that the resources required to perform tasks are available in finite amounts. As difficulty is increased, more resources are required to maintain consistent performance. If available resources are sufficient, performance will remain constant. If they are not, performance will degrade on a single task or on one or more concurrent tasks. Since some tasks have been found to interfere with each other more than others, Wickens (1980) and Wickens and Liu (1988) suggested that the performance of concurrent tasks depends on the pattern of requirements that the two tasks share in common. Thus, performance decrements on one task may occur in the presence of some concurrent tasks, but not others, due to the pattern of resource competition between the two tasks. However, it is difficult to assess workload from the measures of performance that are obtained. If two tasks do not compete for the same resources, or if sufficient resources are available to perform both, performance is maintained but workload is increased. If two tasks do compete for the same, insufficient resources, performance on one or both tasks will degrade, but workload may not be affected.

Performance is also influenced by task schedule; almost any task can be performed well if unlimited time is available and even easy tasks may become impossible to perform if the time available is reduced below a critical point. Thus, measures of performance will reflect workload only in the region where sufficient, but not unlimited, time is available for task performance; as time pressure is increased within this region, measures of performance will show a decrement even though additional effort is exerted.

In both theoretical and applied research, it is always assumed that operators attempt to respond immediately and perfectly to task demands. If they do not, then these measures lose their meaning. Even for simple tasks, an operator may choose to emphasize speed at the expense of accuracy, or vice versa. This may result in two competing estimates of his performance.

In most operational tasks, there is considerable flexibility in when and how task requirements are accomplished. Operators may try to maintain acceptable performance on each of two concurrent tasks (by time sharing or rapidly switching between them), emphasize one task at the expense of the other, or perform the tasks sequentially. In fact, delaying the performance of one (less important) task in favor of another (more important) task might well represent an optimal strategy. In addition, most tasks do not have be performed perfectly. Rather, a vehicle must be kept within certain boundaries (not perfectly lined up) and many discrete tasks can be performed at the operator's discretion, as long as they are completed by a deadline. Thus, operators' task performance strategies determine when and how they focus their attention on individual tasks. This, in turn, determines which measures of performance are the most appropriate for analysis and the magnitudes of performance decrements that will be recorded for specific task components. Strategies adopted to minimize workload might result in poor performance on one element of a complex task, but, if the strategies are part of a more global workload-management strategy, they can result in better overall performance. Alternatively, very good performance, often thought to index low workload, might also reflect the exertion of extreme effort and very high workload. "Extreme effort" might be defined as a level of exertion that can be sustained for only a short interval of time.

Primary Task Measures for Simple Tasks

Despite these problems, several types of performance measures seem to covary with other measures of workload sufficiently reliably that they might be considered. For example, in laboratory research, consistent relationships have been found between an increase in reaction time (due, for example, to an increase in the number of remembered items, alternative choices, or arithmetic operations) and other workload measures (e.g., Mosier & Hart, 1985). For target acquisition tasks, a reliable relationship has been found between increased capture time (as target size decreases or its distance increases) and subjective ratings (e.g., Hart, Shively, Vidulich, & Miller, 1986). For tracking tasks, including simulated flight, increased tracking error (associated with an increase in the difficulty of the forcing function, the number of controlled axes, or the order of control) generally correlates with an increase in subjective ratings and decrements in secondary task performance (e.g., Bortolussi, Hart, & Shively, 1989; Kramer, Sirevaag, & Braune, 1987; Vidulich & Wickens, 1986).

Primary Task Measures for Complex Tasks

Interpretation of performance measures is quite complicated when many measures are available; different measures taken concurrently might suggest very different levels of workload. Since the units of measurement, frequency of occurrence, and priority vary across task components, it is not always easy to combine multiple performance measures into a summary figure of merit for performance which might, in fact, reflect workload. Two approaches to such combinations are possible: (1) The use of multivariate analyses (i.e., principle components analysis) that reveal the most important variables in discriminating among levels of workload, and (2) A linear combination of all of the measures of primary task performance that the investigator feels are important, such that measures of "good" or "bad" performance, respectively, are consistently given the same sign. Although, in theory, all measures that are recordable could be combined, there are some important qualifications to using some measures. For example: (1) Errors generally provide a poor index of workload; slips and blunders occur as often in low workload situations as high (Morris & Rouse, 1988). However, the activities required to resolve an error can contribute to a subsequent increase in workload; (2) Operators must devote at least some attention to a task for its performance to reflect overall workload; (3) The measure must be recorded with adequate frequency and accuracy; and (4) The operators' inputs must not be masked or delayed by the system. Figures 9-2d and 9-2e depict explicit examples of situations where higher workload is associated with better, rather than worse primary task performance in a complex system.

Some elements of complex primary tasks do appear to provide consistent and reliable indices of workload. For example, operators' abilities to estimate the passage of time generally degrades as overall workload (as indexed by other measures of workload) is increased in laboratory, simulation, and flight research. Altitude control variability often increases in simulation and flight experiments as other task demands are increased. Communications are often delayed and abbreviated as overall task demands increase.

Summary: Primary Task Measures

While measures of performance on the task of interest can be used to estimate momentary variations in workload and the degree to which a specific level of effort can be sustained, the problems outlined above suggest caution in their use. In fact, Gopher and Donchin (1986) concluded that direct measures of task performance are usually poor indicators of mental workload because they do not reflect the resource investments prompted by changes in task demands and do not diagnose the source of load. Furthermore, a review of the field recently conducted for the Army Research Institute (Lysaght et al., 1989) concluded that primary task measures "should not be generally treated as appropriate for assessments of overall workload." Taken together, these recommendations provide a strong note of caution about using primary task measures to evaluate workload, although they must always be obtained to determine whether or not the operator was able to accomplish the task; other measures of workload are virtually impossible to interpret without such information. Thus, while primary task performance measures can evaluate the "bottom line" – Can the task be done? – they may not reflect the workload that an operator experienced in achieving that level of performance. As task demands are increased, operators are often able maintain a consistent level of performance (which would suggest no difference in workload), even though the workload "cost" of doing so may be greater (if they responded to the demands of the task by exerting additional effort) or less (if they responded by adopting a more efficient task-performance strategy). Conversely, performance decrements accompanying an increase in task demands may either reflect a constant or reduced level of effort (i.e., the operator did not choose to devote more resources to the task or adopt a more efficient strategy) or the upper limit of his capabilities (i.e., the operator did not have any additional resources available and a more efficient strategy was not possible).

Secondary Task Measures

Because measures of performance on the task of interest may not provide an adequate estimate of workload (if task demands remain within the operator's capabilities, more effort may be exerted to maintain consistent performance), yet objective measures are desirable, the use of "secondary" tasks was developed as an alternative approach. Operators are instructed to maintain performance on the primary task and use their "reserve capacity" to perform the additional tasks. The secondary tasks impose a sufficient additional load so as to exceed the operator's capabilities. The level of performance on these tasks is used as an indirect measure of the resources demanded by the primary task; secondary task performance degrades as primary task demands increase. Thus, secondary tasks can provide a useful measure at the low end of the workload continuum, where primary task performance measures are insensitive.

Attractive as the concept seems to be, in theory, secondary tasks are not without some problems. For example, if primary task demands are fairly high, particularly in an operational environment, the operator may simply abandon the secondary task in order to maintain acceptable performance on the primary task, making the secondary task measure ineffective. Initially, it was thought that secondary task "yardsticks" could be developed and used to compare the workload of a variety of primary tasks (Ogden, Levine, & Eisner, 1979). However, the data suggest that secondary tasks are selectively sensitive, depending on the pattern of requirements they share with the primary task, again suggesting the existence of multiple resources (Wickens, 1980, 1984). Each task requires a specific pattern of resources. Concurrent tasks that require similar (insufficient) resources will interfere with each other, while those that require different resources will not. Because interference results in performance decrements on one or both tasks, the assumption that "secondary" tasks are always performed after "primary" task requirements have been met is not always supported. Another possible problem is that most secondary tasks occur at discrete intervals. Thus, their performance reflects workload at the times they were introduced, rather than the workload throughout a period of time. If secondary task presentations are time-locked to significant events in the primary task, however, they can provide more precise information than can a more global measure (Kantowitz, Bortolussi, & Hart, 1987). If operators modify their primary-task performance strategy when a secondary task is present, the information that a secondary task provides about primary task workload becomes ambiguous.

A careful task analysis is particularly important in using secondary tasks; tasks must be selected that require the same resources necessary for performance of the primary task. Since every task requires several types of resources, performance on several tasks that require different patterns of resources can provide converging information about primary task workload. In addition, using secondary tasks that have graded levels of difficulty can provide a more accurate estimate of primary task workload.

Most secondary tasks are relatively simple activities for which the input (visual, auditory) and output (verbal, manual) can be presented precisely and measured directly and accurately. The intervening cognitive processes are predicted by psychological models and inferred from variations in the speed and accuracy of performance. Most tasks were originally designed for purposes other than workload assessment (e.g., to test theories of human performance, memory, and attention), however, others represent simplified versions of "real-world" tasks. The following describes several of the secondary tasks which have received the widest application:

Reaction Time

Reaction time tasks generally include a visual or auditory stimulus presented during performance of a primary task. The operator responds by pressing a button or making a verbal response. Multiple-choice tasks are more sensitive than single-choice tasks, as they impose some information processing and response selection load. Increased response time and errors index an increase in primary task workload (e.g., Bortolussi et al., 1989; Kantowitz et al., 1987). Depending on the modality of input (visual/auditory) and output (verbal/manual) selected, this task can be designed so as to require the same resources also required by a specific primary task.

Monitoring

Monitoring tasks generally require the operator to pay consistent attention to one sensory modality (visual/auditory) and respond when a particular event occurs (verbally/manually) or maintain a running count (which adds a memory requirement). Increased response time, misses, and false alarms index an increase in primary task workload (e.g., Kramer, Wickens, & Donchin, 1983). Again, this task can be implemented so as to compete for the resources also required by the primary task.

Time Estimation

Timing tasks require the subject to produce a specific interval of time (usually in the range of 1-20 seconds) by manually pressing a button or verbally indicating its beginning and end. As attention is drawn away from timekeeping, the length of produced durations increases (e.g., Hart, McPherson, & Loomis, 1978). Although the demands of this task are almost entirely on central processes, it has been found to be sensitive to variations in perceptual/motor load as well.

Mental Arithmetic

Mental arithmetic tasks also require primarily cognitive resources. The difficulty of the task can be manipulated by increasing the number of digits or the number of operations that must be performed (e.g., Roscoe & Kraus, 1971). Performance is measured by response latency and the accuracy of verbal, written, or typed responses.

Memory Search

Memory search tasks require the operator to remember one or more letters, number, or words (the memory set) and then to respond whether subsequent stimuli (probes) were members of the memory set or not. As memory set size increases, or as concurrent primary task demands increase, response time generally increases (e.g., Wickens, Hyman, Dellinger, Taylor, & Meador, 1986). This task imposes loading on short-term memory, as well as perceptual and response resources.

Tracking

Tracking tasks can provide a continuous index of primary task workload, although they are often impractical in operational environments. Different difficulty manipulations (e.g., forcing function amplitude and bandwidth, order of control, and number of axes controlled) create different patterns of interference with concurrent primary tasks (e.g., Jex & Clement, 1979). In general, error about a target value increases as the difficulty of a concurrent task is increased.

Embedded Tasks

One problem with secondary tasks in operational environments is that they tend to be considered unimportant and uninteresting by the operators and their performance may be terminated altogether. A solution to this problem is to present embedded secondary tasks which are designed to appear as a natural and integral part of the task of interest. This method of presenting a secondary task substantially improves user acceptance (in operational situations) and increases the likelihood that operators will attempt to perform them. Many of the tasks described above can be modified so as to integrate them with an operational task. For example, a memory search task can be designed as a response to a radio call sign. The time at which a particular activity is performed can be treated as a time production. Discrete responses to alternative display configurations can be evaluated as reaction time or monitoring tasks, and so on. In addition, there are many components of complex tasks whose performance reflects overall workload levels, such as communications, which can be singled out for analysis.

Summary: Secondary Task Measures

Performance on a concurrent, additional task can index the reserve capacity remaining after primary task performance and provide diagnostic information about the specific resources required. However, secondary task performance may interfere with or change primary task performance. Thus, this measure may be inappropriate for many operational situations unless presented as embedded elements of the primary task. In addition, not all secondary tasks are equally sensitive to the specific types of demands imposed by different activities. A lack of secondary task decrement can never be interpreted with certainty as evidence of low workload; it might just indicate that the resources required by that secondary task were not depleted by the performance of that primary task. Thus, the use of several secondary tasks, each of which demands different combinations of resources, will provide a more accurate assessment of the sources and magnitudes of primary task workload. A thorough task analysis and a strong theoretical basis is necessary to select and implement secondary tasks successfully.

Physiological Measures

Measurable, involuntary changes in such measures as heart rate, eye blink rate, pupil diameter, respiration, blood pressure, electrical activity of the brain, and so on may accompany variations in the physical and mental demands of a task (Hancock, Meshkati, & Robertson, 1985; Wierwille et al., 1986). In addition, since emotional stress may accompany an increase in workload, measures of arousal provide an indirect indicator of workload. Thus, many physiological measures have the potential for providing an objective and unobtrusive indication of operators' responses to the demands placed on them. In addition, these measures rarely require an overt response, and, thus, do not interfere with task performance. Finally, many physiological indicators can be monitored and measured continuously, thereby providing a fine-grained analysis of subtle, momentary changes in workload.

There are two classes of physiological measures: (1) measures of emotional and physical activation (e.g., heart rate, pupil size), and (2) measures of mental and perceptual processing (e.g., event related potentials and direction of gaze). Measures of voluntary actions (such as hand or leg movement) are not generally considered to be "physiological" measures.

Although the selection of a particular physiological measure to index workload is rarely based on sound theoretical considerations, most of the measures have a sound basis in physiology. The general expectation is that heart rate will increase, heart rate variability will decrease, pupil diameter will increase, the pattern of eye blinks and eye movements will change, and the amplitude of event-related potentials (in response to a secondary task probe) will decrease as workload is increased. However, different measures are specifically sensitive to different types of workload. For example, heart rate is particularly responsive to stress and physical effort, while other measures of cardiovascular functioning and event-related potentials reflect mental effort. The following briefly describes a few of the physiological measures that seem most useful:

Heart Rate

Measures of the operator's heart rate provides an integrated index of the overall effect of task demands and the operator's emotional response to them. Heart rate is particularly sensitive in situations where stress and responsibility play an important role. For example, heart rate typically increases during takeoff and landing in simulation and flight research for the pilot responsible for controlling the aircraft, but does not for other pilots in the cockpit (e.g., Hart & Hauser, 1987). Heart rate is generally less useful in low-workload situations and non-operational environments. The variability of beat-to-beat intervals also indexes workload; normal heart rate irregularities are suppressed as task demands, particularly cognitive demands, are increased (e.g., Derrick, 1988; Vicente, Thornton, & Moray, 1987). Although this measure can be computed without any additional cost, it has not proven to be as reliable a measure as is heart rate. Heart rate is easy to record in almost any environment using portable devices.

Pupil Diameter

Variations in the diameter of the operator's pupil provides an accurate index of cognitive load (e.g., Beatty, 1982). However, this measure is quite difficult to obtain reliably in operational environments, due to its sensitivity to ambient illumination. In general, the finding has been that pupil diameter increases as cognitive workload is increased.

Measures of Eye Function

Most modern systems present a substantial amount of information visually and operators may control the flow of this information consciously or unconsciously through blinking, redirecting their gaze, or changing their focus. Thus, measures of these functions can provide useful information about visual and cognitive workload. The timing, frequency, and duration of eyeblinks have been found to reflect workload-related phenomena (e.g., Stern & Skelly, 1984); blinking is inhibited until information is obtained and tends to occur after decisions are made. Information about where the operator is looking, and for how long, and transition patterns among displays provide valuable information about task-performance strategies (e.g., Harris, Glover, & Spady, 1986); operator's abilities to monitor an added visual display can indicate the visual monitoring and processing, resources required by the primary task, fixation durations may decrease, and scan patterns may change under high workload. A modification of this technique, useful in Army combat or vehicle control environments, is to calibrate workload by the reduction in peripheral scanning or head movements that result when central load increases. This may be estimated by head movements in conjunction with peripheral eye movements. The primary drawbacks of measures of eye function are the difficulty of implementing them in operational environments and interpreting the data (e.g., people do not always "see" what they are looking at and can process information from the periphery).

Event-Related Potentials (ERP)

ERPs have been proposed as an objective index of mental workload because of the obvious connection between the brain and behavior (Gopher & Donchin, 1986; Donchin, Kramer, & Wickens, 1986). The occurrence of sensory events is related to predictable patterns of electrical activity in specific parts of the brain that can be measured by electrodes placed on the scalp. The general practice is to average the data recorded from several presentations of similar stimuli so that the ERP "signal" can be detected in the "noise" of ongoing brain activity; this is possible because the latencies and amplitudes of ERP waveforms are similar for the same types of events. Generally, cognitive workload is reflected in the amplitude of a positive component of the ERP waveform that occurs between 300-500 milliseconds after presentation of a stimulus; ERP amplitudes increase as mental effort increases and are reduced if cognitive resources are also demanded by other concurrent tasks. Although relatively difficult to implement in an operational environment, Kramer et al. (1987) found reliable relationships among event-related potential amplitudes, performance on a primary, simulated flight task, secondary task performance, and subjective ratings. Futhermore, Kramer et al. (1983) used the ERP to index workload changes as training progressed in a complex task. This measure provides a direct index of information processing activities, and, because it is time-locked to a specific event, it can provide very precise information about the workload at that time. It cannot, however, reflect continuous workload variations and is more expensive and difficult to implement and analyze than heart rate or rating measures. Finally, event-related potentials are subject to artifacts (e.g., eye movements or blinks) which are themselves related to workload and require special filtering to remove.

Summary: Physiological Measures

Some physiological measures can index the physical and emotional responses that accompany an increase in workload, while others index cognitive load. Many vary with sufficient frequency to be sensitive to momentary shifts in workload. Most can be obtained with minimal primary task interference. They provide "objective" information that is quantified in physical units. Despite these advantages, however, their use in operational environments has met with only limited success, and, with the possible exception of heart rate, no single measure covaries with other measures of workload or performance consistently enough to be recommended as the only measure of workload. As part of a battery of measures, however, they can provide useful, unobtrusive information that may be absent from subjective ratings or measures of primary or secondary task performance.

HOW ARE MEASURES SELECTED AND IMPLEMENTED?

Several expert systems have been developed to aid in selecting and applying available assessment techniques. The Workload Consultant for Field Evaluation (WC FIELDE) was developed at NASA (Casper, Shively, &

Hart, 1987) to serve as an aid in formulating the research question, assessing the research environment, evaluating the relevant task dimensions, and selecting the most appropriate measures. This system includes an extensive data base that describes the measures, evaluates their strengths and weaknesses, summarizes previous applications, and suggests how to implement them. A second system was developed by Analytics, Inc., under contract to the Army Research Institute for use during all stages of system specification, design, test and evaluation, and operation for military system acquisitions. The Operator WorkLoad KNowledge-based Expert System Tool (OWLKNEST) includes predictive techniques as well as assessment techniques in its data base (Harris, Hill, & Lysaght, 1989). It provides a brief summary and evaluation of each measure and predictive technique.

In selecting the measure(s) that will be used, the following should be considered: (1) the focus of the research question, (2) the grain of analysis that is required, (3) the probable level and sources of workload, (4) the importance of diagnostic information, and (5) practical constraints imposed by the research environment. Different measures are particularly sensitive to specific workload dimensions and magnitudes, and they may either summarize the overall effects of task performance on the operator or provide diagnostic information to isolate the causes or consequences of a specific problem. Depending on which measure is selected and how it is implemented, the data may provide a fine-grained analysis of the time-varying characteristics of the operator's behavior and experience or an integrated summary across sources of workload and time. Finally, some measures can be obtained in almost any environment with minimal cost and interference, while others are difficult to obtain in the field or require expensive equipment and special training. In practice, there is usually a trade-off between the time and effort devoted to workload assessment and the quality and precision of the information that is obtained.

Formulating the Question

The potential causes (e.g., mission requirements, equipment design, environment) and consequences (e.g., performance decrements, physical and emotional stress) of inappropriate levels of operator workload encompass a broad range of factors. Thus, a formal statement of the research question will ensure that the most appropriate measures are selected and that the obtained evidence does, in fact, answer the original question. The following represent a sample of the questions that might motivate a workload analysis: (1) Are sustained or momentary workload levels consistent with acceptable system performance? Operator's emotional and physical well being? (2) Could crewmembers perform additional duties? (3) Should crew duties be reassigned to distribute workload more evenly? (4) Could the crew complement be reduced? (5) Will adequate performance be possible under degraded weather? Reduced visibility? Battlefield conditions? If system components fail? (6) Will the addition of automated subsystems reduce the workload of existing crewmembers? Allow a reduced crew complement? or (7) What is the best design alternative? Visual/auditory displays? Manual/vocal controls? Multifunction/integrated/separate display formats? Different measures are particularly sensitive to different aspects of workload and are, thus, better able to answer different workload-related questions:

Sustained vs. Momentary Workload

If the goal of a workload analysis is to estimate average or total workload associated with performing a particular task, then any form of rating scale might be appropriate, as would overall performance on the most salient measures, or average heart rate. If a more fine-grained analysis of momentary variations in workload are required, then subjective ratings would be inappropriate, as would global performance measures. However, momentary variations in heart rate or performance on a continuous control task, secondary tasks (and associated ERPs) timed to coincide with specific periods of interest, and the pattern of eye movements and eye blinks might be appropriate.

Reserve Capacity

If the research question seeks to identify operators' potential abilities to perform a task for a longer duration or under more extreme environmental conditions or to assume additional responsibilities, then measures of primary task performance will provide little information. Prospective subjective ratings might be used to index the likely workload under such hypothetical conditions. Secondary tasks would provide the most sensitive measure; the level of performance operators are able to achieve on one or more secondary tasks (by maintaining a constant level of effort and performance on the primary task and by employing all of their available resources) can index the resources required by the primary task, and thus by inference, the amounts and types of resources still available to perform additional activities. ERPs associated with the presentation of secondary task stimuli would provide additional useful information.

Emotional and Physical Consequences

Heart rate and multidimensional subjective ratings (such as SWAT and NASA-TLX) provide the most sensitive indices of the stress often associated with inappropriate levels of workload. In addition, heart rate and NASA-TLX also provide an indication of the physical demands placed on an operator. Primary and secondary task performance and ERPs would provide little information.

Specific Source of a Workload Problem

Multidimensional rating scales, such as SWAT and NASA-TLX, and a battery of secondary tasks are the best approaches for diagnosing the specific cause of a workload problem. Depending on the method chosen, the user might determine that excessive time pressure, heavy visual or manual demands, excessive mental effort, or emotional stress was the primary cause of an overall increase in workload or a performance decrement. In addition, the pattern of primary task performance decrements might also provide diagnostic information.

Comparison Among Design Alternatives

Any subjective rating scale can provide information about the workload associated with alternative designs, however multidimensional ratings scales will provide more diagnostic information. Secondary tasks designed so as to require similar resources to those also required in using the design alternative would prove useful information as well. Primary task measures of performance with the alternative systems would be essential. Patterns of eye movements and visual fixation durations might be particularly useful in evaluating alternative visual displays. ERPs and response latency might be used to index how quickly operators were able to notice or extract relevant information.

Research Environment

Although studies performed in the actual system, or an adequate simulation of it, provide the most accurate assessments of workload, it is possible to obtain useful information through research conducted in similar systems, part-task simulators, or the laboratory using more abstract representations of the activities that will be performed when it is not feasible to conduct the study using the actual system. In fact, it is often possible to obtain more precise information by evaluating the impact of specific elements of a task in isolation. However, it will be difficult to assess the influence of environmental factors in any other environment than the real thing. Different measures will be more or less appropriate depending on the constraints of the environment in which an assessment is performed.

Laboratory

In the laboratory environment, it is feasible to implement any of the measures described above. Here, the decision of which measure to use might be based on the equipment and funds available, the techniques with which the experimenter has expertise, the design of the experimental tasks, and the specific research question being addressed. Generally, both primary and secondary task measures can be obtained with great precision and ratings can be timed so as not to interfere with task performance. Although ERPs have been used successfully in this environment, heart rate is generally insensitive.

Simulation

In simulation research, large quantities of primary task performance data are generally available. Here, the problem is one of selecting which measures are relevant and summarizing and integrating the information provided by multiple measures. Subjective ratings are generally appropriate, as are most physiological techniques. Secondary tasks may be somewhat more difficult to implement, depending on the flexibility of the simulation hardware and software and the creativity of the experimenter. As the complexity of the activities being simulated increases, correlating the presentation of secondary tasks, ERPs or patterns of eye movements to the operator's activities can become quite difficult.

Field Tests

In the operational environment, accurate measures of performance are often not available, and it may be difficult to implement secondary task measures. However, attempting to obtain such measures would be worth the effort. Particularly in field tests, presenting secondary tasks as embedded elements of the overall task is particularly important. In this environment, subjective ratings are the most commonly used measure, although the experimenter must exercise care that the rating procedure does not interfere with task performance, thereby compromising safety. Heart rate measures are generally useful in an operational situation and can be obtained unobtrusively and with little difficulty using portable recording devices. ERPs, eye movements, eye blinks, and pupil diameter measures are generally not appropriate, due to practical constraints.

Estimating Imposed Demands

A preliminary analysis of objective task requirements, the equipment that will be used, and the environment in which the task will be performed is necessary to select the most appropriate measures. This analysis, which might be quite informal, defines: (1) the magnitude of task demands, and (2) the most salient workload dimensions. In addition, this analysis can aid in identifying the performance levels operators will be expected to achieve and practical opportunities for introducing workload measures.

Magnitude of Task Demands

Since different measures are particularly effective when overall workload is generally low, moderate, or high, a preliminary analysis will increase the probability that measures will be selected and applied effectively. *A priori* assessments of imposed task demands generally catalog the number of activities that an operator will be expected to accomplish and the precision with which they must be performed. Task demands can be approximated by computing the information that must be processed, the number of decisions that must be made, the number and frequency of responses, the difficulty of mental calculations and transformations, and the predictability of events. The difficulty of almost any task can be altered by reducing the time available for performance; workload can be approximated by comparing the time required to complete subtask components to the time available. Task scheduling also influences workload; subtasks performed in sequence may impose relatively low workload, whereas unacceptably high workload may occur if they must be performed concurrently. The workload associated with the scheduling of task components can be approximated by evaluating the complexity of schedules operators must remember or develop, the frequency with which multiple tasks must be performed simultaneously, and the degree to which concurrent tasks require similar or different resources.

In general, secondary tasks are most appropriate when primary task demands are low to medium. Primary task measures are most appropriate when task demands are moderate. Subjective ratings can be used across the range of task difficulties. Heart rate is more effective for relatively demanding tasks, particularly when stress or responsibility are involved. Eye movement data and ERPs may be difficult to interpret for very complex tasks.

Sources of Task Demands

The system resources available (e.g., controls, displays, automatic subsystems, other crewmembers, support personnel) define (or limit) the task-performance strategies available to an operator. The workload impact of these factors can be approximated by assessing the types of information that will be available, the number of different information sources that must be monitored and integrated, the location and handling characteristics of controls, and the availability of automatic subsystems.

Analyzing the human resources that will be required to perform a task is particularly important in designing and implementing secondary tasks. To be most effective, secondary tasks must require the same resources that are required to perform the primary task. In addition, a thorough analysis of the primary task might suggest opportunities for introducing embedded secondary tasks. In highly automated systems, primary task measures will provide little information about workload. However, primary task measures may prove an excellent source of information when the operator is required to make frequent, measurable inputs. Multidimensional rating scales are most appropriate when assessing tasks with many, overlapping sources of workload; subscale ratings can identify variations in the composition of workload over time.

Implementing Workload Measures

The quality of information provided by a workload measure depends entirely on the quality of the study in which it was obtained. Although experimental control becomes more difficult as the research environment moves from the laboratory to simulation and field test, the additional effort required to conduct a "clean" experiment is always worth its cost.

Training

Some techniques can be administered adequately by relatively inexperienced technicians (e.g., rating scales) while others require substantial sophistication and experience (e.g., ERPs). Thus, both experimenters and subjects must be familiarized with the assessment procedures before experimental data are collected. For example, the terms used in rating scales must be defined clearly so that raters will adopt the definitions and criteria that the experimenter intended. Operators must receive advance training with secondary tasks to ensure they have reached asymptotic performance; if they have not it will be difficult to determine whether variations in secondary task performance reflect primary task workload or increased expertise with the secondary task itself. Because operators' strategies and skills change as they gain experience with a task (thereby changing their workload), it is essential that the subjects in a workload experiment are similar (in terms of experience and skills) to the people who will be the operational users of the system.

Experimental Control

Counterbalancing the order in which different conditions, tasks, displays, etc., are presented will distribute the effects of training (on the primary task) and experience (with the evaluation procedure). If, for example, the most difficult task is always presented last, the effects of increased familiarity or improved skills (a potential source of workload-reduction) may result in erroneous conclusions. In addition, if identifying the workload associated with a specific aspect of a complex system is the goal of a workload analysis, it is essential that all other sources of workload variability are held constant across measurement intervals.

Comparing Measures Across Tasks

A major concern is that it is difficult to compare the workload of two tasks directly if different measures were obtained. The most obvious problem is that units of measurement and scaling might be different. This problem might be solved by standardizing the scores or expressing them in terms of percent change. However, since different measures are sensitive to different aspects of workload, it may be inappropriate to compare them directly. In addition, individual differences in the use of rating scales, resting heart rates, and level of experience on a task may create large differences in absolute levels between experiments that do not, in fact, represent true workload differences. And, apparently similar measures of performance may in fact have different meanings, reflect differences in systems (rather than the behavior of the human operator), or task-specific instructions and priorities across different tasks. Thus, workload measures are most useful in assessing relative differences in workload within a specific context, and less useful for comparisons across tasks and experiments.

Quantification and Analysis

In summarizing workload analyses and making recommendations, measurement variability must be reported as well as average differences. Formal statistical analyses take between-subject variability into account when determining the statistical significance of average differences. However, when data are available for only a few operators, a situation common in many operational situations, statistical treatment of the data is nearly impossible. In this case, it is imperative that the ranges of values are reported as well as the averages. Finally, just because different measures can be expressed with great numerical precision, this does not imply that the information they provide is equally precise. For example, secondary task reaction times might be recorded to the nearest millisecond, while there may be only three possible values for a rating, however, the sensitivity of both measures might be such that neither can do more than distinguish between high and low workload. However, precision of measurement, rather than precision of expression, is important; inaccurate or "noisy" data certainly limits the faith that can be placed in the information provided.

Interpretation

To be useful, workload measures must be translated into terms that are meaningful to the end users. Reporting that ratings of "25" and "35" were obtained for System A and B is not particularly meaningful. However, reporting that ratings for "System A were 25 percent higher than for the reference system" puts this information in context and compares it to a known quantity. Again, note that the example is couched in relative terms; it is still very difficult to make absolute statements about workload (e.g., "This task requires 75 percent of the operator's capacity"; "Momentary workload was excessive 25 percent of the time"). Even "objective" measures that are reported in physical units, as opposed to obviously subjective values, do not, in fact, convey information that is as concrete and absolute as a similar statement about fuel consumed or exceeded oil pressure limits. The ranges of acceptable and unacceptable workload simply are not known. Identifying these points or ranges has proven to be difficult because sources of workload vary among tasks, different people respond to the same objective demands by adopting different strategies and exerting different effort, and individuals' abilities to cope with excessively low or high workload differ.

WHY DO MEASURES DISSOCIATE?

Often, workload practitioners and system designers are confronted by a lack of agreement or instances of dissociation between different measures of workload, or between measures of workload and performance (e.g, Vidulich, 1988; Vidulich & Wickens, 1986; Wickens & Yeh, 1988): (1) The correlation between workload measures is either not significant or is negative. For example, people (or systems) that appear to have higher workload by one measure are shown to have lower workload by another; or (2) A system shown to have significantly higher workload by one measure, may have lower workload (or better performance) by another measure. An understanding of the theory of workload, as outlined earlier in the chapter, as well as an understanding of the nature of the measures used, may make many of such dissociations interpretable.

Figure 9-2 presented a number of examples which are interpreted within the framework of the performance resource function. Consider both Figures 9-2d and 9-2e, for example. In Figure 9-2d, circumstances that produced better performance – as a consequence of allocating greater effort – also led, naturally, to higher workload. In Figure 9-2e, the circumstances that resulted in higher workload (the use of the less "natural" keyboard interface) were different than those that resulted in better performance (the faster response of the keyboard system).

Given that both workload and performance are influenced by many different factors, and given also that some measures are affected by factors that are not directly related to workload, it is not surprising that measures may dissociate. For example, the performance of certain secondary tasks might be disrupted by structural interference with the primary task (e.g., the eyes cannot look in two directions at the same time), while heart rate measures might reflect physical exertion. If the amount of physical load and the effect of structural interference are not equivalent in two systems that are compared, it is not surprising that these measures would not coincide.

To this form of dissociation may be added a note of caution about the use of correlations in comparing workload measures. Correlations may be computed between people, between systems or between people and systems, depending upon what is defined as a "case" (e.g., the data point in a scatter plot upon which a correlation is based). In the case of correlations between people, the data tell us whether the people who find a system difficult to operate (as indicated by a higher subjective rating, for example), are the same people who show low performance on the system or respond to an increase in difficulty by a measurable increase in heart rate. The major problem with the use of these correlations is that there are so many differences between people (e.g., resources available, skill on a particular task, resource-investment strategy, and willingness to use the high or low ends of a subjective rating scale). Since each of these factors can influence different measures in different directions, low correlations are not surprising (and high correlations are not always interpretable). The correlations between systems (averaged across people) are more informative; they demonstrate whether systems that appear to have low workload (based on one measure) are the same systems that appear to have low workload (based on another measure). Here, correlations tend to be more meaningful as long as the unique characteristics of the different measures are taken into account. Correlations of the third type, in which variability between people and between systems is combined, are often uninterpretable; they mix two entirely different sources of variability to produce a single number.

HOW CAN WORKLOAD BE PREDICTED?

Up to this point, the focus of this chapter has been on workload assessment; a system is designed and its figure of merit is established. In contrast, models for workload prediction are designed to make projections about the workload imposed by a system (or a system modification) which does not yet exist. Should a helicopter be designed to support a one crew or two crew flight deck? Will the requirement to operate a communications system in degraded lighting make workload unacceptable? Will the workload reduction created by introducing an automated function be significant? To answer questions such as these, designers would like to turn to predictive models; it is far more efficient and cost effective to resolve workload problems on paper, in advance, than to wait until the first system has been fielded and then solve a problem that arises by a change in design. Many predictive models are described in Elkind, Card, Hochberg, and Huey (1989), McMillan (1989), Phatak (1983), and Wickens (1989a, 1989b) and they will not be reviewed in detail here. However, an important feature of all of these models is that they do not explicitly predict either primary task performance or resource requirements, but leave the distinction between the two purposefully unspecified. This is because, as we have noted, workload predictions for the high-demand range focus on changes in performance, while workload predictions for the low-demand range focus on changes in reserve capacity. However, the state of the art of prediction is not yet sufficiently precise to specify within which range a prediction will fall. All that is offered is a general predicted figure of merit for a system, or a time line which shows how this figure of merit might fluctuate over time.

Typically, these models are employed for one of two purposes: to make a prediction of absolute workload (e.g., whether the workload imposed on a single pilot flying a particular mission will be excessive) or to make predictions of relative workload differences among design alternatives. The foundation of most predictive workload models is a task analysis (in which the mission is decomposed into specific activities that the operator will perform) and a time line analysis which computes workload based on the percentage of time the operator will be busy performing relevant tasks, as shown in Figure 9-3 (Parks & Boucek, 1989). Usually, some value, such as 80 percent, is chosen to define a "red line," above which workload is considered to be unacceptably high. Task performance times are estimated by reference to standardized tasks in a data base, by empirical research, or expert opinion.

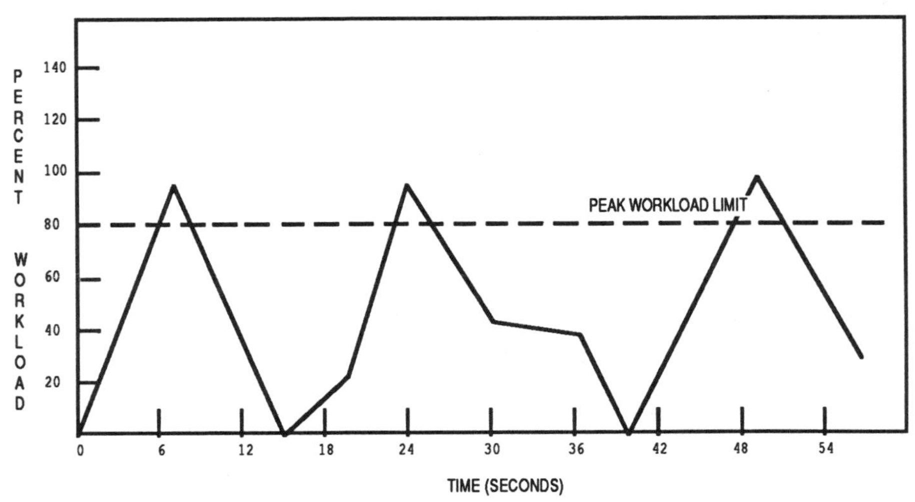

Figure 9-3
Example of Workload Time History Profile as Produced by TLAP

Two advances to time line analysis, which incorporate information about task difficulty and task interference, have added greater precision to the predictions. McCracken and Aldrich (1984) and Aldrich, Szabo, & Bierbaum, (1989), have argued for the importance of coding the resource demands of different activities in the time line, such that the workload of two easy tasks carried out concurrently might be less than (or at least not double) the workload of a single difficult task. Aldrich et al. developed a table lookup coding scheme for defining the demand values of different activities on four 7-point scales (visual, auditory, cognitive, physical load). Recent validation of predictive workload model assumptions suggests that this feature is an extremely important component of useful models (Wickens, Larish, & Contorer, 1989). For example, time-sharing vehicle control and vocal communications is fairly easy for well-trained operators, but time-sharing reading procedural instructions with communications is not. North and Riley (1989) and Wickens, Harwood, Segal, Tkalcevic, and Sherman (1988) developed predictive workload models that attempt to consider competition for multiple processing resources, as well as demand levels and time lines.

A final feature of some predictive models that is of considerable value in forecasting real world operating conditions is the ability of these models to dynamically schedule task performance according to momentary demands. A static time line, such as that shown in Figure 9-3 in which task analysis establishes the moment at which tasks will be performed, does not take into account the fact that an operator may choose to postpone the performance of a demanding task if it occurs when workload would otherwise be excessive. Models such as the Human Operator Simulator (HOS) (Harris, Iavecchia, Ross, & Shaffer, 1987; Harris, Iavecchia, & Bittner, 1988), MicroSaint (Laughery, Drews, & Archer, 1986), and Procedure Oriented Crew Model (PROCRU) (Corker, Davis, Papazian, & Pew, 1986) have incorporated this dynamic process of activity rescheduling.

To provide accurate workload predictions, a computational model must address not only single task demand levels and performance times, but also penalties for concurrent task conflicts. By combining such information, the model can identify workload peaks for specific workload dimensions, or averages across dimensions, at a given point in time or averaged over intervals of time, and extrapolate the consequences of such overload points to decrements in performance. Existing models vary in the procedures they offer for task decomposition and analysis, taxonomies of task-related and behavior-related functions, grain of analysis, availability of standardized data bases of subjective or objective values for task performance times and load levels, output format, and the degree to which workload values are related to the degree or type of subsequent performance decrements. Furthermore, the scientific basis for and empirical evaluation of these models differ substantially.

In principle, it should be possible to develop predictive models that are sufficiently accurate that empirical assessments are not also required. In practice, however, predictive models neither can, nor should, be used alone without empirical assessments to verify their accuracy, given the capabilities of existing techniques. The predictions provided by these models should not be treated as established fact, but rather as suggested guidance. In particular, the value of predictive models is that they can make rough predictions of figure of merit which, if not 100 percent reliable, are at least considerably better than chance (or intuition). Furthermore, these models can also often objectively identify points of potentially high workload and diagnose their cause. These points can then serve as the focus for in-depth empirical evaluation, using the workload assessment techniques that were described in the section entitled "How is Workload Assessed?"

Workload predictions are not an adequate substitute for an empirical evaluation; it is the interaction between a particular operator or team of operators and the demands imposed by the task, equipment, and environment that determines the workload experienced and system performance achieved, not task demands alone. The effort that operators can (or will) exert and the task-performance strategies that they adopt determine the workload they actually experience and the level of system performance they can achieve. It is this interaction that is the focus of workload assessment.

CONCLUSION

Workload is an important, integrative concept that determines the abilities of human operators of complex systems to accomplish mission requirements, given the equipment and training that are provided and the organizational and environmental constraints that are placed on them. Workload can be measured with considerable of accuracy, however, workload prediction is a much more difficult and less precise process. A variety of subjective rating scales, measures of primary and secondary task performance, and physiological indicators have been developed, tested, and used to aid designers, manufacturers, and users in quantifying the effects of task requirements on the operators. Because each measure is especially sensitive to different workload causes and consequences, the results obtained with different measures may not covary. However, recent research in the field is focused on clarifying the underlying causes of such dissociations and formulating a model of workload/performance trade-offs to aid in interpreting the results of workload analyses, identifying workload criteria, and improving the accuracy of workload prediction.

REFERENCES

Aldrich, T. B., Szabo, S. M., & Bierbaum, C. R. (1989). The development and application of models to predict operator workload during system design. In G. McMillan (Ed.), *Applications of models to system design* (pp. 65-80). New York: Plenum Press.

Beatty, J. (1982). Task-evoked pupillary responses, processing load, and the structure of processing resources. *Psychological Bulletin, 91*, 276-292.

Bittner, A. C., Byers, J. C., Hill, S. G., Zaklad, A. L., & Christ, R. E. (1989). Generic workload ratings of a mobile air defense system (LOS-F-H) (Technical Memorandum 1). To appear in *Proceedings of the Human Factors Society 33rd Annual Meeting.* Willow Grove, PA: Analytics, Inc.

Bortolussi, M. R., Hart, S. G., & Shively, R. J. (1989, February). Measuring moment-to-moment pilot workload using synchronous presentations of secondary tasks in a motion-base trainer. *Aviation, Space, and Environmental Medicine*, 124-129.

Byers, J. C. (1989). *Workload assessment of the pedestal mounted stinger (PMS)* (Technical Memorandum 7). Willow Grove, PA: Analytics, Inc.

Byers, J. C., Bittner, A. C., & Hill, S. G. (1989). Traditional and raw task load index (TLX) correlations: Are paired comparisons necessary? *Advances in Industrial Ergonomics and Safety* (Vol. 1). London: Taylor & Francis.

Byers, J. C., Bittner, A. C., Hill, S. G., Zaklad, A. L., & Christ, R. E. (1988). Workload assessment of a remotely piloted vehicle (RPV) system. *Proceedings of the Human Factors Society 32nd Annual Meeting* (pp. 1145-1149). Santa Monica, CA: Human Factors Society.

Casper, P. A., Shively, R. J., & Hart, S. G. (1987). Decision support for workload assessment: Introducing WC FIELDE. *Proceedings of the Human Factors Society 31st Annual Meeting* (pp. 72-76). Santa Monica, CA: Human Factors Society.

Cooper, G. E., & Harper, R. P. (1969). *The use of pilot ratings in the evaluation of aircraft handling qualities* (NASA TN D-5153). Washington, DC: National Aeronautics and Space Administration.

Corker, K., Davis, L., Papazian, B., & Pew, R. (1986). *Development of an advanced task analysis methodology and demonstration for Army-NASA aircrew/aircraft integration* (Report No. 6124). Cambridge, MA: Bolt Beranek and Newman, Inc.

Derrick, W. (1988). Dimensions of operator workload. *Human Factors, 30*, 95-110.

Donchin, E., Kramer, A. F., & Wickens, C. D. (1986). Applications of brain event-related potentials to problems in engineering psychology. In M. G. H. Coles, E. Donchin, & S. Porges (Eds.), *Psychophysiology: Systems, processes, and applications* (pp. 702-718). New York: Guilford Press.

Elkind, J., Card, J., Hochberg, J., & Huey, B. (Eds.). (1989). *Human performance models for computer-aided engineering*. Washington, DC: National Academy Press.

Ericsson, K. A., & Simon, H. A. (1980). Verbal reports as data. *Psychological Review, 87(3)*, 215-251.

Gopher, D., & Donchin, E. (1986). Workload - An examination of the concept. In K. Boff, L. Kaufman, & J. P. Thomas (Eds.), *Handbook of Perception and Human Performance* (pp. 41-1 - 41-49). New York: Wiley & Sons.

Hancock, P. A., & Meshkati, N. (Eds.). (1988). *Human mental workload.* Amsterdam, The Netherlands: North Holland Press.

Hancock, P. A., Meshkati, N., & Robertson, M. M. (1985). Physiological reflections of mental workload. *Aviation, Space, and Environmental Medicine, 56(11),* 1110-1114.

Harris, R. M., Hill, S. G., & Lysaght, R. J. (1989). OWLKNEST: An expert system to provide operator workload guidance. *Proceedings of the Human Factors Society 33rd Annual Meeting* (pp. 1486-1490). Santa Monica, CA: Human Factors Society.

Harris, R. M., Iavecchia, H. P., Ross, L. V., and Shaffer, S. C. (1987). Micro-computer human operator simulator (HOS-IV). *Proceedings of the Human Factors Society 31st Annual Meeting* (pp. 1179-1183). Santa Monica, CA: Human Factors Society.

Harris, R. M., Iavecchia, H. P., & Bittner, A. C. (1988). Everything you always wanted to know about HOS micromodels but were afraid to ask. *Proceedings of the Human Factors 32nd Annual Meeting* (pp. 1051-1055). Santa Monica, CA: Human Factors Society.

Harris, R. L., Glover, B. J., & Spady, A. (1986). *Analytic techniques of pilot scanning and their application* (NASA TP-2525). Washington, DC: National Aeronautics and Space Administration.

Hart, S. G. (1986). Theory and measurement of human workload. In J. Zeidner (Ed.), *Human productivity enhancement: Training and human factors in systems design* (Vol. I, pp. 396-456). New York: Praeger.

Hart, S. G., & Hauser, J. R. (1987). Inflight application of three pilot workload measurement techniques. *Aviation, Space, and Environmental Medicine, 58 (5),* 402-410.

Hart, S. G., McPherson, D., & Loomis, L. L. (1978). Time estimation as a secondary task to measure workload: Summary of research (NASA-CP-2060). *Proceedings of the 14th Annual Conference on Manual Control* (pp. 693-712). Washington, DC: National Aeronautics and Space Administration.

Hart, S. G., Shively, R. J., Vidulich, M. A., & Miller, R. C. (1986). The effects of stimulus modality and task integrality: Predicting dual-task performance and workload from single task levels (NASA-CP-2428). *Proceedings of the 21st Annual Conference on Manual Control* (pp. 5.1-5.18). Washington, DC: National Aeronautics and Space Administration.

Hart, S. G., & Staveland, L. E. (1988). Development of NASA-TLX (Task Load Index): Results of empirical and theoretical research. In P. A. Hancock & N. Meshkati (Eds.), *Human mental workload* (pp. 139-183). Amsterdam, The Netherlands: North Holland.

Hill, S. G., Byers, J. C., Zaklad, A. L., Bittner, A. C., and Christ, R. E. (1989).

Prospective workload ratings of LOS-F-H mobile & defense missile system (Technical Memorandum 2). Willow Grove, PA: Analytics, Inc.

Hill, S. G., Zaklad, A. L., Bittner, A. C., Byers, J. C., & Christ, R. E. (1988). Workload assessment of a mobile air defense missile system. *Proceedings of the Human Factors Society 32nd Annual Meeting* (pp. 1068-1072). Santa Monica, CA: Human Factors Society.

Jex, H. R., & Clement, W. F. (1979). Defining and measuring perceptual-motor workload in manual control tasks. In N. Moray (Ed.), *Mental workload: Its theory and measurement* (pp. 125-179). New York: Plenum Press.

Kantowitz, B. H., Bortolussi, M. R., & Hart, S. G. (1987). Measuring pilot workload in a motion base simulator: III. Synchronous secondary task. *Proceedings of the Human Factors Society 31st Annual Meeting* (pp. 834-837). Santa Monica, CA: Human Factors Society.

Kramer, A. F., Sirevaag, E. J., & Braune, R. (1987). A psychophysiological assessment of operator workload during simulated flight missions. *Human Factors, 29 (2)*, 145-160.

Kramer, A. F., Wickens, C. D., & Donchin, E. (1983). A analysis of the processing requirements of a complex perceptual motor task. *Human Factors, 25*, 597-621.

Laughery, R., Drews, C., & Archer, R. (1986). A micro-SAINT simulation analyzing operator workload in a future attack helicopter. *Proceedings of the IEEE National Aerospace and Electronics Conference* (pp. 86-92). New York: Institute of Electrical and Electronics Engineers.

Lysaght, R. J., Hill, S. G., Dick, A. O., Plamondon, B. D., Linton, P. M., Wierwille, W. W., Zaklad, A. L., Bittner, A. C., and Wherry, R. J. (1989). *Operator workload: Comprehensive review and evaluation of operator workload methodologies* (TR 851). Alexandria, VA: US Army Research Institute for the Behavioral and Social Sciences.

Mallery, C. J., & Maresh, J. (1987). Comparison of POSWAT ratings for aircraft and simulator workload. In R. Jensen (Ed.), *Fourth International Symposium on Aviation Psychology* (pp. 644-650). Columbus: Ohio State University.

McMillan, G. R. (Ed.). (1989). *Applications of models to system design.* New York: Plenum Press.

McCracken, J. H., & Aldrich, T. B. (1984). *Analyses of selected LHX mission functions: Implications for operator workload and system automation goals* (Technical Note ASI479-024-84). Fort Rucker, AL: Army Research Institute Aviation Research and Development Activity.

Moray, N. (Ed.). (1979). *Human mental workload: Its theory and measurement.* New York: Plenum Press.

Moray, N. (1988). Mental workload since 1979. In D. J. Osborne (Ed.), *International Reviews of Ergonomics, Vol. 2* (pp. 38-64). London: Taylor and Francis.

Morris, N. M., & Rouse, W. B. (1988). *Human operator response to error-likely situations in complex engineering systems* (NASA CR-177484). Washington, DC: National Aeronautics and Space Administration.

Mosier, K. L., & Hart, S. G. (1985). Levels of information processing in a Fitts Law task (NASA CP-2428). *Proceedings of the 21st Annual Conference on Manual Control* (pp. 4.1-4.15). Washington, DC: National Aeronautics and Space Administration.

Norman, D., & Bobrow, D. (1975). Data limited and resource limited processing. *Cognitive Psychology, 7*, 44-60.

North, R. A. and Riley, V. A. (1989). W/INDEX: A predictive model of operator workload. In G. R. McMillan (Ed.), *Applications of models to system design* (pp. 81-99). New York: Plenum Press.

O'Donnell, R. D., & Eggemeier, F. T. (1986). Workload assessment methodology. In K. Boff, L. Kaufman, & J. Thomas (Eds.), *Handbook of Perception and Human Performance*, (Vol. 2, pp. 42-1 - 42-49). New York: John Wiley & Sons.

Ogden, G. D., Levine, J. M., & Eisner, E. J. (1979). Measurement of workload by secondary tasks. *Human Factors, 21(5)*, 529-548.

Parks, D. L., & Boucek, G. P., Jr. (1989). Workload prediction, diagnosis and continuing challenges. In G. R. McMillan (Ed.), *Applications of models to system design* (pp. 47-63). New York: Plenum Press.

Phatak, A. V. (1983). *Review of model-based methods for pilot performance and workload assessment* (Report for NASA Contract NAS2-11218). Mountain View, CA: Analytical Mechanics Association, Inc.

Reid, G. B. (1985). Current status of the development of the subjective workload assessment technique. *Proceedings of the Human Factors Society 29th Annual Meeting* (pp. 220-223). Santa Monica, CA: Human Factors Society.

Reid, G. B., & Colle, H. A. (1988). Critical SWAT values for predicting operator overload. *Proceedings of the Human Factors Society 32nd Annual Meeting* (pp. 1414-1418). Santa Monica, CA: Human Factors Society.

Reid, G. B., & Nygren, T. E. (1988). The subjective workload assessment technique: A scaling procedure for measuring mental workload. In P. A. Hancock & N. Meshkati (Eds.), *Human mental workload* (pp. 185-213). Amsterdam, The Netherlands: North Holland.

Roscoe, S., & Kraus, E. (1971). Pilotage error and residual attention. *Navigation, 20*, 267-279.

Roscoe, A. H. (Ed.). (1987). *The practical assessment of pilot workload* (AGARD-AG-282, pp. 78-82). Neuilly-sur-Seine, France: Advisory Group for Aerospace Research and Development.

Roscoe, A. H., & Ellis, G. A. (in press). *A subjective rating scale for assessing pilot workload in flight. A decade of practical use* (Technical Report). Bedford, England: Royal Air Force Establishment.

Stassen, H. G., Johannsen, G., & Moray, N. P. (1988). *Internal representation, internal model, human performance, and mental workload* (EPL- 88-01). Paper presented at the International Federation of Automatic Control. Urbana-Champaign: University of Illinois, Department of Mechanical and Industrial Engineering.

Stern, J. A., & Skelly, J. J. (1984). The eye blink and workload considerations. *Proceedings of the Human Factors Society 28th Annual Meeting* (pp. 942-943). Santa Monica, CA: Human Factors Society.

Vicente, K. J., Thornton, D. C., and Moray, N. (1987). Spectral analysis of sinusarrythmia: A measure of mental effort. *Human Factors, 29 (2)*, 171-182.

Vidulich, M. A. (1988). The cognitive psychology of subjective mental workload. In P. Hancock & N. Meshkati (Eds.), *Human mental workload* (pp. 219-229). Amsterdam, The Netherlands: North Holland Press.

Vidulich, M. A., & Wickens, C. D. (1986). Causes of dissociation between subjective workload measures and performance. *Applied Ergonomics, 17*, 291-296.

Wickens, C. D. (1980). The structure of attentional resources. In R. Nickerson (Ed.), *Attention and performance*, (Vol. VIII, pp. 239-257). Englewood Cliffs, NJ: Erlbaum.

Wickens, C. D. (1984). Processing resources in attention. In *Varieties of attention* (pp. 63-102). New York: Academic Press.

Wickens, C. D. (1989a). Models of multitask situations. In G. McMillan (Ed.), *Applications of models to system design* (pp. 259-273). New York: Plenum Press.

Wickens, C. D. (1989b). Resource management and time sharing. In J. Elkind, S. Card, J. Hochberg, & B. Huey (Eds.), *Human performance models for computer aided engineering* (pp. 180-202). Washington, DC: National Academy Press.

Wickens, C. D., Harwood, K., Segal, L., Tkalcevic, I., & Sherman, W. (1988). TASKILLAN: A simulation to predict the validity of multiple resource models of aviation workload. *Proceedings of the Human Factors Society 32nd Annual Meeting* (pp. 168-172). Santa Monica, CA: Human Factors Society.

Wickens, C. D., Hyman, F., Dellinger, J., Taylor, H., & Meador, M. (1986). The Sternberg memory search task as an index of pilot workload. *Ergonomics, 29*, 1371-1383.

Wickens, C. D., Larish, I., & Contorer, A. (1989). Predictive performance models and multiple task performance. *Proceedings of the Human Factors Society 33nd Annual Meeting* (pp. 96-100). Santa Monica, CA: Human Factors Society.

Wickens, C. D., & Liu, Y. (1988). Codes and modalities in multiple resources: A success and a qualification. *Human Factors, 30 (5)*, 599-616.

Wickens, C. D., & Weingartner, A. (1985). Process control monitoring: The effects of spatial and verbal ability and current task demand. In R. Eberts

(Ed.), *Trends in ergonomics and human factors* (pp. 25-32). Amsterdam: North Holland Publishing Company.

Wickens, C. D., & Yeh, Y. Y. (1988). Dissociation of performance and subjective measures of workload. *Human Factors, 30 (1)*, 111-120.

Wierwille, W. W., Casali, J. G., Connor, S. A., & Rahimi, M. (1986). *Evaluation of the sensitivity and intrusion of mental workload estimation techniques. Advances in man-machine systems research* (Vol. 2, pp. 51-127). Greenwich: JAI Press, Inc.

PART 

SYSTEMS INTEGRATION METHODOLOGIES

Methods of systems integration take on special meaning for an organization's decision makers when they provide comprehensible answers to trade-off questions between product requirements, cost limits, and people constraints. Several advances in system integration methods can be presented collectively as illustrated in the Figure 1 for Part III. For military systems the Cost Benefit Analysis links together the principal product drivers: Reliability (RAM) and Performance Modeling (Complex Environment Models) – with the four primary cost drivers – manpower, R&D, procurement, and training. Human Factors Engineering and Manpower, Personnel and Training (MPT) Methodologies, along with Training and Training/Aiding Trade-off Analyses make up principal people data inputs for an overall *MANPRINT Analysis Integration Methodology*. The chapters presented in this part describe various concepts which collectively make-up the state of the art for systems integration which fully considers the people variables as part of the systems being integrated. It will be obvious that much R&D work remains to be done by government, industry, and universities before people, cost, and product data can be integrated by systems engineers as smoothly as hardware and cost data are today. Nevertheless, progress on integrating people into analysis methodologies has been rapid and much is already available.

Chapter 10, *Acquisition Decision Process*, by Shields, Johnson, and Riviello. describes how MANPRINT can become part of the routine process for procuring complex systems and other products. Shields, Johnson, and Riviello use the Defense Department acquisition process to illustrate the technical, managerial, and economic complexity of including human performance as part of total systems performance requirements. They lead the reader step-by-step through the acquisition process pointing out examples of MANPRINT issues to be addressed, relevant decision making information, and useful analytical tools for critical milestone decisions. Shields Johnson, and Riviello also provide several examples of similar activities and techniques applicable to the process of acquiring products in the commercial world.

The following chapter, Chapter 11, discusses *Complex Environment Models in Systems Integration*. Parry, Collins, and Van Nostrand state that systems development combat models are increasing in popularity in the military as a cost-effective means of evaluating competing systems candidates. Their use ranges from the selection of technologies early in the conceptual phase of systems development to the testing of candidate systems without ever building a prototype. Because the combat model is often the only analytical tool used to provide performance parameters early in systems design, human factors design influence must be achieved through the complex environment modeling community. The complex environment model is also frequently used to interact dynamically with the development of other key design parameters such as reliability, availability and maintainability (RAM) through cost-benefit trade-off analyses.

Ramifications of the combat model approach to non-military commercial applications are also discussed.

Booher and Hewitt in Chapter 12 describe in some detail many of the *MANPRINT Tools and Techniques* available and how they apply to systems design. They point out that these MANPRINT analytical techniques are particularly important:

1. in identifying high resource costs and human error potential at very early stages of preliminary system design;
2. for influencing the design considerations of human physiological and cognitive requirements;
3. in converting human performance requirements into engineering specifications;
4. in developing and validating designs that meet total system performance requirements; and
5. in providing essential data for decision makers throughout the materiel acquisition process.

There is probably no analysis area more difficult than the integration of training and training devices and other techniques (like aiding) for enhancing human performance in complex systems. Oneal with *Integration of Training Systems and Analysis*, Chapter 13, describes the MANPRINT concept for realistically bringing training costs, training performance requirements and training technologies into the systems integration process. With emphasis toward integration of training costs and performance requirements Oneal discusses systems integration opportunities which result from the desirable trend toward acquiring complete training systems rather than training equipment. She is optimistic about the far reaching impact of this trend on institutional management and organization for training both in industry and government.

Chapter 14, *Training and Aiding Personnel in Complex Systems*, by Rouse, focuses on the central human performance trade off between providing the knowledge and skills which give personnel the potential to perform (through training) versus enhancing performance directly (through aiding). Alternative approaches to training and aiding are discussed and several methods for performing trade-off analyses are considered.

The final step in the acquisition process before a system or product is ready for military fielding or commercial customer purchase is operation testing and evaluation by the actual user. Hennessy in Chapter 15, *Practical Human Performance Testing and Evaluation*, discusses both the theoretical and practical limitations in operational measures of human performance. Although MANPRINT has made significant changes in the way operational tests and evaluations are planned and executed in the military, there is still a major problem facing the test and evaluation communities assigned with the responsibility of acquiring useful operational human performance data. As a start to resolve this problem Hennessy suggests a MANPRINT Hierarchical Performance Measurement Model as an approach to relate user's performance to system effectiveness and suitability.

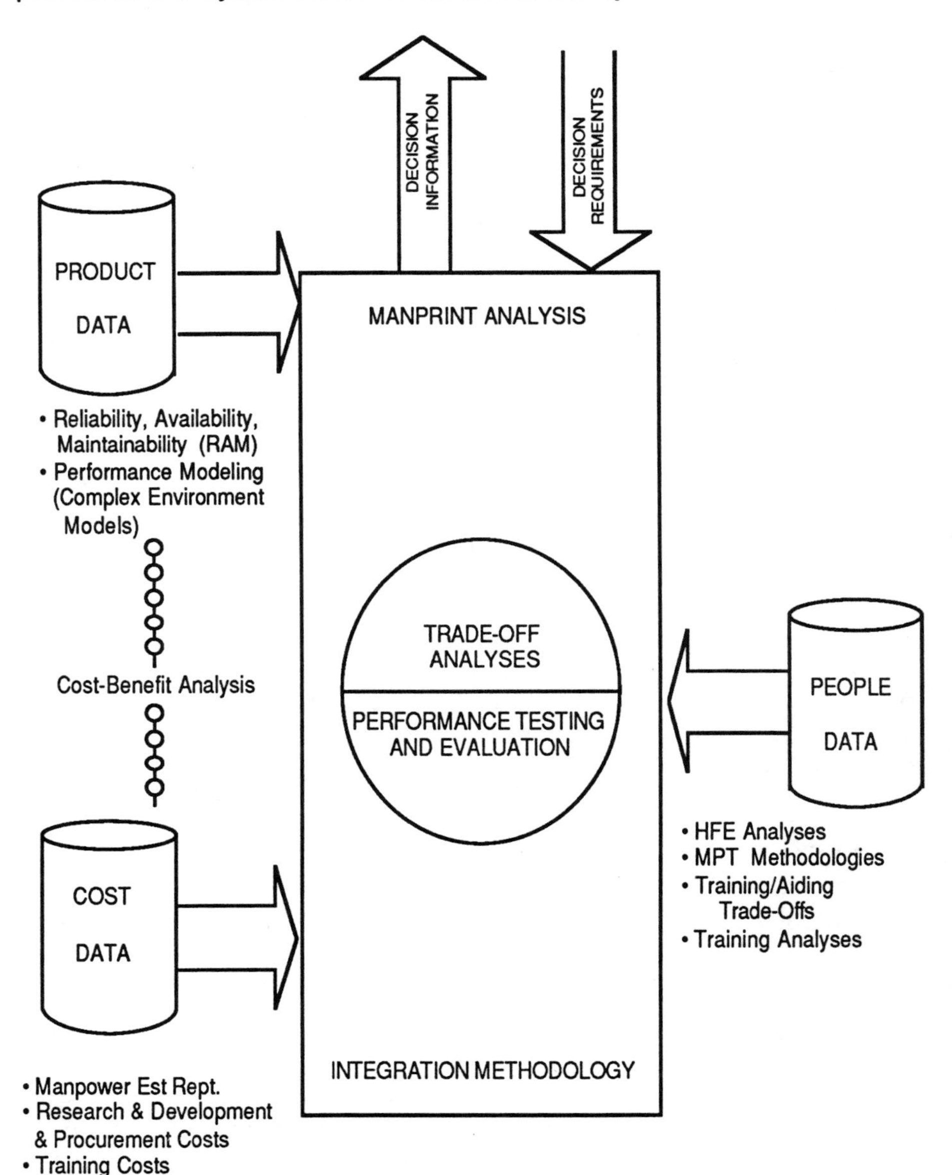

Part III, Figure 1
Systems Integration Methodologies

CHAPTER 10

THE ACQUISITION DECISION PROCESS

Joyce L. Shields
Kenneth M. Johnson
Robert N. Riviello

ABSTRACT

The acquisition of major systems is a complex process in any environment. In the Defense Department it is further complicated by an intricate set of interactions between the government and private industry. The roles of the buyer (the government) and the builder (private industry) are defined in lengthy and numerous regulations and directives. It is the purpose of this chapter to highlight the key issues and decisions in the systems acquisition process as they are influenced by Manpower and Personnel Integration (MANPRINT) domain concerns. This chapter will use the Defense Department's acquisition process as an example of the technical, managerial and economic complexities of including human performance as part of total systems performance requirements. This model is typical of all large, government bureaucracies. This chapter will also discuss and compare this process to the important issues in acquiring major systems in the private sector.

INTRODUCTION

Properly harnessed, new technology can enhance performance in every aspect of an organization. However, there are many examples of where sophisticated, high-tech systems were misused, or sat idle, because the critical human factor was overlooked during the design phase. Frequently, requirements for training, operation, maintaining and repairing new systems are far in excess of available resources.

In the mid 1980s, the U.S. Army instituted the Manpower and Personnel Integration (MANPRINT) program which imposed consideration of human factors engineering, manpower, personnel, training, health hazards and system safety on the entire systems acquisition process. Recent initiatives from the Department of Defense (Directive Number 5000.53) and Congress (Title 10, U.S. Code) underscore the importance that the MANPRINT program has placed on these issues. These directives require the military departments to attend to MANPRINT issues in the systems acquisition process. This process is the focus of this chapter. The objective of this chapter is to provide a framework for understanding the information provided in subsequent chapters.

The problems and challenges that have been described in the Defense Department that drove the Army to create MANPRINT exist to the same degree in the private sector. Product designers have pushed technology out to the frontier, but it's way ahead of the consumer. The consumer is not the only one affected by these advances. A national survey of mechanics found that many advances in automotive engineering are turning out to be burdens in car-repair shops. They reported that it is more difficult to determine what's wrong, and therefore, more difficult to repair today's cars. The design of high technology machinery and systems often require people to adjust to advanced hardware and software. The examples range from computer dashboards in cars, cameras and VCRs to automated manufacturing systems, robotics and office automation.

But U.S. industry is changing. U.S. manufacturers are once again discovering that good design is a key to industrial competitiveness. This comes after relegating design to the backseat in the 1970s. Industry has relearned that good design is more than skin deep. It must be reliable, easy and economical to operate and service, and simple to manufacture. Simplicity of assembly is a key to slashing costs while still producing higher quality products. If any single trend stands out in American design today, it is simplicity. The designer is being forced to match products to their customers. The similarities in the systems acquisition process in the private and public sectors will be discussed in the final section of the chapter.

Objectives of the Defense Systems Acquisition Process

The principal objective of the Defense systems acquisition process is to acquire and deploy effective systems in response to an identified deficiency or threat, or to capitalize on technology breakthrough, and thereby increase total force effectiveness. The objectives of the acquisition management decision makers are to influence and approve a systems acquisition program at key milestones and to provide the information necessary to program and budget for the implementation of the system. This acquisition process is designed to develop, procure and field a totally integrated and supportable system of technology, people and organizations, and to insure that the complex system meets its cost, schedule and performance goals.

The Systems Integration Challenge

The ultimate goal of addressing MANPRINT issues during the development and acquisition of systems is to maximize the likelihood that the system will meet its performance requirements. The central objective of the program is to influence system design so that the system makes best use of the human resources that are available. Key decision makers must understand the importance of MANPRINT issues and design engineers must be willing to work with MANPRINT practitioners to integrate these considerations into the engineering process. The underlying concept is that the MANPRINT objectives must be considered from a total system perspective.

As shown in Figure 10-1, the system must be evaluated on varying levels of interaction beginning with the man-machine interface through the mission-organization interface. At the man-machine level, the concerns focus on soldier performance issues and the likelihood that the soldier will be able to perform all tasks to standard under operational conditions. At the unit performance level, the interoperability of systems and support issues are addressed. At the total system performance level, MANPRINT evaluates the interaction of the equipment (hardware and software), the manpower and personnel requirements, the organizational structure and the operating environment, and the likelihood that all of the subsystems will interface correctly to achieve total system performance.

In order to make a difference in the design and development of a system, the critical MANPRINT issues must be identified and addressed within the context of the process through which systems are developed, tested and deployed. MANPRINT information must be presented in a compelling,

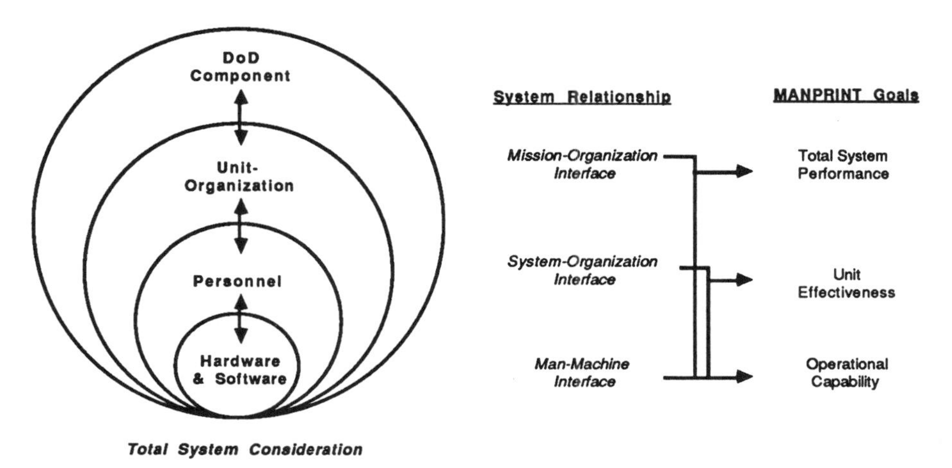

Figure 10-1
MANPRINT Goals and Total System Performance

timely manner throughout the acquisition cycle. In other words, MANPRINT concerns must be treated exactly like all of the other performance and cost parameters during the acquisition of a system. To do so, the MANPRINT practitioner must understand the systems acquisition process and manage the planning, budgeting and implementation activities necessary to field a new system with all of the people and equipment necessary to optimize total system performance.

THE DEFENSE SYSTEMS ACQUISITION PROCESS

The systems acquisition process at its simplest level is illustrated in Figure 10-2. Simply stated, the government determines what is needed, why it is needed, what it should do, how much it can cost, how and who will design and build it. Private industry determines how to design and build the system based on the government requirements and specifications. However, the acquisition process is not simple. It consists of a complex set of actions, functions, decisions and procedures designed to acquire an enhanced capability. These policies, procedures and responsibilities for the acquisition of systems are well documented in government directives and

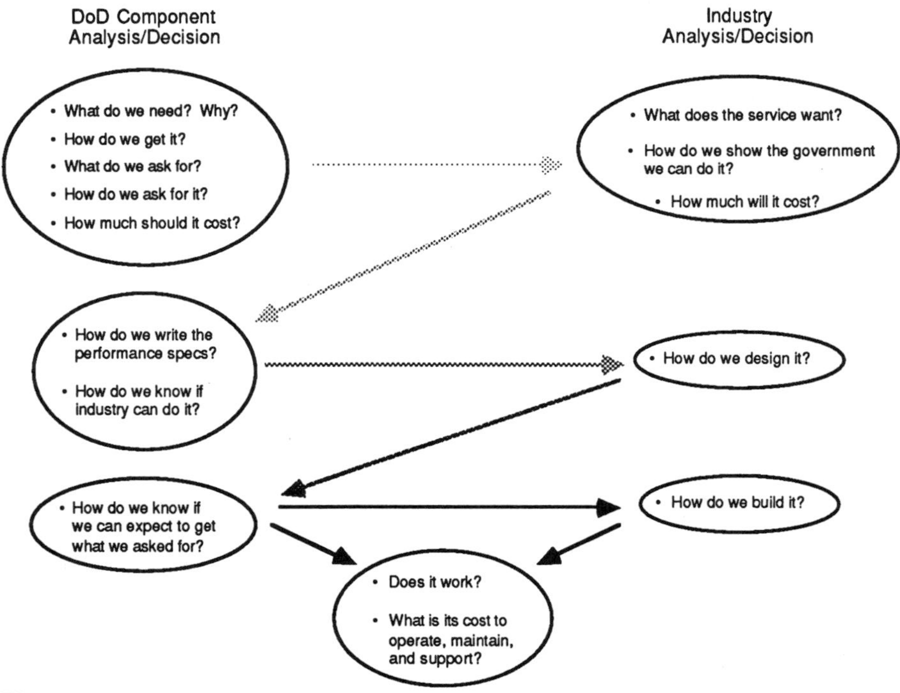

Figure 10-2
Conceptual View of the Acquisition Process

regulations (e.g., Department of Defense Regulation 5000.1, Army Regulation 70-1). There is some one hundred and fifty feet of documentation describing this process housed in the Library of Congress!

Overview of the Defense Systems Acquisition Process

The Defense systems acquisition process (for conceiving, developing, acquiring and fielding new systems) is formalized in Department of Defense (DoD) Directives and has been tailored by each of the services through
The phases shown in Table 10-1 are for the traditional, full development

Table 10-1
Description of the Activities in the Systems Acquisition Process

• Preconcept

Prior to formally entering the Acquisition Process, the government defines needs for new or improved systems and identifies opportunities and constraints.

• Concept Exploration/Definition

The purpose of this phase is to identify and explore conceptual alternatives and select concept approaches for further development.

• Concept Demonstration and Validation

In this phase, preliminary designs are verified and may include prototype hardware and software demonstrations and early testing with troops; trade-offs among system characteristics are evaluated and system design specifications are developed.

• Full Scale Development

During this phase, the system design is finalized, tested, evaluated and documented.

• Production and Initial Deployment

In this phase sustained production of the system is initiated, personnel and operational units are trained, associated support equipment is acquired and distributed, and logistical support is provided.

acquisition and will be used to illustrate the principles of integrating people, organizations and technology in the acquisition process. There are many tailored applications and versions of this process such as the Army's Streamlined Acquisition Program and Nondevelopmental Item Programs, as well as additional activities in each phase. Each of the services has a similar process as directed by DoD Directive 5000.1. The issues highlighted in this chapter apply to all forms of the acquisition process, to include acquiring major systems in private industry. The primary focus is on how to manage the development, design and purchase of systems so that the user will be able to operate, maintain and support the system in a manner that meets system performance and cost requirements.

MANPRINT Information Required in the Systems Acquisition Process

The consideration of MANPRINT issues throughout the acquisition cycle is no longer just good design practice; it is a requirement. At the overall level all decision makers should consider the questions listed in Table 10-2 as they plan and evaluate alternative concepts, decisions and designs.

Table 10-2
Key MANPRINT Issues

Decision makers should ask: Will this concept/design increase the likelihood that the:

- Total system will meet performance requirements?
- Manpower will be sufficient to operate, maintain, and effectively support the total system?
- Personnel will have the skills and abilities to perform the tasks necessary to operate, maintain, and support the system?
- Personnel with the right skills, abilities, and training will be at the right place and at the right time to properly operate, maintain, and support the system?
- Individuals will perform the required tasks correctly under all operational and environmental conditions?
- Operation, maintenance, or support of the system will not result in safety or environmental hazard problems?

The new DoD Directive 5000.53 formally establishes that Manpower, Personnel, Training, and Safety (MPTS) issues such as those included in the MANPRINT program, must be considered, assessed, refined, documented and reported at each phase of the acquisition. The DoD Directive establishes policy, assigns responsibilities and prescribes procedures for the integration and implementation of MPTS considerations throughout the systems acquisition process.

Compliance with these directives requires a MANPRINT process of data collection, analysis, design, testing and verification. Each phase builds upon information available from existing or predecessor systems, previous designs and earlier phases, gaining in specificity as the developmental process unfolds. The key MANPRINT objectives within each of the phases is summarized in Table 10-3. The MANPRINT activities will be described in the next section.

MANPRINT Activities in the Systems Acquisition Process

The principal function of the MANPRINT activities within the systems acquisition process is to generate the information necessary to influence design and to manage the acquisition, training, and distribution of the work force needed to operate, maintain and support the total system. Both the government (buyer) and industry (builder) must execute a series of planning, organizing, managing and analysis activities. The general flow of these MANPRINT activities is illustrated in Figure 10-3.

This section reviews these key MANPRINT activities within and across the acquisition phases. On a general level, MANPRINT activities must be integral to all acquisition activities in order to influence the design and development process. Many of the activities illustrated in Figure 10-3 are repeated in each acquisition phase (e.g., competitive solicitation and system evaluation); the levels of detail, however, will increase until the design and development of the production model is finalized. Key MANPRINT activities include planning, requirements formulation, solicitation and source selection, and design and validation.

MANPRINT Planning

For the government, the MANPRINT planning process should begin during Preconcept activities. The objective of early planning is to identify the MANPRINT issues and decisions that must be addressed during the acquisition of the system and to plan for MANPRINT activities and analyses for the remainder of the acquisition cycle. The initiation and development of a MANPRINT Management Plan provides the required documentation to plan, organize, and manage the MANPRINT effort (including required

Table 10-3
MANPRINT Objectives By Phase

1. Preconcept

The principal MANPRINT objective in this phase is to influence the consideration of acquisition alternatives and requirements through the identification of known or projected manpower, personnel, training, safety, health hazard or human factors constraints, opportunities and goals. In this phase, the MANPRINT goals, constraints and opportunities are identified and documented.

2. Concept Exploration and Definition

During concept exploration, the primary objective is to identify and evaluate the characteristics of the alternative system design concepts that drive MANPRINT requirements and costs or that may lead to human performance problems. In thisphase, the key issues, test and evaluation criteria, and design specifications and trades are identified.

3. Concept Demonstration and Validation

In this phase, the principal objective is to influence the system design specifications through an assessment and evaluation of the preliminary design studies and trade-off analyses. A secondary objective is to develop the MANPRINT planning factors necessary to field and support the system.

4. Full Scale Development

The principal objectives of this phase are to determine if MANPRINT goals and constraints have been addressed adequately, to identify and resolve any outstanding human performance or system safety issues, and to implement the actions necessary to field the system.

5. Production and Initial Deployment

During this phase, the objectives are to verify that MANPRINT considerations have been fully integrated, to identify MANPRINT issues that need to be considered in future improvements, and to evaluate lessons learned.

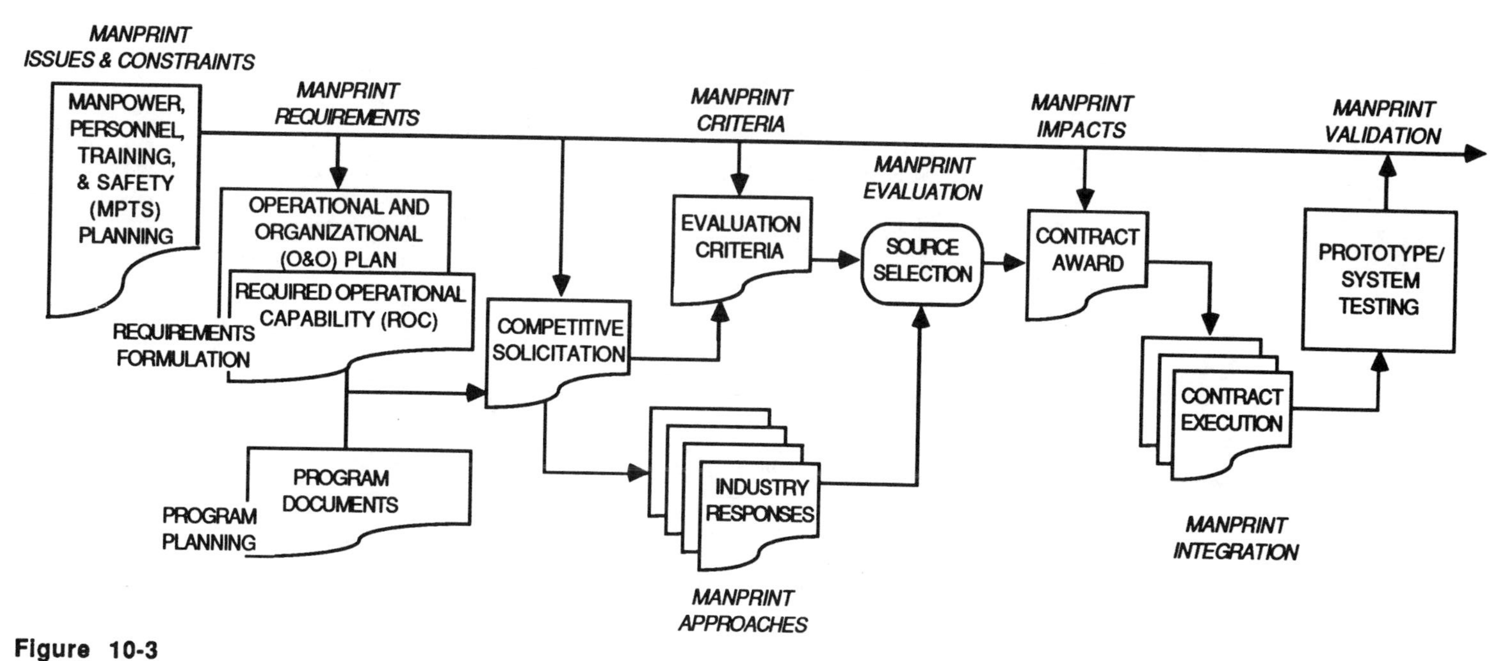

Figure 10-3
MANPRINT Activities in the System Acquisition Process

analytical support) for a system. The early focus of MANPRINT activities should be to identify previous problems, performance deficiencies, MANPRINT opportunities, required improvements, and lessons learned that have affected predecessor system performance. The planning aspects of MANPRINT focus on identification of issues that are potentially critical to successful development of the new system, determination of appropriate analytical tools and related data bases to address those issues, and programming the analytical support required to ensure that requisite information will be available to key decision makers in a timely manner.

As the system develops, from early concepts through production prototypes, the issues will be validated and refined as data become available to support a reorientation away from existing or predecessor system information to the specific system being developed.

MANPRINT Requirements Formulation

The formulation of the requirements for a new system begins during Preconcept and continues to be refined through the Demonstration and Validation Phase. A precept of an effective MANPRINT program is that human performance must be integrated with hardware/software requirements to define the total system performance envelope. In government acquisitions, this is accomplished by embedding MANPRINT issues in the system requirements documents and program planning documentation such as the Required Operational Capability (ROC) and Operational and Organizational Plan (O&O). This requires that a cross-walk of information takes place throughout the acquisition process.

Requirement documents state the minimum essential performance requirements that the government has determined will satisfy the deficiency. These identified requirements form the basis for subsequent solicitation packages which define what industry must provide in terms of system performance. These documents address MANPRINT issues in addition to hardware/software requirements in order to adequately define the total system performance goals for the emerging system. Individual domain or element issues, identified in the MANPRINT Management Plan, are integrated into the hardware/software capability statements to provide a cohesive description of the system.

Program planning documents (such as the Integrated Logistics Support Plan and the Test and Evaluation Master Plan) also include MANPRINT elements. Supportability and testing requirements are derived from the overall MANPRINT goals constraints.

MANPRINT Solicitations and Source Selection

The solicitation to industry requires a translation of government requirements into contractual language that can convey to both management and design engineers what the government is seeking in terms of minimum system performance levels.

Included in the solicitation package is how MANPRINT will be evaluated during the source selection process. The emphasis that the government places on MANPRINT will, to a large extent, guide industry's approach to integrating MANPRINT in the design and development process.

During the source selection process, the government's evaluation of proposals will include a determination of how well industry understood what the government was seeking in terms of MANPRINT and how well industry responded with an integrated plan to address the total system performance goals.

Once the contract is awarded, a government-industry partnership is formed and industry becomes an extension of the government's effort to ensure that total system performance can be achieved and sustained.

MANPRINT Design and Validation

Industry plays the primary role in the integration of MANPRINT in the design and development of a new or improved materiel system. Theirs is the challenge of translating human performance requirements into system designs. Some of the challenges that must be addressed in this process are the:

- Ability to convert human performance requirements into engineering specifications.
- Ability to design to the cognitive abilities of the work force.
- Ability to assess the cognitive work load of designs.
- Quantification of human performance and cost factors.
- Ability to assess human error potential at preliminary design stages.

Early in the design process, use of modeling, simulations, and mockups provide the initial opportunities for assessing the MANPRINT impacts of alternate design approaches. As the concept matures, subassembly and component prototypes are developed and opportunities for early user testing provide increasingly reliable data on the human performance impacts of the total system. By Full Scale Development, validation of the MANPRINT contributions to the total system performance can be assessed during full-up user testing.

IMPLEMENTING THE MANPRINT EFFORT

In the previous sections we have discussed the global MANPRINT issues to be addressed, decisions to be made, questions to be answered, and the

MANPRINT activities within the systems acquisition process. Using this information as the process which must be accomplished, this section uses an Army example to illustrate the activities in each phase.

Preconcept

The primary objectives of Preconcept MANPRINT activities are to identify the key issues and decisions that must be addressed in the systems acquisition process and to plan for MANPRINT activities and analyses for the remainder of the acquisition cycle. The role of MANPRINT activities during Preconcept is to provide insights into the previous problems, performance deficiencies, required improvements, and lessons learned that have affected past system performance. These previous data points will be used to lay the foundation of predecessor data upon which to build the framework for the modified or replacement system. If the opportunity to maximize system performance is to be realized, MANPRINT should play an active role in this early stage.

For example, The Chief of Infantry, U.S. Army, identified a significant combat deficiency in light infantry forces medium-range antiarmor capabilities. Due to a projected increase in enemy armor capability and performance deficiencies in the current weapon, the need for a new man-portable antiarmor weapon system was determined. The following paragraphs trace the identification and management of key issues through the acquisition phases. In the Preconcept phase, MANPRINT practitioners analyzed antiarmor projected mission requirements, doctrine, emerging technological information, current and projected force availability, and lessons learned from the existing antiarmor weapon system. Doing so, they identified numerous MANPRINT issues that were critical to successful development of the new system. Examples of issues are seen in Table 10-4.

These are only a few of the MANPRINT issues that were determined for this system. The identification of these issues became the basis for determining the degree to which MANPRINT would play in this particular acquisition and provided the foundation for the development of the MANPRINT goals and constraints which guided that role.

Concept Exploration and Definition (CE) Phase

During this phase in the antiarmor program, a number of analyses were performed to estimate MANPRINT impact on possible courses of action and to refine the performance requirements of the antiarmor weapon. Some of these were a human factors engineering/portability study, induced

Table 10-4
MANPRINT Issues Identified in Preconcept

- Soldier performance: Could the expected Infantry user identify targets, correctly range, fire the weapon within short time limits and perform these engagement tasks correctly to ensure a 90 percent probability of hit? If unable to do so, soldier survivability would be unacceptable. Predecessor performance information indicated significant man-machine interface problems precluding anything near this hit probability.

- Portability: Due to manpower constraints, this system had to be one-man portable; a two-man crew was not acceptable. The weight of the weapon had to be such that one man could carry it in the expected operational environment over distances expected by tactical requirements.

- Health hazards: Doctrine required employment of the weapon from enclosure areas (concealed in a building or foxhole). The capability of the soldier to fire more than one shot from an enclosure was defined as a requirement. Propellant gases present a hazard to the operator.

- Training: Information showed the predecessor training program was unsatisfactory and not achieving soldier performance desired. Additional training time was not considered affordable. A solution, whether better training devices or an embedded practice system, was required.

environment study, target acquisition predictor study, initial health and safety assessments, and cost and operational effectiveness analysis. Research data and information from industry indicated several different, technological approaches were possible. MANPRINT contributed greatly to define the performance requirements that selected industrial companies that would be asked to demonstrate in the next phase. Table 10-5 shows how the issues identified in Preconcept influenced Concept Exploration activities and requirements. These, along with many other MANPRINT factors, were embedded into the Demonstration/Validation Request For Proposal. Competing companies had to show their plans to fulfill these requirements in order to be selected to proceed to the next phase.

Table 10-5
Issues in Concept Exploration

- Soldier performance: The cognitive aptitude (in terms of Armed Forces Qualification Test [AFQT] scores) and physical capabilities (e.g., 20/20 correctable vision) of the Infantry user who must correctly accomplish operational tasks were determined; the range estimation function was allocated to the equipment; time limits in terms of seconds per engagement were specified. Soldier performance was made a specific part of the overall system performance formula included in the technical specifications.
- Portability: Based on the existing load of the fully equipped user, a 20.5 kilogram weight limit was imposed and a carrying strap and handle required.
- Health hazards: The requirement to fire more than one round from an enclosure was reaffirmed; laser protection for the user was mandated.
- Training: A 40 hour institutional and 6 hour/quarter sustainment training constraint was established; embedded and thermal viewer training devices and a training round were identified as requirements; performance standards for the training program were defined requiring 95 percent successful performance demonstrated after training.

Concept Demonstration and Validation (Dem/Val) Phase

During Demonstration and Validation of the antiarmor weapon system, intensive Army/industry interaction took place. Three contractor teams, each offering a different firing technology, completed prototype development, contractor testing, contractor demonstrations, limited soldier demonstrations, health and safety assessments, human factor/portability demonstrations, and technical prototype shoot offs. Army MANPRINT practitioners analyzed each prototype to determine MANPRINT impacts and constraint compliance. The results of these analyses along with the efforts cited by the contractors in their proposals for Full Scale Development contributed to selection of one company's system for Full Scale Development. The issues evaluated during this phase are shown in Table

Table 10-6
MANPRINT Issues Evaluated During Demonstration/Validation Phase

- Soldier performance: Assessed expected performance using soldiers representative of the projected user population (limited because of safety factors of prototypes); estimated soldier error probability; determined resolution of performance issues cited earlier; and evaluated each prototype for source selection input.
- Portability: Tested prototype size, weight and configuration for acceptability.
- Health hazards: Performed assessment to determine impact on the soldier and identify further hazard issues.
- Training: Evaluated proposed training packages to determine if within affordability constraints and to estimate training effectiveness for soldier skill development.

10-6. Based on these evaluations, MANPRINT was a significant factor in the decision of which contractor to select to proceed into Full Scale Development.

Full Scale Development

During Full Scale Development, the Army/industry interaction is again intense. Industry makes final design decisions and the Army assesses the data provided and results of their operational testing. At this time, this phase has not been completed on the antiarmor system. Action is planned to track what was documented in earlier phases to ensure system configuration and performance results meet all defined criteria. These criteria include MANPRINT considerations in almost all aspects. Barring areas of technical infeasibility, human performance and supportability requirements for the antiarmor system should meet or exceed the requirements defined in order to eliminate the light infantry combat deficiency identified at the start of the systems acquisition process. Table 10-7 contrasts the predecessor system with some of the improvements already realized with the new antiarmor system.

Table 10-7
Antiarmor System Improvements

Current antiarmor system:

- Difficult for gunners to acquire and track targets. Limited lateral movement allowed by launch position (gunner serves as "third leg of tripod"). Limited lateral movement makes it difficult to hit a moving target, the most likely on the battlefield.

- Difficult for gunner to make fine adjustments when tracking because he is forced to rely on the use of larger muscle groups and postural adjustments when tracking.

- Sighting system forces the gunner into an awkward position that is difficult to maintain.

- Weight shift caused by missile leaving the launch tube can lead the gunner into either under or over compensating his tracking. Weight loss at back of the launch tube (which rests on the gunner's shoulder) can cause gunner to inadvertently direct the missile into the ground.

- Firing signature (noise, smoke, dust) and exposure while tracking the target make the gunner vulnerable to enemy suppressive counter fire. Heat and noise of launch are stressful, if not threatening, to the gunner.

- Separate day and night tracking systems must be used. Normally one is carried by the gunner and one by the assistant gunner. Use of separate trackers poses the likelihood that one will not be available when needed.

New antiarmor system:

- Gunner does not guide the missile and does not have to track the target. He acquires and locks on the target using TV-like thermal imagery. Target is automatically tracked during missile flight by a seeker in the missile's nose. Gunner training does not have to include target tracking.

- Automating the missile guidance process reduces demands imposed on the gunner, lowers his exposure time, and allows for an increased rate of fire. Immediately after firing, the gunner can take cover, move or fire again.

- Incorporates a two-stage rocket motor. This permits a "soft launch" with a much reduced heat, noise and signature impact on the gunner. Second stage rocket firing takes place during missile flight.

- Command Launch Unit combines the functions of both day and night tracker. This reduces the gunner's load, eliminates the need for sight assembly training, and depends only on the activation of a switch for high resolution thermal imaging.

Production and Initial Deployment

Additional MANPRINT issues may be identified during the fielding and deployment after the antiarmor weapon system is in the hands of actual users in field units. MANPRINT issues will be determined as these units take the weapon into the operational environment for field training and attempt to maintain gunner proficiency with only limited unit firing possible. Follow-up studies will take place to determine lessons learned and document supportability problems. This information will be incorporated into any future changes to the system or, in the long-run, into development of a replacement system as has been shown in this example.

ACQUISITION DECISION PROCESS IN THE PRIVATE SECTOR

It is apparent that the demands of the customers in the private sector are similar to the demands of the customers in the Defense Department. However, the roles of the buyer and builder differ in the private sector. Figure 10-4 presents a simplified version of the *acquisition process* from a

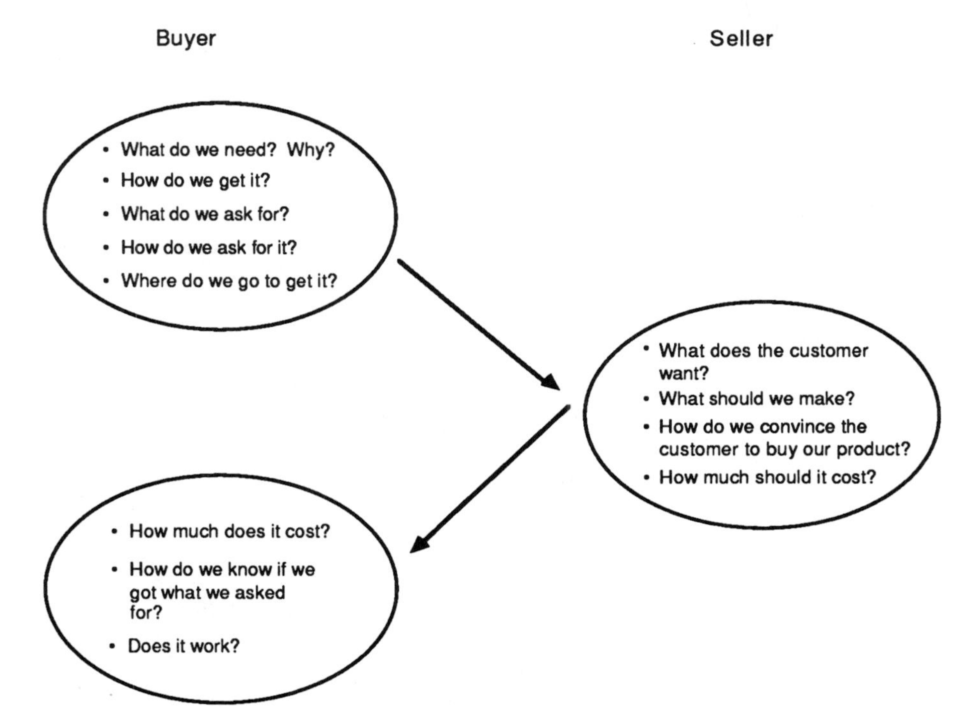

Figure 10-4
Private Sector Acquisition Process

private sector perspective. The *acquisition issues* are similar, but the *acquisition process* in the private sector has not been codified to the same degree. In the private sector, the average customer does not write specifications for the design of new products. However, buyers in large industrial and service companies are learning that they must attend to the integration of people, organizations and technology as they invest in high-technology machinery and systems. As they acquire complex systems such as computer-assisted manufacturing systems or computer-based, satellite-distributed training systems, they must consider how multiple hardware and software systems will work together and how the systems will be operated, maintained and supported. Industry can write specifications for new systems or evaluate the capability of vendors to deliver systems that match the capabilities of the work force and organization. For example, in acquiring a new automated manufacturing system, the buyer should evaluate alternative systems on the key issues shown in Table 10-8. The introduction of automated technologies will lead to the redesign of tasks, jobs and work environments. The evaluation of alternative systems must be analyzed in the context of the overall structure of the automated factory.

Attention to the consideration of the full range of issues in the interaction of people, organizations and technology is key to successful implementation of advanced technology. This is a new focus in business planning and human resource management. The tools, data bases and

Table 10-8
Key Human Resource Issues

Decision makers should ask: Will this concept/design increase the likelihood that the:

- Total system will increase productivity?
- Size of the work force will be sufficient to operate, maintain, and effectively support the total system?
- Current work force will have the skills and abilities to perform the tasks necessary to operate, maintain, and support the system?
- Individuals will perform the required tasks correctly under all operational and environmental conditions?
- Operation, maintenance, or support of the system will not result in safety or environmental hazard problems?

approaches used by the Defense Department and described in succeeding chapters of this book can be adapted by the private sector to assist in this process. These approaches describe both the more traditional human factors or micro-ergonomic tools and data bases, as well as the application of new macro-ergonomic tools used to evaluate the integration of people, organizations and technology.

CONCLUSIONS

The Defense systems acquisition process takes place in a complex political, bureaucratic and design environment. This chapter has presented the key issues and decisions in this acquisition process as they are influenced by data bases described in the following chapters. Through this understanding, the MANPRINT practitioner can present information in a compelling, timely manner throughout the acquisition cycle. The practitioner will be able to manage the planning, budgeting and implementation activities necessary to field a new integrated system of people, organizations and technology necessary to optimize total system performance.

REFERENCES

Army Regulation 602-2 (1987, May 18). *Manpower and Personnel Integration (MANPRINT) in the Materiel Acquisition Process.* Washington, DC: Department of the Army.

Army Regulation 70-1 (1988, October 10). *System Acquisition Policy and Procedures.* Washington, DC: Department of the Army.

Department of Defense Directive 5000.53 (1988, December 30). *Manpower, Personnel, Training, and Safety (MPTS) in the Defense System Acquisition Process.* Washington, DC: Department of Defense.

CHAPTER **11**

COMPLEX ENVIRONMENT MODELS IN SYSTEMS INTEGRATION

Sam Parry
Dennis D. Collins
Sally J. Van Nostrand

ABSTRACT

This chapter describes complex environment models and their application in systems integration, with a particular emphasis on military applications because of the rich background provided by wargames and combat modeling. Although combat models have been utilized extensively by military decision makers for more than 30 years, their potential in systems integration is only recently becoming realized. High resolution models are becoming especially important decision aids as pressures mount to make major commitments to design concepts without building and testing prototypes. MANPRINT factors have been generally unrepresented in combat modeling even though it is obvious that people play extensive roles in any operation as complex as war. The reasons for and the impact of this omission on system development models are discussed. A proposal is made for improving systems integration by including greater human factors representation in complex environment models, such as the military combat model, Janus. Finally, a paradigm is presented for the development of human factors data which can be most useful in complex modeling.

BACKGROUND

Complex environment models have their origins in attempts to predict the outcome of perhaps the most complex of all operational environments, that of man at war. Philosophers of war have long recognized the need to recognize the unique role of people in that environment. Carl von Clausewitz (1832/1984, p. 184) in the 19th century stressed:

> The effects of physical and psychological factors form an organic whole which, unlike a metal alloy, is inseparable by

> chemical processes. In formulating any rule concerning physical factors, the theorist must bear in mind the part that moral factors may play in it; otherwise he may be misled into making categorical statements that will be too timid and restricted, or else too sweeping and dogmatic.[1]

Modern combat models have evolved from nineteenth century German wargames. The original German wargame protocol, for example, called for the color blue to represent "friendly" forces and red to represent the enemy. This protocol has become universal, with almost all measures of effectiveness being shown in both "red" and "blue". Wargame usage for tactical planning continued to grow through the early Twentieth Century such that by the beginning of World War II, wargames were in use by virtually all major powers. The Japanese attack on Pearl Harbor was extensively wargamed, and the early versions of United States World War II operations in Europe and the Pacific (called the Rainbow Plans) were wargamed by the War Department in the 1920s and 1930s (Greenfield, 1960; Prange, 1981). Wargame usage has branched from pure operational prediction to tactical training devices and item-level modeling for systems design.

It was the marriage of the wargame and the computer which produced the modern version of the combat model. Combat simulations with computers are a modern method of trying to understand war. As stated by Schroth (1989, p. 2):

> In a combat simulation the essence of combat is distilled and specifically formulated as mathematical representations of war and battles. The methodology to distill this essence is one of systematic study, where assumptions are made until a mathematical relationship is established, and then an iterative process of removing assumptions and rewriting the relationships begins. This process continues until time, money, or the ability to replace assumptions with data prevent the process from continuing.

The automated wargame offered the designer of weapons systems the opportunity to evaluate candidate systems concepts much earlier in the systems design process creating a new use for the wargame. By the late 1960s, for example, the software for the Army Small Arms Requirements Study (ASARS) model contained over 18,000 lines of FORTRAN IV computer code and could show rudimentary performance differences between rifle candidates (ASARS, 1970). The modern combat models used by the military have now grown into systems development models which show the candidate system's performance in its operational environment.

Systems Development Implications

A systems development combat model is an analytical tool used to view how a conceptual candidate for systems selection will perform in actual combat. The intent of systems modeling is to facilitate the development and selection of the best possible system by allowing early design intervention. In simple terms, each systems candidate is given movement and performance directions which are intended to approximate its expected performance. These directions are then placed in a wargame which represents its combat environment. Combat modeling can be described as analogous to the game of chess (see Figure 11-1). Imagine that each chess piece represents a candidate system. The chess piece movement rules represent the systems performance parameters and the board represents the operational environment. If detailed records were kept of the chess game's progress one would begin to get a picture of the relative value of each piece. The chess game could be structured to hold constant all variables such as strategy, initial position and player (i.e., act as dependent variables). New pieces with differing performance characteristics would be introduced to become the independent variable. Computerized versions of the chess game are particularly appropriate for the combat model analogy. It would be relatively easy to develop a systems design computer chess model by automating the human player, substituting a chess piece variant, and running the game repeatedly to show the impact of the variant. "Knight Modification 1.1" for example, might be capable of moving three spaces forward before a diagonal move. Multiple runs of the automated chess game

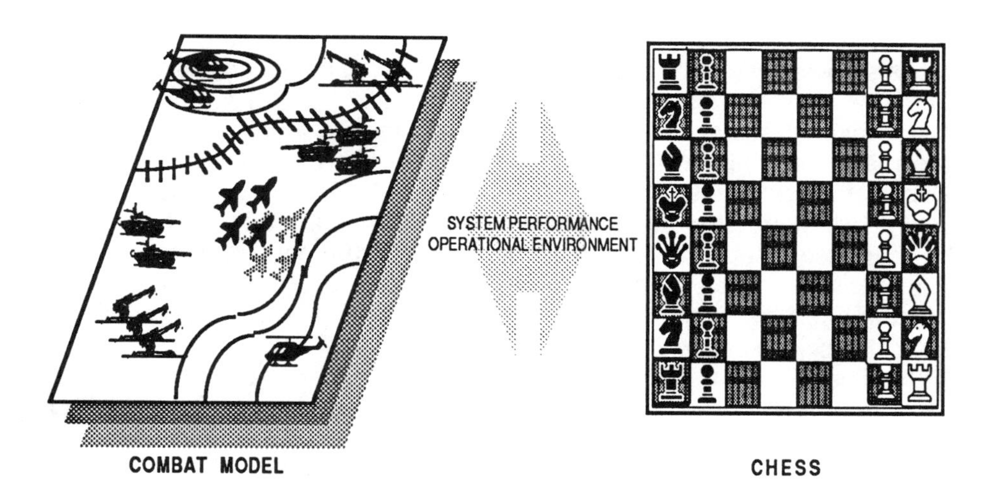

Figure 11-1
The Chess Analogy

would show the impact of "knight mod 1.1" on the outcome of the game. If this process were repeated many times, a statistically significant picture of the relative value of each chess piece would emerge. The systems development combat model is simply a more elaborate version of the chess game.

Systems development combat models are used to show a system's expected performance in its operational environment. Most importantly, they compare the relative operational value of candidate conceptual systems. In modeling terms this type model is called a "high resolution" model because it shows its systems in detail. Combat models used for tactical training and strategic analysis often aggregate systems (tanks, airplanes, trucks, etc.) into groups called units. The aggregated combat model is usually called a "low resolution" model, and is not often used for systems analysis because the performance of individual systems is not observable.

Human Performance Parameters

Because the systems development combat model grew from analytical communities which were oriented to tactics and engineering, the representation of human performance parameters in the evolution of combat models was rarely considered. This omission has caused some analysts to complain of lack of *realism* in combat models. "When I look at our combat models, what do they represent? Almost totally it is the weapon systems and other tools of war – the physical elements, the scenarios, and the environments. Where are the real combatants? The human performance never seems to suffer from fatigue, or battle, just as the modeled weapon's performance rarely suffers from nonincapacitating battle damage or from wearing out" (Murtaugh, 1987, p. 22). Moreover, in historical analyses of battle, it is often human factors which are found to be the major determinants of victory or defeat. (As was the conclusion of Dupuy and Hammerman (1980) regarding the 1973 Arab-Israeli Wars).

By failing to properly consider the human component of systems performance, people have an assumed value of 100 percent effectiveness. It is generally accepted, even among combat modelers, that this assumption has the effect of exaggerating systems performance, and accelerating the tactical pace of a battle.

Modelers do not, of course, believe people's performance is perfect. One argument as they see it is that people's performance is simply too difficult to quantify. According to Gropman (1987, p. 1), " . . . many of the war gamers among us pursue the quantification of psychological or human military quantities which are unquantifiable . . . we do not argue here that psychological elements are unimportant; on the contrary they are extremely important, but also extremely variable . . . their uncertainty, their variability, their inconsistency and their lack of utility in modeling future conflict should make one reluctant to expend resources pursuing them." There is some merit in this argument since human performance parameters are generally less well defined than hardware performance parameters, and no clear consensus has as yet emerged as to how human factors should be modeled. But this argument is shallow if people do have a significant influence on the outcome of the battle and the model is to be relied on by decision makers in that regard. What has actually happened is more subtle. By omitting human factors from both "red" and "blue" forces, the tactical/engineering modeler tended to deal with the amorphous area of human factors through a *balanced omission*. That is, since neither side considered human factors, the effect was balanced. The people effects were assumed canceled out so far as any tactical or engineering conclusions drawn from the model's output. But as noted above, through a balanced omission of human factors, human performance was actually being modeled as 100 percent effective. In early tactical wargames and engineering models, this approach was acceptable because the computers of the day were functional only in aggregated, low resolution modeling. Low resolution models provided valuable tactical insights, but little design information about specific systems. Engineering models were also simple: tank "blue" fired at tank "red" in a straightforward duel format. But more recently, at least one analyst (Van Nostrand 1986, 1988) has argued effectively that human performance parameters like stress and fatigue can and should be included in low resolution models used for group (crew, unit, theater) decisions as well.

As wargames became automated, the ability to conduct high-resolution simulation allowed the tactical and engineering modeling of actual systems in dynamic combat (high resolution modeling). Automation of wargames, however, has also made omission of human factors both unnecessary and problematic. The continuation of balanced omission is no longer necessary because of the analytical power of advanced data systems. The continuation of balanced omission is now problematic because the new analytical power of advanced data systems provides decision makers with an extraordinary insight into notional combat, but a combat devoid of any human representation. The question has become: How do we deal with the possibility of a very sophisticated but wrong answer?

The Janus Combat Model

One combat model which has begun to work with human factors parameters is the Janus model. The Janus model was developed in the late 1970s at Lawrence Livermore National Laboratory in Livermore, California. An initial use of Janus was an analysis of the "neutron bomb" which the Carter administration sought to develop. Janus was unique at the time in its ability to provide high-quality color graphics on its display. Janus' quality visual output capability quickly made it popular with decision makers. As the national laboratories began to market their technologies in the early 1980s Janus was adopted as both a training development and systems development model by the United States Army's Training and Doctrine Command (TRADOC). Subsequent army modifications to the model have tailored it to the analysis of army missions and systems. Recent use of the Janus model by the Army element at the Naval Postgraduate School in Monterey, California, for example, has led to the demonstration of Janus as a valuable tool in the analysis of joint-service systems such as the air-launched cruise missile in support of air and ground conventional combat (Balaconis, 1989).

In 1988 the Janus model was selected as a pilot for a human factors integration project. This project, called the Janus-Human Factors Project is sponsored by the Department of the Army and conducted at the United States Army element at the Naval Postgraduate School at Monterey, California.

The Janus model is particularly suited to the chess analogy because, like the chess game, it has human players. The human players usually act as commanders of units rather than moving specific systems, but they will influence the model much like a chess player will influence the game. This human-interactive feature makes Janus particularly well suited for the tactical training of military leaders. Janus also allows the amount of human interaction to be varied. Human interaction would usually be limited in a systems development use of Janus in order to minimize the variances introduced by human players. Systems development models often restrict human intervention in order to control human variations and permit statistically significant iterations (called runs in software terminology). Multipurpose models like Janus are not statisticians models of choice, but are frequently used because they can be set-up quickly and are "transparent" in that the internal influences of the analytical combat are easily understood by decision makers. The chess analogy holds here, also. It would be much easier to understand the results of several videotaped chess games than to decipher the numeric results of several hundred chess games played internally within a computer.

The chess analogy can also be extended to how each chess piece functions on the board. Like combat models, the chess game has a built-in assumption that human performance is 1.0. The knight, for example, represents both horse and rider. The knight "system" (man and horse) is played as a perfect system. If we were to convert the chess board knight to a combat model system, we would degrade the horseman's ability to see the opposing player's chess pieces (acquiring a target) and his ability to correctly identify the opposing chess pieces (identifying-friend-or-foe).

In Janus runs, systems can be seen moving about the battlefield and firing. The host computer tracks the performance of each piece and aggregates these performances on preselected "measures of performance." It is important to note that these measures of performance not only reflect the performance of systems directly, but also show the performance of the overall force. A system's contribution to the success of the force is often considered a greater indicator of value than its isolated, individual performance. In a private sector context, this would be like being able to model a new product in a market of the future against a competitor's future products and show not just the product's performance, but also its value within the company's product line and its contribution to the performance of the company as a whole.

ACQUISITION PROCESS

Systems development combat models are increasing in popularity as a cost-effective means of evaluating competing military systems. Their use ranges from the selection of technologies early in the conceptual phase of systems development to the testing of candidate systems without ever building a prototype. Because the combat model is often the only analytical tool used to provide performance parameters early in systems design, human factors design influence must be achieved through the combat modeling community. The combat model is also frequently used to interact dynamically with the development of other key design parameters such as reliability, availability, and maintainability (RAM) through cost-benefit trade-off analyses.

Technology Selection

Combat models have become powerful determinants in systems design and selection. They provide a quantification of performance in the cost-performance analyses which take place early in systems design. Combat models are useful in early concept development, when systems are only analytical "notions" and traditional human factors influence is weakest. The decisions made in the early stages of design predetermine the commitment of resources later in the design process where human factors normally interfaces (Figure 11-2). Without human factors intervention during this "notional" analytical period of design, both the technology and design selections may be frozen, leaving only a brief opportunity for intervention through an ergonomic review of what will essentially be a final product.

Figure 11-3 illustrates a historical example of the role of combat models in decision making. Looking at the left of Figure 11-3, one sees the enemy submarine of 1941 as the problem to be solved. Solutions to this problem could be found in several candidate technologies. At the time, potential solutions were an airship, a fixed-wing aircraft or a surface sea vessel. If

Figure 11-2 is overlaid on Figure 11-3, it can be seen how important it is for people factors to be considered far earlier than when human factors are traditionally involved. The *opportunity* for greatest design influence was in the technology selection area. After the aircraft technology was selected, the systems developer was faced with yet another selection from the various aircraft designs available. Once again, the *opportunity* for design influence was much greater than the next step in the process.

In the final step, man was matched with the selected design in what we call man-machine interface, or classic human factors engineering. This example helps to illustrate one of the differences between traditional human factors engineering and MANPRINT. Even in its earliest stages, human factors engineering was not a major player until a systems concept was relatively fixed. MANPRINT, however, is involved in the optimization of the systems performance through human-centered analysis throughout the entire acquisition decision process (see Shields, Johnson, and Riviello, Chapter 10). The modern human factors professional should participate in all phases of these systems analysis models.

The point of the technology and design selection area discussions in this chapter is that those selections are made today using the output of combat models. Technology and design selection decisions are made based on candidate performance in a military complex environment model. Each candidate is evaluated on its ability to contribute to the competitive power of the overall force against the estimated opposing force in the predicted environment of the future. If combat models had been available for analysis of the 1941 antisubmarine problem, the preconcept modeling would have been conducted about 1926. The "problem" of the enemy submarine would have been a 1926 intelligence estimate of the likely German and Japanese submarine capability fifteen years in the future. The technologies and design for solution would have been based on the emerging technologies and designs of 1926. While the strategic planning processes of the 1920s were not as structured as they are today, planning for a war in the Pacific did begin in 1921. Virtually all possible scenarios were considered:

> For about fifteen years, the strategic concepts embodied in the ORANGE Plan formed the basis for most American war planning. Variations of the plan were prepared and discussed at length. Every conceivable situation that might involve the United States in a war with Japan, including a surprise air attack on Pearl Harbor, was carefully considered and appropriate measures of defense were adopted (Greenfield, 1960, p. 15).

Complex environment models now provide the systems developer the ability to view a conceptual future in such high resolution that systems and even subsystems are observable. To the degree we can see the general trends of our future, we can model that future in a complex model.

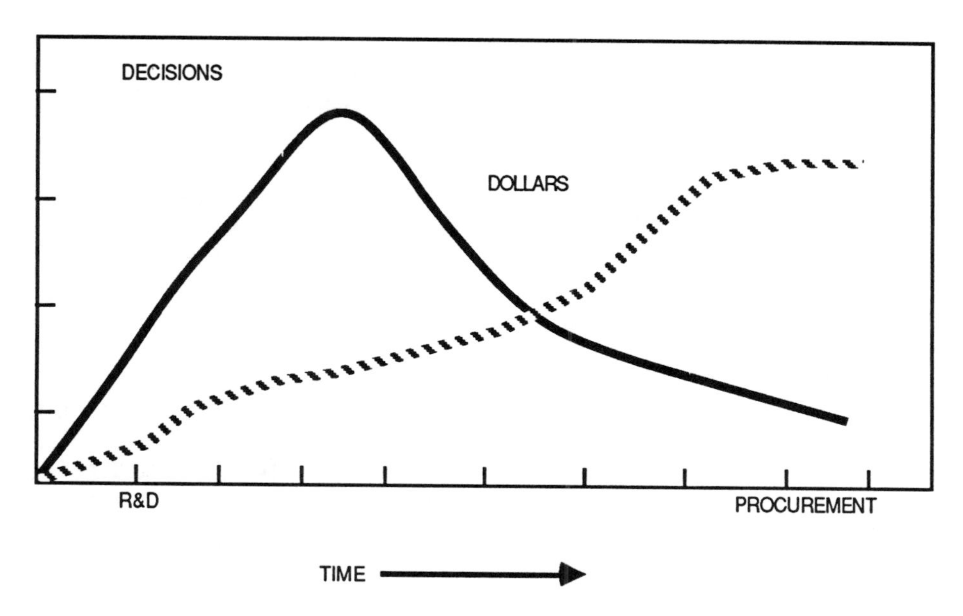

Figure 11-2
Decisions and Dollars: Early Decisions Drive Later Expenditures

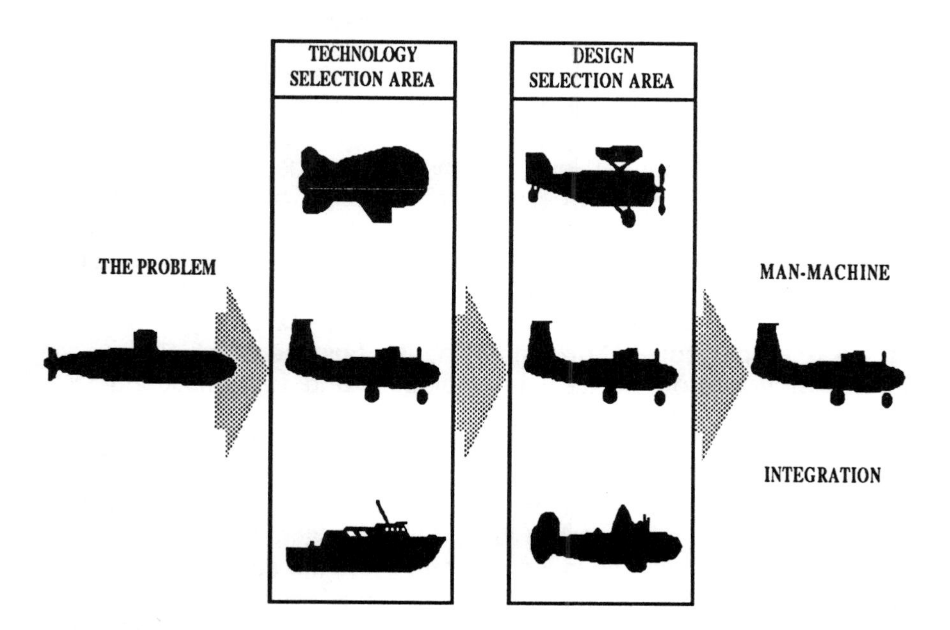

Figure 11-3
1941 Antisubmarine Warfare Example

Role of Analytical Programs

The conceptualization of any modern system requires early integration with its operational environment. The requirement for early systems integration is particularly important for military systems which are unique in that they must function in an environment which is extraordinarily hostile. Survival in this environment is frequently the principal mission of the system and also its principal measure of effectiveness. It is the analytical merger of the conceptual system with its operational environment which defines both the objective and importance of military combat modeling.

Current versions of systems development models are virtually all computer resident. Because of the complexity of systems development, plus the requirement for many repetitions, modern combat models are best suited for an automated environment. Combat models differ from computer-aided design/computer-aided manufacturing (CAD/CAM). CAD/CAM is used to conceptualize and manufacture a specific system. A systems development combat model, on the other hand, is used to demonstrate a system's performance in its anticipated wartime environment performing against its probable enemy. Combat Models are also unique in that both the system and the wartime environment are required to be speculative in order to estimate the probable reality at the time the system will actually perform its battlefield mission. Speculation concerning the future for the "red", or enemy force often involves highly classified intelligence estimates. This secrecy compounds the air of mystery which surrounds combat models.

Modern data systems provide the capability to view systems operational performance early in design, allowing elimination of candidate concepts well before they leave the drawing board. This relatively new capability to observe "draft" or "notional" systems inside a model of an operational environment presents not only new powers of design, but new problems as well. For example, entire technology options and systems design concepts can be eliminated before they have had a fair evaluation. Traditional human factors engineering begins when the concept of a system is sufficiently firm to permit the design of at least a mock up of the man-machine interface such as a cockpit simulator. The combat model, however, has allowed the selection of first order military technologies and systems candidates completely inside the notional reality of a computer.

In general, the advantages of automated complex modeling outweigh the disadvantages. This is true, however, only if the human performance factors can be modeled as well. The approach envisioned is as follows. Using the 1941 Antisubmarine example of Figure 11-3, a macro design for human factors influence is illustrated in Figure 11-4. Here the submarine becomes simply the problem-to-be-solved and the cost-benefit (Cost and Operational Effectiveness Analysis [COEA]) and the reliability, availability, and maintainability (RAM) analyses of each area influence flow into the analyses which influence design. There are several key documents (not shown in

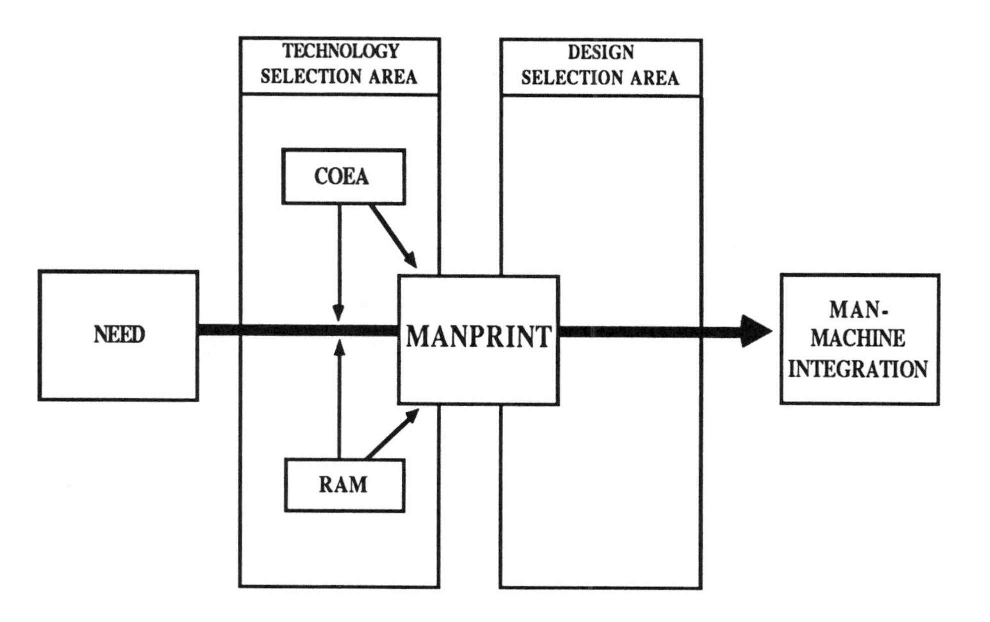

Figure 11-4
MANPRINT Analysis Integration Methodology

Figure 11-4) in this process which drive systems trade-offs and determine design. One of these, the Operational Mode Summary/Mission Profile (OMS/MP) is the first document to quantify a conceptual system solution to a problem (Thurmond, 1988). The OMS/MP converts the concept into numbers and becomes the quantitative basis for the development of both RAM and COEA analyses. The MANPRINT approach to quantitative design influence is through early involvement and quantification of human factors. The process begins with the OMS/MP, flowing through RAM and COEA processes, through source selection and, finally, into testing and system fielding.

A series of methodologies exist which are useful in assessing and predicting operator workload (OWL). Described in more detail in Booher and Hewitt, Chapter 12 (also see Christ, Bulger, Hill, and Zaklad, 1990), the OWL methodologies present the complex environment modeling community with an excellent start-point for the modeling of human operators.

HIGH IMPACT HUMAN VARIABLES

In order to be cost effective, those tasks frequently allocated to the human in systems which also appear to influence the outcome of combat should be selected for early study. Tasks not frequently allocated to humans (i.e., frequently automated) are not likely to provide support for the next level of data collection. Analysis of tasks which do not influence the outcome of combat will weaken the case for inclusion of human factors in models. As the process matures, high impact human tasks will be demonstrable. Two general areas offer promise as a start point:

Cognitive Performance

Man's brain is unique. We can build sophisticated controls to imitate man's motor skills in driving a vehicle, but we cannot yet build a computer with even the simplest of man's cognitive driving skills (Sworder & Haaland, 1989). There are a number of combat environmental factors which diminish a human operator's cognitive task performance. For example, there is a compounding relationship between combat stress and human behavior. Combat fatigue in the form of sleep loss causes stress; the demands of continuous combat aggravate the causes of sleep loss, and so on. From sleep decrement alone, Van Nostrand (1988) estimates the average combat officer will degrade to less than 50 percent of capacity performance within 13 days. The compounding impact of combat stress may be even greater when group and leadership situations are considered. Fear is another important human emotion in combat not only because it increases stress, but because it has an extraordinary ability to degrade people's willingness (and therefore likelihood) to perform critical systems tasks. By its very nature, combat is a hostile environment intended to diminish human performance. Temperature extremes, primitive living conditions, and the increased incidence of illness and non-combat injury also have a major impact on the human operator's ability to perform his systems-critical tasks.

Added to all other aspects of the combat environment are the human operator stresses created by the system itself. The system may generate excessive noise, toxic fumes and vibration, for example. Inadequacies in the ergonomic, man-machine interface may also diminish the human's ability to function. The sum of these factors which degrade human performance may be much greater than the sum of their parts. When we discuss the measurement and model replication of human systems components, we may be seeing the root of an exponential function, i.e., multiple, reinforcing stresses may degrade human performance much more than a single or isolated stress. None of these combat conditions are new or unique. Man was never expected to perform at 100 percent in combat . . . except in the models we use to select combat systems.

A major finding of the Army's Yellowstone National Forest Firefighting Study was a significant depression and hostility level in the soldiers participating in firefighting (King, Fatkin, & Hudgens, 1989). These findings of depression were considerably higher than those measured in military field exercises; they were, in fact, "higher than had ever been measured outside the laboratory" (see Figure 11-5). While the underlying causes of the

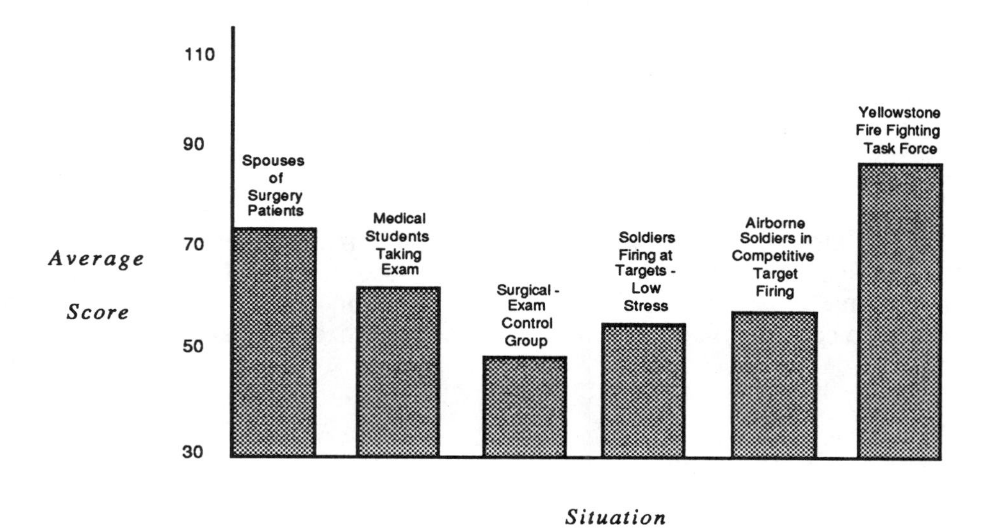

Figure 11-5
Depression Rates

psychological phenomenon observed in the Fire Fighting Study are not clear, their magnitude suggests the cumulative exponential impact of multiple stresses.

Visual Performance

One of the most common human-assigned systems tasks is the implementation of cognitive skills through vision. Very few systems allow the identification of friend-or-foe, for example, without human review of the system's decision parameters. Man's visual function may be direct or through visual review of a systems screen. Because the danger of human operator blinding by directed energy weapons is increasing, the near term trend will be toward indirect viewing (electronics offer a "fuse" to protect the operator's eyes). Ground based combat systems are uniquely dependent upon human vision for certain critical functions, however. Air and sea systems do not face the difficult background clutter problems of ground based systems. Because the background clutter presented around ground targets is a significant technological difficulty for automated target seekers, human visual performance is likely to remain a systems critical task in combat modeling. As such, human visual performance replication in combat modeling offers an excellent opportunity for model enhancement. Frequently modeled visual subtasks are:

- Target Acquisition
- Target Identification
- Target Tracking

Target Acquisition

Finding targets, particularly ground targets, remains a critical, human-allocated systems function. This function is also hampered by diminished human performance. Degradation of this function in combat modeling is rare. If a computer target is available, the computerized attacker always sees it.

Target Identification

Commonly referred to in the military as identify-friend-or-foe (IFF), this is one of the most difficult tasks in modern warfare and one of the most difficult to automate. The recent United States Navy experiences with the USS Stark and USS Vincennes incidents bring this point home vividly. Virtually all combat experience and military exercises show that accurate IFF is a serious problem which will worsen as systems gain the ability to acquire targets farther than targets can be identified. More important to the systems designer, however, is the failure to model this systems critical task which is always allocated to a human, even if only after the fact. When a fielded system fails to properly identify-friend-or-foe, no one will accept "the computer did it."

Target Tracking

Target tracking (performed after acquisition and identification) continues to be allocated to people in systems where tracking technology cannot meet the systems' demands. This is particularly the case in long range ground antiarmor weapons. Automation of target tracking is limited by the weight of portable systems and the costs of high quantity systems with missile mounted trackers. Ground systems must also overcome, as mentioned earlier, a severe background clutter. Target tracking is, therefore, another critical systems function likely to be allocated to the human operator's visual capabilities, and a good candidate for early integration into combat models.

Visual aptitude for many of the same military tasks seen in combat models has been measured (O'Neill, Batten, Woontner, 1988). An aptitude for target acquisition, identification, and tracking, for example, has been investigated as a discriminator for personnel selection and assignment. Figure 11-6 illustrates a test stimulus for target detection designed by the

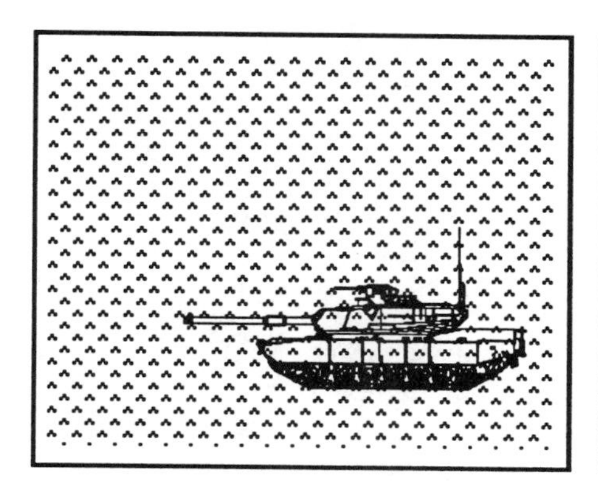

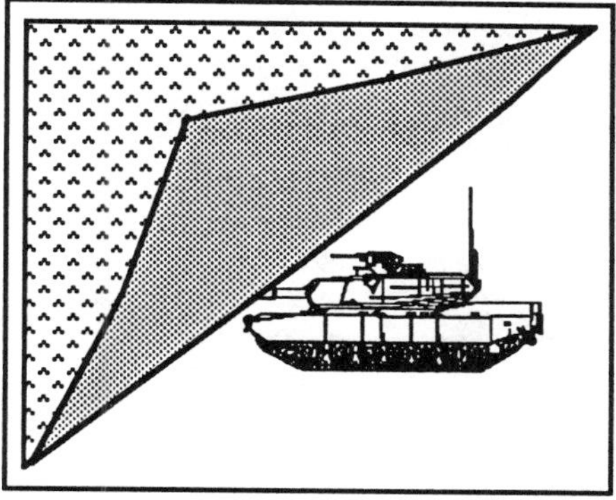

A: BACKGROUND (NOISE) B: TARGET

Figure 11-6
Target Detection Aptitude Test

Department of Behavioral Sciences and Leadership at the U.S. Military Academy. Subjects were shown target silhouettes against a camouflage background. The authors were able to show that certain individuals demonstrated an unusual aptitude for target detection. The relative wealth of data in this area of human performance offers both the combat modeling and personnel research communities an opportunity for mutual benefit.

IMPROVING COMBAT MODEL USEFULNESS

In order to make existing and future combat models more useful for systems integration, a scheme is needed to bring combat model structure and human performance data together. A suggested paradigm for the integration of human performance variables in combat models is illustrated in Figure 11-7. A first step is to identify the combat models most often used in the design and selection of systems. Care must be exercised, however, not to select models solely on the basis they are easily modified for human performance factors. Each model needs to be examined for impact on systems development. Systems development models typically are sophisticated engineering development models which *do not* lend themselves to human performance integration.

Second, select those systems for study which require "man-in-the-loop" for optimal functioning. Good candidates for study are those systems which depend upon people for the performance of critical functions. The intent, early in a MANPRINT program, is to pick those systems for study which are likely to show the importance of human factors, even when only limited human performance is modeled.

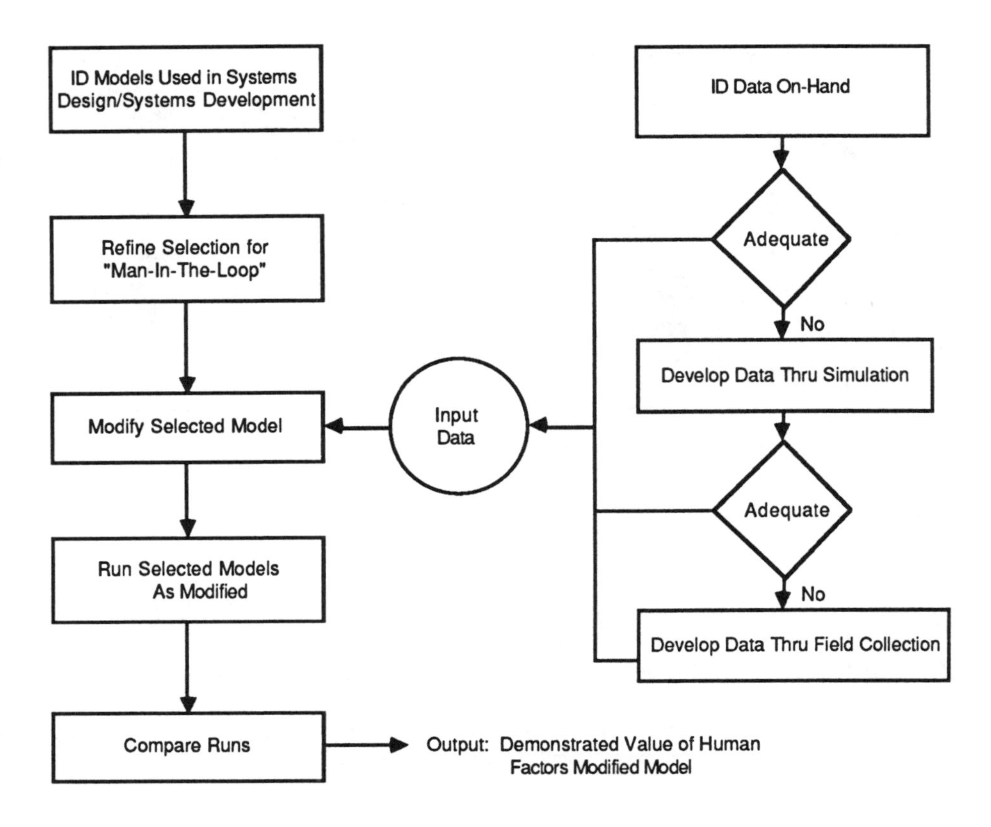

Figure 11-7
A Paradigm for the Integration of Human Factors in Combat Models

Third, select human systems tasks which are currently modeled by implication (i.e. man as 1.0) and for which data can be obtained, such as acquire target, identify target, lock-on target and fire. When systems are conceived, their designers allocate some tasks to man, some to the machine and some to both man and machine. In highly sophisticated design processes using elaborate task analyses this process is formal. More often however, it is informal. A combat aircraft, for example, might acquire a target automatically through the system, depend on its operator for correct identification and attack decision, then return control to the system for attack launch and execution. Selection of human-critical tasks will, like the first step, increase the likelihood that human variance will have an independent variable impact on model outcome.

Fourth, modify the selected model to allow replication of the discrete human functions selected. Actual model algorithm modification need not be complex. The initial modifications are required only to demonstrate that the human tasks selected do, in fact, influence the outcome of the analysis as shown by the measures of effectiveness. Modifying complex models to show the more discrete human functions such as suppressed action due to fear or diminished target acquisition due to cognitive overload is within our current capability. Some models already represent these functions to some extent (Janus(T) Documentation, 1985; Hughes, 1984).

Fifth, run the model with the human factors modifications using the best available data.

Sixth, compare the model output (systems exchange ratios, force exchange ratios, etc.) between the basic combat model and the human factors modification. At this point human performance can be observed in a quantified fashion which is both understandable and acceptable to the senior engineering design community.

Seventh, demonstrate the value of human factors algorithms in combat modeling through any significant differences between the basic and human factors modified model.

Prioritize Human Factors Data for Models

Using those critical systems tasks which are frequently assigned to humans (identify friend-or-foe, for example), develop a plan for the collection of human performance task data. Referring again to Figure 11-7.

First, search for existing data with high human factors and engineering community acceptance. In other words, use what we have. The data provided through Crew Systems Ergonomics Information Analysis Center (CSERIAC) (see Van Cott, Chapter 16) and other human performance data bases (see Haas & Laine, Chapter 17) are very applicable. This approach is particularly important early in the effort when the needs for combat modeling data are ill defined. Data collected without a good understanding of how they will be used are likely to be wasted. As the process matures, the personnel data development and modeling communities will develop an understanding of one another's needs and a protocol for data communication will evolve.

Second, develop data through the use of cost effective means such as developmental tests in training simulators. Since personnel data formats for combat models are likely to evolve, the costly process of test or field developed data is likely to be ineffective due to inevitable changes. The new family of aircraft and vehicle simulators offers an excellent opportunity to collect human performance data for combat model input. (Chung et al., 1988; Gaver & O'Muircheartaigh, 1989; Gaver & Parry, 1989; Geiger, 1989; Siegal, Wolf, & Schorn, 1981)

Third, develop data through field operations research. An excellent example of this concept was the Fire Fighting Task Force study sponsored by the United States Army's Concepts Analysis Agency in Bethesda,

Maryland (Van Nostrand, 1989). The Fire Fighting Task Force studied the psychological impact of stress caused by United States Army Infantry units fighting the Yellowstone National Forest fire in 1988. This type of effort not only generates data for use in modeling, but contributes to our understanding of combat theory.

Finally, loop early data development back to the human factors modified model algorithms in order to demonstrate human factors as an independent variable in the outcome of combat and document those human variables which warrant further developmental research. As human factors data is provided to combat model algorithms, the formats for data will become apparent, and a human factors/combat modeling data protocol will emerge.

TRENDS FOR THE FUTURE

The trend toward increased use of combat models for systems design and selection will continue. The ability to look first at competing solution technologies, and then review the operational performance of various technology designs in a model provides senior management with an extraordinary capability. When coupled with CAD/CAM, complex environment modeling provides a systems preview from concept through manufacturing, fielding and employment. Even the test and evaluation of complex systems is becoming so expensive that models are an attractive alternative. Many field operational tests are now modeled or heavily augmented with model based analysis. The M1 Tank and the F15-18 Fighter series were selected from the operational testing of several competing prototypes, but we may never see this approach again. The B2 bomber was designed and operationally tested entirely within an algorithm. The B2 may have benchmarked the design process of the future.

Although combat models are growing in popularity, they are likely to continue to omit human factors. Combat model proponents have a somewhat justified view of their critics as romantics who wax philosophically about the value of such human traits as leadership, morale and courage on the battlefield, but cannot quantify these dimensions in order that they be shown as "independent variables" in the outcome of analytical combat.

As we look to the future, several trends are emerging:

Combat models will become combat multipliers. Combat models are analogous to business accounting system models. They have provided military managers with estimates of their organization's performance, and are increasingly capable of providing real-time estimates of strategic and tactical situations. Military decision makers have an increasingly valuable decision aid in combat models. The battle won by the better model is just over the horizon.

Our understanding of the theory of combat will be enhanced. Our current understanding of combat is anecdotal, and it appears that the extraordinary stress of the experience makes accurate communication of combat reality difficult. How does combat flow? Do linear functions apply? Is it exponential, or chaotic? The Army has taken an important step forward in answering these questions in its efforts to compare the Janus (T) model to actual field exercises at the National Training Center (NTC) at Fort Irwin, California.

Combat models are being used more as on-line decision and training aids for senior management, enhancing their acceptability. Most military leaders are now trained in tactics and strategy using combat models. Students at the United States Army War College are trained using the same Janus (T) model that they will see again should they receive systems development assignments. Combat models have become the principal form of combat reality for the post-Vietnam era military leader.

The evolution of complex environment modeling. Military combat models have evolved to such a level of sophistication that they are now microcosms of reality, albeit a conflict reality. The field may be on the threshold of an expansion into other sectors. In the complex model family, the combat model has a close-cousin in the econometric model. Econometric models have been used for policy formulation for some time, although they appear to have been limited, for the most part, to the public sector (U.S. Long-term forecast model, 1988). Complex models may not have reached the private sector because of their inordinately expensive software development costs. Because they are intended to act as future predictors, these models also require a lengthy return-on-investment period. It would appear that there would be good applications in other areas, particularly in corporate long-range planning.

The MANPRINT community must respond to the growing use of models for systems design and selection by participating in the development and management of complex environment models. The output of a complex system model is a function of its input and its algorithms. By "building the tunnel from both ends," both will be influenced. Involvement after-the-fact will not allow optimization of the human dimensions of the system, putting the human factors proponent in the unenviable position of perennial "nay-sayer."

ACKNOWLEDGMENTS

For their assistance in preparing this chapter, the authors wish to thank Mr. Eugene Visco, Headquarters, Department of the Army, and U.S. Army Major Hirome Fujio, U.S. Army Personnel Command.

NOTE

[1]von Clausewitz, Carl, ON WAR, ed. and trans. Michael Howard and Peter Paret. Copyright (c) 1976 by Princeton University Press.

REFERENCES

ASARS I - Army small arms requirements study (1970). Fort Benning, GA: U.S. Army Combat Developments Command, Infantry Agency.

Balaconis, R. J. (1989, June). *Integrated strike warfare high fidelity simulation: Cruise missiles and TACAIR support of airland battle.* Monterey, CA: Naval Postgraduate School Thesis, Naval Postgraduate School.

Christ, R. E., Bulger, J. P., Hill, S. G., and Zaklad, A. L. (1990). *Incorporating operator workload issues and concerns into the systems acquisition process: A pamphlet for Army managers*. Alexandria, VA: U.S. Army Research Institute for the Behavioral and Social Sciences.

Chung, J. W., et al. (1988). *SIMNET M1 Abrams main battle tank simulation software description and documentation (Revision 1)*. Cambridge, MA: BBN Systems and Technologies Corporation for the Defense Advance Research Project Agency (DARPA), Tactical Technology Office, Arlington, VA.

Dupuy, T. N., & Hammerman, G. M. (1980). *Soldier capability – Army combat effectiveness (SCACE): Volume III, Historical combat data and analysis*. Fort Benjamin Harrison, IN: U.S. Army Soldier Support Center.

Gaver, D. P., & O'Muircheartaigh, I. G. (1989). *Latent factor models and analyses for operator response times.* Unpublished paper. Monterey, CA: Naval Postgraduate School.

Gaver, D. P., & Parry, S. (1989). *Human factors in combat models.* Research Proposal for 1989. Monterey, CA: Naval Postgraduate School.

Geiger, R. E. (1989). *Experimental design and analysis of M1A1 commander/gunner performance during CONOPS using the U-COFT.* Master's thesis in progress. Monterey, CA: Naval Postgraduate School.

Gropman, A. (1987, December). On pursuit of the holy grail. *Military Operations Research Society Phalanx.*

Greenfield, K. R., (Ed.) (1960). *Command decisions.* Washington, DC: Office of the Chief of Military History, Department of the Army.

Hughes, W. P. (1984). *Military modeling.* Alexandria, VA: Military Operations Research Society.

Janus (T) documentation (1985). White Sands Missile Range, NM: U.S. Army Training and Doctrine Command.

King, J. M., Fatkin, L. T., & Hudgens, G. A. (1989). Soldier Performance and Combat Stress, U.S. Army Human Engineering Laboratory, Aberdeen,

MD. In S. J. Van Nostrand (Ed.), CAA Study Report SR-89-10. Bethesda, MD: U. S. Army Concepts Analysis Agency.

Murtaugh, S. A. (1987, December). Of human behavior and actions as an essential ingredient for realism in combat modeling. *Military Operations Research Society Phalanx.*

O'Neill, T. R., Batten, D. A., & Woontner, S. (1988). *What the gunner's eye tells the gunner's brain: III Predicting target detection performance* (Human Sciences Laboratory Report No. 88-1). West Point, NY: United States Military Academy.

Pfeiffer, M. G., Siegel, A. I., Taylor, S. E., & Shuler, L., Jr., (1979). *Background data for the human performance in continuous operations guidelines* (Technical Report 386). Alexandria, VA: U.S. Army Research Institute for the Behavioral and Social Sciences.

Prange, G. W. (1981). *At dawn we slept.* New York: McGraw-Hill.

Schroth, T. F. (1989). *An introduction to human factors and combat models. Master's thesis in C3.* Monterey, CA: Naval Postgraduate School.

Siegel, A. I., Wolf, J. J., & Schorn, A. M. (1981). *Human performance in continuous operations: Description of a simulation model and user's manual for evaluation of performance degradations* (Technical Report 505). Alexandria, VA: U.S. Army Research Institute for the Behavioral and Social Sciences.

Sworder, D. D., & Haaland, K. S. (1989). A hypothesis evaluation model for human operators. *IEEE Transactions on Systems, Men and Cybernetics* (Vol. SMC-19). New York: IEEE.

Thurmond, P. E. (1988). *MANPRINT analysis methodology: Victory through design.* Research paper prepared for the Department of the Army. Frederick, MD: Path Corporation.

U.S. long-term forecast annual model (1988). Bala Cynwyd, PA: Wharton Econometrics Forecasting Association.

Van Nostrand, S. J. (1986). *Model effectiveness as a function of personnel [ME = f (PER)]* (Research Paper CAA-SR-86-34). Bethesda, MD: U.S. Army Concepts Analysis Agency.

Van Nostrand, S. J. (1988). *Including the solider in combat models* (Executive Research Report S73). Washington, DC: Industrial College of the Armed Forces, Fort McNair.

Van Nostrand, S. J. (Ed.) (1989). *Fire fighting task force* (CAA Study Report SR-89-10). Bethesda, MD: U.S. Army Concepts Analysis Agency.

von Clausewitz, C. (1984). *On war* (M. Howard and P. Paret, Trans. and Eds.). Princeton, NJ: Princeton University Press. (Original work published 1832).

CHAPTER 12

MANPRINT TOOLS AND TECHNIQUES

Harold R. Booher
Glen M. Hewitt

ABSTRACT

Analytical tools and techniques for human performance prediction and assessment exist for each of the MANPRINT domains. MANPRINT brings the various human performance methodologies under a common umbrella so that the analysis findings are (a) coordinated among the domains, and (b) can be integrated in a timely manner with total system analytical methodology used by top management decision makers, system designers, managers and implementors, and system support planners and trainers. This chapter addresses (a) the background important to understanding the value of MANPRINT tools; (b) the use of MANPRINT tools in acquisition decisions; (c) MANPRINT domain tools and the current state of the art of some of the better methodologies, e.g., operator workload (OWL), HARDMAN (Hardware and Manpower) III, and advanced human factors engineering (HFE) technologies; (d) factors in tool identification and selection; and (e) some research and development (R&D) issues and future needs for MANPRINT.

BACKGROUND

Human performance analytical models and techniques for each of the MANPRINT domains have been available for military and industrial applications for at least twenty years (Meister, 1976). Yet their effect on system design has been unimpressive. The domains of human factors engineering (HFE), systems safety, and health hazards have had more effect than manpower, personnel, and training (MPT), because of their historical and practical closeness to engineering, and, in the case of HFE, because of the greater availability of tools. Even with the more advanced technologies of HFE, extensive use has not been made by system integrators (see Price, Chapter 6). There are a number of reasons for the failure to apply MANPRINT tools and techniques. Before the Navy HARDMAN (Hardware and Manpower) methodology was developed, most human resource MPT models were not designed to support decisions for specific systems; rather they were directed toward the routine process of managing human resources (Rostker, 1984). These earlier human resource models were used to determine how many people would be recruited and promoted, and how much they would be paid each month, but were unable to predict how many people, or what unique personnel knowledges, skills, or aptitudes would be needed to operate and maintain a weapon system once placed into operation.

Another reason for the lack of use of MANPRINT techniques is simply the myriad of techniques in one stage of development or another which have been or are being prepared to aid the engineer and system integrator. Fleger, Permenter, and Malone (1988) identified 113 advanced tools in HFE alone. Lysaght et al. (1989) reviewed over 1500 technical and research reports in evaluating operator workload methodologies; and, the first phase of a Air Force effort to define a Manpower, Personnel, Training, and Safety (MPTS) analytical system for the weapon system acquisition process examined more than 75 MPTS tools and 45 data bases (Shields, Rossmeissl, & Johnson, 1989). Still, another reason for their disuse is that these tools and data bases are not generally tailored to the system integrator or to specific user's needs. Moreover, many are "presently inconsistent, incomplete, and often unquantified" (Gentner, 1989).

Understanding the value of human performance tools and techniques requires an appreciation of two critical human factors concerns. These are (1) individual differences, and (2) limits in performance capacity – operator workload.

Individual Differences

The utility of MANPRINT tools depends upon their sensitivity to individual differences. While it is obvious that people are different, expert researchers and engineers fail to account *how* different and conduct experiments and design systems around "the average man." Nearly all experimental studies concentrate on comparisons of group "means" at the expense of obscuring individual differences (using terms such as "within group variance" or "error variance") (Hanser, 1989; Jones, 1966; Rimland, 1983; Snow, 1986). Kaplan and Miles (1981), for example, report how a high performance jet aircraft was designed for the average person resulting in an ejection system that would break the legs of larger pilots. The variability of human anthropometrics (the most obvious of individual differences) is often not a consideration until late in the development process. (See Price, Chapter 6, for many other examples of this tendency to ignore obvious human differences.) Illustrating the magnitude of individual differences, Williams

(1956) long ago reported huge variations even in something so basic as the heart. In displaying human samples of the right atrium, for example, he notes that the forms vary so widely "one might doubt they are from the same species." Rimland, after examining individual comparisons within groups conducting such widely disparate activities as sports, typing, school work, warfare, and computer programming, concludes that in the areas of human behavior and performance, "the range of variation in even small groups rarely falls below 200 percent, and differences in the thousands of percents are not uncommon" (Rimland, 1983). Moreover, performance differences are not reliably reduced solely through training. People do not become similar with training or experience; differences broaden in proportion to complexity of task (Rimland, 1983; Farr, 1990).

One very critical challenge for MANPRINT tools to aid in systems design is for human beings to be "considered not only as interacting elements to be integrated into the system, but also as elements varying on many dimensions that contribute significantly to system variance" (Sorensen, 1983). This interaction becomes complex in the development of effective MPT tools because of the time and cost involved in relating "predictive validity" of individual aptitudes (as measured by test scores) with actual performance on tasks. The interaction between humans and systems becomes even more complex because of the relationship between jobs and abilities (see Figure 12-1). Matching people from expected manpower pools in the *future* to jobs resulting from projected new systems conceived and designed *now* is one of the many uses of MPT tools.

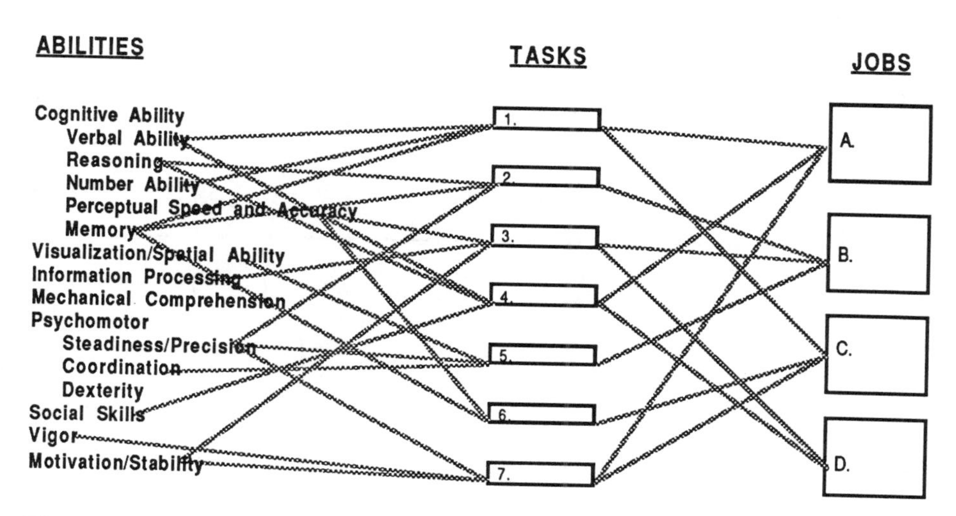

Figure 12-1
Relationships Among Abilities, Tasks, and Jobs
(Hanser and Arabian, 1989)

Limits in Performance Capacity – Operator Workload

As technology becomes increasingly integrated into operations systems, trends are evident in the transfer from physical workload imposed on people, to cognitive (or mental) workload. As designs require operators to perform more supervisory, monitoring, and overseeing functions, hardware and software do more of the physical work while operators spend more time analyzing, planning, and checking for failure or emergency conditions. That is, in high technology systems, the operator's job may shift from a "line" function to performing a "staff" function. Unquestionably, physical workload can induce fatigue or stress which degrades performance, but recent MANPRINT efforts in system design have concentrated on cognitive (or mental) workload. Workload from cognitive tasks includes perceptual and psychomotor tasks and, in many instances (such as in warfare or firefighting), it is nearly impossible to sort out the physical component from the cognitive.

Cognitive workload has a complex relationship with performance. As shown in Figure 12-2, it seems to be an inverted U relationship. Performance can be adversely affected at low (boring or desensitizing) levels of workload as well as at overload levels. There is also a general interval of workload where, as task demands increase, operators are able to increase their efforts thus maintaining a performance level in the acceptable range. At some threshold, however, increases in workload will cause performance to drop catastrophically.

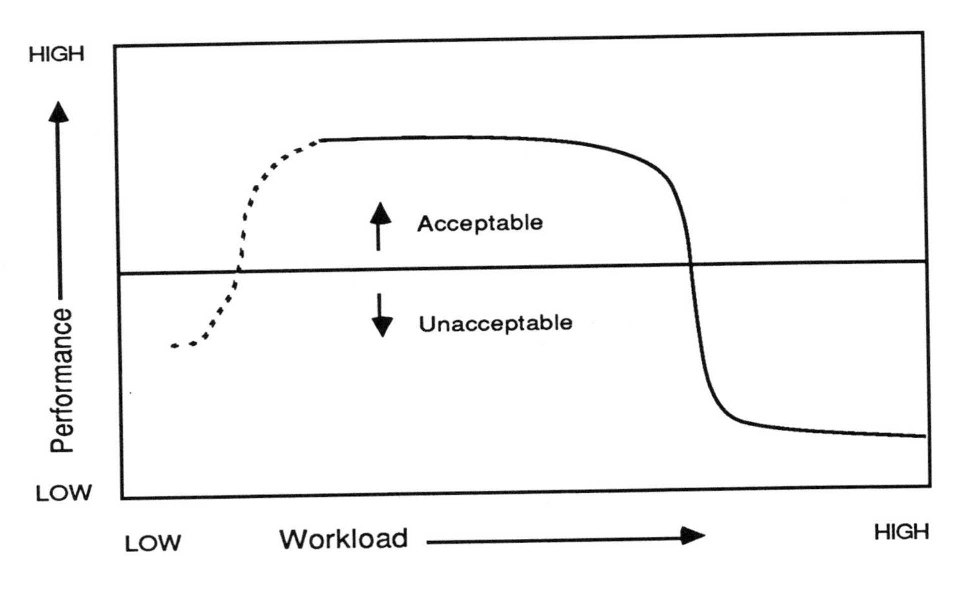

Figure 12-2
The Hypothetical Relationship Between Workload and Performance

To understand cognitive workload in high technology systems, a simple illustration may help (Christ, Bulger, Hill, & Zaklad, 1990). Place your watch in front of you and keep track of the time to do the following three tasks:

1. recite the alphabet
2. count from 1 to 26
3. now do both together, intertwining the alphabet with the counting (A-1, B-2, etc.), saying the answers.

If you actually got all the way through the combined task, you are unusual. Most people give up about G-7 or H-8. It is difficult because of the load on memory. Now, try the double task again, but write down the answers as you go. Performance is improved by reducing the mental load.

Some insight into workload has come from recent research by the Army (Christ et al., 1990, p.10). Several general conclusions have emerged:

- Workload is relative. It depends on both the external demands and the internal capabilities of the individual. Also, it can vary over time for an individual.
- Workload causes individuals to react in different ways. Workload is related to but not the same as the individual's performance.
- Workload involves the depletion of internal resources to accomplish the work. The higher the workload, the faster the resources are depleted.
- There are a diversity of task demands and a corresponding diversity of internal capabilities to handle these demands. Persons differ in the amount of these capabilities that they possess.

The Army MANPRINT program for workload research accepted a general definition of workload as the *relative capacity to respond*. This working definition implies both an amount of spare capacity and the ability and willingness of the operator to use that capacity in the context of the specific personal and environmental situation.

With a better appreciation of some of the human factors concerns that MANPRINT tools are designed to address, it is now possible to examine how MANPRINT tools are used in acquisition decisions.

MANPRINT TOOLS IN ACQUISITION DECISIONS

The principal objective for MANPRINT tools (Table 12-1) can be summarized as providing methods to get human performance data into the decision process in a way sufficiently quantitative that decision makers can make meaningful and reliable tradeoff decisions. The application of MANPRINT methodologies are critical at each phase of the materiel acquisition decision process, but, the earlier the data is provided and acted upon, the better.

Table 12-1
Objectives for MANPRINT Methodologies

1. To identify high resource costs and human error potential at very early stages of preliminary system design.
2. To influence the design considerations of human physiological and cognitive requirements.
3. To convert human performance requirements into engineering specifications.
4. To determine goals and constraints for the MANPRINT program.
5. To develop and validate designs that meet total system (man/machine) performance requirements.
6. To provide essential data for the decision body throughout the materiel acquisition process.

In order to be most useful, the various existing techniques need to be tailored to answer questions posed by the different decision makers at each system acquisition stage. As suggested by Shields, Johnson, and Riviello (Chapter 10, Table 10-2), a set of generic questions should be asked by the decision makers to identify issues throughout the system acquisition cycle. As examples, they suggest asking: Will this concept/design increase the likelihood that the total system will meet performance requirements? Or, will the anticipated manpower will be sufficient to operate, maintain, and effectively support the total system?

Figure 12-3 shows a simple model of a military system acquisition process and the MANPRINT influence on decisions. The questions, posed by Shields, Johnson, and Riviello (Chapter 10), must be addressed within the context of acquisition process phases to best influence design. Table 12-2 lists examples of the kind of issues that MANPRINT tools and data bases should be designed to address.

Currently, the most comprehensive decision modeling approach is that of the Air Force in its Integrated Manpower, Personnel, and Comprehensive Training/Safety (IMPACTS) program (Potempa & Gentner, 1988). The first phase of a three-phase effort has been completed. The Air Force approach uses a functional modeling procedure which graphically

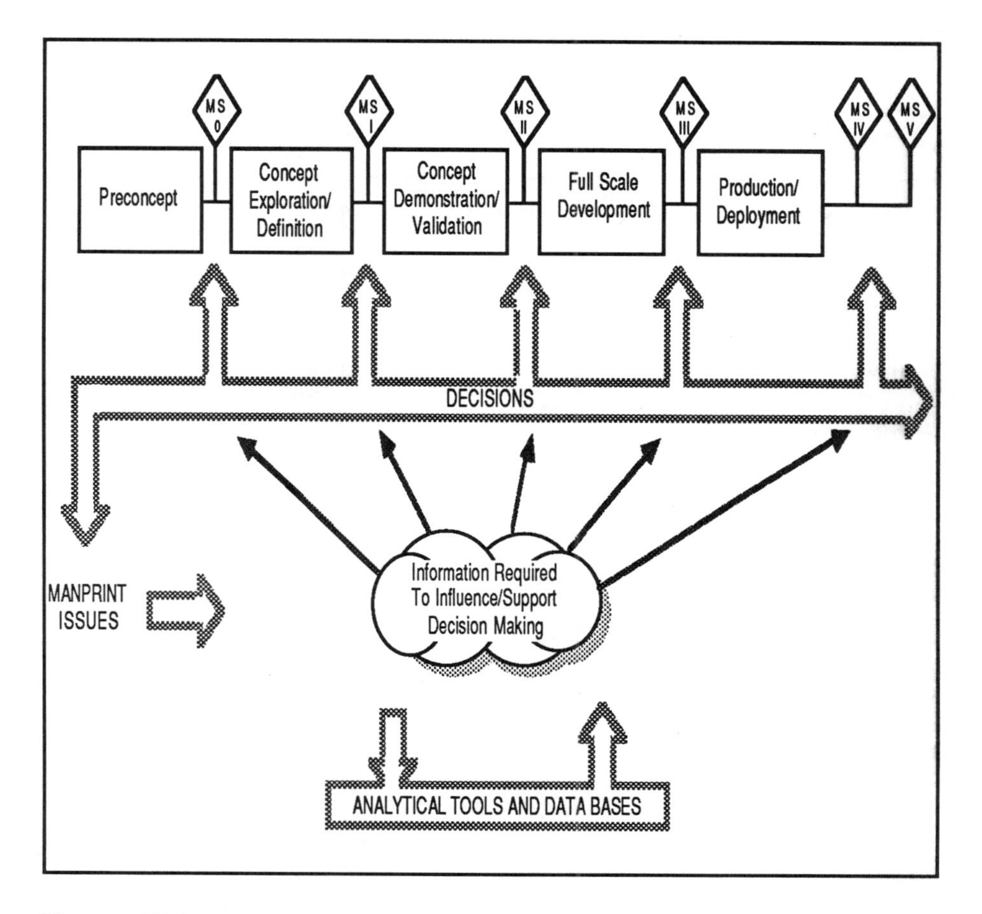

Figure 12-3
MANPRINT Issues, Decisions, Analytical Tools, and Data Bases in the Acquisition Process

describes complicated processes in terms of a hierarchy of functions or activities (Shields, Rossmeissl, & Johnson, 1989). The first level represents an entire system or process as a single modular unit (with inputs and outputs). The single unit is broken down further and further into modules and submodules. In this approach to weapon system acquisition phases, the single unit represents the total process; the next level addresses the five stages already shown (Figure 12-3 and Table 12-2). The third level of detail can be shown by using the Demonstration/Validation (DEM/VAL) Phase as an example. During this phase, the next lower decomposition level consists of five subactivities:

Table 12-2
Examples of MANPRINT Issues by Acquisition Phase

Pre-Concept Exploration	Concept Exploration	Demonstration and Validation	Full Scale Development	Production and Deployment
What are the MANPRINT goals and constraints that should be included in the functional capability requirements? What conceptual changes in the new system may require changes in the numbers and types of personnel employed in the predecessor system or systems?	What are the MANPRINT impacts of the conceptual approaches and how do the MANPRINT issues impact upon the approaches? What are the components of the baseline comparison system (BCS) and what is the MANPRINT profile of the BCS? Do MANPRINT/System concept impacts require a refinement to the operational and maintenance concept? What MANPRINT data to be provided by the contractor should be called out in the DIDS and CRDLS?	What MANPRINT impacts should be incorporated in updating the requirements, program, and decision documents? What MANPRINT trades require documentation? What level of personnel performance can be achieved with the system and does that level meet the requirements? What MANPRINT impacts require changes to the program funding? Have revisions to the MANPRINT description been integrated into the Decision Concept Paper and have MANPRINT decisions been reflected in the Manpower Estimate Report?	What MANPRINT decisions and trades should be incorporated into the production model? How much emphasis should be placed upon MANPRINT during source selection and what should be the MANPRINT selection criteria? Will a complete training system be ready for concurrent fielding? Have any safety issues been identified during system testing and what is their nature?	What MANPRINT issues, concerns, or opportunities have been Identified during system fielding and deployment? Have MANPRINT issues been identified that require modifications to the production system?

(1) award and execute DEM/VAL contracts
(2) evaluate and select alternative concept prototypes
(3) develop/update program documents
(4) develop RFP and solicit proposals for FSD
(5) conduct acquisition Milestone II decision revision

Proceeding deeper and deeper down the hierarchy Shields and associates reach a lowest level decomposition. For example, under (2) above, "evaluate and select alternative concept prototypes," activities like "update personnel baseline, evaluate skill requirements, determine Air Force Systems Command structure and skill levels, conduct personnel tradeoffs and initial early manpower and personnel actions" are the lowest level actions analyzed. Using the graphic model of the activities required to produce an effective system, the next effort analyzed and documented the MANPRINT issues that needed to be addressed (and, consequently, the MANPRINT questions to be asked). Figure 12-4 shows a sample page of

What are the MANPRINT requirements that should be integral to the system design specifications? Are MANPRINT issues being addressed as system issues? What are the results of subsystem/component prototype testing? Are there MANPRINT impacts?

- Is MANPRINT integrated into the design process?

  - •• What MANPRINT goals, constraints and processes should be highlighted in the post award conference?

  - •• What are the appropriate contractor MANPRINT trades? What level of sensitivity analysis is conducted to determine MANPRINT trades?

    - -- What is it that the government really wants? How will this alternative affect MANPRINT?

    - -- What guidance has been provided? What are the goals and constraints that should be considered in the design process?

    - -- Which configurations meet both performance and MANPRINT goals and constraints? What are the costs? What are the risks?

    - -- How does this subsystem fit in the total system? What impact does it have on MANPRINT?

  - •• What changes need to be made to government planning documents? What program feedback is required due to MANPRINT planning? What initial projections (demographics, etc.) require updating?

  - •• Are the contractors using an organizational structure and structured process to integrate MANPRINT into the design? Are the contractors using MANPRINT techniques during the design? Is the contractor MANPRINT information appropriate? Are there design issues that could change the MANPRINT description? What is the appropriate allocation of tasks to hardware/software operational, maintenance, and support personnel?

- What are the MANPRINT impacts of the alternative designs and how does MANPRINT impact the design? What design aspects should be selected for full scale development?

  - •• Are technical tests structured to support early MANPRINT data collection? Does the system design/subsystem prototypes support MANPRINT goals? What are the MANPRINT impacts of the design approach? What are the MANPRINT impacts of the subsystem prototype?

    - -- What cost and feasibility studies should be conducted to influence the functional baseline? How does the functional baseline influence future MANPRINT requirements?

    - -- Do MANPRINT impacts of the functional baseline require a change to the operational and maintenance concept?

Figure 12-4
MANPRINT Decisions/Issues in the Demonstration/Validation Phase

some of the relevant MANPRINT decisions/issues for the Demonstration Validation Phase. (A complete list of typical decisions and issues by system acquisition stages is provided in Rossmeissl, Akman, Kerchner, Faucheux, Wright, and Shields, 1990.)

Having completed the above efforts, the Air Force then identified all MPTS tools and data bases which appear useful in addressing the catalog of issues and questions. The initial analyses reviewed the 75 tools and 45 data bases for certain characteristics. The data base characteristics included (a) when it will be available in the Weapon System Acquisition Program, (b) what information it will contain and how the content will change as the acquisition process changes, (c) how an individual can access and use it, and (d) what other data bases will it interface with. Analytical tool characteristics included (a) what it does, (b) how it is used, (c) what data bases are needed to support it, (d) what are the interlinkages with other MPTS tools, (e) what is tool validity under specified conditions, and (f) how adequate is the tool (i.e., strengths and weaknesses). (A full listing of the tools and data bases are available from either Rossmeissl et al. at HAY Systems, Inc. or the Air Force Systems Command.)

MANPRINT TOOL CLASSIFICATION

Recently a number of studies have identified, reviewed and assessed the various MANPRINT tools and techniques (Fleger et al., 1988; Rossmeissl et al., 1990; Bogner, Kibbe, & Laine, 1988; Lysaght et al., 1989). In addition to classification by system acquisition phase, some of the more useful classification criteria of tools and techniques are:

- MANPRINT domains
- Type of activity (design, analysis, evaluation, etc.)
- Task based vs. comparability
- Manual vs. software
- Simulation vs. statistical
- Analytical vs. empirical

Human Factors Engineering Tools

While one useful classification scheme for systems integration analyses is by MANPRINT domains, the domains are not equally represented by analytical tools. The domain best represented is human factors engineering (HFE). Fleger et al. (1988) provide the most complete survey of advanced HFE tool technologies. Using an 8-point classification scheme (acquisition phase, tool activity, class, type, role, application, status, and cost) they have developed a database which includes 88 HFE tools.

Table 12-3 lists the highest rated tools for four phases of the materiel acquisition process (based on six trade-off criteria – availability, accessibility, adaptability, utility, training, and mobility). (The complete data base is available from the U.S. Army Human Engineering Laboratory.) A cost estimate (low, medium, high) was also provided. The tools recommended for government procurement were those which are operational with good adaptability and demonstrated utility, that fall toward the low to moderate end of the cost spectrum.

A tool to integrate HFE with the other MANPRINT domains is currently under development. Entitled IDEA (Integrated Decision/Engineering Aid), it is designed to be an automated, productivity enhancement system for conducting HFE and MANPRINT interface analyses (Westerman et al., 1989). IDEA seems especially promising as an easy to use expert system which (when completed) can identify and access all the relevant HFE information provided by technical reports, standards, and guidelines. Its goal is to provide a focal point and repository for all HFE and other MANPRINT domain interface information regardless of origin, to encompass HFE design, analyses, and evaluation requirements of all materiel acquisition phases.

MPT Tools and Techniques

MANPRINT has placed considerable interest on MPT tool and technique development, because they are needed to influence the "macro" decisions (manpower costs, personnel aptitudes, training resources) early in systems acquisition and tool availability was limited to costly, slow manual techniques (like the original Navy HARDMAN). The MPT technique void is being filled as new products, resulting from increased emphasis on MPT, become available.

Table 12-4 describes special features of four of the most frequent types of analytical modeling methods, i.e. *simulation, statistical, comparability* and *task based.* Table 12-5, which lists a large number of MPT methods, provides for each: a brief description, applicable domain(s), applicable acquisition phase(s), type of analytical technique, type of tool, and availability.

Safety and Health Hazard Tools

Although not specifically labeled as such, much of systems safety and health hazards assessment can be covered by the human factors engineering tool technologies. Several of the analytical tools (Fleger et al., 1988) were specially designed for safety and health application. There are also some useful data bases maintained for these domains (Rossmeissl et al, 1990). Table 12-6 lists common sources of safety and health hazards.

Table 12-3
Best Advanced HFE Tools (Fleger et al., 1988)

NAME	DESCRIPTION	ACTIVITY / ROLE	ACQUISITION STAGE	COST
CADAM/ADAM (Anthropometric Design Aided Mannequin) & EVE (Ergonomic Value Estimator) CADAM Inc. Lockheed Missiles & Space Co. Sunnyvale, CA (McGuiness et al., 1986; Hickey et al., 1985)	Man-model generates 2-D engineer drawings that can be viewed from three angles.	Design/technician access to equipment during operation and maintenance.	DEM/VAL FSD	Moderate
CAPRA (Computer-Aided Probabilistic Risk-Assessment) Essex Corp. Alexandria, VA	Hardware reliability model- integrates hardware starting with machine operation. Based on tasks broken into micro-motions categorized by difficulty level.	Design, T&E; copiers, maintenance analysis	FSD	Moderate
CAR (Crewstation Assessment of Reach) Analytics Inc. Willow Grove, PA (Morrissey et al., 1985; Harris et al., 1982)	Link man-model and an adjustable workspace model for assessing pilot anthropometric data.	Design/reach evaluation, panel design	FSD	Moderate
DART (Data Analysis and Retrieval Technique) Douglas Towne Redondo Beach, CA	Breaks scenario into component tasks and analyzes the workload associated with tasks.	Analysis, workload, T&E, Front-end Analysis	DEM/VAL FSD	Moderate
	Presents analysis of hand motions associated with task actions.	Manufacturing and assembly, commercial and military aircraft	DEM/VAL FSD	Moderate
GEO MOD (Geometric Modeling Tool) Hughes Aircraft Co. Fullerton, CA	Tool for developing workstation around a man-model. Produces 2-D blueprint, system modifications on screen.	Design/development of aircraft cockpits, workstations	FSD	Low
GRASP (Graphical Robot Applications Simulation Package) Dept. of Production Engineering & Production Management Nottingham University Nottingham, England (McGuiness et al., 1986)	Improve safety features within a robot installation; robot operating zones, man-interation with robot.	Design/non-aviation, commercial applications		

Table 12-3 *(continued)*

NAME	DESCRIPTION	ACTIVITY / ROLE	ACQUISITION STAGE	COST
HF - Robotex (Human Factors - Robotics Expert System) White Oak Laboratory Naval Surface Weapons Center Silver Spring, MD (McGuiness et al., 1986)	Assist in application of HF principles, data and techniques to robotics systems design. Purpose is to allow designer a rapid and efficient search of the HF knowledge base.	Design/robots	FSD	Low
Micro SAINT (Micro-Systems Analysis of Integrated Networks of Tasks) Micro Analysis Design Boulder, CO (Laughery, 1984)	Microcomputer version of SAINT. Simulates activities of human operations within complex systems. Primary used by human factors specialists - numerous user-interface features.	Analysis/building and executing task network models of human operators	Concept Eval. DEM/VAL FSD Prod. Imp.	Moderate
SIMWAM (Simulation for Workload Assessment and Modeling) Carlow Associates, Inc. Fairfax, VA (Kirkpatrick & Malone, 1984)	Micro-computer-based task network modeling technique for assessmentof operator workload and performance effectiveness in man-machine systems.	Analysis,T&E/ applied to aircraft carrier air operations, surface ship air detection and tracking area and surface/sub-surface area.	DEM/VAL FSD Prod. Imp.	Low
WORG (Workspace ORG) Office of Naval Research Arlington, VA (Pulat, 1983a)	Part of Multiman-machine Work Area Design and Evaluation System (MAWADES). Prepares graphic layout of several work-stations within a workspace.	Design/workstation arrangements facility design	FSD	Moderate
WOSTAS (Workstation Assessor) Office of Naval Research Arlington, VA (Pulat, 1982, 1983b)	Accepts mission oriented task requirements. Generate alternative scheduling schemes of tasks to workstations through application of scheduling and line balancing concepts.	Analysis, task allocation, workload/ probabilitic branching to allow operator to assume alternative tasks to prevent bottlenecks.	FSD	Moderate
ZITA (Zero Input Tracking Analyzer) Norman Walker Associates Maryland	Method of predicting shifts in behavior as result of workload- induced stress. Tests persons tracking ability under stress.	Analysis/workload and stress as it affects tracking skills.	Concept Eval. DEM/VAL	Low

Table 12-4
Analytical Modeling Methods

Simulation:

Simulations use models for mimic, study, and experimentation of dynamic systems. They provide valuable insight into system sensitivity to MANPRINT issues and offer low cost hypothesis testing on actual or proposed systems. Simulations have been successfully applied to electrical, optical, aerospace, biological, psychological, acoustic, and various other types of complex systems. Simulations, which may also involve analog and analog-digital (hybrid) processes, have become increasingly powerful analytical tools as a result of the high technology computers.

Statistical:

Mathematical modeling is universally recognized as a valid, stand-alone method to analyze a system. Mathematical analytical techniques have been developed as a result of simulation and system design analysis as well as an adjunct to them. The development of algorithmic descriptions of a dynamic system necessarily entails uncertainty that can be adequately represented by statistical methods. Interactive adjustment of mathematical parameters leads to greater model improvement and increased understanding of system MANPRINT implications.

Comparability:

Comparability analysis provides one of the most readily available methods of understanding a system design impact. This approach involves decomposing the system into subsystems and subfunctions for which data on predecessor or similar subsystems and subfunctions already exist. Through a rational process of comparison new versus old (especially where a full task analysis is unavailable), MANPRINT domain sensitivity to the design can be determined, even early in the acquisition process.

Task-based:

For those systems, subsystems, or components that have been analyzed sufficiently to identify human performance tasks or subfunctions, an analysis based upon the aggregation of these tasks or subfunctions offers the greatest opportunity for detailed design review. Such evaluations that assess both cognitive and physical workload, especially in a time-sequence analysis, support considerations for functional allocation of tasks, man-machine interface complications, and MANPRINT requirements.

Table 12-5
MANPRINT Methods for Manpower, Personnel, and Training

METHOD	DESCRIPTION	Domain	Acquisition Phase(s)	Analytical Techniques	Type Tool	Availability
Addressing Manpower, Personnel and Training (MPT) Issues in Human Factor Engineering Analysis (HFEA) (Bogner, Kibbe, & Laine,1990)	Method (flow chart and narrative instructions) for preparing MPT portions HFEA. Input: ASVAB profiles, training and performance data critical tasks. Output: Matrix of soldier performance related to ASVAB composite scores and training resources consumed.	MPT	CE-PD	S	H/G	C
AIM (Authoring Instructional Material) (Rossmeissl et al., 1990)	Training system upgrade for production and life-cycle maintenance support of instruction and instructional materials to enhance personnel training. Introduces state-of-art technology and advanced computer-based technologies into current system.	T	CE-PD	NA	S	N
AMCOS (Army Manpower Cost System) (Bogner, Kibbe, & Laine,1990)	PC - based, used to forecast manpower costs for life cycle of proposal weapon system. Input: Manpower by grade by MOS. Output: Manpower costs by year and budget appropriation category	M	CE-PD	S	S	C
BRAT (Budget/Readiness Analysis Technique) (Rossmeissl et al., 1990)	Provide link between support resources (manpower, Line Replaceable Units (LRUs), and Support Equipment) and weapon system readiness. Useful in translating support resources to level of readiness comparing alternative support concepts and operational procedures.	M		S	S	C

LEGEND

Domains:
- M = Manpower
- P = Personnel
- T = Training
- HFE = Human factors engineering
- SS = System safety
- HH = Health hazards

Acquisition Phase(s):
- CE = Concept exploration
- D/V = Demonstration/validation
- FSD = Full scale development
- PD = Production/Deployment

Analytical Techniques
- Si = Simulation based
- S = Statistical based
- C = Comparability based
- T = Task based

Type Tool
- H/G = Handbook/guide
- DB = Database
- S = Software

Availability
- C = Current
- N = Near Term

Table 12-5 *(continued)*
MANPRINT Methods for Manpower, Personnel, and Training

METHOD	DESCRIPTION	Domain	Acquisition Phase(s)	Analytical Techniques	Type Tool	Availability
CODAP (Comprehensive Occupational Data Analysis Programs) (Rossmeissl et al., 1990)	Set of computer programs to automate, process, organize, and report occupational data. Useful in occupational analysis for evaluating, updating enlisted classification structures and redesigning jobs.	P		S	S	C
COMBIMAN (COMputerized BIomechanical MAN-Model) (McDaniel & Hofmann, Chapter 7)	Three-dimensional interactive computer-graphics model of aircraft pilot, evaluate physical accommodation of pilot to existing or conceptual crew station designs; performs fit analysis, visibility analysis, reach analysis and strength for operating controls with arms and legs.	P HFE SS	FSD PD	Si	S	C
CREW CHIEF (McDaniel & Hofmann, Chapter 7)	A CAD man-model that simulates an aircraft maintenance technician designed for use by aerospace manufacturers to improve maintainability and supportability	P HFE	CE-FSD	Si	S	C
CREWCUT (Bogner, Kibbe, & Laine,1990)	Models, predicts, and assigns workload estimates to crew tasks to simulate the effects of workload on crew and system performance.	M	CE-PD	Si	S	N
CRDS (Crew Requirement Definition Subsystem & Methodology (Bogner, Kibbe, & Laine,1990)	Stand alone component of SORD finds and graphically displays key data points of personnel-system performance spectrum (e.g., longest or critical task sequence, crew loading)	MPT	CE-PD	T	S	C
CRM (Cognitive Requirements Model) (Rossmeissl et al, 1990)	Provides systems developers with information on cognitive demains of system tasks. Non-automated, based on 12 rating scales of cognitive factors, gives estimated skill requirements with overall measures of task difficulty.	P	CE-PD	S T	H/G	C
CROSSWALK (Haas & Laine, Chapter 17)	Identifies the relationship between equipment and the occupations that operate and maintain it.	MPT	CE-PD	NA	DB	C

Table 12-5 *(continued)*

METHOD	DESCRIPTION	Domain	Acquisition Phase(s)	Analytical Techniques	Type Tool	Availability
ECA (Early Comparability Analysis) (Bogner, Kibbe, & Laine, 1990)	Non-computerized tool (12 steps) using "'lessons learned" approach to identify deficiencies resulting from poor task performance and identifies ways of making tasks easier.	MPT	CE-PD	NA	H/G	C
EAM (Electronic Aids to Maintenance) Impact on Weapons System Availability (Bogner, Kibbe, & Laine,1990)	Used to determine level of performance of built in test (BIT) and built in test equipment (BITE) for specific system availability level and determine opportunities for improvement of system performance by changing BIT/BITE performance, maintenance doctrine, or policy.	M	CE-PD	S	S	C
Embedded Training (ET) Guidelines & Procedures (Bogner, Kibbe, & Laine,1990)	Method for MANPRINT domains training for all phases of system acquisition.	T	CE-PD	NA	H/G	C
FOOTPRINT (Haas & Laine, Chapter 17)	Utilizes existing data bases to develop the MPT profile of a Military Occupational Specialty.	MPT	CE-PD	NA	S DB	N
HARDMAN III (See this chapter for details; also Bogner, Kibbe, & Laine, 1990)	Set of six interrelated PC-based tools to assist analysts to develop MPT estimates for requirements and constraints early in acquisition process (SPARC, M-CON, P-CON, T-CON) and to evaluate designs during later phases (MAN-SEVAL, PER-SEVAL).	MPT	CE-PD	Si C S T	DB S	N
HOS IV (Human Operator Simulator) (Bogner, Kibbe, & Laine, 1988)	Partially menu driven, simulation-based approach to evaluate man-machine interface design. Unique combination of user friendliness with complex system simulator and complex human processing micro-models. Input: Man-machine interface design and task analysis. Output: Detailed design interface evaulation data	MPT HFE	CE-PD	Si T	S	C

Table 12-5 *(continued)*
MANPRINT Methods for Manpower, Personnel, and Training

METHOD	DESCRIPTION	Domain	Acquisition Phase(s)	Analytical Techniques	Type Tool	Availability
JASS (Job Assessment Software System) (Bogner, Kibbe, & Laine,1990)	Addresses issues in MANPRINT personnel domain. Estimate aptitude requirements (cognitive, perceptual, psychomotor) of operations and maintenance tasks. Taxonomy of 40 aptitudes (not in ASVAB terms).	P	CE-PD	T	S	C
LCOM (Logistics Composite Model) (Rossmeissl et al., 1990)	Large scale Monte Carlo simulation model that relates sortie generation to variety of support resources including manpower, spare parts, and facilities; used extensively in supportability, maintainability, and trade-off analyses; can be used with comparability analysis to estimate new system maintenance requirements and impact of support limitations.	M	CE-PD	Si	S	C
MANCAP II (Manpower Capabilities)	At various unit levels (up to Division size) models operational availability, maintenance manpower requirements, and organization configurations result-ing from usage failures and combat damage.	M	CE-PD	Si	S	N
MANPRINT Handbook for Request for Proposal (RFP) Development, AMC PAM 602-1 Defense Technical Information Center, Bldg. 5, Cameron Station, VA 22314 (AD A 188321)	Provides guidance with examples on how to include MANPRINT require-ments in RFPs for any phase of major system development program.	MPT HFE SS HH	CE-FSD	NA	H/G	C
MANPRINT Handbook for Conducting Analysis of the Manpower, Personnel & Training (MPT) Elements for a MANPRINT Assessment Headquarters, U.S. Training & Doctrine Command, ATTN: ATCD-SP, Fort Monroe, VA 23651-5000	Analysis Guide for MPT topics as part of a MANPRINT assessment; includes underlying theory, technical and administrative steps.	MPT	CE-FSD	NA	H/G	C

Table 12-5 *(continued)*

METHOD	DESCRIPTION	Domain	Acquisition Phase(s)	Analytical Techniques	Type Tool	Availability
MANPRINT Handbook for Non-Developmental Item (NDI) Acquisition AMC PAM 602-2 Headquarters, U.S. Army Materiel Command, ATTN: AMCDE-AQ, 5001 Eisenhower Avenue, Alexandria, VA 22031	Aids in establishing key MANPRINT issues in each NDI acquisition phase; includes Independent Evaluation Plan, Market Investigation, and NDI procurement solicitation.	MPT HFE SS HH	CE-FSD	NA	H/G	C
MANPRINT in Test and Evaluation (Bogner, Kibbe, & Laine,1990)	Quantitative method for predicting manned system performance, given sample data which describes human performance and hardware and software reliability	MPT HFE HH SS	CE-PD	S	H/G	C
MANPRINT Practitioners Guide HQDA Pamphlet 602-2	Supplements AR 602-2; provides guidance and references necessary to apply MANPRINT throughout the acquisition process.	MPT HFE SS HH	CE-PD	NA	H/G	C
MANPRINT Primer, Defense Technical Information Center, Bldg. 5, Cameron Station, VA 22314 (AD A 197681)	Provides conceptual basis for U.S. Army's MANPRINT program and "how to" guidance for incorporating MANPRINT in the materiel acquisition process.	MPT HFE SS HH	CE-PD	NA	H/G	C
Manufacturers MANPRINT Management Plan (MMMP) Expert System (Bogner, Kibbe, & Laine, 1990)	Computer-based expert system for guiding industrial practitioners in the development of the Manufacturers MANPRINT Management Plan (MMMP). From series of questions about specific system and its applications, select appropriate report format and specific passages for inclusion.	MPT HFE SS HH	CE-PD	NA	S	N

Table 12-5 *(continued)*
MANPRINT Methods for Manpower, Personnel, and Training

METHOD	DESCRIPTION	Domain	Acquisition Phase(s)	Analytical Techniques	Type Tool	Availability
MANRRS (MANPRINT Reference Retrieval System)	Consists of a data base and a retrieval program to assist in locating MANPRINT reference materials.	MPT HFE SS HH	CE-PD	NA	DB	N
MRA (MANPRINT Risk Assessment)	Provides a list of potential MANPRINT issues and overall system risk assessment	MPT HFE SS HH	CE-PD	S	S	C
Methodologies for Planning Unit and Displaced Equipment Training (Bogner, Kibbe, & Laine,1990)	Useful in determining the most effective (population trained) and efficient (resources consumed) training method. Output is training plan for organization scheduled to receive a new system; plan includes training schedule for each organization and resources required.	T	CE-PD	NA	S	C
MIST (Man-Integrated Systems Technology) HARDMAN II) (Bogner, Kibbe, & Laine,1990)	Early MPT estimation based on comparability analysis, most useful Pre-milestone I, but applies up to Milestone III.	MPT	CE-FSD	C	S	C
MSDS (Manpower Standards Development System) (Rossmeissl et al., 1990)	Automates development of manpower standards outlined in AFM 25-5; provides powerful analysis tools including bivariate and multivariate regression, task-ratio analysis, category size analysis, and work count analysis.	M	CE-PD	S	S	C
Operations and Maintenance Requirements Simulators Methodology Model (Bogner, Kibbe, & Laine,1990)	Dynamically simulates and links system mission capability with system maintenance and supply support concepts of resources. Output can be used to quantify and study relationship among hardware design, manpower availability, personnel capability and mission capability.	M	CE-PD	Si	S	C
OWLKNEST (Christ, 1989)	A microcomputer-based tool that provides guidance in selecting the most appropriate technique for assessing operator workload for developing Army systems.	MPT	CE-PD	NA	S	C

Table 12-5 *(continued)*

METHOD	DESCRIPTION	Domain	Acquisition Phase(s)	Analytical Techniques	Type Tool	Availability
RIA TAS (Requirements Identification and Technology Assessment Summary)	5-25 year look at potential human deficiencies; formal Air Force process for weapons system development.	MPT HFE SS HH	CE-PD	NA	NA	C
S^3 (Speciality Structuring System) (Rossmeissl et al., 1990)	Research to address MPT in acquisition process. Designed to establish base-line comparison system, establish MPT goals, and conduct trade-offs among M, P, and T issues encountered during acquisition process. Uses LSAR as task level input data and user defined MPT interrelationships.	MPT	CE-PD	C	S	N
SIMNET-D (Simulation Network) (Bogner, Kibbe, & Laine,1990)	Technology demonstrations supported by Government Defense Advanced Research Projects Agency (DARDA); modular crew compartments(Abrams M1 Tank and Bradley M213 Fighting Vehicle) for a company team have been simulated and networked to operate simultaneously on common simulated terrain. Can be used in addressing research issues in MANPRINT domains.	MPT HFE SS	CE-PD	Si	S	C
SORD (Systematic Organizational Design) (Bogner, Kibbe, & Laine,1990)	Method of specifying composition of Army units and combination of similar units necessary to form larger organizational structures in response to requirements of new high-technology subsystems, changing doctrinal conceptsof mission requirements and personnel and materiel constraints.	MPT	CE-PD	Si	S	C
SUMMA (Small Unit Maintenance Manpower Analysis) (Rossmeissl et al., 1990)	Integrated MPT modeling and analysis method; microcomputer-based decision support tool; uses task level input data from LCOM; evaluates impact of specialties restructuring. Useful for examing ways of organizing aircraft maintenance work into job specialties and for integrating MPT issues in context of Air Force operational environment.	MPT	CE-PD	Si	S	C

Table 12-5 *(continued)*
MANPRINT Methods for Manpower, Personnel, and Training

METHOD	DESCRIPTION	Domain	Acquisition Phase(s)	Analytical Techniques	Type Tool	Availability
Supply Support Methodology and Model (Bogner, Kibbe, & Laine,1990)	Estimates manpower and personnel (supply) required to support a given mission profile for an emerging weapon system. Stand alone component of (MANCAP) model (LHX program only).	MP	CE-PD	S	S	C
TASCS (Training Analysis Support Computer System) (Rossmeissl et al., 1990)	Instructional System Design (ISD) tool implement on a PC; used by instructional system designers and subject matter experts to define, analyze training requirements, design supporting training systems.	T	CE-PD	T	S	C
TCA (Training Contract Action Data Base) (Bogner, Kibbe, & Laine,1990)	Provides information on contractual activities for training aids and devices reported by Federal agencies.	T	CE-PD	NA	DB	C
TDS (Training Decision System) (Rossmeissl et al., 1990)	Air Force training programming and planning system. Uses information from job tasks performed by airmen, combined with assignment information and Air Force training capacities, to determine cost-effective training options.	T	CE-PD	T	S	N
TEDB (Training Equipment Data Base) (Bogner, Kibbe, & Laine,1990)	Provides a data base on training equipment and terminology	T	CE-PD	NA	DB	N
TOSS (Task Analysis Work Load [TAWL] Operator Simulator System) (Bogner, Kibbe, & Laine,1990)	Performs data base management functions for task analysis information to develop operator workload prediction models.	MP	CE-PD	T Si	S	C
TRANSFORM (Training for Maintenance) (Rossmeissl et al., 1990)	PC-based support of ISD process and interface with LSAR data base; avoids ISD specialist manual searches through contractor provided LSAR output reports.	T	CE-PD	NA	S DB	C
TWS (Taxonomic Workstation System) (Bogner, Kibbe, & Laine,1990)	A PC-based, relational data base with a library of 14 taxonomies facilitating job and task analysis.	MPT	CE-PD	T	S DB	C

Table 12-6
Potential Safety and Health Hazards
(Adapted from Leibrecht, 1988)

MECHANICAL FORCES

The potentially hazardous mechanical forces found among Army systems include acoustical energy (noise), vibrations, shock, and trauma. These hazards tend to occur together as characteristics of engines, drive trains, tracks and wheels, transmissions, rotors, guns/cannons, and munitions – common components of Army vehicles and aircraft.

Noise, steady state: intermittent, sustained, narrow or wide band. Arises from generating, transmitting, and converting power; drive elements interacting with ground or air; generation of sound; gas or fluid flow-friction; steady combustion.

Noise, impulse: blast, impact, repetitive or nonrepetitive. Arises from propellant combustion; detonation of explosives; sudden release of pressure; forceful impact.

Vibration: high or low frequency, linear, rotational, intermittent or sustained. Arises from generating, transmitting, and converting power; drive elements interacting with ground or air; resonance dynamics, induced changes or oscillations in system attitude or position.

Shock: acceleration, deceleration, force loading. Arises from system impact (crash, collision, hard landing); system recoil; sudden aircraft displacement due to air turbulence; windblast, parachute opening.

Trauma: blunt, sharp, musculoskeletal. Arises from objects or components impacting soldier; weapons blast or recoil; shattering of components or materials; limb or head flail due to vehicle/terrain interaction; airblast; musculoskeletal overload.

CHEMICAL SUBSTANCES

Chemically active compounds frequently occur in basic system construction (e.g., paints, sealants, adhesives), routine operations and logistical support (e.g., fuels, coolants), maintenance (e.g., solvents, cleaning agents), and special functions (e.g., fire/flame suppression, decontamination) and are readily identifiable as such. In contrast, there is another family of substances generated by normal system operations, usually byproducts of engine combustion and weapons combustion. These primary substances and byproducts occur as liquids, gases, and solids:

Liquids: stable, volatile, enclosed, open. Associated with fueling, maintaining, and repairing systems; systems salvage and disposal; pest and plant control; decontamination; generation of obscurants; sewage handling and tretment. Common types include fuels, lubricants, coolants, hydraulic fluids; solvents, cleaning agents; paints, adhesives; pesticides, herbicides, defoliants; decontamination solutions.

Gases: vapors, fumes. Arise from vaporization of liquids listed above; engine or weapons combustion; compressed gas; air filtration; electric motors; welding; flame/fire suppression.

Solids: coatings, aerosals, mists, dusts, smoke, particulates. Arise from system-environment interaction; burning materials; generation of smokes/obscurants; construction activities; blasting; welding, brazing, soldering; cutting, grinding, and sanding of metals, plastics, wood; decontamination; pest and plant controls; air filtrations.

Table 12-6 *(continued)*
Potential Safety and Health Hazards
(Adapted from Leibrecht, 1988)

BIOLOGICAL SUBSTANCES

Hazardous biological substances result from contamination or infiltration of systems by disease-causing microorganisms residing in the earth's environment. Common types include bacteria, viruses, parasites, Rickettsia, and fungi. These organisms may grow (or at least survive) wherever there is a "reservoir" containing a hospitable medium such as water or nutrified liquid.

RADIATION ENERGY

Common types of radiation which accompany Army systems include visible light, infrared, ultra-violet, radiofrequency energy, laser energy, and ionizing radiation. System or subsystems designed for special functions, especially of an electrical or electronic nature, frequently give rise to these types of energy. The basic forms and generic sources of each type of radiation are summarized below:

Visible light, high intensity: artificial, natural, transient, sustained. Generic sources: search lights, landing lights, strobes, high intensity lamps, light amplification devices, cathode ray tubes, natural sunlight, highly reflective surfaces, laser reflection, gas torches, nuclear flash.

Infrared: sustained, transient. Generic sources: heating elements (such as those used in food preparation equipment and space heaters), gas torches, soldering equipment, electronic repair equipment.

Ultraviolet: near and far UV, artificial, natural, transient, sustained. Generic sources: ultraviolet lamps, gas torches, gas discharge tubes, natural sunlight (varies with season, altitude, etc.).

Radiofrequency energy: microwaves, millimeter waves, transient, sustained. Generic sources: telecommunications systems, radar systems, microwave ovens.

Laser energy: pulsed, transient, sustained. Generic sources: rangefinders, target designators, training simulators, sensor-targeted countermeasure systems, material processing systems.

Ionizing radiation: transient, sustained. Generic sources: high-voltage electronics, X-ray equipment, radioluminescent materials, nuclear weapons, depeleted uranium munitions.

ENVIRONMENTAL EXTREMES

On the training range and battlefield, environmental factors such as temperature, humidity, wind, and altitude interact with combat systems and their operations. In their extreme forms and combinations, these factors may threaten the soldier's health. The Army is concerned with three categories of environmental extremes – ambient heat, ambient cold, and oxygen deficiency.

Ambient heat: convection, radiant, natural, artificial, transient, sustained. Arises from environmental heat, sunlight; heat-generating systems and subsystems; human metabolism.

Ambient cold: natural, artificial, transient, sustained. Arises from environmental cold, ice; cooling subsystems.

Oxygen deficiency: natural, artificial, transient, sustained. Arises from high altitude (terrestrial, airborne); oxygen displacement in confined spaces; systems which constrain breathing.

Safety and health hazard concerns are generally not regarded as performance enhancers; rather they are considered to impose limits on system design to prevent degradation to human performance. As potential sources of degraded system performance, the methodological approach to incorporating them into product design differs significantly from that of MPT and HFE.

Tool Example 1: Controlling Operator Workload (OWL)

The Army Research Institute project for controlling operator workload (OWL) has identified factors that contribute to increased workload and has produced for system designers an evaluation of existing methodologies for assessing operator workload (Lysaght et al., 1989).

Operator Workload Measurement Techniques

Operator workload measurement techniques fall within two categories: (1) analytical techniques, and (2) empirical techniques. As illustrated in Figure 12-5a, analytic techniques include comparison techniques, mathematical models, expert opinion, task analyses, and simulations. Empirical techniques (Figure 12-5b) include primary task measures, subjective methods, secondary task techniques, and physiological techniques applied during system simulation, prototyping, or testing. Both analytical and empirical techniques can be useful in prediction as well as assessment. Both analytical and empirical techniques can be useful once an operator and some actual hardware, system prototype, or system simulator are available. However, during early system concept exploration, when soldier-in-the-loop assessments are not possible but the consequences of the soldier-system interface must be predicted, analytical tools are all that can be implemented to assess OWL.

The workload measurement techniques currently available can satisfactorily be used in every stage of the acquisition process. System acquisition managers are advised, however, that no one technique will tap all potentially relevant aspects of OWL for a given system in a particular set of environmental situations. It is, therefore, recommended that a battery of OWL assessment techniques be used; an OWL assessment program should be implemented that will utilize the strengths of different types of techniques and assessment procedures. (Christ et al., 1990)

There are also automated tools available to assist the MANPRINT, human factors, or workload analyst in choosing the most appropriate techniques to use when looking at workload. The Army Research Institute has developed OWLKNEST (Operator Workload Knowledge-based Expert System Tool); the National Aeronautics and Space Administration has created the WC

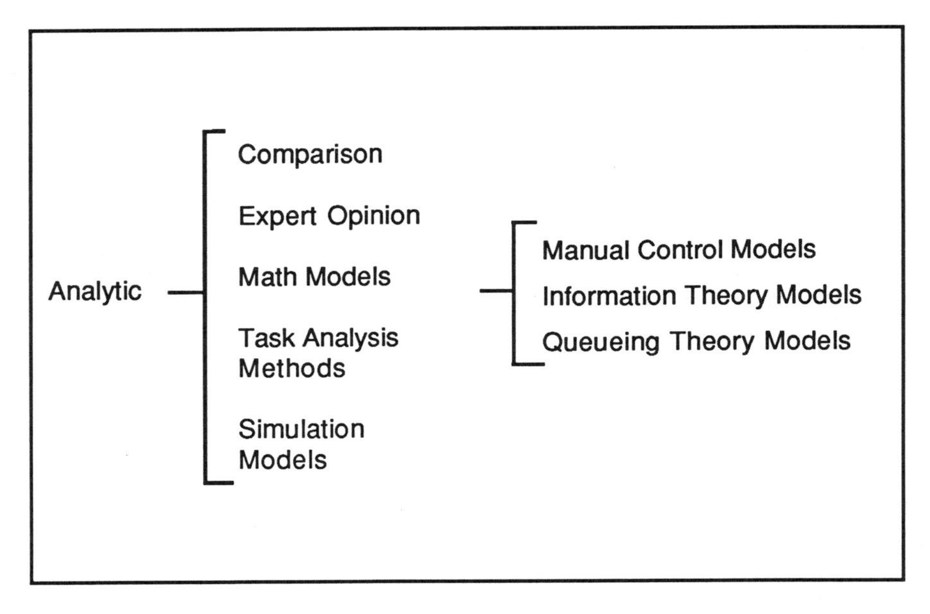

Figure 12-5a
Analytical OWL Prediction and Assessment Techniques

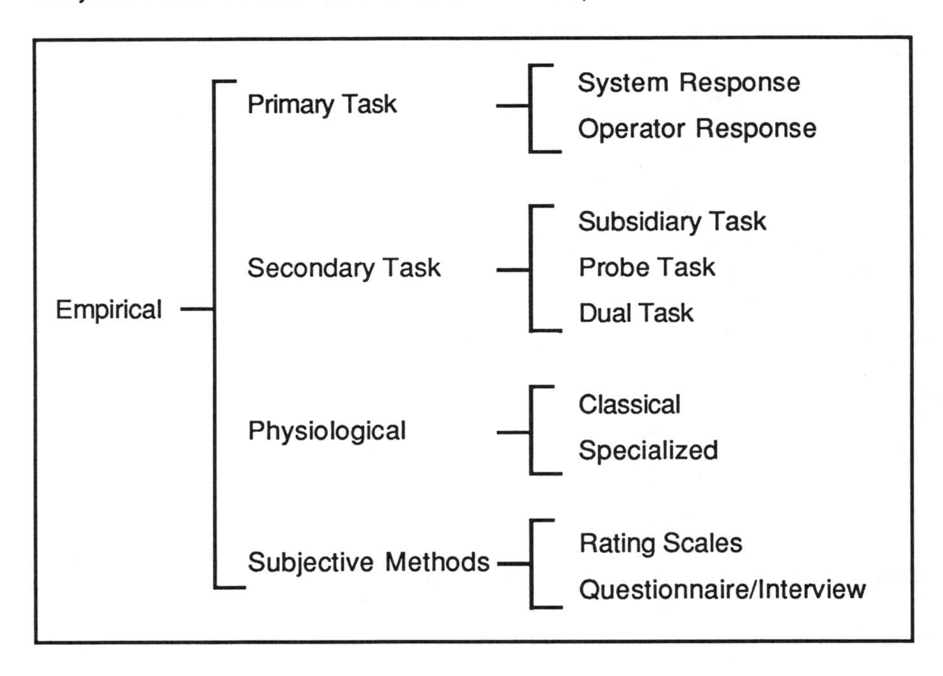

Figure 12-5b
Empirical OWL Prediction and Assessment Methods

FIELDE (Workload Consultant for Field Evaluations) expert system. OWLKNEST is oriented to workload evaluation under field conditions and provides information about measures that can be used for both prediction and evaluation throughout system development. Some of the specific technologies available are listed in Table 12-7. WC FIELDE addresses evaluation techniques developed primarily for use in controlled environments. For more information on these automated tools, see Casper, Shively, and Hart (1986, 1987); Harris, Hill, and Christ (1990); and Hill and Harris (1989).

Application in System Acquisition Process

Some of the specific applications planned by the Army are: (1) forecast the impact of operator workload on the design and performance of new Army systems; (2) effectively allocate workload-imposing tasks among soldier, hardware, and software components of the systems; (3) assess the

Table 12-7
OWLKNEST Operator Workload Techniques

ANALYTICAL	EMPIRICAL
Closed Questionnaires	AHP
Comparability Analysis	Bedford
Delphi Interviews	Blink Rate
Human Operator Simulator	Choice RT Secondard Tasks
McCracken-Aldrich Task Analysis	Closed Questionnaires
MicroSaint	Embedded Secondary Tasks
Open Ended Questionnaires	Evoked Potentials
Prospective -OW	Eye Movement
Prospective SWAT	Heart Rate
Prospective TLX	Heart Rate Variability
SIMWAM	Modified Cooper-Harper
Structured Interviews	NASA-TLX
TAWL	Open Ended Questionnaires
Tr/Ta Task Analysis	OW
Unstructured Interviews	Pupil Measures
Zaklad/Zachary Task Analysis	Steinberg Memory Secondary Tasks
	Structured Interviews
	SWAT
	Time Estimation Secondard Tasks
	Type 1 Primary Measures
	Type 2 Primary Measures
	Unstructured Interviews

influence of workload factors on the organization design of Army units; and (4) establish procedures for the selection, classification, and training of soldiers to effectively cope with operator workload in operational situations.

Figure 12-6 shows that a mechanism exists within the Army for integrating OWL-related activities into the system acquisition process. The Army already requires that MANPRINT be incorporated into all stages of the materiel acquisition process (Army Regulation 602-2). Furthermore, it has been shown that OWL issues and concerns can be related to the MANPRINT domains. Consequently, activities associated with assessing and controlling OWL can be specified in the System MANPRINT Management Plan (SMMP), the document required for the implementation of MANPRINT in Army systems.

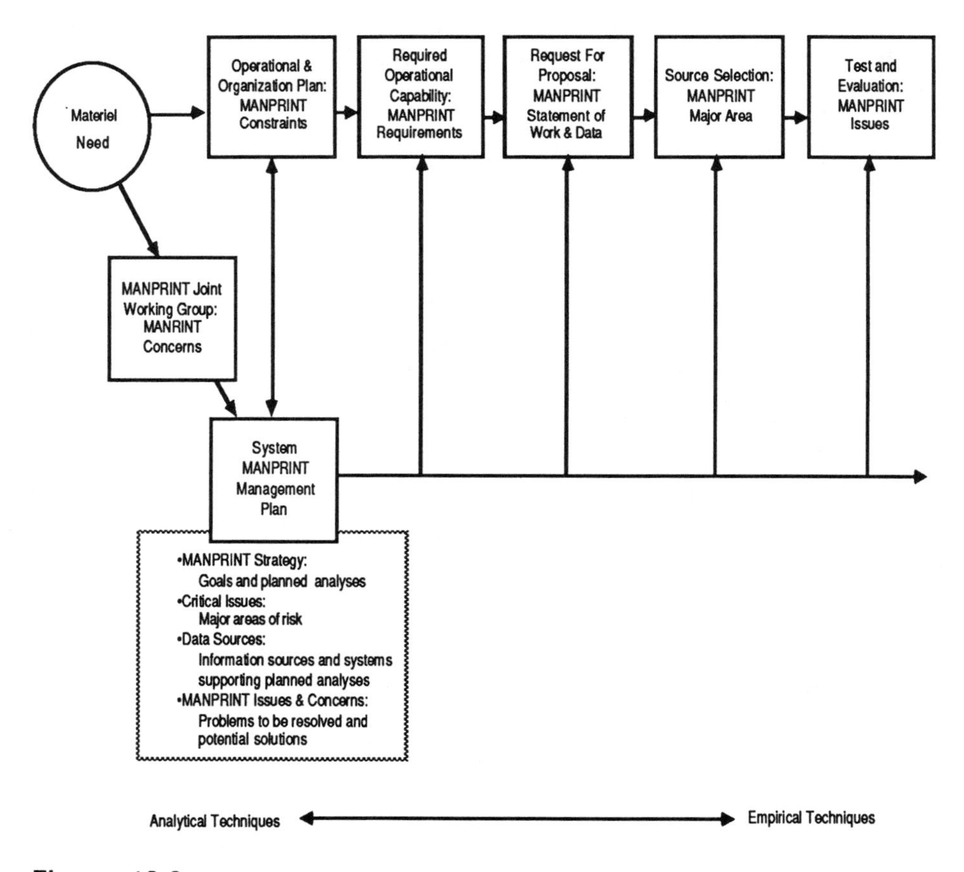

Figure 12-6
Incorporating OWL in the Acquisition Process

Keeping in mind the approach described in detail by Shields and associates for asking critical questions for decisions, OWL-related questions include the following (Christ et al., 1990):

- Is there reason to believe OWL will be a problem?
- During which portions of the mission scenario are OWL problems likely to emerge?
- What will be the nature of expected OWL problems?
- Will these problems become more severe as scenarios become more realistically rigorous?

The answers to these questions are not necessarily obvious or easy to determine. The systems integrator must be reminded that few systems operate with no human interface (Price, Chapter 6). Moreover, the tasks allocated to the operator may be simple one by one, but in combination can be considerably more difficult.

Tool Example 2: HARDMAN III

The Hardware and Manpower (HARDMAN) III Project was initiated to resolve analytic deficiencies of the past (see Table 12-8) and accomplish three

Table 12-8
Common Limitations of MPT Analytic Techniques

Sensitivity:

Lack measures of sensitivity to mission and operational scenarios.

Hardware Dependence:

Require predecessor systems for system comparisons of similarity. Most methods do not apply to classes of systems (e.g., armored vehicles, air defense systems, communication equipment).

Data:

Lack necessary data support to influence early decisions in the materiel acquisition process.

Flexibility:

Do not provide sufficient parameter ranges for MANPRINT domain impact; limited ability to conduct "what ifs."

Cost:

Too costly and too cumbersome (requiring time, manpower, dollars, and large mainframe computers to conduct analyses).

objectives (Kaplan & Holman, 1989): (1) influence system design to improve accommodation of projected manpower, personnel, and training constraints; (2) evaluate systems in development to determine manpower, personnel, and training requirements for acceptable system performance and availability; (3) enhance the tools and techniques available to MANPRINT practitioners. The HARDMAN III decision support system developed by the Army Research Institute offers a set of integrated analytical tools using a new conceptual base. The concept involves defining manpower, personnel, and training constraints for a developing system and assessing the impact of these constraints on human task performance and overall system performance before design is begun. Design features, which include ease of iteration, personal computer-based, and built-in data libraries, make it a promising capability for MANPRINT analysts.

The HARDMAN III assessment methodology consists of six independent but mutually supporting modules (see Figure 12-7):

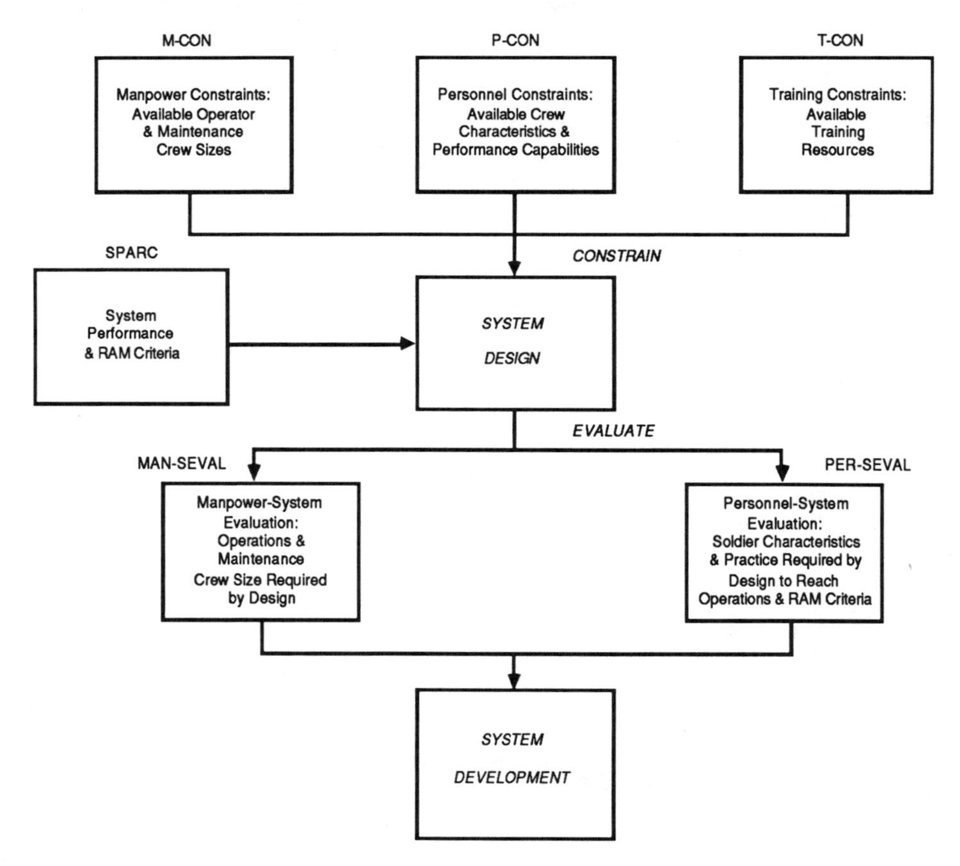

Figure 12-7
Hardware and Manpower (HARDMAN) III Analysis

• *SPARC (System Performance and RAM Criteria) Module*: Aids the user in decomposing a mission description into functions and subfunctions and then provides system performance criteria (e.g., time, accuracy) at the subfunction level required to meet mission criteria. SPARC sets the minimum acceptable system performance requirements and specifies the reliability, availability, and maintainability (RAM) requirements for the system at the subsystem level. It lets hardware and software designers know what the manned system will have to do to be minimally acceptable.

• *M-CON (Manpower Constraints) Module*: Provides an estimate of the Army-wide and per system manpower that is likely to be available to support a developing system. Use of this module involves setting manpower constraints, evaluating the sensitivity of the system performance to these constraints, and adjusting as necessary. M-CON addresses quantitative manpower constraints for maximum crew sizes by system and total operator and maintenance manpower available at pay grade, military occupations specialty (MOS), and maintenance organizational level of detail (e.g., direct support, general support). This method is based on manpower availability predictions of the MOS due to replacement of systems already in the force.

• *P-CON (Personnel Constraints) Module*: Provides an estimate of personnel aptitude constraints. Distributions of personnel characteristics (e.g., armed forces entrance test scores, education, language levels, and gender) are displayed for operations and maintenance crews projected for future years. P-CON gives the performance expected at the task level based on Project A Data Base (see Table 12-9) and the MOS cutoff score required for criterion performance. This tool aids in predicting the availability of soldiers with desired abilities and relates these characteristics to soldier performance.

Table 12-9
Project A Data Support of HARDMAN

Project A is a research data collection effort by the Army Research Institute that has produced a longitudinal data base matching Armed Services Vocational Aptitude Battery (ASVAB) test scores with job performance for 10,000-15,000 individuals. HARDMAN III uses the data to predict task performance of the available soldier population, the effects of operating and maintaining weapons systems with varying quality of soldiers, and soldier aptitude requirements for existing, developing, and non-developmental (NDI) systems. (See Haas & Laine, Chapter 17)

• *T-CON (Training Constraints) Module*: Provides an estimate of the training constraints most likely to affect the system. It identifies the probable training time and the training devices for assigned operator and maintainer functions at the subsystem level. With a comparison based prediction, T-CON supports early estimation of training supportability and evaluates the training consequences of the mission and operational requirements of a new system. T-CON aids system design by providing a training prediction so as to avoid an interface design that requires unavailable training.

• *MAN-SEVAL (Manpower-System Evaluation) Module*: Evaluates the quantitative manpower requirements at the completion of the system design phase and prior to the development of any system prototypes. MAN-SEVAL identifies the jobs associated with each design, and the tasks, the number of operators and maintainers, and the occupational specialty and skill level associated with each job. MAN-SEVAL deals with operator manpower through a combination of modeling and workload analysis. It deals with maintainer manpower by modeling the relationship between maintenance manpower and system availability. It allows the service to determine manpower requirements and compare them to manpower availability.

• *PER-SEVAL (Personnel-System Evaluation) Module*: Assists analysts in identifying the aptitudes of personnel needed to support a particular design. It evaluates the contractor's design by identifying what level of personnel characteristics is needed to meet design system performance requirements with fixed amounts of training and under the specific conditions in which the system tasks will be performed. The components of PER-SEVAL include (1) performance shaping functions that predict performance as a function of personnel characteristics and training; (2) stressor algorithms that degrade performance to reflect the presence of critical environmental stressors; and (3) operator and maintainer simulation models (using *Microsaint* logic) that aggregate the performance estimates of individual tasks and produce estimates of system performance. This tool can be used to predict performance of all aptitude levels of soldiers operating or maintaining the system under conditions specified by the user.

TOOL IDENTIFICATION AND SELECTION FACTORS

For systems integrators faced with identifying and/or evaluating MANPRINT issues in a timely and cost-effective manner, it is important they have the most appropriate tools available. The tools of the highest quality are those which have sufficient information to answer the users' questions. In addition, they need to be accessible, highly flexible, provide rapid response, and easy to use. In order for the systems integrator to assess the quality of potential techniques, the quality of the data sources which the tools use would also need to be known. Chapters 16 and 17 discuss several of the human performance data banks which are available for the MANPRINT tools and will not be discussed further here. This chapter provides the practitioner with a door to the currently existing MANPRINT domain techniques and MANPRINT integration tools which interface with other analytical methodologies. The chapter references identify a large number of potential useful tools for the MANPRINT domains and for integration with other systems acquisition analytical tools. Choosing to use a tool may be assisted by a six-factor decision table (see Figure 12-8).

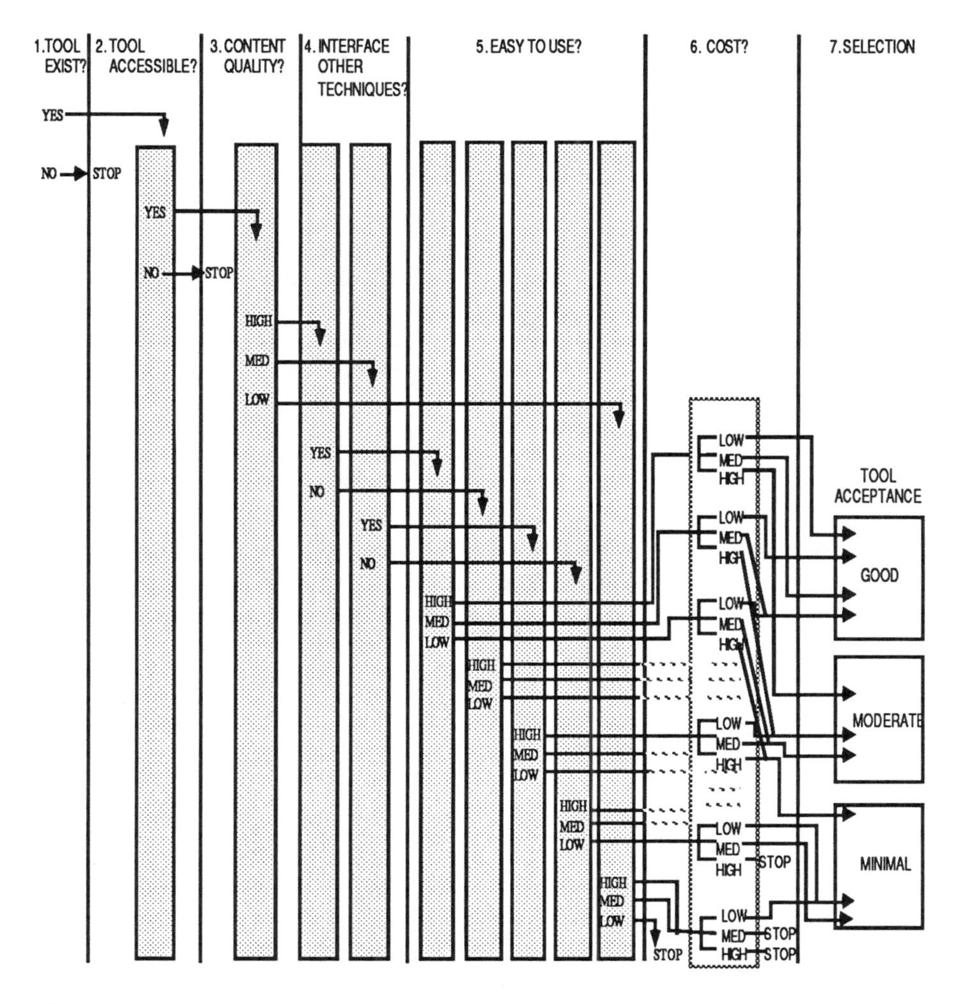

Figure 12-8
MANPRINT Tool Selection Criteria and Decision Table

Does a Tool Exist?

There may be any number of instances where a tool does not exist for the practitioner's question. Prior to HARDMAN and predecessor system comparability analyses there essentially were no tools which could assess the impact of a system design on future manpower, personnel or training domains. There are currently few techniques categorized as system safety tools. On the other hand, as discussed earlier, there are a large number of useful operator workload assessment and prediction methodologies. Other than this chapter, the primary sources for determining tool existence include the authors of the various surveys and investigators at the various government laboratories or contractor and academic facilities.

Is the Tool Accessible?

Gaining a high degree of confidence that a technique is adequately accessible requires consideration of the following: Is the tool available to the general public or is it company proprietary or otherwise unavailable? For example, Fleger et al., (1988) indicate CHESS (Crew Human Engineering Software System) is a highly useful tool and has the capability to assess reach of female operators and mixed-sex populations, but it is proprietary to Boeing. Is it fully completed, verified, and validated, or is it in a R&D stage? Many of the tools in Table 12-5 are listed as near term, meaning they should be available for use sometime in 1990 or early 1991. There are other tools, however (not listed in the chapter's tables) which hold great promise for the future, but are really not yet ready for application. The Army-National Aeronautics and Space Administration (NASA) Air Crew/Aircraft Integration (A^3I) program at NASA-Ames Research Center is developing a human factors computer-aided engineering facility to evaluate alternative designs of helicopter cockpits (Elkind, Card, Hochberg, & Huey, 1989), but it is still four years away from meeting the criterion of "accessible." Another tool very much needed for the MANPRINT interface is the Air Force RAMCAD (Reliability, Availability, Maintainability/Computer-Aided Design), but it is not expected to be ready for use until the fall of 1991.

Is the Content Quality Good?

This is often the most difficult factor to determine. First consider the quality of the source data. Second, the tool itself can be evaluated by (a) its general intent from published descriptions of the tool and (b) its ability to provide the expected information asked for by the user. The classification schemes discussed in the earlier section can be useful in this determination. For example, if a tool is described as appropriate only for full scale development, it probably has low content quality for pre-concept stage issues. Assessments by experts or past users provide the most dependable means of determining content quality for the new user.

Does the Technique Interface with Other Tools?

There are generally two problems to consider here. First, is the information from the tool designed to interface with other analyses? The interaction with other MANPRINT domains may affect its ability to influence total system performance issues. Second, are there software interface implications? Is the tool operable on a computer system that is generally available? Can the technique communicate with other software to be used during the development of the system?

Is It Easy to Use?

The original HARDMAN methodology provides an example of a tool which meets the first three criteria above, but is valuable even though difficult to use and costly. To assure wide application, the tools need MANPRINT philosophy applied to them as well. Some considerations are:

- Easy use of the information system
- Quick response
- Flexibility to model at different levels of specificity
- Built-in job aids and variable help levels
- Reasonable training time for new users
- Mobility or portability of tool hardware/software

Is Its Cost to Operate Reasonable?

Cost can be evaluated against the above criteria in choosing among several alternatives; but in many instances where only one tool is available, cost becomes the sole criterion.

The HARDMAN family of tools provide an example of how the six factors may be considered together in determining how acceptable any particular tool is for MANPRINT application. Referring to Figure 12-8, all three HARDMAN tools (HARDMAN I, HARDMAN II [also called MIST], and HARDMAN III) exist and are accessible for use in evaluating MPT issues. HARDMAN I would be evaluated as medium (MED) content quality, if, for example, we were to use it in the Pre-Concept stage. It is not evaluated as HIGH because it has the limitation of being based on a comparability analysis. It is not designed to interface with other techniques (NO), is not easy to use

(LOW) and is HIGH cost. This leads to a selection recommendation not to use (STOP). MIST would be found to have the same quality (MED), same interface with other techniques (NO), but is MED ease of use and MED cost. This is an improvement in cost and usability over HARDMAN I, placing it somewhere in the minimal to moderate range of tool acceptance. HARDMAN III will exist and be accessible in the near term (1990). It is rated HIGH on content quality because it is task analysis based and therefore not totally dependent on predecessor system information. It is designed to interface with other techniques (YES), relatively easy to use (MED), and LOW cost. This leads to the very good range of tool acceptance.

ISSUES AND NEEDS

Without tools and techniques to bring human behavior and performance information into the processes of the decision maker and the designer, there can be little influence on systems design, development, and operation. Although many tools and techniques are available, numerous issues remain to be addressed by new technology and research.

System Acquisition MANPRINT Integration Process

In 1987, an Air Force acquisition panel consisting of representatives from six Air Force systems command product divisions prioritized their MPT technology needs (Gentner, 1989). Their top three needs were: (1) Integrated MPT Data Base, (2) Current Technology MPT Analysis System, and (3) MPT Requirements Estimation Aid. The Army's prioritized list in the past has indicated similar needs. Studies of methodology generally agree with these recommendations:

Integrated Data Base

Integrated data bases are needed to communicate MPT issues to engineers and to facilitate participation in the trade process. Integration of data bases is also needed to provide input to the various MPT tools and to close any gaps among the different MPT domain data systems. A further expansion would include all the MANPRINT domains as part of an overall integrated human performance data base (or network of data bases).

Current Technology Analysis System

The Air Force panel stressed the need for an MPT analysis system to become operational as soon as possible, even if the technology still has bugs to work out. The Rossmeissl et al (1990) report provides the requirements for writing functional specifications to acquire this capability for the Air Force. The Army HARDMAN III is the most fully developed tool to meet some of this need in the immediate future. HARDMAN III and Air Force Technology Analysis System development should be closely considered for an overall system acquisition MANPRINT integration technique.

Requirements Estimation Aid

The Air Force plans to further the state of the art by developing requirement estimation aids which can be used to set MPT goals at the outset of a program. It would incorporate MANPRINT research findings; operations, maintenance, and deployment concepts; and other features such as the design characteristics of prime support and training equipment in projecting MPT requirements. Since HARDMAN III is believed to provide the current state of the art, the plans to analyze it along with the other the most promising tools and data bases identified in Rossmeissl et al (1990) should provide clear guidance for completing new tools and improving existing ones.

MANPRINT Domain Tools

Safety and Health Hazards

System safety and health hazard tools development lag that of the other domains. This gap should be filled. While HFE and MPT tools also apply to system safety and health hazard domains, both Air Force and Army needs-identification efforts have noted that there are few viable methods for early influence of system design in these two domains. There are, for example, no dependable methods to predict increases in number of tasks, task times, additional training, and additional personnel screening requirements due to potential safety and health hazard concerns of weapon systems. One notable safety and health tool quantifies the hazard of carbon monoxide in terms of human (soldier) performance (Steinberg & Neilson, 1977). Gentner (1989, p. 9) notes that "every additional safety and health hazard requires additional man hours to adapt to or 'work around' and be trained for its danger and methods to deal with the hazard. In addition, certain substances are more toxic to some people than others, and selected screening may be warranted to prevent contact with these chemicals." Standard methods are needed to extract and relate relevant existing safety and health hazard data to new system proposals. Methods are needed to predict safety and health hazard induced MPT consequences in new systems in order to reduce the impact on individual and system performance.

Human Factors Engineering

Studies indicate more HFE tools are needed and their application expanded. Advanced tools needed by human factors engineering specialists (Fleger et al., 1988) in prioritized order are:

(1) Computerized Workload Prediction Tool: Ideally it would integrate measures of cognitive workload with physiological performance predictors to yield objective measures of performance, accurately predict workload across a wide spectrum of job assignments, have good face validity, and be accepted by engineers.

(2) Generic Expert System: Connect artificial intelligence (AI) research with HFE data to allow non-HFE designer capability to solve problems related to systems design and to use HFE tools to consider changes in the design and development process.

(3) Other Microcomputer Based Tools: Microcomputer HFE data base compendiums; User Control Interface (UCI) rapid prototyping software; additional CAD programs and anthropometric man-models; automatic operational sequence diagrams.

HFE tools should expand to broader application areas. Of the 88 advanced tools in the Fleger et al. data base, about 64 percent were developed primarily for military applications. Even within the military, the applications are primarily aviation oriented. In fact, most of the tools, military and commercial (72 percent), have some aerospace application. Although there is much in aviation tools that are relevant elsewhere, the translation to non-aerospace continues to be a challenge. Direct applications of military HFE tools to non-military situations are even more challenging. After the Three Mile Island nuclear power accident, several years of research were required to build useful nuclear power plant control room design aids even though military control room and aviation crew station HFE tools were available. There is, however, some obvious movement to develop tools that have more general commercial, non-aerospace application. Thirty percent of the advanced HFE tools in the Fleger et al. data base were developed specifically for areas like transportation, nuclear power plants, manufacturing, and general business applications. In fact, although only one of their "best" tools had no aviation application, 25 percent were developed specifically for non-military use.

Manpower

The military has a considerable number of manpower estimate methods and models, but they are not used consistently. Some existing manpower models have been combined and coordinated, but they are usually applied one at a time, sequentially. A thorough scientific comparison of the various manpower models is needed, especially with a goal of improving utility. Those of greatest utility need cost modification estimates for adaptation to early acquisition applications. Selected models should be modified to meet early acquisition needs. The early phases of this effort have begun (Rossmeissl et al, 1990).

A useful and validated model exists for aircraft maintenance manpower estimation (Logistics Composite Model [LCOM]). This model does not, however, cover depot-level maintenance and total training manpower. To be complete aircraft manpower models need to be expanded to cover all manpower affected by new system projections. More importantly for the military is the need for non-aircraft manpower models which would cover all operator, maintenance (field and depot) and training requirements.

Personnel

Some generally agreed upon needs reflected in personnel tool development are:

- Methods to project demographic effects on future forces (7-20 years) must be able to match requirements to aptitudes by military specialty.
- Better quantification basis for personnel aptitudes and training requirements (or options) – very helpful would be a measure of commonality of skills – in both existing and emerging technology.
- Methods to optimize military specialty structures to support new eras of technology.

Training

The needed developments in the training domain are:

- An early training requirements methodology. Research products (such as a functional task cluster analysis) are needed to explore alternative ways to describe the nature of jobs and job related training for emerging systems *before* the detailed *full scale development* (FSD) task analysis.
- Research is needed to find ways to model training planning and pipeline optimization options (beginning in Demonstration/Validation) through the fielding of systems. Models should be able to project training course length, resources, and costs by option. An emerging data support issue for such models is the relationship between job aids and training methods (See Rouse, Chapter 14).
- Methods are needed to speed training systems development. There has been a long recurring problem of training not being delivered before

fielding systems. Methods are needed to develop training for emerging systems as they are being developed. It may be feasible to identify essential training development and update information from government and industry's CALS applications.

• Prototype Intelligent Maintenance Trainer. Most of manpower training time is in maintenance training, yet the state of the art in training systems has been applied almost entirely to pilot and operator crew simulations and training systems. State-of-the-art maintenance training technology, which would include advanced job aid and training technology in an intelligent delivery system, needs prototype development.

MANPRINT and Concurrent Engineering

MPT analyses to influence design are primarily "after-the-fact" in response to design concepts. To be considered as equal factors in the design process, engineers need to have a better understanding how MPT issues specifically affect system design. MPT should be translated (similarly to the way anthropometric data is already) for use in their CAD system. Also, MANPRINT needs to become part of the larger concurrent engineering approach as well as philosophically consistent with Total Quality Management (Booher & Fender, Chapter 2).

In achieving this relationship, significant research issues for MPT emerge. Standard definitions for MPT factors should be developed; their relationships more comprehensively determined, and critical MPT factors integrated into the product design process. There is a need for greater quantification of these factors. This implies more research into how to format data into engineering design and program manager "symbology" for easy interpretation.

General Research Issues

Much remains to be done in the research of MANPRINT methods and techniques. Included are the following research endeavors:

• *Domain Trade-offs*: Research has only recently diversified enough to provide the much needed emphasis upon methods and techniques in all six of the MANPRINT domains. This progress leads to the even more difficult issue of relating the effect of one domain upon another. Achieving the integration goals of MANPRINT include continuing research to identify trade-offs among MANPRINT domains.

• *Validity*: Validating the techniques that have been developed is itself a large research agenda (e.g., validation for different scenarios, environments, and classes of equipment and systems).

• *Complexity*: While considerable study on human performance has been done where one or two tasks or one or two situations are involved, little research has been done on multiple tasks in multiple types of situations. Research results that identify the multitask effects in different situations will assist in (1) making task allocation decisions, (2) predicting overall job performance, (3) assessing system and individual performance sensitivity to varying degrees of task difficulty, (4) determining the boundaries of human workload, (5) determining individual time-sharing abilities, and (6) understanding task prioritization processes (Lysaght et al., 1989).

• *OWL and System Performance*: Research in operator workload requires emphasis on (1) the ability to assess workload in operationally significant environments, (2) improving the understanding of the relationship between workload and operator and system performance, and (3) identifying cost-effective methods for reducing the impact of OWL on organizational, system, and individual performance (Christ, Hill, Zaklad, Bittner, & Linton, 1989). Table 12-10 identifies potential areas of future research.

• *Individual Differences*: Understanding the individual's capabilities to perform various types of tasks, the knowledge necessary to bear on a problem, and the human resources (e.g., behavioral, cognitive, sensory) necessary to achieve various levels of performance will continue to be one of the most elusive, yet fruitful fields of research to provide man-machine integration and to improve total system performance. Further, identifying population parameters that adequately describe the range and distribution of individual differences is as important as understanding the role that certain characteristics play in predicting performance.

This chapter provides ample evidence that there is no lack of tools, techniques, or data resident within the systems development community at large. However, it is equally demonstrable that the distribution, validation, maintenance, and support of tools and data is insufficient. Areas needing greater emphasis include:

• Improving methods by which to translate research findings and lessons learned into meaningful engineering design constraints and guidelines.

• Facilitating the distribution and training of new techniques and related data bases to the user community.

• Determining priority applications for the use of pertinent technical tools.

• Acquiring sufficient amounts of standardized electronic equipment to support the community of users.

• Enhancing the process of providing government furnished information on tools, techniques, and data bases to the industrial community for their use during system design.

• Improving the aggregation and sharing of government, academic, and industrial research to improve system engineering techniques and data.

Table 12-10
Needed Workload Research

Methodologies:	Validation of a greater number of assessment methodologies.
Environmental Impact:	Assessment in greater diversity of environments.
Type Systems:	Assessment of more types of systems/ more operator tasks and functions.
Data Support:	Improved data bases for workload evaluation.
Early Prediction:	Analytical tools to be used earlier in the system development.
Extremes of Workload:	Greater understanding of "underload" as well as "overload."
Source Diagnosis:	Greater understanding of the source of operator overload.
Individual Differences:	Increased understanding of the influence of individual differences on reactions to workload extremes.
Technique Assessment:	Enhanced ability to assess the value of workload measure techniques so as to capitalize upon the most lucrative and most productive.

Basic or Exploratory Research Questions

Gentner (1989) lists a number of longer-range research questions that need answered:

(a) What data are needed to answer the essential MPT questions in the acquisition process; which analytic models need to be used and what data are needed to drive them?

(b) Which critical MPT data are missing and how can they be most cost-effectively researched and/or developed?

(c) If data are normally collected in the wrong form, grain size, etc., can they be transformed into suitable data for analysis, or would it be more economical to collect the data in a different way? If the data are transformed, how much fidelity is lost?

(d) Can artificial intelligence be used to link these diverse data bases and make them more useful?

(e) Can symbology and hypermedia help communicate MPT data to engineers?

(f) Can a facility be developed for the integrated MPT data base that allows the MPT analyst to "build" a comparable MPT system by taking subsystems from several different comparable weapon systems?

(g) Can these data be linked through concurrent engineering, unified life cycle engineering, or the CALS to influence design or speed the acquisition process?

Non-Military, Non-Aerospace Applications

Doppelt (1987) in discussing the findings of a workshop in the psychology of system design, reiterates the long time complaint of the generally poor application of relevant research data to engineering developments. He notes that the military, academia, and industrial research literature is literally "burgeoned with data . . . suggestive of applications to the human-system design problem . . . but these data are not presently factored into design decision making" (p. 3).

He suggests several reasons for this non-use of data; lack of practical usefulness; choice of format and media; personal habits and styles of design engineers; perceived cost/benefits; and deluge of already available information. Boff (Chapter 19) discusses this at length. Doppelt's complaint was directed primarily to the military and aerospace because most of the data available were collected specifically for those applications. But an even more crucial issue for national public health and safety is how poorly such data is exploited in other areas. Perrow (1984) concludes in his examination of high-risk technologies that the airways system "has become very safe, as safety goes in inherently risky systems," but with marine transport for example "the opposite" is true; it is a system "that induces errors through its very structure." The nuclear industry has now awakened to the need for human factors, but are still way behind in its application relative to aerospace. The chemical industry and the medical world have yet to discover even 1950s human factors technology.

Some of the more immediately obvious applications would draw from the HFE methodologies described in this chapter, but the other manufacturing and operational industries can benefit from the advances in MPT as well. Some of these are suggested in Booher (1990). Also, Weddle (1986) notes that these Department of Defense developed products can:

• Aid product engineers to assess design impact on likely user populations.

• Allow trade-off analyses which identify design changes to improve the system-work force fit.

• Be used to evaluate alternative production strategies to assure the total production process can (with people-in-the-loop) actually meet production and quality goals.

• Help determine the profile of skills and skill level requirements for new manufacturing equipment or operating process change.

Weddle suggests that the benefits of such analytic tools applied to human resource planning could be especially useful for timely development and implementation of training programs or perhaps in supporting union negotiations regarding a planned work force change as a result of introducing new technology.

SUMMARY AND CONCLUSION

The principal goal of MANPRINT tools and techniques is to get critical human performance data into the acquisition decision process as early and as quantitatively as possible. MANPRINT tools and techniques are in general available such that in the hands of MANPRINT domain specialists high quality trade-off decisions can be made, both among MANPRINT domains and with MANPRINT as a whole against total system performance and cost. HFE tools are more fully developed than the other domains, but sufficient MPT tools and techniques exist (or will exist(in the near term) to make issue decisions on MANPRINT domains with as much confidence as those on hardware and software. Advanced HFE technologies, HARDMAN III, and Operator Workload techniques are extremely usable and inexpensive tools. Tables 12-4 and 12-5 provide a quick look at the state of the art for HFE and MPT. Safety and health hazards have few tools of their own for directly influencing design, and keyed to acquisition decision stages. These domains are, generally dependent on being represented through the other available techniques. Extensive research and development is still needed, to insure that MANPRINT advances along with other technology and that identified gaps are filled. Suggestions are made regarding (a) complete system acquisition integration process, (b) specific needs within MANPRINT domains, (c) MANPRINT and concurrent engineering, (d) general analytical and study issues, (e) basic or exploratory long-range research questions, and (f) non-military, non-aerospace applications.

ACKNOWLEDGMENTS

For their assistance to preparing this chapter, the authors wish to thank Richard E. Christ, John L. Miles, Jr., Marilyn Sue Bogner, Jonathan Kaplan of the Systems Research Laboratory and Jane M. Arabian of the Manpower and Personnel Research Laboratory, all of the U.S. Army Research Institute; Lawrence Hanser of the RAND Corporation; Paul Rossmeissl of Hay Systems, Inc.; Thomas Malone of Carlow Associates; and Dean Westerman of U.S. Army Human Engineering Laboratory.

REFERENCES

Bogner, M. S., Kibbe, M., & Laine, R. (1990). *Directory of design support methods.* Developed for Department of Defense Human Factors Engineering Technical Group (Designing for the User Subgroup). Washington, D.C.: Headquarters, Department of the Army, Office of the Deputy Chief of Staff for Personnel, MANPRINT Directorate.

Booher, H. R. (1990). MANPRINT implications for product design and manufacture. *International Journal of Industrial Ergonomics.* (to appear)

Casper, P. A., Shively, R. J., & Hart, S. G. (1986). Workload consultant: A microprocessor-based system for selecting workload assessment procedures. *Proceedings of the IEEE Conference on Systems, Man, and Cybernetics.* Piscataway, NY: IEEE Service Center.

Casper, P. A., Shively, R. J., & Hart, S. G. (1987). Decision support for workload assessment: Introducing WC FIELDE. *Proceedings of the Human Factors Society 31st Annual Meeting.* Santa Monica, CA: Human Factors Society.

Christ, R. E., Bulger, J. P., Hill, S. G., & Zaklad, A. L. (1990). *Incorporating operator workload issues and concerns into the system acquisition process: A pamphlet for Army managers.* Alexandria, VA: U.S. Army Research Institute.

Christ, R. E., Zaklad, A. L., Bittner, A., Jr., Hill, S., & Linton, P. (1989). The Army Operator Workload (OWL) Program: Review and prospects. *Proceedings of the Human Factors Society 33rd Annual Meeting,* Vol. 2, pp. 1471-1475. Santa Monica, CA: Human Factors Society.

Doppelt, F. F. (1987). Introduction and overview.. In W. B. Rouse and K. R. Boff (Eds.), *System design.* New York: North Holland.

Elkind, J. I., Card, S. K., Hochberg, J., & Huey, B. M. (Eds.) (1989). *Human performance models for computer-aided engineering.* Washington, DC: National Academy Press.

Farr, B. J. (1990). Personal communication. Alexandria, VA: U.S. Army Research Institute.

Fleger, S., Permenter, K., & Malone, T. (1988). *Advanced human factors engineering tool techniques.* Aberdeen Proving Ground, MD: U.S. Army Human Engineering Laboratory.

Gentner, F. C. (1989). *Manpower, personnel, and training integration technology needs – Tools for IMPACTS.* Dayton, OH: Air Force Systems Command.

Hanser, L. M. (1989). Personal communication. Santa Monica, CA: RAND Corporation.

Hanser, L. M., & Arabian, J. M. (1989). Personal communications. Santa Monica, CA: RAND Corporation and Alexandria, VA: U.S. Army Research Institute.

Harris, R., Bennett, J., & Stokes, J. (1982). Validating CAR: A comparison study of experimentally-derived and computer-generated reach envelopes. In R. E. Edwards & P. Tolin (Eds.), *Proceedings of the Human Factors Society 26th Annual Meeting* (pp. 969-973). Santa Monica, CA: The Human Factors Society.

Harris, R. M., Hill, S. G., & Christ, R. E. (1990). *Handbook for operating the OWLKNEST Technology (HOOT)* (TR 2075-5C). Willow Grove, PA: Analytics, Inc.

Hickey, D. T., Pierrynowski, M. R., & Rothwell, P.L. (1985). *Man-modeling CAD programs for workspace evaluations.* Downsview, Ontario: Defense and Civil Institute of Environmental Medicine.

Hill, S. G., & Harris, R. M. (1989). OWLKNEST: A knowledge-based expert system for selecting operator workload techniques. In A. Genaidy and W. Korwowski (Eds.), *Computer-aided design: Applications on ergonomics and safety.* London: Taylor and Francis.

Jones, M. B. (1966). Individual differences. *Acquisition of skill. Conference on acquisition of skill.* New York: Academic Press, Inc.

Kaplan, J. D., & Holman, C. (1989). *HARDMAN III decision support system.* Alexandria, VA: U.S. Army Research Institute.

Kaplan, J. D., & Miles, J. L., Jr. (1981). Human factors in weapons design: The performance gap. *Concepts* (Journal of Defense Systems Acquisition Management), IV(4), pp. 76-89.

Leibrecht, B. (1988). Health hazards: Their habits and haunts. *MANPRINT Bulletin, Vol. II, No. 6.*

Kirkpatrick, M. III, & Malone, T. B. (1984). Development of an interactive microprocessor based workload evaluation model (SIMWAM). In M. J. Alluisi, S. DeGroot, and E. A. Alluisi (Eds.), *Proceedings of the Human Factors Society 28th Annual Meeting* (Vol. 1, pp. 78-80). Santa Monica, CA: The Human Factors Society.

Laughery, K. R., Jr. (1984). Computer modeling of human performance on microcomputers. In M. J. Alluisi, S. DeGroot, and E. A. Alluisi (Eds.), *Proceedings of the Human Factors Society 28th Annual Meeting* (Vol. 2, pp. 884-888). Santa Monica, CA: The Human Factors Society.

Lysaght, R., Hill, S., Dick, A., Plamondon, B., Linton, P., Wierwille, W., Wherry R., Jr., Zaklad, A., & Bittner, A., Jr. (1989). *Operator workload: Comprehensive review and evaluation of operator workload methodologies.* Alexandria, VA: U.S. Army Research Institute.

McGuinness, J., Wagner, J., Nicholas, J. M., & Rhoads, C. J. (1986). Expert system design for application of human factors in robotics (Contract No. N60921-85-C-0252). Alexandria, VA: Person-System Integration.

Meister, D. (1976). *Behavioral foundations of system development.* New York: John Wiley & Sons.

Morrissey, S. J., Herring, B. E., & Gennetti, M. G. (1985). Applicability of using the CAR-II model in design and evaluation of multioperator workstations with shared controls. In R. W. Swezy, T. J. Post, L. B. Strother, and M. G. Knowles (Eds.), *Proceedings of the Human Factors Society 29th Annual Meeting* (Vol. 2, pp. 698-699). Santa Monica, CA: The Human Factors Society.

Perrow, C. (1984). *Normal accidents.* New York: Basic Books.

Potempa, K., & Gentner, F. C. (1988). *Manpower, personnel, training and safety in the Air Force weapon systems acquisition*. Brooks Air Force Base, TX: Human Systems Division.

Rimland, B. (1983). Human individual differences. In R. C. Sorensen (Ed.), *Human individual differences in military systems.* San Diego, CA: Navy Personnel Resarch and Development Center.

Pulat, B. M. (1982). *A computer aided workstation assessor for crew operators: WOSTAS* (ONR Contract No. N00014-81-C-0320). Greensboro, NC: North Carolina A & T State University. (NTIS No. AD-A116 045)

Pulat, B. M. (1983a). *Computer aided techniques for crewstation design work space organizer-WORG: Workstation layout generator-WOLAG* (ONR Contract No. N00014-81-C-0320). Greensboro, NC: North Carolina A & T State University. (NTIS No. AD-A132 981)

Pulat, B. M. (1983b). A computer aided workstation assessor for crew operations-WOSTAS. In A. T. Pope & L. D. Haugh (Eds), *Proceedings of the Human Factors Society 27th Annual Meeting* (Vol. 2, pp. 887-891). Santa Monica, CA: The Human Factors Society.

Rossmeissl, P. G., Akman, A., Kerchner, R., Faucheux, G., Wright, E., & Shields, J. L. (1990). *Analysis of manpower, personnel, training, and safety during the acquisition of Air Force Systems: Requirements and capabilities.* Washington, DC: Hay Systems, Inc.

Rostker, B. (1984). Human resources models: An overview. In W. P. Hughes, Jr. (Ed.), *Military modeling*. Alexandria, VA: Military Operations Research Society.

Shields, J. L., Rossmeissl, P. G., & Johnson, K. (1989). *Manpower, personnel, training, and safety in the Air Force weapons systems acquisition process*. Dayton, OH: National Aerospace and Electronics Conference.

Snow, R. E. (1986). Individual differences and the design of education programs. *American Psychologist, 41(10)*, pp. 1029-1039.

Sorensen, R. C. (1983). Human individual differences in military systems. San Diego, CA: Navy Personnel Research and Development Center.

Steinberg, S., & Neilson, G. A. (1977). A *proposal for evaluating human exposure to carbon monoxide contamination in military vechicles* (Technical Memorandum 11-77). Aberdeen Proving Ground, MD: U.S.

Army Human Engineering Laboratory.
Weddle, P. D. (1986, July). Capturing the benefits of high technology. *Personnel Administration.*
Westerman, D. P., Malone, T. B., Heasly, C. C., Kirkpatrick, M., Eike, D. R., Baker, C. C., & Perse, R. M. (1989, October). *HFE/MANPRINT IDEA: Integrated Decision/Engineering Aid.* Aberdeen Proving Ground, MD: U.S. Army Human Engineering Laboratory.
Williams, R. J. (1956). *Biochemical individuality.* New York: John Wiley & Sons.

CHAPTER **13**

INTEGRATION OF TRAINING SYSTEMS AND ANALYSES

Judy A. Oneal

ABSTRACT

Training people in the use, maintenance, and support of complex operational systems and equipment is well recognized in both military and commercial applications as one of the most important elements of system performance. Too often efforts to provide effective training either come after the design and development of the operational system or overemphasize the hardware and technology aspects of training devices. In either case, the expected benefits from training for the costs invested have been disappointing. The Systems Approach to Training (SAT) has long been recognized as beneficial to making training programs more effective, but even in areas like weapons systems development where SAT is applied extensively, total integration has not occurred. MANPRINT offers an opportunity for the development of completely integrated training systems through a more comprehensive approach to the conduct of training analyses. This chapter describes (1) the background leading up to the MANPRINT approach applied to training system development; (2) a conceptual model and methodology for integrating training systems and analyses on a timeline matching operational equipment design and development; and (3) the effect of MANPRINT (through integrating training system development and analyses) on operational equipment and training system development.

BACKGROUND TO INTEGRATED TRAINING SYSTEMS

The development of complex training systems has come under increasing scrutiny as weapon systems and other operational equipment have become more complicated, resources more constrained, and system performance more critical. Training systems currently being developed should accommodate four major considerations: (1) common causes of failure, (2) the impact of cost, (3) a systems approach to training, and (4) the need to integrate training analyses.

Common Causes of Failure

Modern training systems are composed of high technology training equipment, sophisticated software, courseware, trainers, and trainees. Planning and producing these modern training systems may entail a development process that is as complex as the corresponding process for the operational equipment that training supports. While training systems have been developed successfully for military and industrial applications, poor system performance as a result of inadequate training attests to the need for further improvements in training systems planning, development, and implementation. For example, in spite of the tremendous investments in training for commercial pilots, 65 percent of all airplane crashes in the past thirty years is attributed to crew error with inappropriate training cited as the most frequent cause (Stockton, 1988). And, in 1984, when a major producer of high quality paper invested heavily in equipment-oriented training to match plant operation expansion, production decreased. Lost revenues totaled in the millions of dollars. A subsequent investigation blamed inadequate training for the loss of essential paper plant production skills (Oneal, 1989). And, a detailed examination of IBM's 900 million dollar corporate training program (4 percent of the corporation's operating budget) identified three severely deficient areas: (1) too little direct job-related training, (2) no systematic training design or central planning, and (3) a lack of modern training technology (Galagan, 1989).

These deficiencies and other challenges are well documented in the literature on military training (Beecher, 1988; Hays & Singer, 1988; Hughes, 1986; Oneal, 1988; U.S. Congress, 1987), on commercial training (Feldman, 1985; Naisbitt, 1982), and on private and public education (Becker, 1982; Lusterman, 1977). Hays and Singer (1988), for example, stress that failure to coordinate training subsystem decisions degrades the total system capabilities. Too often training decisions that are mutually dependent (such as instructor to student ratios, required student matriculation, training length, number of pieces of equipment/training aids used, source of training, and presentation methods) are decided in isolation. Feldman (1985) projects that the impact of the information and technological revolutions on the conduct of work and training in factories and offices will be far greater than that of the industrial revolution, radically altering our training agendas and methods. And, Becker (1982) concludes that the American educational system and industry must systematically adapt the job to the worker produced by our public education.

The primary causes of modern training failures fall into two categories: (1) overemphasis on training hardware at the detriment of its functional utility, and (2) lack of a systematic process for integrating training decisions with supporting analyses. Factors that need to be considered in attempts to meet these training challenges on large scale systems are summarized in Table 13-1.

Table 13-1
Training Development Pitfalls

User Interface Complexity

Simulator complexity should not equate to user-interface complexity. Trainer and trainee abilities to operate simulators must limit the degree of interface complexity.

Interdependency of Training System Processes

Decision makers must be discouraged from the tendency to oversimplify the interdependencies associated with developing a training system. Majchrzak (1988), in the context of factory automation, outlines the complexity involved in planning and coordinating among such diverse areas as corporate goals, organizational structure, job structure, career progression, personnel aptitudes, motivation, abilities, learning strategies, teaching methods, and budget – all which may significantly impact training decisions.

Reflexivity

Effective training systems possess two elements of learning: (1) methods to teach employees, and (2) methods to learn from employees so as to identify new opportunities for improvement and to evaluate the effectiveness of the application of new technology. Training systems developers should not forget to incorporate the second element as well as the first. (Carnevale & Schultz, 1988).

Rapid Technology Changes

Early advances in technology may be so rapid as to outstrip the ability of the user to recognize or utilize its potential. As technology matures, identification of users and uses emerge. Increased appreciation for the technology and its application results in improved identification to materiel developers of intended capabilities and expected results. Failure to allow sufficient technological maturity may result in investments in high risk training technology with resources too limited to realize the full potential of the technology or to capture the most advantageous applications. (White, 1988)

Encumbering Fidelity

Simulators and other training devices should be designed to produce the degree of physical fidelity that is needed for training. Depending on the tasks to be trained and the skills, knowledge, and ability of the trainee, greater reality is not necessarily better for training. (Baum, Riedel, Hays, Mirabella, 1982; Garris, Mulligin, Dwyer, Moskal, 1987).

Cost Constraints

Until the 1970s, cost constraints had minimum impact on military training. Training was supported by spare parts and ammunition manufactured and stockpiled during the previous three decades. At the end of the Vietnam War, two significant types of training cost increases emerged. First, because old stockpiles had been depleted, training programs began paying for training equipment, spares and ammunition. Second, the focus on force modernization allocated a larger part of the budget to new weapon system acquisition, thereby limiting training budgets and training capabilities (Beecher, 1988). These cost increases added to other cost pressures (such as longer training for complex equipment, and requirements for larger maneuver areas) caused the military to seek more cost-effective training programs (Demers & Kitfield, 1989). A systematic approach to training development led to enhanced training effectiveness.

Systems Approach to Training

One of the most fundamental and ubiquitous changes in training in recent years was the development of the Systems Approach for Training (also called Instructional Systems Development). Developed for the Department of Defense (Bransen, Rayner, Cox, Furman, King, & Hannum, 1975), the Systems Approach to Training provides a systematic process by which training needs are identified and training products developed to meet those needs. Its five phases include: (1) analysis of training needs; (2) design of training programs, courses and materials; (3) development and validation of the materials; (4) implementation of the training in various settings and locations; and (5) evaluation of training in order to measure and improve learning. Methodologies similar to SAT are used by the Federal Aviation Administration, National Aeronautics and Space Administration, Office of Management and Budget, and Nuclear Regulatory Commission. Commercial training organizations have also benefitted from adopting a systems approach to training (Dick & Carey, 1985; Vaught, Hay, & Buchanan, 1985). Procedures and techniques of a systems approach to training generally include the activities depicted in Table 13-2.

Need for Integrated Analyses

Widespread use of SAT has been credited with improving the development of training systems in a variety of applications. But, while the SAT procedures have been validated, the development of training systems suffers from a number of problems (Hays & Singer, 1988): (1) lack of resources sufficient to employ SAT fully; (2) lack of guidance on how to

Table 13-2
Common Activities in the Systems Approach to Training

- Define the training requirement and the need for a training system.
- Describe the functions of each component of the training system and how components are interrelated.
- Develop the training system specification in terms of training and equipment objectives and parameters.
- Identify test and evaluation criteria for the training system.
- Develop alternative approaches and conduct trade-offs.
- Design and develop the training system. Test and select the best alternatives.

follow SAT's complex procedures; (3) failure to adequately understand the SAT's broad perspective and wide applicability; (4) inadequate attention to training behavioral models (e.g., focusing on simulation technology or hardware rather than on instructional technology); and (5) failure of design engineers to focus on the objective of improving job performance.

To help resolve these problems, a change in the training system acquisition process is needed which would systematically integrate training and all human-component analyses for training decisions. The MANPRINT program provides new concepts for institutionalizing these changes by adhering to such goals as: (1) matching the training systems acquisition to the materiel acquisition cycle; (2) achieving broad trade-offs early in concept stages among the elements of equipment design, training, manpower, and personnel; and (3) designing complete training systems considering the human component and the training environment as part of the system. Reaching these goals is largely accomplished by integrating training analyses into the equipment development process.

INTEGRATING TRAINING ANALYSES

The training community strives to develop training systems which fully integrate with the operational equipment, the operational environment, and the trainers and trainees. To achieve this objective, the elements of a *complete training system* (consisting of Training Management, Training Delivery, and Training Support subsystems) must be thoroughly coordinated. How well these three subsystems (Figure 13-1) are coordinated with the weapon system depends upon the degree to which support analyses and evaluations influence training and weapon system decisions. How well these training analyses influence the decisions on training and weapon systems depends on (1) the type of training acquisition strategy adopted to match to the operational system procurement, (2) the characteristics of the training analyses conducted, and (3) the process, procedures, or model used for integrating the analyses. (Chapter 10, Shields, Johnson, & Riviello, outlines the acquisition cycle for operational systems and may be used to follow the acquisition of a training system during an operational system procurement.)

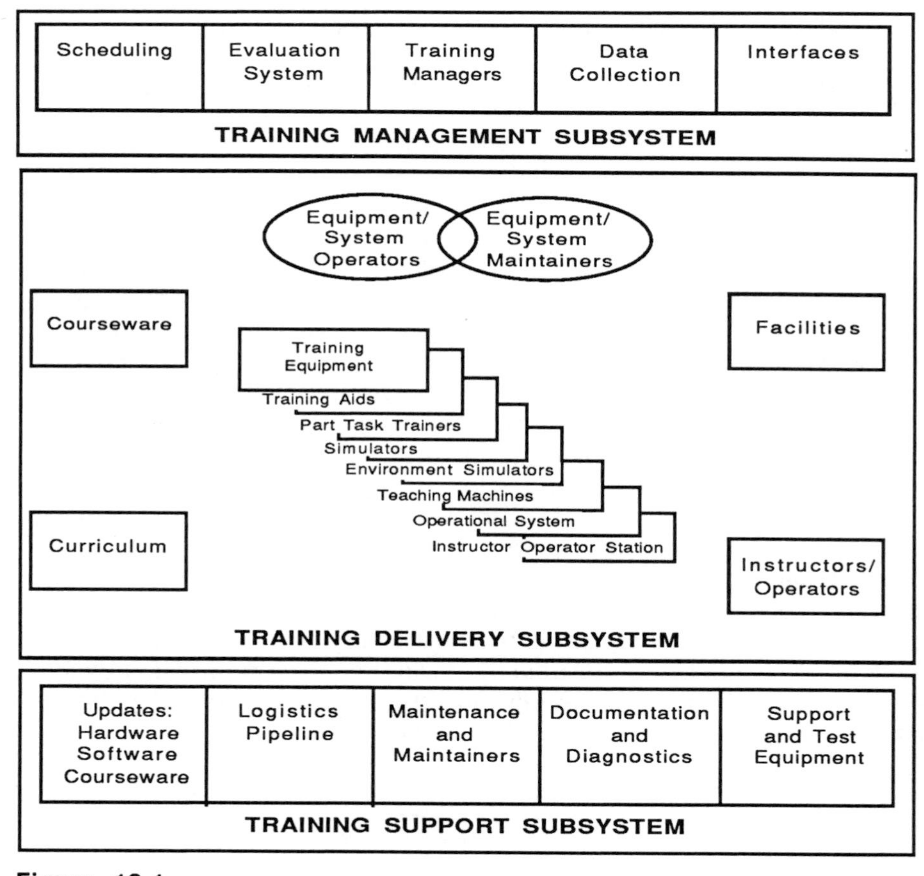

Figure 13-1
Elements of a Complete Training System

Training Acquisition Strategies

The various strategies for acquiring training can be grouped into four broad procurement categories (Figure 13-2): (1) complete training systems acquired concurrently with their operational system; (2) complete training systems acquired noncurrently with their operational system (phased); (3) partial training systems acquired either separately from or concurrently with their operational system; and (4) training equipment acquired sequentially or concurrently with their operational subsystems.

These acquisition strategies apply to industrial training as well as military training. Operational equipment acquisitions in industry range from multimillion dollar plant openings, or expansions involving the procurement of numerous and varied pieces of state-of-the-art equipment, to the addition of a single piece of manufacturing equipment. Each acquisition has training implications; and, the degree to which training and operational equipment

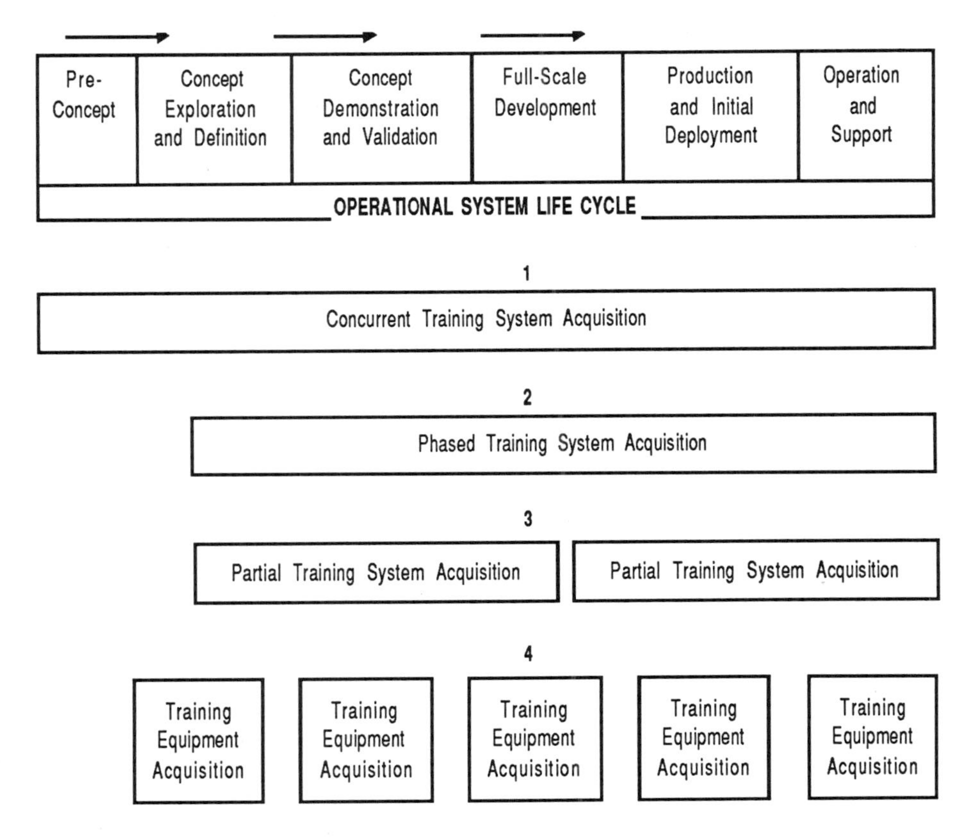

Figure 13-2
Four Acquisition Strategies for Training Systems

acquisitions are mutually supportive offers industry opportunities for training systems integration similar to those in the military.

An integrated training system can be achieved with any of the four acquisition strategies. Those training systems whose acquisition closely parallels the development of their operational equipment provide the most effective and timely training with minimal downtime once the equipment is operational. Majchrzak (1988) recognized the importance of timing in industrial training by stressing the coordination of training decisions with other plant activities. She emphasized the potential conflict between peak training activities and operational start-up activities if training plans have not been developed concurrent with equipment selections. Without congruity between training development and equipment development, productivity and organizational effectiveness are certain to suffer.

Training Analyses

The development steps of a training system follow a logical pattern from need definition through implementation and improvement (Figure 13-3). Each step is supported by various types of training analyses. These training analyses fall into three categories (Figure 13-4): (1) requirements analyses,

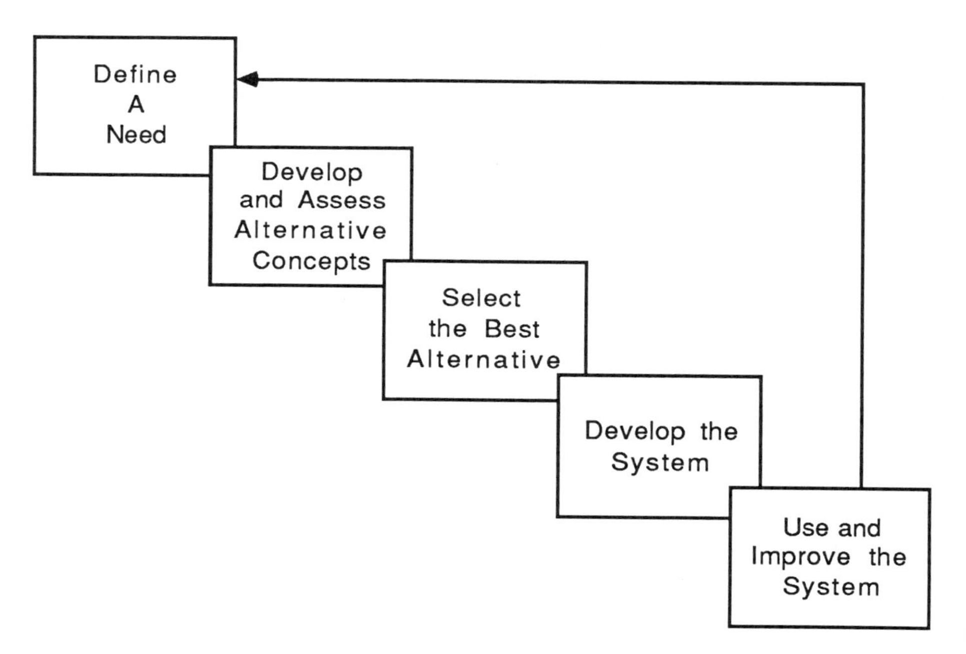

Figure 13-3
Training System Development Steps

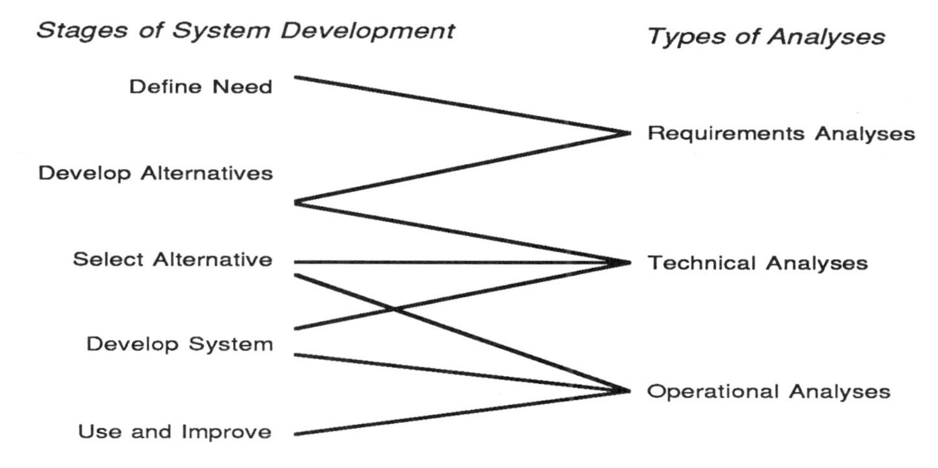

Figure 13-4
Training Analyses Supporting System Development

(2) technical analyses, and (3) operational analyses. Requirements analyses are conducted to determine the specifications necessary to achieve the desired capabilities. They may include review of predecessor systems as well as evaluation of new system alternatives. Technical analyses assist in the development and selection of alternatives and in the detailed development of the system. They provide identification of trade-offs and the technical feasibility of achieving expected results. Operational analyses measure the sensitivity of the integrated solution. They may assist in the selection of alternatives, evaluation of the development of the system, and assessment of its fielded capabilities.

Conceptual Model for Integrating Training Analyses

To ensure that training analyses achieve the expected results on the decision process, their application must follow a structured process. A conceptual model for integrating training analyses into equipment design and development in accordance with the MANPRINT philosophy is illustrated in Figure 13-5. This model, developed for military systems, conforms to Department of Defense documentation and regulations including Army Regulation 602-2 (1987), Army Regulation 700-127 (1987), TRADOC Pamphlet 350-30 (1978), TRADOC Regulation 350-7 (1985), and AMC/TRADOC Pamphlet 70-2 (1987). The model's application need not be restricted to military training systems. It provides a framework for integrating training analyses into the decision cycle, avoiding disjointed and repetitive efforts.

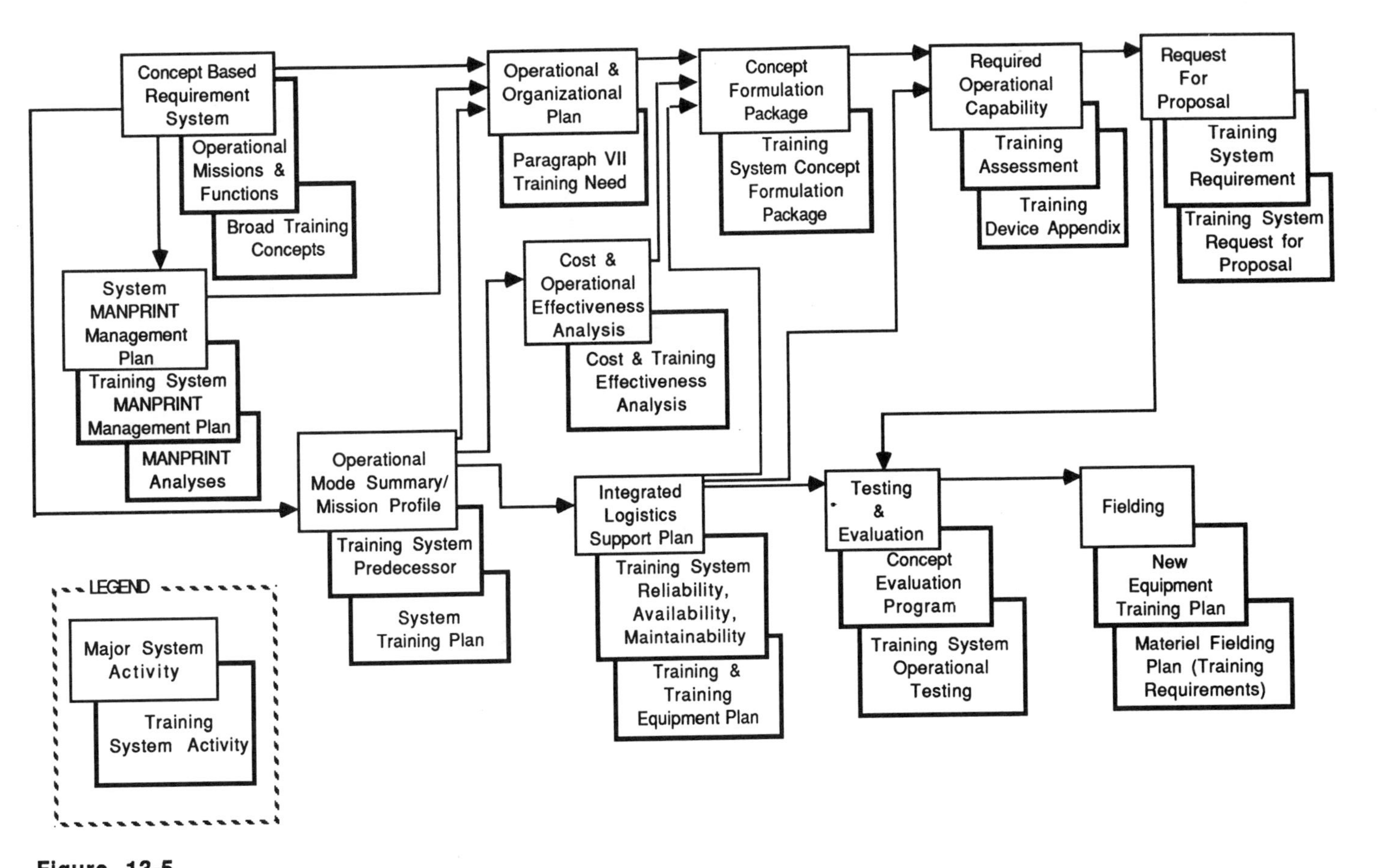

Figure 13-5
Key MANPRINT Activities for an Integrated Training System

MANPRINT analysis for Army training systems begins with the Concept Based Requirement System and the evaluation of battlefield function deficiencies. Concepts relating to operational missions and training concepts are developed. As the operational equipment requirements for the weapon system emerge, broad training requirements, concepts, and constraints are clarified. The System MANPRINT Management Plan provides a planning and management guide to insure that MANPRINT issues are recognized, analyzed, and documented during the weapon system's and training system's life cycle (especially during early stages). A System Training Plan is initiated to plan for development of training courses, training support products, and training facilities. Using predecessor system information, the System Training Plan, and weapon system mission concepts (weapon system Operational Mode Summary/Mission Profile), the training need is incorporated into the documentation for operational concepts (the weapon system's Operational and Organizational Plan). Structured trade-off studies in the form of cost and operational effectiveness analyses (supported by Integrated Logistics Support Plans that document training reliability, availability, and maintainability [RAM] and equipment plans) are initiated. These trade-off studies formulate the training system concepts and associated MANPRINT issues (documented in the Concept Formulation Package). The training and training devices requirements are specified in the weapon system Required Operational Capability (ROC) documentation. It is here that the needs for new equipment training, extension training, sustainment training, embedded training, and training aids are identified.

Upon system acquisition approval, the requirements are released to industry by way of Requests for Proposal. These requirements are stated not only as training and equipment engineering requirements, but also as MANPRINT requirements for the training system (for example, how the training system will be manned and supported and what its performance factors are to be).

During system development, MANPRINT analyses for the training system are updated as a result of ongoing test and evaluation of both the weapon system and the training system. During and after fielding, users assess training in order to improve the existing training system or identify possible enhancements for future systems. This completes the cycle back to concept analysis. (For simplification, feedback loops are not shown on Figure 13-5).

TRAINING INTEGRATION THROUGH MANPRINT

Acquiring an integrated training system through the MANPRINT approach entails coordinating the various elements of the materiel acquisition process (MAP) with the training analyses done during acquisition. MANPRINT provides the umbrella under which the training system acquisition achieves this coordination. The benefits attributed to the MANPRINT approach include three: (1) MANPRINT assures analyses are conducted with the operator, maintainer, and support personnel in mind; (2) there is a clear definition of when specific analyses are to be considered in order to influence acquisition and other decisions; and (3) the inclusion of human considerations will occur early in the decision cycle. The decision cycle steps follow.

Define a Need

The Army continuously analyzes its missions and force capabilities in order to identify shortfalls in countering a military threat. The Concept Based Requirements System (Figure 13-6) identifies potential solutions to the deficiencies. Priority solutions include (in descending order): (1) doctrine, (2) training, (3) organization, and (4) materiel requirements. Training is the second priority for problem solution and may also be implicated in the other solutions.

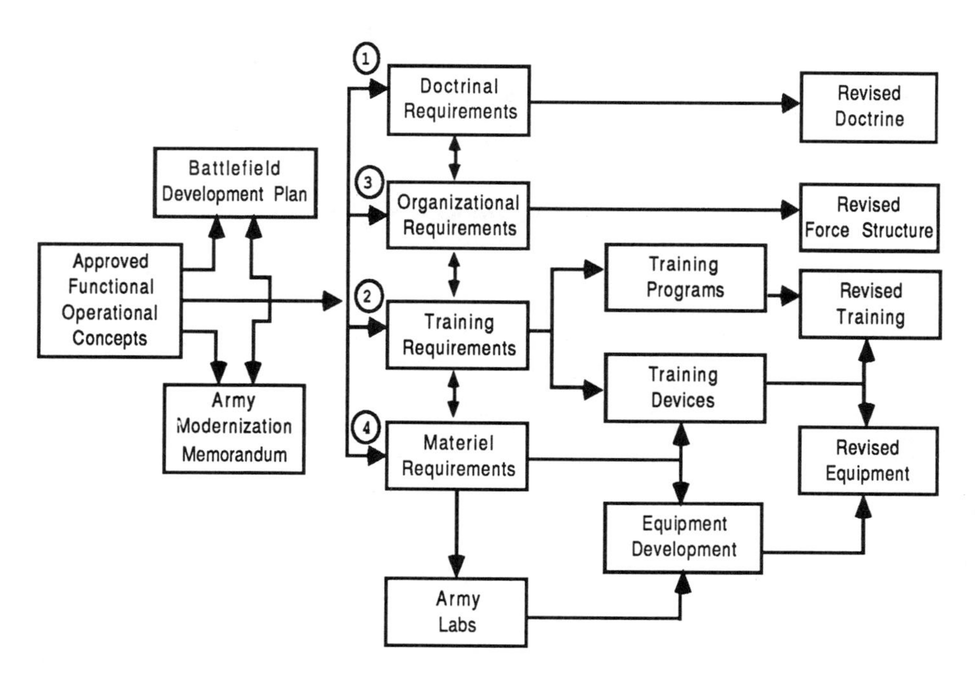

Figure 13-6
Concept-Based Requirement System

Operations Research/Systems Analysis. Within the broad spectrum of operations research/systems analysis (ORSA), the Army conducts training analyses to identify deficiencies and assess priorities. General Maxwell Thurman, former Army Vice Chief of Staff and former Commander of the Training and Doctrine Command, describes the role of ORSA in identifying and selecting solutions to battlefield deficiencies (Thurman, 1989, p. 5):

> ORSA is a bridge for problem solving – it provides a science-based capability to address problem areas. It is an intellectual art that couples imaginative and innovative alternatives within a scientifically grounded discipline. Analysis produces an understanding of phenomena related to the decision process and contributes to understanding the effect of alternative solutions. Analysis tells us what works well and what needs to be fixed. It provides the priorities for correcting problems and identifies cost-effective solutions. Once a solution is recommended, analysis provides the associated downside risks, the driving assumptions, and the weak points in the argument.

Thurman further illustrated the impact of integrated analysis on training solutions. In a recent analysis for the Tube-launched, Optically-tracked, Wire-guided (TOW) missile system, operational testing determined that TOW gunners were taking their first shot hundreds of meters below the expected launch range of the missile. The potential performance of the weapon system was not being realized. The analysis determined that commanders failed to position the missile where long-range launch was possible. New doctrinal and training concepts resolved this problem. Analysis also revealed several problems with the training equipment (the laser engagement system): (1) the laser failed to replicate the TOW guidance system; (2) smoke, fog, and dust degraded the effectiveness of the training system more than the actual missile; and (3) training tracking times were unrealistic (causing the gunner to be exposed to enemy fire during actual firings much longer than necessary). Changes in both the training strategy for the TOW and changes in the actual training equipment were offered to improve overall system performance.

Similar operation research analyses can also focus attention on training and other problems in industry. On a factory production floor, for example, it may be observed that low production output of one piece of equipment is affecting the entire production line. The problem may be an equipment deficiency, a personnel skills or training shortfall, too few operators or maintainers, a labor morale problem, or a combination of these and other factors. Only through a rigorous and thorough operational analysis can optimal solutions be found.

Develop and Assess Alternative Concepts

Once the need and possible solutions have been identified, further analysis helps to assess alternatives. Training analyses integrated with doctrinal, organizational, and equipment analyses help provide the information to develop and assess alternative concepts. If the solution is a materiel one and involves the acquisition of a major weapon system with its supporting training system, the military initiates analyses for the new systems. Table 13-3 lists the analyses having the most impact upon the development of the training system at this stage. An explanation follows.

Comparability Analysis. Analysts collect and evaluate relevant human performance and training effectiveness data on the predecessor and baseline systems as they apply to the definition of the new training system. Training analysts explore available training technologies and methodologies and conduct any required additional research. If there is no predecessor system, the analyses rely on basic and applied research and interpolate from these findings. In both cases, the analyses result in estimates for the new system training impacts and requirements.

Early Comparability Analysis. These analyses are distinguished by the stage at which they are conducted (i.e., during the rudimentary beginnings of the system). Early (in the acquisition process) training comparability analyses evaluate a hypothetical training system called a baseline comparison system using components from existing systems or, if necessary,

Table 13-3
Analyses Affecting Training Decisions

Analyses
Comparability Analysis
Early Comparability Analysis
Personnel Trade-offs and Constraints
MANPRINT Analyses
Manpower and Personnel
Training
Human Factors Engineering
Safety and Health Hazards
Cost and Training Effectiveness Analyses
Integrated Logistics Support Analyses
Reliability, Availability, Maintainability (RAM)
Human Performance RAM

from the research base. This baseline system provides the information needed to estimate system performance factors and to refine the projected training system's functional and resource requirements. This early analysis is repeated as necessary, incorporating empirical data as weapon system and training system developments progress.

Personnel Trade-offs and Constraints Analyses. These analyses identify trade-offs between parameters affecting personnel for the weapon system and the training system itself. Key sources of information for these analyses include: (1) weapon system descriptions, (2) training system descriptions, and (3) user personnel descriptions (in quantitative and behavioral terms). User personnel descriptions for the weapon system are documented in the weapon system Target Audience Description (TAD) and feed the TAD for the training system. The training system TAD expands the weapon system TAD by including the managers, instructors, maintainers, and support personnel for the training system. Like comparability analyses, training system target audience descriptions help provide baselines using data from the military and civilian service personnel systems and incorporate projections for the future population.

Issues surfaced by these trade-off analyses address every major aspect of the personnel support that may affect training, including mobilization requirements, personnel turbulence, qualitative availability, assignment and distribution schemes, and other personnel policy implications. The results of personnel trade-off analyses influence decisions on other entities of the training system, such as the training equipment design and the training system support elements (including configuration management, logistics pipeline, maintenance procedures, facilities, and the like). Evaluating the interrelationship of these decisions identifies constraints upon the training system and insures incorporation of personnel estimates into the earliest equipment and training concepts.

MANPRINT Analyses for the Training System. MANPRINT analyses consist of detailed front-end activities to supplement the broad preliminary training requirements that have been identified. These front-end activities provide detailed training requirements and alternatives through the Systems Approach to Training process. Basic stages of the analyses (Figure 13-7) include: (1) identifying training deficiency; (2A) conducting job, task and skill analysis; (2B) performing media analysis; (3A) developing training concept/macro and micro strategy; (3B) determining training alternatives, and (4) conducting a training effectiveness analysis. MANPRINT analyses determine constraints on the training system in each domain and formulate data to feed the trade-offs for the Concept Formulation Package (see Figure 13-5). As analyses in each MANPRINT domain are conducted for the weapon system, the training system MANPRINT analyses build on the weapon system analyses and are integrated with them. The MANPRINT analyses, described below, are repeated, as required, for the life of the training system.

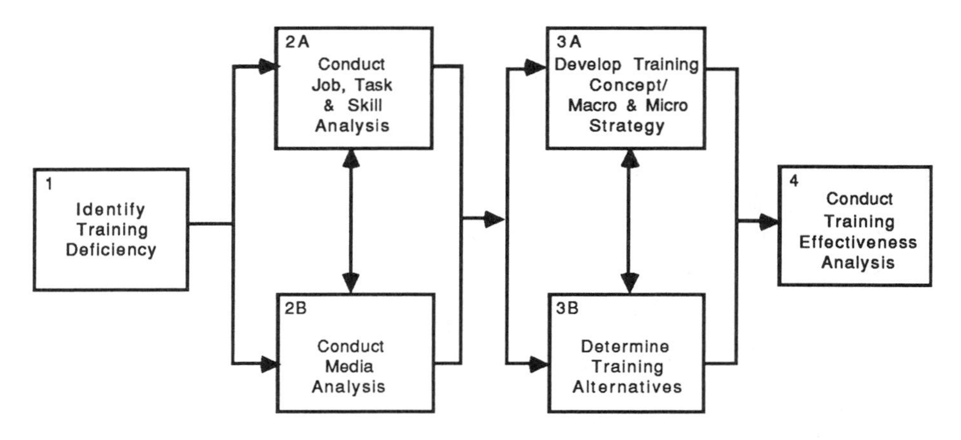

Figure 13-7
Stages in Integrated Training Front-End Analyses

MANPRINT Manpower and Personnel Analyses. Manpower and personnel requirements for the training system include consideration of the instructors, operators, maintainers, and support personnel needed by the training delivery, training management, and training support subsystems. Decisions concerning the type of training system to be acquired affect the quantity, availability, quality, and organization of these personnel in the training base and in operational units.

MANPRINT Training Analyses. Training requirements analyses are of two types: (1) Mission, and (2) Support. Mission training requirements analyses define the training tasks for the weapon system operators and maintainers. Support training requirements analyses identify the requirements for training the instructors, operators, maintainers, and support personnel for the training system itself. These training requirements become the basis for subsequent training curriculum design and development.

MANPRINT Human Factors Engineering Analyses. The human factors engineering analyses for the weapon system are an integral part of the training system analyses. The detailed procedural information developed for the weapon system provides information used by training analysts to structure training programs and to detail task performance factors. Human factors analyses for the training system address three major areas: (1) the trainees' ability to function within the training system, (2) the instructors' and managers' ability to use the instructor/operator station and management subsystem, and (3) the maintainers' access to equipment and use of procedural guides and documentation. Workload analyses and other related man-in-the-loop analyses and simulations identify potential human factors engineering problems for instructors and operators. MANPRINT emphasizes the need for human factors engineering analyses for the training system and for identifying trade-offs among other domains. Issues identified by these analyses are translated into subsequent test and evaluation issues.

MANPRINT Safety and Health Hazard Analyses. Information concerning safety and health hazards for the training system are based on the weapon system analyses. Additional analyses are required to identify potential training equipment hazards (such as those associated with sophisticated laser technology). Safety and health hazard issues may vary significantly from those of the weapon system. The recent surge of embedded training adds another dimension to safety issues which must be identified and tested early during training system design and development (Finley, Alderman, Peckham, & Strasel, 1988). Embedded training software must not interfere with operation of the weapon system and must allow quick and safe transition from a training mode to an operational mode.

MANPRINT Analyses Example. The training analysis for the Army's Advanced Antitank Weapon System-Medium (AAWS-M) illustrates the importance of MANPRINT analyses to training systems development. (The AAWS-M is being developed as a replacement for the M-47 Dragon weapon system which has been plagued with performance and training problems since its development in the 1970s.) In a 1989 study, Dyer, Lucariello, and Heller compared the results of an early AAWS-M training requirements analysis to an integrated MANPRINT analysis. The MANPRINT analysis caused significant changes in the AAWS-M training system, including tasks to train, training time, location of training, method of training, and embedded training equipment design. Table 13-4 provides some examples:

Cost and Training Effectiveness Analysis. The Army uses analyses in the six MANPRINT domains to support the Cost and Training Effectiveness Analysis. A Cost and Training Effectiveness Analysis is a subset of a Cost and Operational Effectiveness Analysis for a weapon system (see Figure 13-5). As such, it provides training system cost data for use in assessing the related acquisition, manpower, personnel, and training costs for candidate weapon system designs. The Cost and Training Effectiveness Analysis also develops and consolidates cost and effectiveness data concerning training system alternatives. The Army conducts the Cost and Training Effectiveness Analyses at various stages of the weapon system acquisition process. Detailed design and testing of the system updates earlier comparability studies with quantifiable data. As the training system design matures, logistics specialists initiate and document more detailed supportability studies in the training annexes to the Integrated Logistics Support Plan. Procedures for conducting Cost and Training Effectiveness Analyses are documented in TRADOC Pamphlet 11-8, *Studies and Analysis Handbook* (1985).

Integrated Logistics Support (ILS) Analyses. The management of Integrated Logistics Support analyses is documented in an ILS Plan. The plan includes those analyses to be conducted for evaluating weapon system support and MANPRINT plans for the training system (Army

Table 13-4
MANPRINT Analyses and AAWS-M Training System

Tasks Trained:

The original analysis concluded that all Dragon tasks were trained either in a centralized school setting or in the unit (on-the-job) and that a 40-hour program of instruction at the school would train all critical AAWS-M tasks. The MANPRINT analyses showed that several tasks were not being trained at all and that an 80-hour program of instruction would be needed for AAWS-M training.

Task Difficulty:

It was originally assumed that since the AAWS-M design would simplify the firing task, all AAWS-M firing skills would be simple to train and sustain. The MANPRINT analysis identified antiarmor related tasks that were difficult to train (such as target recognition). The training system architecture was changed to resolve this shortcoming.

Cognitive Demands:

A significant change in the training resulted from a shift from high motor skill requirements (Dragon) to high cognitive demands (AAWS-M) which represented a need for the training of more difficult and complex tasks.

Training Device Design:

Early assumptions concluded that the improved training device design would eliminate many of the difficulties in acquiring skills associated with the Dragon, such as tracking skills and launch performance under environmental stress (noise, smoke obscuration, debris). However, the original analysis failed to consider unit turbulence of both gunners and their trainers, and the inability of training devices to compensate for the lack of gunners' entry level skills as well as the lack of trainers' instructional and technical skills. The MANPRINT analyses concluded that (1) training devices should be designed with self-instructional capabilities, and (2) unit training programs must focus on initial skills training, or the schools must train all critical AAWS-M tasks (requiring all gunners to attend formal schooling).

Regulation 700-127, 1987). Early in the acquisition process, these MANPRINT plans are very general, consistent with the level of maturity of the system's design. As more training system information emerges, the Training and Training Equipment Plan is developed to identify support requirements for the training system (including facilities, support personnel and equipment, technical and training manuals and documentation, spare parts and maintenance services).

Reliability, Availability, Maintainability (RAM). Reliability, availability, and maintainability estimates influence logistics support analyses for the training system as they do for the weapon system. However, consideration to training system usage parameters (e.g., component usage rates) may differ significantly from the operational system as a result of embedded training and other training factors.

Human Performance in RAM. MANPRINT objectives extend the concept of RAM to incorporate the "human-in-the loop". This is often referred to as human performance RAM and refers to the performance parameters that may be expected for the operator, maintainer, and supporter of the weapon and training system. Additional research in this area must be accomplished to develop the necessary human performance analytic techniques (see Booher & Hewitt, Chapter 12) and to incorporate these concepts in cost modeling and simulations (see Parry, Collins, & Van Nostrand, Chapter 11).

Select the Best Alternative

The process of selecting the best training alternative involves using all analyses previously accomplished and a series of interrelated trade-off analyses: (1) major trades, (2) concept formulation package support analyses, and (3) required operational capability training analyses.

Major Trades. These evaluations examine the relationship between equipment and support system design alternatives and the cost of alternatives. Each of the major trades looks at alternatives among such entities as training, personnel, and hardware design, or at alternatives within such entities as alternative training media or variations in personnel aptitude criteria.

Concept Formulation Package (CFP). The CFP (Figure 13-5) documents the trade-off analyses and their results, and provides the analytic rationale for the training system concepts which have been considered. The Concept Formulation Package provides MANPRINT analyses, cost/benefit assessments, and supporting technical documents prior to demonstration and validation of the selected training concept. The CFP is updated only if major changes in the basic system occur.

Required Operational Capability. A Training Assessment is completed as part of the overall MANPRINT Assessment and provides in the Required Operational Capability (ROC) the training strategy as well as the need for system training devices and embedded training. The Training Device Appendix provides the operational, technical, logistical, and cost information necessary to develop, procure, and manage the system training devices and includes a MANPRINT Assessment for each device.

Comprehensive documentation of specific training system requirements and the degree to which they are satisfied by alternative systems should be developed as part of the rationale for selecting any training system. The documentation serves three purposes: (1) it provides information needed to cost, schedule, and budget for all components of the training system, (2) it insures that training requirements at both the training system level and the detailed task level are met, and (3) it provides an historical account of the rationale and issues affecting the selection.

Following selection of the best training system alternative, training analysts reevaluate all six MANPRINT domains, but with a narrower focus. As modification of the equipment itself becomes less feasible, analysts are limited to examining trade-offs among the training support systems. MANPRINT analyses during this phase focus on the goals established for each domain, assessing the impact upon total system performance.

Since the MANPRINT approach focuses on early analyses with the trainees, operators, and maintainers in mind, the training system alternative optimizes design from a people-equipment perspective. Fewer training system modifications result. Also, since the rationale for training trade-off decisions early in the program have been well-substantiated, this data is available for later system modification should that be required. Once the decision makers have selected the best alternative for the weapon system and the training system, training system requirements are translated into a Request for Proposal (Figure 13-5).

Develop the System

The Request for Proposal (RFP) communicates training system requirements to industry. For complex equipment, there may be several related Requests for Proposal for training equipment. These requirements are evaluated, tested, and refined as the system is developed.

Following the Request for Proposal and as the training system becomes more clearly defined, the purpose of analyses increasingly supports test and evaluation of intended concepts. Test and evaluation of a training system is based on new and previous analyses. Test and evaluation begins during early development and continues for the life of the system. Evaluators and analysts incorporate MANPRINT issues into the Concept Evaluation Program and into Operational Test and Evaluation. MANPRINT issues may also be used in product improvements for a fielded training system.

Concept Evaluation Program. The Concept Evaluation Program entails an analysis that provides information concerning the operational feasibility of a training concept or system. The program provides for quick reaction, inexpensive testing early in the acquisition process. The Concept Evaluation Program often uses models, simulations, mockups and prototypes for testing, all of which lend themselves to testing MANPRINT issues for an integrated training system.

Operational Test and Evaluation. Formal operational test and evaluation reviews assess the total training system along with its weapon system in an operational environment. They validate training tasks and their performance criteria as well as the performance of the total training system (including its equipment and courseware). Supportability tests validate quantitative and qualitative manpower and personnel requirements, maintenance and support concepts, training and technical documentation, software, human factors, and workload requirements in order to influence the design of the system. Tests also validate the reliability, availability and maintainability of the training equipment with man-in-the-loop considerations.

MANPRINT activities conducted early in the training system design process review operational test and evaluation issues, often using individuals chosen from local high school or college populations, informally representing trainees, instructors, operators, and maintainers of the training system. In later, formal operational test and evaluation reviews (but prior to fielding), representative samples from the actual target Army population participate in the test process. Developed properly, an integrated training system procurement involves the training system in weapon system operational testing. Further integration of analyses continues as the system is being used in an operational setting.

The early user testing which took place in 1988 for the Advanced Antitank Weapon System (AAWS-M) demonstrates integration of MANPRINT in the test process. Decay of operator skills severely degraded system performance for the Dragon (the AAWS-M predecessor system). No similar decay was evident when the Army tested AAWS-M trainees and retested them six weeks later. Incorporating training lessons learned from the predecessor system (for example, by redesigning AAWS-M to reduce the number of switches) improved the new training system (Janosko, 1989).

Use and Improve the System

The realization of a complete integrated training system using modern MANPRINT concepts on training is only in its infancy. The services plan fully integrated training systems for weapon systems such as the Air Force C-17 airlifter, the Army light assault helicopter (LHX), the Air Force Advanced Tactical Fighter, and the Army Heavy Force Modernization track vehicles. In these programs, MANPRINT training issues are being identified early in the development process, and related analyses are being conducted to verify and correct these issues. The results of fielding tests and changes in user-specific requirements are incorporated into other MANPRINT activities to improve total system performance. Users and analysts study these MANPRINT issues to determine trade-offs between weapon design, training system configuration, and user-specific needs. Follow-on systems, product improvements, and future training system changes benefit from these training-related MANPRINT activities during and after fielding.

ISSUES AND CONCLUSION

The MANPRINT program for integrated training systems continues to evolve. Because the program is evolving, several critical issues will be considered in the next few years including: (1) alternative concepts for integrated training systems, (2) advancements in analytical techniques, and (3) the impact on industry.

Alternative Concepts for Integrated Training Systems

The training community has defined only a limited number of concepts for integrated training systems. As the government and industry acquire the first complete integrated training systems and conduct related MANPRINT analyses, the benefits of alternative concepts at the training system and training subsystem levels will be determined. The costs and benefits of alternative acquisition and training strategies, training delivery concepts, and training system operation and support alternatives should also be determined. These benefits can only be determined empirically, after integrated training systems have been acquired and fielded. This empirical data will provide information to (1) refine the definition of an integrated training system, and (2) determine the most cost-effective means of acquisition.

Advancements in Analytical Techniques

Advancements in integrating multidisciplinary analyses and availability of training system support data have just begun. The military has initiated a much needed program to streamline acquisition procedures. This program, to be maximally effective, must influence the conduct, availability, and utility of analyses and data. The acquisition of integrated training systems offers a unique opportunity to streamline training-related analyses and data and to coordinate training system analyses with operational equipment analyses (Figure 13-8). Use of emerging computer technology and data base management techniques will enhance the effort. As the capability to identify trade-offs among acquisition costs, life cycle costs, and alternative training system concepts improves, "best guess" analyses of the past will be replaced with cost/benefit analyses that better integrate with MANPRINT analyses within each discipline and domain.

Impact on Industry Capabilities

The impact of integrated training systems upon industry capabilities is also just beginning to emerge. The most significant changes will be realized in

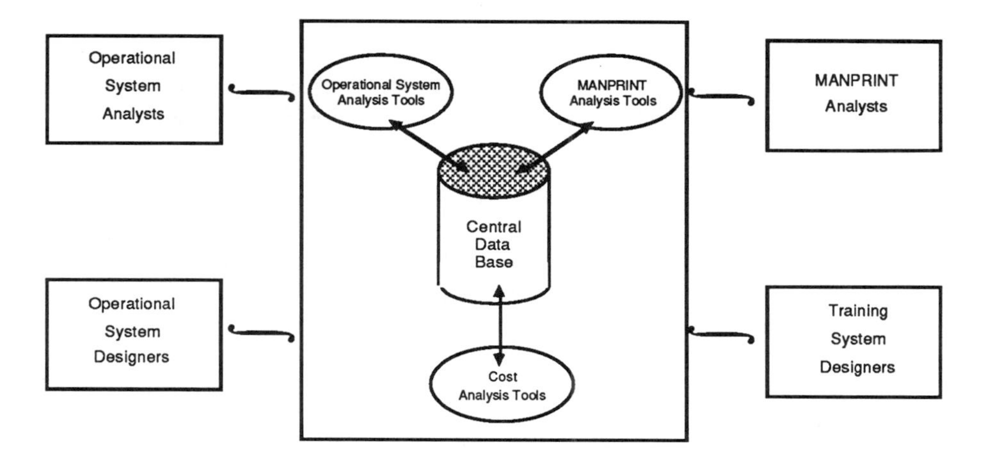

Figure 13-8
Integrated Analyses Provide for Concurrent and Consistent Operational System and Training System Data

the areas of: (1) organizational planning, (2) training development capabilities, (3) competition, and (4) personnel staffing.

Organizational Planning: Managers responsible for implementing changes to equipment (within the manufacturing plant as within the defense community) must change organizational planning to accommodate the concepts inherent in acquiring integrated training systems. This entails understanding the comprehensive implications of new equipment and new training. Planning functions must be structured, staffed, and directed to capture the opportunities (and avoid the pitfalls) of an integrated training system.

Training Development Capabilities: Larger corporations are becoming more involved in the training business (although there are few companies that provide a full array of training services from early requirements and supportability analyses to training development, training equipment development, and training implementation). Larger corporations have three team-building options in regard to obtaining capabilities for developing integrated training systems: (1) develop the capabilities internally (2) subcontract training development, or (3) acquire training development capabilities through acquisition. As larger companies recognize the operational efficiencies and economic benefits of possessing training system capabilities, competition will increase.

Competition: The trend to acquire integrated training system capabilities has led to increased competition among training companies. Mergers, acquisitions, and reorganizations within the training industry will continue (Beyers, 1988; Hull, 1988).

Personnel Staffing: The implications of integrated training systems on personnel staffing affect the hiring schedule, job description, supervisory design, and skills required of personnel within the organization that develops as well as within the organization that acquires new equipment and an integrated training system.

CONCLUSION

The performance of complex operational systems and equipment is dependent upon effective and timely training development. MANPRINT offers the opportunity to develop completely integrated training systems through a comprehensive approach to conducting and incorporating training analyses in the acquisition process. The result is the fielding of training systems completely compatible with the operational equipment, the support systems, and with the human dimensions of complex operational systems.

REFERENCES

AMC/TRADOC Pamphlet 70-2 (26 March 1987). *Materiel acquisition handbook*. U.S. Army Materiel Command/U. S. Army Training and Doctrine Command.

Army Regulation 602-2 (17 April 1987). *Manpower and personnel integration (MANPRINT) in the materiel acquisition process*. U.S. Department of the Army.

Army Regulation 700-127 (18 January 1987). *Integrated logistics support*. U.S. Department of the Army.

Baum, D. R., Riedel, S., Hays, R. T., & Mirabella, A. (1982). *Training effectiveness as a function of training device fidelity* (ARI Technical Report 593). Alexandria, VA: U.S. Army Research Institute for the Behavioral and Social Sciences.

Becker, R. J. (1982). Education and work: A historical perspective. In H. F. Silverman (Ed.), *Education and work*, Part II (pp. 1-14). Chicago: University of Chicago Press.

Beecher, R. G. (1988). Strategies and standards – an evolutionary view of training devices. *Proceedings of the Tenth Interservice/Industry Training Systems Conference*, 528-553.

Beyers, D. (1988, July/August). Contractors take over training and one another. *Military Forum*, 104-109.

Bransen, R. K., Rayner, G. T., Cox, J. L., Furman, J. P., King, F. J., & Hannum, W. J. (1975). *Interservice procedures for instructional systems development* (5 volumes). Fort Monroe, VA: U.S. Army Training and Doctrine Command.

Carnevale, A. P., & Schulz, E. R. (1988, November). Technical training in

America: How much and who get it. *Training and Development Journal*, 18-32.

Demers, W. A., & Kitfield, J. (1989, July). Training in Europe: fenced in. *Military Forum*, 46-53.

Dick, W., & Carey, L. (1985). *The systematic design of instruction*. New York: Scott Foresman.

Dyer, J. L., Lucariello, G., & Heller, F. H. (1989, July/August). Implications of system test data for training resource decisions. *MANPRINT Bulletin*, 8-9.

Feldman, M. (1985). The workplace as educator. In M. D. Fantini & R. L. Sinclair (Eds.), *Education in school and nonschool settings* (pp. 102-113). Chicago: Chicago Press.

Finley, D. L., Alderman, I. N., Peckham, D. S., & Strasel, H. C. (1988). *Implementing embedded training (ET): Volume 1 of 10: Overview* (Research Product 88-12). Alexandria, VA: U.S. Army Research Institute for the Behavioral and Social Sciences.

Galagan, P. A. (1989, January). IBM gets its arms around education. *Training and Development Journal*, 34-41.

Garris, R. D., Mulligin, C. P., Dwyer, D. J., & Moskal, P. J. (1987). *An experimental analysis of level of graphics detail and method of cue presentation on locator task performance for computer-based training and job aiding* (NTSC IN87-034). Orlando, FL: Naval Training Systems Center.

Hays, R. T., & Singer, M. J. (1988). *Simulation fidelity in training system design: Bridging the gap between reality and training*. New York: Springer Verlag.

Hughes, R. G. (1986, November). Aircrew training: The relative contribution of major training system resources to future readiness. *Proceedings of the Eighth Interservice/Industry Training Systems Conference*, 197-204.

Hull, S. (1988, November/December). A new owner. *Military Forum*, 5.

Janosko, T. (1989, June 7). *The role of test and experimentation in systems acquisition*. A briefing presented to the Military Operations Research Society. Fort Leavenworth, KS.

Lusterman, S. (1977). *Education and industry*. New York: The Conference Board.

Majchrzak, A. (1988). *The human side of factory automation*. San Francisco: Jossey-Bass Publishers.

Naisbitt, J. (1982). *Megatrends*. New York: Warner Books, Inc.

Oneal, J. (1988). Eliminating forced technology in military training – A conceptual model. *Proceedings of the Tenth Interservice/Industry Training Systems Conference*, 340-351.

Oneal, J. (1989, April). Personal notes. Nashua, NH: Oneal Brooks Associates.

Stockton, W. (1988, March 29). Trouble in the cockpit. *New York Times Magazine*, p. 39.

Thurman, M. R. (1989, March). Analysis counts. *Phalanx*, 4-8.

TRADOC Pamphlet 11-8 (1985, February). *Studies and analyses*. U.S. Army Training and Doctrine Command.

TRADOC Pamphlet 350-30 (1978). *Interservice procedures for instructional systems development*. U.S. Army Training and Doctrine Command.

TRADOC Regulation 350-7 (1985, November). *Systems approach to training*. U.S. Army Training and Doctrine Command.

U.S. Congress, House (1987, October 21). *Duplicative threat simulators waste millions and compromise testing of vital weapons* (House Report 100-529). 100th Congress, 2nd Session.

Vaught, B., Hay, F., & Buchanan, W. (1985). *Employee development programs – An organizational approach*. Westport, CT: Quorum Books.

White, W. J. (1987). The great divide. *Proceedings of the Ninth Interservice/Industry Training Systems Conference*, 99-105.

CHAPTER **14**

TRAINING AND AIDING PERSONNEL IN COMPLEX SYSTEMS

William B. Rouse

ABSTRACT

This chapter is concerned with enhancing human performance in complex systems. Discussion focuses on the central trade-off between providing the knowledge and skills that give personnel the potential to perform (via training) vs. augmenting performance directly (via aiding). Alternative approaches to training and aiding are discussed. Several methods for performing trade-off analyses are considered. The nature of training vs. aiding trade-offs are discussed in the broader MANPRINT context.

INTRODUCTION

Human performance in complex systems can be enhanced in several ways. One approach is *selection* which involves recruiting people with abilities, aptitudes, and attitudes appropriate for the roles, jobs, and tasks to be performed. While this chapter does not directly deal with selection, this topic is returned to when the broader manpower, personnel, and training (MPT) context within which this chapter fits is discussed.

Another way to enhance performance is *training*. Training can be defined as the process of managing people's experiences so that they gain the requisite knowledge and skills that give them the potential to perform. The extent to which this potential is created depends on the nature of the training experiences and the aptitudes and abilities of the personnel being trained.

The third way involves enhancing performance directly via *aiding*, which can range from task procedures, to more sophisticated decision support, to complete automation that personnel only monitor. These approaches usually are oriented toward improving performance or making difficult tasks easier. In some cases, aiding is central to being able to perform tasks at all.

This chapter discusses alternative approaches to training and aiding. Primary alternatives are discussed in terms of where and when they are most applicable. Also discussed is the fundamental trade-off between training and aiding – how much should be invested in creating the potential to perform (via training) vs. augmenting performance directly (via aiding)? In other words, how much intelligence should you put in people's heads vs. how much should you put in machines?

Depending on one's philosophy toward people and automation, either extreme might be staunchly defended in this trade-off. However, as will be illustrated, the trade-off is much more subtle and complex than can reasonably be addressed with simplistic strategies. The real issue, therefore, involves determining the best mix of *training and aiding*. Appropriate resolution of this issue is central to achieving the levels of system integration and performance espoused by the MANPRINT philosophy.

APPROACHES TO TRAINING

Choices among alternative approaches to training depend on the training objectives of interest, the population to be trained, and the resources available. Objectives usually relate to gaining a particular set of knowledge and skills so as to enable desired behaviors. Skill can be viewed as proficient application of knowledge gained by practice and repeated use. From this point of view, the critical issue, beyond the need for practice, concerns knowledge requirements. These requirements determine the most appropriate approaches to training.

It is useful to characterize knowledge requirements in terms of system knowledge and operational knowledge (Johnson, Maddox, Rouse, & Kiel, 1985; Fath & Rouse,1985). Operational knowledge (Table 14-1) denotes information about the way in which tasks are performed. System knowledge (Table 14-2) refers to information about the equipment system with which operators and maintainers perform their tasks.

The contrast between Tables 14-1 and 14-2 can be summarized by noting that operational knowledge concerns how to work the system while system knowledge relates to how the system works. This distinction has important implications for choosing among training methods. This distinction does not, however, imply that operational knowledge and system knowledge are mutually exclusive.

In general, analysis of operational knowledge requirements leads to determination of system knowledge requirements. In other words, analysis of how to work the system leads to determining the extent to which personnel need to know how the system works. The results of such an analysis tend to be quite different depending on whether the personnel of interest are maintainers, operators, managers, or designers. For example, knowledge of principles and theories tends to be more important for designers than maintainers.

Table 14-1
Types of Operational Knowledge

LEVEL	TYPES OF KNOWLEDGE		
	"WHAT"	"HOW"	"WHY"
Detailed/ Specific/ Concrete	Situations (What Might Happen)	Procedures (How To Deal With Specific Situations)	Operational Basis (Why Procedure Is Acceptable)
↓	Criteria (What Is Important)	Strategies (How To Deal With General Situations)	Logical Basis (Why Strategy Is Consistent)
Global/ General/ Abstract	Analogies (What Similarities Exist)	Methodologies (How to Synthesize And Evaluate Alternatives)	Mathematical Principles/ Theories (Why: Statistics, Logic, Etc.)

Table 14-2
Types of System Knowledge

LEVEL	TYPES OF KNOWLEDGE		
	"WHAT"	"HOW"	"WHY"
Detailed/ Specific/ Concrete	Characteristics of System Elements (What Element Is)	Functioning of System Elements (How Element Works)	Requirements Fulfilled (Why Element Is Needed)
↓	Relationships Among System Elements (What Connects To What)	Co-Functioning of System Elements (How Elements Work Together)	Objectives Supported (Why System Is Needed)
Global/ General/ Abstract	Temporal Patterns of System Response (What Typically Happens)	Overall Mechanism of System Response (How Response Is Generated)	Physical Principles/Theories (Why: Physics, Chemistry, Etc.)

Table 14-3 characterizes a range of alternative approaches to training. The methods noted as "passive" in this table tend to be instructor-dominated in that the trainee passively consumes knowledge when these methods are used. In contrast, the "active" methods tend to be trainee-dominated in that the trainees actively utilize knowledge, hopefully in a manner that has been designed to clarify, reinforce, and extend operational and system knowledge. Thus, to a great extent, active training methods can be viewed as carefully planned surrogates for actual experience.

Table 14-3 indicates that some methods are better for imparting system knowledge than operational knowledge, and vice versa. The classifications shown are obviously very approximate. However, these classifications are consistent with the traditional distinction between "in-the-head" and "in-the-hands" training. Further, as is shown below, these classifications emphasize the need to design training programs with significant "active" components.

The training methods in Table 14-3 can be differentiated further on the basis of their relative effectiveness and efficiency. Effectiveness is defined as the degree to which a method can successfully support the acquisition and retention of the desired type of knowledge and/or skill. Efficiency is defined in terms of the time and resources required to achieve this success - to be effective.

Table 14-4 presents an assessment of the relative effectiveness and efficiency of the methods in Table 14-3. Passive training methods are both effective and efficient for imparting all aspects of system knowledge. Active methods are effective, but inefficient for system knowledge, and actual experience is very inefficient.

Considering operational knowledge, passive methods are not effective. Active methods and actual experience are effective, but incur varying levels of inefficiency. Of course, the general inefficiency of actual experience is partially compensated for by the fact that personnel can do other useful things as they are gaining operational knowledge.

Tables 14-1 through 14-4 provide a four step process for synthesizing a mix of training methods into an integrated and comprehensive training program. First, overall training objectives (task behaviors and/or performance requirements) are decomposed into operational knowledge requirements using Table 14-1. For example, for an entry-level maintenance job, operational knowledge requirements are likely to be in the categories of situations, criteria, and procedures. The results of an analysis using Table 14-1 would be lists of context-specific knowledge in each of these three categories.

Using Table 14-2 as a guide, step two is to determine system knowledge requirements. For the entry-level maintainer, system knowledge requirements are likely to be in the categories of characteristics and relationships, with perhaps some knowledge in the category of functioning. The overall assessment of operational and system knowledge requirements

Table 14-3
Methods for Acquisition of Knowledge and Skills

EMPHASIS	METHODS		
	"PASSIVE" TRAINING	"ACTIVE" TRAINING	ACTUAL EXPERIENCE
System Knowledge	Classroom Lecture	Equipment Mockups	Induced Malfunctions
↓	Classroom Discussion	Flat Panel Simulators	Practice With Real Equipment
↓	Video and Films	Part-Task Simulators	On-The-Job Apprenticeship
Operational Knowledge	Laboratory Demonstrations	Full-Scope Simulators	On-The-Job Responsibility

Table 14-4
Effectiveness and Efficiency of Alternative Methods

PRODUCT	METHODS		
	"PASSIVE" TRAINING	"ACTIVE" TRAINING	ACTUAL EXPERIENCE
"WHAT"/ "HOW" (System)	Effective And Efficient	Effective But Inefficient	Effective But Very Inefficient
"WHAT"/ "HOW" (Operational)	Very Ineffective But Efficient	Effective And Efficient	Effective But Inefficient
"WHY" (System)	Effective And Efficient	Effective But Inefficient	Ineffective And Very Inefficient
"WHY" (Operational)	Ineffective But Efficient	Effective But Inefficient	Effective And Very Inefficient

can be performed by instructional system developers and domain experts for jobs where much previous experience is available. For jobs where there is not a strong baseline, considerable engineering analysis may be needed to determine knowledge requirements.

The third step is to consider and choose among alternative training methods. The nature of the knowledge requirements are used to choose among methods based on the guidance in Tables 14-3 and 14-4. For the entry-level maintainers, operational knowledge requirements focus on "what" and "how," and tend to be greater than system knowledge requirements. Table 14-4 indicates that active training methods are likely to be the best choice for this operational knowledge, while passive methods are a good choice for the system knowledge.

Choosing among the active methods requires assessing the strengths and weaknesses of specific instances of the training methods in Table 14-3. This can be accomplished using compilations such as provided by Johnson, Maddox, Rouse, and Kiel (1985). Issues include size of trainee population, availability of training devices and real equipment, and budgets available. This assessment of alternative training methods should include consideration of newly available training technologies (e.g., Pstoka, Massey, & Mutter, 1988).

The fourth step is to integrate the mix of methods chosen into a coherent training program. This includes considering how different levels of training device fidelity can best be combined to achieve training objectives (Rouse, 1982). Also central to this step are the analyses associated with developing an overall instructional system (see Gagne, Briggs, & Wager, 1988, for a detailed discussion). Before investing in such an integration effort, it is often useful to consider alternative approaches to achieving desired task behaviors and/or performance objectives – perhaps aiding (which can be less expensive) can decrease training requirements.

APPROACHES TO AIDING

As noted earlier, in contrast to training, which focuses on providing the knowledge and skills which yield the potential to perform, aiding focuses on augmenting task performance directly. Since the focus is on tasks, rather than knowledge and skills, it makes sense to organize approaches to aiding in terms of tasks supported by these approaches (Rouse, 1986).

A task taxonomy has been devised based on a series of efforts to develop aiding systems for command and control (Rouse & Rouse, 1983) and nuclear power (Rouse, Kisner, Frey, & Rouse,1984). Subsequent uses of this taxonomy for applications in manufacturing (Rouse, 1988) and design information systems (Rouse, Cody, & Frey, 1989) have enabled extension and refinement of the ways in which this taxonomy is used.

This task taxonomy, shown in Table 14-5, is premised on task requirements being composed of at least four different classes of activities.

Table 14-5
General Set of Human-Machine Tasks

EXECUTION AND MONITORING
1. Implementation of plan.
2. Observation of consequences.
3. Evaluation of deviations from expectations.
4. Selection between acceptance and rejection.
SITUATION ASSESSMENT: INFORMATION SEEKING
5. Generation/identification of alternative information sources.
6. Evaluation of alternative information sources.
7. Selection among alternative information sources.
SITUATION ASSESSMENT: EXPLANATION
8. Generation of alternative explanations.
9. Evaluation of alternative explanations
10. Selection among alternative explanations.
PLANNING AND COMMITMENT
11. Generation of alternative courses of action.
12. Evaluation of alternative courses of action.
13. Selection among alternative courses of action.

Execution and monitoring is the predominant class. For routine situations where observations are in reasonable agreement with expectations, execution and monitoring continue uninterrupted.

However, when deviations from expectations are large, task requirements expand to include *situation assessment*, which involves both *information seeking* and *explanation*. Information must first be sought as a basis for assessing the situation, and then, an explanation that "fits" the information must be recognized or devised.

Task requirements then shift to *planning and commitment*. A course of action for dealing with the assessed situation must be recognized or devised. This process is likely to include explicit, but not necessarily analytical, evaluation of the consequences of pursuing a particular course of action.

The relative frequency of the tasks in Table 14-5 depend on the nature of the job of interest (e.g., maintenance vs. operations vs. management vs. design), as well as the expertise and experience of the individuals of interest. For example, the tasks of the entry-level maintainer discussed earlier tend to fall primarily in the routine execution and monitoring category. In contrast, a higher-level maintainer is more likely to deal with subtle and complex problems requiring much situation assessment, and perhaps some planning.

Three characteristics of task classification are of particular importance because they provide very useful means for identifying aiding alternatives. First, note that most of the tasks in Table 14-5 involve generation, evaluation, and selection among alternatives. Second, emphasis is on alternatives in terms of interpretations of deviations, information sources, explanations, and courses of action. Third (not obvious in Table 14-5), the range of these tasks implies a wide variety of types of aiding, including for example straightforward job performance aids such as procedures, decision support systems for making sophisticated judgments, and automated systems for performing substantial portions of tasks.

An analysis of tasks in this taxonomy yields a 3 x 4 array of action words vs. objects of actions. These are generation, evaluation, and selection vs. deviations, information sources, explanations, and courses of action. This method of organization brings an important degree of structure to the process of identifying appropriate methods of aiding. This process can be illustrated in the context of Table 14-6 which provides a summary of a wide variety of approaches gleaned from hundreds of previous efforts.

For *implementation* (task 1), aiding can monitor activities, perform activities, and possibly adapt its monitoring and performance. Within maintenance tasks, for example, an aid could monitor for execution errors in remove and replace procedures, perform appropriate test procedures, and adapt its monitoring to a change in goals – for instance, switching the goal to assembly removal rather than fault isolation.

Observation (task 2) can be aided by modifying, filtering, interpolating, and extrapolating information. These operations can also be adapted to users' reactions to displayed information. Example observation aids for maintenance include parameter monitoring systems and a wide range of caution, alarm, and warning systems.

Aids for *generation* of alternatives can support three of the tasks in Table 14-5 (tasks 5, 8, and 11). These aids retrieve previous alternatives (e.g., past maintenance procedures) or alternatives with particular attributes (e.g., test procedures not requiring removal). An aid can also employ users' reactions to alternatives initially retrieved as a basis for inferring what users really wanted but may not have crisply specified.

Evaluation aids can be used to support four of the tasks in Table 14-5 (tasks 3, 6, 9, and 12). An aid can be used to assess characteristics (e.g., availability of documentation), correspondence (e.g., consistency of hypothesis and fault symptoms), consequences (e.g., likely repair time), and relative merits of alternatives (e.g., least cost repair). An aid can also adapt its evaluations based on users' reactions to initial evaluations.

Aids for *selection* can also be used to support four of the tasks in Table 14-5 (tasks 4, 7, 10, and 13). Given a set of evaluated alternatives, perhaps obtained via generation and evaluation aids, a selection aid can use a multiattribute criterion function to rank-order alternatives and/or recommend the best choice (e.g., the minimum time troubleshooting procedure). Such

Table 14-6
General Approaches to Aiding

IMPLEMENTATION

1. For a given plan and information regarding the user's actions, an aiding system can monitor plan implementation for inconsistencies and errors of omission and commission.

2. For a given plan and information regarding the user's actions and intentions, an aiding system can implement some or all of the plan to compensate for the user's inconsistencies, errors, or lack of resources.

3. Given information on intentions, resources available, priorities, etc., an aiding system can adapt its monitoring or implementation to reflect, for example, a change in goals.

OBSERVATION

4. For given information, an aiding system can transform, format, and code the information to enhance human abilities and overcome human limitations.

5. For a given set of evaluated information, an aiding system can filter or highlight the information to emphasize the most salient aspects of the information.

6. For a given sample of information, an aiding system can fit models to the information to integrate and interpolate within the sample.

7. For a given user and system constraints, as well as individual differences, an aiding system can adapt transformations, models, etc. (e.g., modify what information is presented and how it is presented).

GENERATION OF ALTERNATIVES

8. For a given situation, an aiding system can retrieve previously relevant and useful alternatives.

9. For a given set of attributes, an aiding system can retrieve candidate alternatives with these attributes.

10. Given users' assessments of suggested alternatives (e.g., via ranking or ratings), an aiding system can adapt its search strategy (e.g., attribute weights or logical operations) to produce new alternatives.

Table 14-6 *(continued)*
General Approaches to Aiding

EVALUATION OF ALTERNATIVES

11. For a given alternative, an aiding system can assess the alternative's a priori characteristics such as relevance, information content, and resource requirements.

12. For a given situation and alternative, an aiding system can assess the degree of correspondence between situation and alternative.

13. For a given alternative, an aiding system can assess (e.g., via simulation) the likely future consequences such as expected performance impact and resource requirements.

14. For given multiple alternatives, an aiding system can assess the relative merits of each alternative.

15. Given users' assessments of evaluation results (e.g., via requests for explanations), a support system can adapt its evaluation in terms of time horizon, statistical measures, etc.

SELECTION AMONG ALTERNATIVES

16. For given criteria and set of evaluated alternatives, an aiding system can suggest (e.g., via optimization) the selection that yields the "best" allocation of human-system resources.

17. Given users' assessments of selections (e.g., via ranking time-variations of criteria, preferences, and evaluations), an aiding system can adapt (e.g., by modifying criteria weights) its processing to provide suggestions that respond to these variations.

an aid could also modify its criteria in response to users' reactions, and subsequently, produce new recommendations.

Identification of methods appropriate for a particular job begins by mapping each of the tasks in the job to one or more of the general tasks in Table 14-5. The next step involves assessing each task in terms of the likely nature of human abilities and limitations. Perceptual and motor limitations associated with implementation and observation tend to be predominant for operations and maintenance tasks. In contrast, limitations for tasks associated with management and design tend to be more cognitive, with emphasis on information seeking, explanation, and planning.

The third step of the identification process involves reviewing the mapping to the general tasks and the nature of task limitations and abilities. This review enables choices among the aiding concepts in Table 14-6. In particular, the mapping to the general tasks indicates which subset of the concepts in Table 14-6 are important. The nature of task limitations and abilities enables choosing within subsets, as well as providing context specificity to the alternatives selected. For example, retrieving previous alternatives in maintenance might be stated as identifying previous faults for similar symptom sets.

The vast support system literature describes theories, design concepts, and evaluative results that relate to one or more of the capabilities summarized in Table 14-6. Review of this literature yields the conclusion that support concepts for selection and observation are common; concepts for evaluation and implementation are not uncommon; concepts for generation are fairly rare. Thus, while the literature on aiding is fairly dispersed, (see, for example, the journals *IEEE Transactions on Systems, Man, and Cybernetics* and *Information and Decision Technologies*; and the book series *Advances in Man-Machine Systems Research*) there is a rich base from which to draw.

TRAINING VS. AIDING TRADE-OFFS

From the foregoing, it should be clear that there is a wide variety of alternative ways to train and aid people. Several factors can affect whether training or aiding, or more frequently some combination, is the most appropriate way to achieve performance objectives (Rouse, 1985, 1987). These factors include characteristics of people, equipment, tasks, organizations, and environments.

Trade-offs are usually pursued in terms of performance, time, and cost. If a particular combination of training and aiding provides a performance advantage compared to alternative mixes, it will be chosen if time and costs are not adversely affected. Quite frequently, performance requirements are fixed. The primary issues then concern enabling personnel to achieve performance objectives as quickly and inexpensively as possible.

Within such analyses, the central contrast between training and aiding is quite clear. Training has the advantage that it can produce more flexible personnel who can cope with changing demands. The disadvantages of training are the recurring time and costs of delivery.

Aiding has the advantage that, with the exception of maintaining the aid, its recurring costs are relatively low. It may, however, have higher capital costs of acquisition. Aiding is likely to have the additional disadvantage of being fairly rigid relative to new demands. However, such technologies as artificial intelligence have the potential to broaden the capability and adaptability of traditional aids.

An interesting interaction between training and aiding concerns the training requirements imposed by aiding. For other than very simple types of aiding, personnel have to be trained to use the aiding appropriately.

These training requirements can be particularly subtle for tasks where the applicability of the aiding varies with situations.

Several research and development (R&D) efforts have focused on developing methods of trading off the advantages and disadvantages of training and aiding as a function of the characteristics of people, equipment, and tasks, as well as life-cycle costs. One of the earliest systematic approaches was the development of structured guidelines for mapping from the above characteristics to training and aiding alternatives, including combinations of training and aiding (Booher, 1978a, 1978b; Foley, 1978). These guidelines take the form of flowcharts which proceduralize trade-off decisions – this tends to get you into the right ballpark, but does not allow fine-grained trade-offs. For example, significant but non-major changes in system complexity via redesign would be unlikely to change the recommendations provided by the guidelines.

Somewhat more recently, the comparability approach, e.g., HARDMAN (Hardware and Manpower) (Weddle, 1986), was developed to enable analyzing the design of a new system in terms of deviations from an existing system. This approach assumes that you are in the right ballpark already, via an existing baseline system. This assumption is probably acceptable for modest updates of existing systems or changes involving well-understood technology.

Irwin, Blunt, and Lamb (1988) have developed a method for codifying experts' opinions on training vs. aiding trade-offs. Using statistical methods, they related experts' judgments to task and equipment variables such as complexity and accessibility, respectively. This approach seems particularly well-suited for initial choices among broad classifications of training and aiding alternatives. Finer-grained design applications require the use of much less aggregated attributes of human performance and equipment characteristics.

Very recently, computational approaches have emerged for analyzing training vs. aiding trade-offs. HARDMAN III (see Chapter 12, Booher & Hewitt) builds upon the comparability approach of earlier HARDMAN efforts, but adds modeling capabilities that enable predicting performance in ways other than baseline comparisons.

In conjunction with an Air Force program called JATAT (Job Aiding/Training Allocation Technologies), Rouse and Johnson (1989) have developed a framework for analyzing training vs. aiding trade-offs that depends on being able to represent the behavior/performance impact of training and aiding within computational models. The computer-based realization of the framework, which is called TRAIDOFF (Rouse, Frey, Wiederholt, & Zenyuh, 1989), potentially provides the basis for integrating a wide variety of computational tools. Many of these modeling tools and computational packages are already available. However, they have not previously been integrated to provide an appropriate mix of methods for formulating and resolving training vs. aiding trade-offs. This ongoing R&D

effort should provide a widely applicable analysis methodology and support package within the next few years.

BROADER CONTEXT OF TRADE-OFFS

Training vs. aiding trade-offs can be formulated using one or more of the methods discussed in the previous section. However, trade-offs cannot be fully resolved using these methods. Resolution requires that trade-offs be placed in the broader contexts of missions, logistics, and costs discussed in other chapters in this book.

Missions and logistics can be considered, to an extent at least, using models such as discussed by Thurman (1989). Costs are more subtle than might be imagined – see, for example Knapp and Orlansky (1983). This is best illustrated by comparing the costs of aiding and training.

Aiding costs tend to occur upstream during system acquisition. Training costs are incurred downstream throughout the system life cycle. Thus, capital costs must be traded off vs. recurring costs. Similarly, but much more subtle, the value of hard assets must be traded off vs. people assets. The subtlety and difficulty is compounded by the fact that different types of monies (i.e., acquisition dollars vs. operations and maintenance dollars) are usually involved in each side of these trade-offs.

Determination of training costs depends on projecting training activities. This involves determining the "what," "when," and "where" of training personnel – what programs, when during career progression, and where delivered. Concepts such as the Training Decisions System (Vaughan et al., 1989) can help with such projections, as well as assist in optimizing the flow of personnel to maintain availability of people with needed proficiencies.

The projection of aiding costs is much more complicated. Beyond initial capital costs, one must plan for the inevitable obsolescence of aiding and the consequent costs of upgrades. Two trends underlie this process. First, system usage inevitably evolves and aiding, therefore, becomes less useful. Second, procedures and knowledge bases inevitably become obsolete and/or require upgrades. Thus, aiding does involve recurring costs, but they are not as predictable as training costs.

Because of the difficulties of projecting costs, current trade-offs tend to go in the direction of training, at least in the sense that training upgrades are viewed as more viable than designing aiding that easily evolves and upgrades. However, this situation is likely to change due to improved technological feasibility, increasing costs of personnel, and decreasing availability of personnel. Consequently, sophisticated aiding systems will be increasingly viable – see, for example, Chapter 8 by Rouse in this book on error tolerant interfaces. As a result, aiding is likely to become the preferred alternative in training vs. aiding trade-offs.

The evolution outlined in the last few paragraphs is driven by costs and technology. Rather than evolve in this manner, a more principled approach is needed that carefully assesses and balances the advantages and disadvantages of training and aiding alternatives. This chapter provides a foundation for such a principled approach.

CONCLUSIONS

Ideally, the perspective should be even broader than presented in the last section. Training vs. aiding trade-offs are only one component, albeit central, of an overall human resources investment strategy. Such a strategy should include consideration of selection, training, aiding, job design, equipment design, and enabling technologies (Rouse, 1990). Consequently, beyond considering the training vs. aiding aspects of performance enhancement, trade-offs should also involve selection in terms of the costs of recruiting and operating the personnel system. Further, selection aspects of trade-offs should include the issues of the availability and recruitability of personnel. Recruitability, as well as retention, are likely to be related to the promise and delivery of quality training of marketable skills. Beyond these issues, supportability should be considered including facilities and other components of "overhead."

Resolving trade-offs within this overall context requires specifying relationships among many attributes of the problem. An obvious question is whether or not data are available to define these relationships. Rouse and Cacioppo (1989) recently studied this issue by surveying government and industry data sources for the information necessary to specify the relationships and parameters with the aforementioned human resources investment model (Rouse, 1990). They concluded that little of the necessary information is directly available. However, they projected that use of sensitivity analyses and model-based selection of data sources might enable substantial progress.

Thus, it is not yet possible, at least in general, to perform human resource trade-offs in the broadest of contexts. However, this chapter has illustrated that rigorous formulation of training vs. aiding trade-offs is possible. Further, this can be accomplished within a systematic process for designing training and aiding alternatives. The availability of such methods and tools for dealing with the training vs. aiding component of system integration is central to realizing the promise of MANPRINT. The conceptual framework presented in this chapter, in addition to available methods and tools, provide the MANPRINT practitioner with a balanced and flexible means for enhancing human performance in complex systems.

ACKNOWLEDGMENT

The author gratefully acknowledges the comments of Dr. William B. Johnson on an early draft of this chapter. Preparation of this chapter was partially supported by the Air Force Human Resources Laboratory under Contract No. F33615-86-C-0545.

REFERENCES

Booher, H. R. (1978a). *Job performance aids: Research and technology state-of-the-art* (Report TR-78-26). San Diego, CA: Navy Personnel Research and Development Center.

Booher, H. R. (1978b). *Job performance aid selection algorithm: Development and application* (Report TN-79-1). San Diego, CA: Navy Personnel Research and Development Center.

Fath, J. L., & Rouse, W. B. (1985). An approach to training for operation and maintenance in large-scale dynamic environments. *Proceedings of 1985 IEEE International Conference on Systems, Man, and Cybernetics*, pp. 532-536.

Foley, J. P. (1978). *Executive summary concerning the impact of advanced maintenance data and task oriented training technologies on maintenance, personnel, and training systems* (Report TR-78-24). Wright-Patterson Air Force Base, OH: Air Force Human Resources Laboratory.

Gagne, R. M., Briggs, L. J., & Wager, W. W. (1988). *Principles of instructional design.* New York: Holt, Rinehart, and Winston.

Irwin, J. G., Blunt, J. H., & Lamb, T. A. (1988). *A training/JPA model: Its utility for the MPT community.* Dayton, OH: Universal Energy Systems, Inc.

Johnson, W. B., Maddox, M. E., Rouse, W. B., & Kiel, G. C. (1985). *Diagnostic training for nuclear plant personnel. Volume 1: Courseware development* (Tech. Rep. NP-3829, Vol. 1). Palo Alto, CA: Electric Power Research Institute.

Knapp, M. I., & Orlansky, J. (1983). *A cost element structure for defense training* (Tech. Paper P-1709). Alexandria, VA: Institute for Defense Analyses.

Pstoka, J., Massey, L. D., & Mutter, S. A. (Eds.). (1988). *Intelligent tutoring systems: Lessons learned.* Hillsdale, NJ: Erlbaum.

Rouse, W. B. (1982). A mixed-fidelity approach to technical training. *Journal of Educational Technology Systems, 11*, 103-115.

Rouse, W. B. (1985). Optimal allocation of system development resources to reduce and/or tolerate human error. *IEEE Transactions on Systems, Man, and Cybernetics, SMC-15*, 620-630.

Rouse, W. B. (1986). Design and evaluation of computer-based decision support systems. In S. J. Andriole (Ed.), *Microcomputer decision support systems* (Chapter 11). Wellesley, MA: QED Information Systems.

Rouse, W. B. (1987). Model-based evaluation of an integrated support system concept. *Large-Scale Systems, 13*, 33-42.

Rouse, W. B. (1988). Intelligent decision support for advanced manufacturing systems. *Manufacturing Review, 1*, 236-243.

Rouse, W. B. (1990). Human resource issues in system design. In N. P. Moray, W. R. Ferrell, & W. B. Rouse (Eds.), *Robotics, control, and society*. London: Taylor & Francis.

Rouse, W. B., & Cacioppo, G. M. (1989). *Prospects for modeling the impact of human resource investments on economic return*. Washington, DC: Department of the Army, Office of the Deputy Chief of Staff for Personnel.

Rouse, W. B., Cody, W. J., & Frey, P. R. (in press). Information systems for supporting design of complex human-machine systems. In C. T. Leondes (Ed.), *Advances in Control and Dynamic Systems* (Vol. 26). Academic Press.

Rouse, W. B., Frey, P. R., Wiederholt, B. J., & Zenyuh, J. P. (1989). *The TRAIDOFF Concept*. Norcross, GA: Search Technology, Inc.

Rouse, W. B., & Johnson, W. B. (1989). *Computational approaches for analyzing tradeoffs between training and job aiding*. Brooks Air Force Base, TX: Air Force Human Resources Laboratory.

Rouse, W. B., Kisner, R. A., Frey, P. R., & Rouse, S. H. (1984). *A method for analytical evaluation of computer-based decision aids* (Tech. Rep. NUREG/CR-3655). Oak Ridge, TN: Oak Ridge National Laboratory.

Rouse, W. B., & Rouse, S. H. (1983). *A framework for research on adaptive decision aids* (Reort. TR-83-082). Wright-Patterson Air Force Base, OH: Air Force Aerospace Medical Research Laboratory.

Thurman, M. R. (1989). Analysis counts. *Bulletin of Military Operations Research, 22*, 1-8.

Vaughan, D. S., Mitchell, J. L., Yadrick, R. M., Perrin, B. M., Knight, J.R., & Eschenbrenner, A. J. (1989). *Research and development of the Training Decisions System* (Report TR-88-50). Brooks Air Force Base, TX: Air Force Human Resources Laboratory.

Weddle, P. D. (1986). Applied methods for human-systems integration. In J. Zeidner, (Ed.), *Human productivity enhancement*, 1, 332-363. New York: Praeger.

SUGGESTIONS FOR FURTHER READING

Gagne, R. M. Briggs, L. J., & Wager, W. W. (1988). *Principles of instructional design*. New York: Holt, Rinehart, and Winston.

Pstoka, J., Massey, L. D., & Mutter, S. A. (Eds.). (1988). *Intelligent tutoring systems: Lessons learned*. Hillsdale, NJ: Erlbaum.

Rouse, W. B. (1984-89). *Advances in man-machine systems research*. (Five volumes) Greenwich, CT: JAI Press.

Rouse, W. B., & Johnson, W. B. (1989). *Computational approaches for analyzing tradeoffs between training and job aiding*. Brooks Air Force Base, TX: Air Force Human Resources Laboratory.

CHAPTER **15**

PRACTICAL HUMAN PERFORMANCE TESTING AND EVALUATION

Robert T. Hennessy

ABSTRACT

The final step in the acquisition process before a system or product is ready for military fielding or commercial customer purchase is operational testing and evaluation. To insure a valid operational test, the MANPRINT approach is to assure that a reasonable sample of representative users actually participate in the test and require that a system can pass or fail for people reliability as readily as for machine reliability reasons. The state of the art for methods to measure human performance in field tests are generally undeveloped compared to those for machine performance. Techniques used in laboratories and other well controlled settings are generally not suitable for field tests, and evaluations of total system performance are usually inconclusive for answering specific questions on human performance. This chapter discusses the theoretical and practical limitations in operational measures of human performance. In doing so, it is suggested that operational test constructors require more use of expert observational techniques and less of automatically recordable, but uninformative data. Finally, a MANPRINT Hierarchical Performance Measurement Model is proposed as a practical approach to relate user's performance to system effectiveness and suitability in a more meaningful way.

INTRODUCTION

Machines do not work alone. Almost all military and commercial equipment and systems require people as operators, maintainers, and supporters. It is essential, therefore, for public health and safety, as well as operational feasibility, that machines be tested and evaluated in their intended operational settings prior to their more permanent or routine usage in those settings. This is particularly necessary when the machines are intended for complex environments like warfare, nuclear power generation, or commercial air flight.

In the military, for example, every new, major item of equipment (prior to full procurement) undergoes a final operational test and evaluation to determine its suitability and effectiveness for its military purpose. Until the mid-1980s, however, attention in testing was almost wholly focused on the equipment. Individuals in charge of operational test and evaluation tended to emphasize what they knew and understood, i.e., equipment functionality, reliability, availability. It was assumed that either the military service users themselves, their managers, and/or the training developers would cope somehow with the systems as designed.

As described in earlier chapters, the MANPRINT philosophy has made major changes in the way operational tests and evaluations are planned and executed (Army Regulation 602-2, Manpower and Personnnel Integration (MANPRINT) in the Materiel Acquisition Process). The importance of human factors engineering, manpower, personnel quality, training, safety, and health hazards are all formally recognized as critical concerns during operational testing and evaluation. In fact, as pointed out by Shields, Johnson, and Riviello (Chapter 10), the testing of systems with people included as part of the system is the crucial final link in the acquisition process chain. Systems can be procured or canceled based on human performance factors as well as equipment factors (Booher, 1988).

There is still a major problem, however, facing the test and evaluation community assigned with the responsibility of acquiring useful operational human performance data. Even though MANPRINT domains are required to be considered in the design, development, and testing process, there is currently no specific requirement for human performance measurement. Without individual and crew performance measurement, there is little chance that latent human factors problems will be revealed. Although system performance measures include the contribution of human subsystem performance, they do not expose design and operational faults, unusual or inappropriate skill demands, poor function allocations and procedures, training deficiencies, and hazards. Military operators and commercial pilots, for example, learn extraordinary skills and go to great lengths to compensate for them. Only the most serious problems, those that are obvious, immediate, and substantial, will be recognized. Those that are subtle, gradual, and serious, but not critical, will go undiscovered.

The purpose of this chapter is threefold: (1) to discuss the unique problems associated with obtaining operational human performance data; (2) to discourage the application of the two most prevalent approaches to human performance measurement (i.e., the measure-people-like-machines approach and the lab-in-the- field approach); and (3) to suggest a more practical way, through expert observational techniques, to obtain useful human performance data in complex operational environments. A hierarchical performance measurement model, useful to MANPRINT practitioners, is described along with an example of how it is currently applied in a real world environment.

HUMAN PERFORMANCE MEASUREMENT IN COMPLEX SETTINGS

It would appear reasonable to assume that the neglect of human performance measurement in operational testing and evaluation could be redressed simply by requiring that it be done. The performance measurement methods could be adopted from other complex settings, such as laboratory research, field training and simulators, where individuals and crews perform multiple tasks in realistic environments, exercise a significant amount of skill, and sessions go on for protracted periods. It turns out, however, that despite the essential need for human performance measurement in many complex environments, there is no model of how it should be done. *A good human performance measurement system has yet to be developed for meaningful tasks being performed in realistic environments.*

The two principal reasons for lack of a prototype human performance system are (1) the measurement problem is difficult, and (2) the two most prevalent approaches to human performance measurement have not proven to be very useful. These are (a) attempting to apply methods suitable for equipment and system measurement to human performance measurement, and (b) extending laboratory methods to the field and other complex environments. A common element in both these approaches is an overemphasis on the importance and value of objective, (i.e., automatically recorded) performance data. This came largely with the introduction of computers into behavioral research settings.

To understand the measurement predicament and the failings of the standard approaches, it is necessary to review the standards for good performance measures for machines and humans. The following section describes these standards and contrasts the characteristics of machines and humans that make it easy for machines but difficult for people to meet the measurement standards. The unsuitability of laboratory methods and shortcomings of "objective" performance measures will be discussed subsequently.

Characteristics of Good Performance Measures

Lane (1986) provides a thorough review of past research and current issues in human performance measurement using air combat maneuvering as the exemplar application area. In describing the complications of human performance measurement, he makes the important point that when there is an expressed interest in human performance, it is often proficiency, i.e., absolute level of skill, that is really of concern. Behavioral researchers are interested in the processes of behavior underlying the observable output. On the other hand, in the military testing context, it is the outcome relative to some standard that is important.

Table 15-1, adopted partially from the organization of Lane's (1986) report, summarizes the essential concerns for good performance measures when measuring either machine system performance or human performance:

I. The *purpose* of the measurement, i.e., the information goals, must be established. For all measurement, its intended use determines what data is to be collected.

II. *Utility* relates to the practicality and economy of obtaining the measurement information. A measure may not be feasible or the cost may be too great.

III. *Credibility* simply means that the performance information is believable.

IV. *Sensitivity* is relative to the information need. Testing in complex situations has the common problem that variability in conditions reduces the sensitivity to an extent that real differences of practical importance cannot be measured.

V. *Separability* means that the important components influencing outcome measures can be disassociated, i.e., the effects of the operator, equipment, and environment can be identified. System outcome performance measures alone contribute little to understanding of the elements and processes that led to the outcome.

VI. *Comprehensiveness* means that nothing important contributing to the performance of interest has been left out. For example, a hardware system may be effective, but it is also important to know if it is reliable; or, crews may perform in a superior manner but may only be able to sustain the level of performance for a short period. It is important to know the whole picture.

VII. *Specificity* relates to the focus of the measure on a particular aspect of performance. That is, individual performance factors can be recognized and quantified. When measures have this characteristic, sources of problems can be diagnosed in detail and selective improvements made.

Measure-People-Like-Machines Approach

The temptation to measure people performance the same way as equipment performance probably stems from the nature of complex operational systems. Such systems are composed of intimate relationships between people and machines with the person viewed, essentially impersonally, as just another subsystem. Furthermore, it is system performance at some level, not component performance, human or otherwise, that is of chief concern to evaluators.

At the other extreme, it is often assumed that the system output is a direct index of individual or crew performance. For example, a dive bombing

Table 15-1
Characteristics Required for Good Performance Measures

I. PURPOSE – Definition of how the information will be used

II. UTILITY

A. Informativeness relative to alternatives
B. Practicality – Time, cost, feasibility
C. Acceptability – Value to intended user

III. CREDIBILITY – Proof of

A. Reliability – Consistency, accuracy, repeatability
B. Validity – Reveals the performance of interest

IV. SENSITIVITY – Discrimination of smallest performance difference of concern

V. SEPARABILITY – Disassociate influences of

A. Operator
B. System
C. Environment

VI. COMPREHENSIVENESS

A. Completeness – All aspects of interest
B. Relevance – High level interpretability

VII. SPECIFICITY – Captures performance components

score can be easily thought of as solely a measure of pilot performance rather than a combination of pilot and aircraft equipment performance. In either case, the result is the same. In the course of most complex operational system tests, exercises, and field training, there appears no reason to the outside observer to treat human performance measurement as a special case. It may be obvious to say that people differ from machines, but it is not necessarily apparent how this influences performance measurement.

There are, however, several crucial differences which need closer examination. Table 15-2 lists differences between machines and people relevant to performance measurement. Machines have specific functions; the functions are achieved from realization of a design that specifies the component parts, what they do, and how they go together. From the design and the laws of physics, the operation or behavior of the machine is

Table 15-2
Differences Between Machines and People

MACHINES:	PEOPLE:
Perform functions	Carry out intentions
Are predictable	Are unpredictable
Have known design	Have unknown design
Definite output	Indefinite output
Few determinants	Many determinants
Repeatable behavior	Variable behavior
Stable over time	Unstable over time
Discrete parts	Integrated

predictable. People have no specific function; the closest analogy is that people have intentions and carry them out. As individuals or in groups, people generally are in a state of intending to carry out their assigned tasks. Their detailed design, particularly the brain and nervous system, is unknown. Consequently, the exact behavior of people is relatively unpredictable compared to machines.

Compared to people, machines are simple. There are only a few external determinants of its functions, and again they are knowable. Machines are somewhat stable over time and, under identical conditions, their functions are repeatable. The machine output is a physical event that another machine can measure. When a machine fails, the design is a path for diagnosis and the suspected component is both accessible and testable.

People are complicated and their behavior reflects this fact. There are many determinants of behavior, internal as well as external, and not all of them are known. People change rapidly with time. A short-term memory disappears in a few seconds, and vigilance deteriorates dramatically after 20 minutes. Fatigue develops within a day. Under identical circumstances behavior will not be quite the same twice. Moreover, a single behavior can have many different corresponding physical events. For example, saying "yes" or "roger," giving a thumbs up, or nodding the head, all signify the same thing. It is not easy to program a machine to measure the single event behind all these different behavioral expressions. When there is a human performance failure, there are no blueprints, only hypothetical constructs, to guide diagnosis of the problem. Isolated testing is not always possible.

Juxtaposing Table 15-1 and Table 15-2 reveals why machine and human performance measurement is so different. For mechanical and electronic systems, many of the concerns are either taken care of by definition or are inherently easy. A machine's design and structure explicitly reveals how components or subsystems relate to one another, what goes in, and what comes out. Data from all the separate elements can be structured with the pattern of the machine's design. It is then relatively easy to establish the contribution of each element to total performance. In operational testing, total system and subsystem measures of performance are fashioned in this manner.

But in the human subsystem, the subelements, internal to the person, are hidden. There is no established or easily knowable structure for how relevant cognitive processes affect each other or what the critical path is for a particular task. The contents of computer memory can be established quickly and confidently. Verifying what a human knows is more complicated. Attempts are made to understand human performance by building a model of the unobservable processes based on the observable performance. Because of the inferential basis for models, the correspondence between the model and the actual behavioral processes are difficult to verify. Conversely, the correspondence between a machine model, i.e., its design, and any specific embodiment is perfect save only for fabrication errors.

The distinctions are simply too great and people too complex to measure people like machines. When human performance measurement has been treated as a unique problem area, it has been in the behavioral laboratory. The topic of adopting these methods to field use is discussed next.

Lab-In-the-Field Approach

At some point in the development of equipment and systems, people are faced with the task of measuring human performance in a complex situation such as an operational test, full-mission simulation, or on the job. When human performance, not system performance alone, is of concern, machine-style measurement is not useful. The most common alternative under these circumstances is to bring the scientific methods of the behavioral laboratory to the complex environment of the field or the simulator.

This approach is valued because it connotes tradition, systematic procedures, quantitative methods, control, objectivity of test and data, and certainty of findings. The logical structure and comprehensiveness of scientific test plans for field or simulation tests inspire feelings of confidence and competence. The experimental design and statistical analyses methods are standard and time honored. Descriptions of the test conditions, the data collection methods, and the analyses to be performed are presented clearly and concisely. How objective data will be obtained by instrumentation is thoroughly detailed along with a list of the data requirements. Ratings and questionnaires may be used, but it is made clear that these are of secondary importance; they will be used only to gain additional insights on the objective data or pickup a few clever suggestions for improvements.

Unfortunately the price of scientific rigor is not always acknowledged by those involved in using and conducting tests of human performance in complex environments. The trouble usually begins with the division of concern and responsibility between the users of the performance information and those who must obtain it. Operationally, the division is between decision making and collecting the data. These roles are not filled only by individuals; groups or whole organizations may be involved. For example, a source-selection board may be the decision maker while an operational test agency or laboratory facility may be the data collector. These roles conflict over what is wanted and what can be delivered. Table 15-3 summarizes the roles of the decision makers and data collectors.

Requirements and Restrictions - The Dilemma

Decision makers, who are the users of the information, usually have more authority than those whose job it is to collect it. Data collectors have to try to comply with the desires of the decision makers. The users generally define their performance information needs qualitatively and do not specify criteria. They may request information within particular domains, such as offensive and defensive performance, but do not give detailed guidance about what measurable events constitute evidence for good or bad performance; it is up to the person responsible for the conduct of the test to work out these details.

However, decision makers usually show some concern about issues of measurement methodology and quality of the data by stating that the performance data should be objective, i.e., not opinions or judgments. They also state that data should be collected under realistic conditions to be certain the test information is valid. At the same time, decision makers typically limit the time and resources available to conduct the test.

The data collector is faced with a dilemma – attempt an ideal, textbook, laboratory style test or recognize reality, the difficulties and impracticalities of using this approach in complex situations. The data collector usually will opt for the *lab style approach* . It is professionally the safest course. Laboratory methods, following the model of the physical sciences, is beyond reproach as a process. It will not be the data collectors fault if reality does not cooperate to deliver the product expected by the decision maker.

Table 15-3
Roles Assumed by Decision Makers and Test Data Collectors in the Measurement of Performance

DECISION MAKER	DATA COLLECTOR
Is the management expert	Is the measurement expert
Has authority and prerogatives	Accedes to decision maker wishes
Determines information needs: Stated qualitatively At system level Without performance criteria	Develops performance measures: Defines measures Specifies data requirements Determines data weightings
Imposes arbitrary constraints: Data should be objective Realistic test conditions	Designs test: Decides what realistic means Lives with imposed constraints
Limits time and resources: Sets schedule Specifies personnel & equipment	Translates measurement data to terms of decision maker's statements of information needs

Impossible Requirements

The foregoing is a common characterization of requirements for performance tests in applied contexts such as operational field tests, research simulations, and training exercises. However, the conflicts between the high-level information desired and the constraints imposed makes it almost impossible to meet them using conventional laboratory methods; the objectives of laboratory and operational testing are too fundamentally different.

Scientific tests are characterized by tight control and limited scope. Control and limited scope are necessary to eliminate unknown influences and to reduce variability. This assures the validity of the findings and increases the sensitivity of the measures. In contrast, the essential character of an operational test setting or complex simulation is the provision of a rich, realistic environment and all its attendant variability. The behavior of interest is complex, meaningful task performance.

Disappointing Results

After the fact of data collection, the fragility of the laboratory approach becomes evident. Underlying the apparent robustness of classic, scientific test plans, there are strong, unstated assumptions. These are (a) the data will be complete and uncontaminated, (b) the test conditions will be constant and knowable, (c) there will be sufficiently low variability of performance data to assure that data reliability can be verified by standard statistical procedures, and (d) there is an established method for inferring high-level performance information from the low-level data collected.

Typical results of the laboratory test paradigm applied to complex environments are disheartening. Table 15-4 lists common results with data obtained from field settings. Almost never is the promised information delivered. Although, as will be discussed later, the subjective data is often quite useful, the objective data has several critical problems:

Missing Data: Data recording failures or cancellation of one or more test trials leave holes in the data set. This is a particular problem when the missing data is part of the documentation of a test condition. For example, it

Table 15-4
Common Results of Applying Laboratory Methods to Field Testing of Human Performance

- Some data are missing

  Data recording failures
  Test plan not completed

- Large proportion of data variability is due to unknown factors and not to controlled test conditions
- Some low-level data variables are found to be statistically reliable; most are found to be insignificant
- The (statistically reliable) objective measures are difficult to interpret in terms of major test questions
- Subjective ratings and expert observational data are consistent across raters and observers and clearly show differences in preferences and perceived effectiveness
- Final conclusions are actually based on subjective measures but a tortuous argument is made to show that the objective data supports the same conclusions

does no good to know that eleven of fourteen weapon firings resulted in hits on the target if the data for range or some other critical parameter of the engagements are missing.

Unknown Factors: In an ideal test, the manipulated conditions will account for almost all of the variation in the data. That is, there will be little "noise" in the data and the measures will be very sensitive. More typically in complex tests, most of the variability is due to unknown factors. Consequently, the measures are insensitive to even substantial differences. It is misleading under these circumstances to say no difference exists among conditions without stating what size difference could be reliably detected in the data. However, it is often the case that the difference required would be ridiculously large.

Inconclusive Findings: After analysis of the individual, low-level data variables, some variables are found to be significantly different among the test conditions. However, many more variables are not found to differ. Having committed to the importance of data reliability, it is very difficult to rationalize results based on statistically insignificant data.

The problem is compounded when the variables that seem to be major performance indicators are not reliable but there are some variables of seemingly secondary importance that are significantly different among the test conditions. For example, the number of aircraft that can be handled by an air-intercept officer, or the speed of processing new threats may not differ as function of display, training, or experience. However, the length of radio messages may differ reliably with changes in all three factors. Given these circumstances, it is difficult to use the data convincingly to support answers to the major questions of the test, e.g., "Are air-intercept officers more effective with the new display?"

Relating the data to the original question is not easy if there is no clear structure of rules for systematically aggregating numerous low-level measures into a global index of merit. It is especially difficult to reconcile data that basically reflect performance trade-offs. For example, it would be tough to reconcile test data on a new Forward-Looking Infra-Red (FLIR) surveillance system that indicates that operators detect a large proportion of targets but also make a large number of false detections.

Exaggerated Virtues of Objective Data

There is a prevalent superstition that when people are the measurement tools, the data produced reflect private standards, judgments, and opinions and will be strongly biased in some unknown way. It is also believed that mechanical data collection is the only way to obtain objective data.

Incorrect Beliefs About Objectivity

Objective data are over valued based on an incorrect definition of objectivity, utility, and applicability to human performance measurement. It is common to believe that objective means automatically recorded. Objectivity is often taken to include validity, i.e., the data come from the intended source.

It is an error to accord uncritically the virtues of objectivity, reliability, and validity to any mechanical measurement method. The distinction between objective and subjective measures is largely illusory since all measurements are based to some degree on human action or judgment (Muckler, 1977). Human participation is inherent in several stages of the measurement process, even for so-called "objective" data: selecting the measures, summarizing the data, analyzing the data, and interpreting the data.

Value of Computers for Objective Measures

As mentioned earlier, the value placed on automatically recorded, objective data probably stems from the introduction of computers in behavioral research. Prior to about 1970, most performance measures, inside and outside the laboratory, were based on observational data recorded by researchers or subject matter experts. The recorded information could be a simple log of time and events or could include some sort of rating of proficiency of performance. Since the 1970s, computers have greatly extended the ability to create simulated task environments and collect data (Lane 1986). More comprehensive tasks could be investigated in more realistic conditions than was previously possible. Early hopes were that the increased data collection capabilities applied to realistic performance would be like applying a microscope to behavior; new levels of detail would be revealed that would make clear the fine-grained structure of performance. Data on performance components could be used to build models and predict performance. Moreover, the automatically recorded data would be truly objective, i.e., not dependent on human acquisition.

The value of objective, computer-collected data is probably a vestige of the hoped-for benefits and is largely mythical. Despite the pervasive attempts to use computers to control and monitor human performance in laboratories, simulators, and in actual systems, there has been little advancement in objective performance measurement. Over the past twenty years, the effort to develop empirical measures have been disappointing (Lane, 1986). In fact, no objective performance measurement system has been demonstrated to be superior to measurement based on human observation. For example, objective methods developed for assessing the proficiency of pilots are not as good as ratings by instructor pilots. Semple and Cross (1982) pointed out that most so-called automated performance measurement systems for aircrew training are really only automatic data collection systems.

Limitations of Objective Data

Data acquired automatically by computer or other instruments have some inherent limitations. First, the data must be from simple, low-level phenomena such as a switch closure. The switch closure may have resulted from a long and complex process of perceiving and deciding (as when the switch is a weapon trigger) but the only thing the data reflects is the state change of the switch. Second, a large amount of low-level data from numerous sources is difficult to analyze and interpret if, as is usually the case, there is no preestablished structure for aggregating the data to yield some higher-level meaning. A third problem with low-level data that stems from the first two is that these data do not capture all relevant aspects of the performance process and therefore have low reliability and validity. Studies that have compared objective and observational data consistently show that subjective measures have greater reliability and validity (Lane, 1986).

Facing Reality

Table 15-4 shows that common results of field testing are that subjective ratings and expert observation data are the most clear cut and useful. In circumstances where objective data are erratic, it is common to find that subjective ratings and questionnaires show clear preferences or greater perceived effectiveness of one alternative over another. For example, pilots consistently exhibit strong preferences for color displays but the performance data show no difference between monochrome and color displays (Reising & Calhoun, 1982).

Conclusions from results in complex environments are often more supportable by the subjective measures than the objective data. However, it is not customary to state this outright. More often, briefings of final results take the form of showing that there are "trends" in the objective data that suggest performance is better under Condition A than Condition B, or that performance nearly met the expected criteria. The lack of statistical reliability is offset with a statement to the effect that the subjective measures support the interpretation of the statistically significant objective data. Therefore, it is concluded, the inference drawn from the objective data is probably correct. In effect, the objective cart is placed is front of the subjective horse.

Recurrence of results like these confirm that using the laboratory approach in operational test and evaluation is simply impractical. At the very least, certain limitations in field performance testing must be recognized. The most important of these is that increasing variability of performance is directly associated with increased realism of a situation. This in turn implies decreased sensitivity of performance measures. Figure 15-1 shows the relationships among credibility, i.e., sensitivity of performance measures; realism, i.e., variability in test conditions; and the amount of data collected, i.e., effort and costs, in any performance measurement effort. For a given level of sensitivity of performance measures, increasing realism requires more data. Similarly, reducing the amount of data either reduces the level of sensitivity of the performance measures or it requires a reduction in realism i.e., variability of the test situation. If you fix any two of the factors in Figure 15-1, the third is determined. The dilemma is that decision makers always want realism but, at the same time, they constrain the time and cost available for data collection. Consequently, the credibility of the performance measures suffer.

Fortunately, there is a way out of the dilemma. This involves a much higher regard for the use of subjective ratings and expert observational data. Automatically recorded data also has a role when its utility is substantiated and not simply assumed.

SUBJECTIVE METHODS FOR HUMAN PERFORMANCE MEASUREMENT

In a strict sense, objectivity does not mean "no human involved." It simply means that the data can be shown to be free of unknown influences and not biased. Data are objective if the collection or measurement instruments do not influence or act selectively. Data are valid if they correspond to what is purported to be measured. These include data obtained by human observation and judgment.

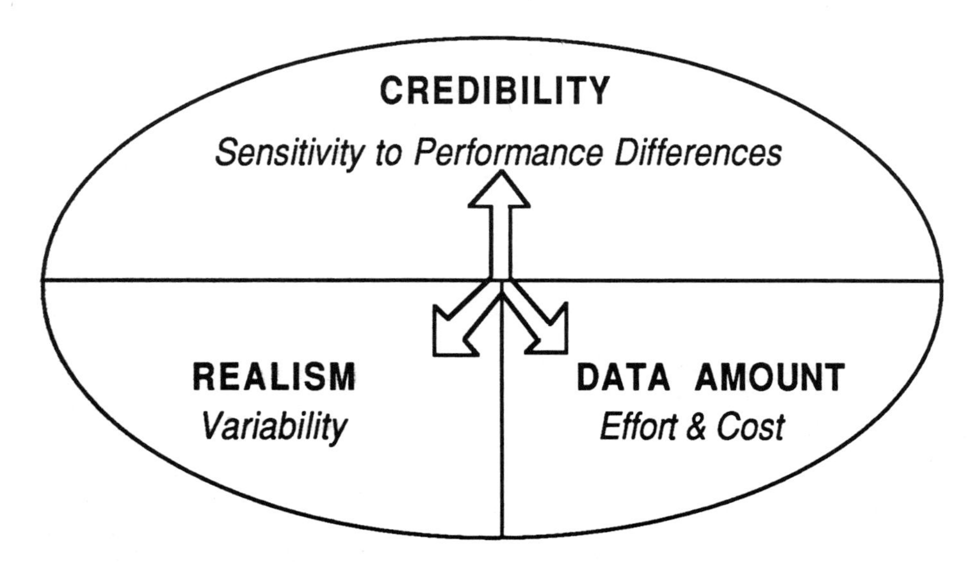

Figure 15-1
Trade-offs in Performance Measurements: Fixing Any Two Determines the Third

High Quality Observational Data

Observational and judgmental data obtained from people is a powerful, undervalued performance measurement technique, well-suited to the operational test setting. Observations and judgments are as objective and valid as performance measures if procedures used to verify the judgments are reliable, unbiased, and the correct aspect of performance is being judged. The requirements for objectivity and validity can be met by pretesting observers to demonstrate that there is intra-observer and inter-observer reliability, i.e., the same data are obtained under the same conditions and by different observers. Some observations, for example, noting when an object first appears, do not require value judgments. Use of multiple observers guards against perceptual errors.

More sophisticated types of observations generally require using subject matter experts who have been indoctrinated and tested on the explicit definitions of criteria and the reporting scale. More or less experience and training is required depending on the nature of the observations. A low-level judgment scale might involve categorizing (simulated) artillery rounds as on-target, close, or distant when the target is an amorphous and changing group of soldiers or vehicles. A high-level judgment, the Cooper-Harper scale of aircraft handling qualities, has been used successfully as quantitative data for many years (Cooper & Harper, 1969).

Measurement Error in Objective and Subjective Measures

For measurement of meaningful performance in complex settings, subjective measures are more likely to be credible and useful than objective measures because measurement error will be much smaller for subjective measures. The error or noise in both objective and subjective measurement of performance comes from four sources: (1) the performance itself, i.e., inherent variability of human behavior; (2) influences of the environmental context; (3) the degree of correspondence between what is measured and what is intended to be measured; and (4) variability in the measuring instrument itself.

Figure 15-2 illustrates the differences in the relative amount of error from the four different sources for objective and subjective measures of human performance. Inherent behavioral variability is independent of the type of measurement. For objective measures, the error due to not accounting for context effects and the inability to directly and fully measure the performance of interest is much larger than the error from the same sources for subjective measures. Objective measures have inherently low variability in the measurement instruments. In contrast, the major potential source of measurement error for subjective measures is the variability of the instruments, i.e., the observers or judges. However, *it is the aggregate of measurement error, not a single source, that ultimately determines the credibility and utility of a measure. Subjective measures of human performance generally will have less total measurement error than objective measures.*

Reliability scores, based on correlations of successive sets of performance scores, are higher for subjective measures than objective measures. Moreover, since the observational scores are more comprehensively representative of the meaningful performance of interest, *they almost always are more valid*, i.e., have better correspondence with what is intended to be measured than automatically recorded data (Lane, 1986).

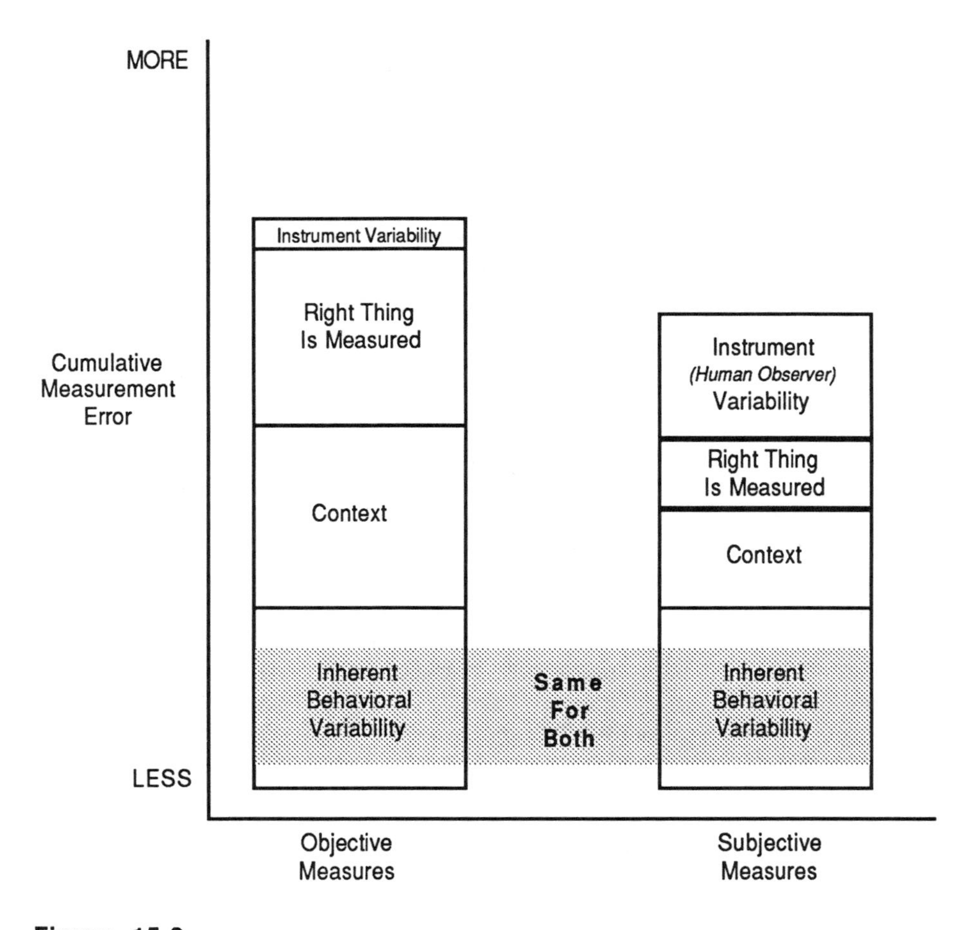

Figure 15-2
Relative Differences in Size of Components of Error in Human Performance Measurement. Total Error is Less for Subjective Measures.

Advantages of Subjective Measures Over Automation

Subjective measurement has several distinct advantages over automated collection of human performance data. The advantages of observational data for operational testing in the field and complicated simulations are listed in Table 15-5 and discussed below.

Economy

Expert observation is an economic method. It usually requires little more than a suitably trained observer to hand record a rating or check mark. The number of events that need to be observed is often small. Observers can filter out spurious data during the collection process. Because the data set is small and free of bad entries, no complex editing and synthesis of the data is necessary.

Direct Measurement of Performance of Interest

Typical events of interest in applied performance measurement are high-order effects, caused by complex and subtle behaviors of operators.

Table 15-5
Advantages of Subjective Measures Over Automatically Recorded Data

- Economical in cost and time
- Directly measure performance of interest
- Context can be taken into account
- Sensitivity of measurement
- Results are available quickly
- Performance of concurrent tasks can be distinguished
- Cognitive tasks can be measured

Counting manipulations of switches and the like does not give an adequate picture of meaningful activities. For example, use of radio equipment and management of several communication channels are activities that are affected by multiple factors and interactions in a realistic setting. The concern during operational test of radio equipment is communications. The questions of interest are whether information flows back and forth in a timely and efficient manner, and are messages complete and accurately sent and received. Performance is revealed largely in content and not by simple aggregation of low-level measures of overt acts or psychomotor control, e.g., control actuations and number and length of transmit switch closures. Observers can interpret content as well as the intentions of the operators and are insensitive to slight variations in sequences of task elements.

Context Taken Into Account

Data obtained by expert observers can be highly resistant to context effects. That is, an observer can take into account, or allow for, situational factors and variabilities that inevitably occur in the real world of operational test. A tank commander may fail to engage a threat on the left because a greater threat just appeared on the right. A pilot in a flight simulator may strike a tree during an otherwise well-executed maneuver because the simulator poorly portrays distance cues. In both of these instances a knowledgeable observer is likely to take these mitigating facts into account and not downgrade performance. On the other hand, if time to engage targets and tree strikes are automatically recorded, the performance measures would be indifferent to the circumstances. Consequently, the automatic performance score will be lower than may be justifiable or desired.

Sensitivity of Measurement

Because the performance of interest is directly measured and people have the ability to take context into account, subjective measures have low variability due to unknown effects. That is, the measures are relatively noise free. Consequently, subjective measures are usually sensitive to small performance differences and more frequently show reliable differences than objective measures. As was illustrated in Figure 15-1, there is a trade-off among credibility, realism, and data collection requirements. Subjective measures allow increased credibility by reducing variability of the measurement process.

Results Are Available Quickly

Subjective data can usually be summarized quickly because there are simply fewer numbers to deal with. Also, very little editing is required since unusual or disruptive events that corrupt performance data are filtered out at the time the data is collected.

In contrast, a massive effort is required to edit, summarize, and analyze objective data from a large-scale field test or simulation. Weeks or months are required before the performance measures of interest are available to the decision makers.

An exception to the rapid availability of subjective data occurs when detailed review and logging of events recorded on videotape are required. This is presently a very labor demanding process with five to ten hours of effort required for each hour of videotape.

Performance of Concurrent Tasks Can Be Distinguished

Observers can readily separate activities of two or more concurrent tasks. This is very difficult to do with automatically recorded data. Moreover, observational data are largely uninfluenced by unimportant departures from strict sequence of performance of task elements, interruptions of a task, or by concurrent performance of tasks.

Cognitive Tasks Can Be Measured

The tasks of operators of modern equipment and systems in the military and industry tend to require a lot of mental work relative to the amount of physical activity. It is almost impossible to obtain useful measures of mental work from objective recording because of its dependency on overt, operator actions. However, trained observers, familiar with the tasks, can reliably interpret performance on tasks that primarily involve perception, attention, decision, or information processing where few operator actions are involved. The level of inferred mental activity will not be very detailed, but at least the kind of activity an operator is engaged in, e.g., seeking information, weighing alternatives, and searching, can be distinguished.

Role of Automation

This is not to say that automatically recorded data has no place in operational testing of human performance in complex settings. They have been over valued, but they are not useless. Pragmatically, whatever data sources are most readily available and pertinent to obtaining the desired human performance information should be used in tests. Both observational and automatically recorded data should be part of a performance measurement plan. It is argued here that observational measures are inherently better for getting information needed for practical decisions. However, automatically recorded data are clearly necessary and useful for documenting timing, accuracy, and sequences of actions and events. Automatically recorded data can also serve their original purpose – providing elemental information to aid in developing and validating models of human performance.

REALISTIC APPROACH TO MEASUREMENT IN COMPLEX SETTINGS

Background Assumptions

Laboratory Methods Will Not Work in the Field

It is common for behavioral scientists to state that the classical scientific, experimental method is the only valid means to obtain useful and credible human performance data from field tests and other complex environments. The logical argument is that the standard scientific method is appropriate and would produce the desired information if only more time was available, more data were collected, and greater control was exercised over the test conditions. In other words, the balance among credibility, realism, and data collection requirements, illustrated in Figure 15-1, should be achieved by reducing realism and collecting lots of data. This is logically correct but, for practical purposes, is a denial of the fact that realism cannot be maintained and controlled at the same time, and the required volume of data is impractically large. The reality of operational test conditions makes it nearly impossible to obtain data in the manner appropriate to a laboratory. Attempting to do so has been likened to trying to read fine print in very dim light (Rosnow & Rosenthal, 1989). Quoted by Lane (1986), Wherry (1957) expressed the futility of such efforts saying, "We don't know what we are doing, but we are doing it very carefully, and hope you are pleased with our unintelligent diligence" (p. 1).

Imperfect Information is Better than No Information

This leads to the practical contention that the responsibility of individuals who do testing is not to conduct a scientifically correct experiment but to provide the most useful information possible consistent with the given constraints. There should be no misunderstanding. The best test is one that produces statistically reliable, objective data. On the other hand, the worst test is one that attempts to do so, but provides only statistically unreliable and/or trivial data that is cited to support uncontrolled opinion data. It is better to plan for and provide data that is imperfect than to completely fail to provide perfect data.

Subjective Measures are Useful

The position argued here is that in highly variable and poorly controlled, i.e., realistic, test situations, data obtained by expert observers is the most reliable and useful source for the information desired by decision makers. The major advantages of expert observational data are that observers can identify complex behavioral events that are difficult or impossible to capture with instrumentation, and that the observational data is not highly sensitive to context factors. The chief disadvantages of observational data are that there is less certainty about exactly what performance is being measured and what the test conditions are that influenced behavior.

With the understanding that laboratory methods will not work, imperfect information is better than no information, and subjective methods are useful, a practical approach for human performance measurement in test and evaluation of systems can be formulated. The approach advocated here includes:

(a) Constructing a Performance Hierarchy
(b) Obtaining Aggregate Weightings
(c) Use of Video Recording for Documentation and Detailed Analyses

Constructing a Performance Hierarchy

In military operational testing and evaluation, system performance measurement is structured hierarchically and developed from the top down. The major questions to be answered are called issues. One or more criteria are associated with each issue that indicate the required level of performance. Measures of Effectiveness (MOEs) are quantitative expressions of the issue criteria, i.e., how well an item of equipment or system performs its combat role. Based on system functions and design, MOEs are analyzed into Measures of Performance (MOPs) that are concretely defined metrics of the component functions that must be performed to achieve the basic system functions described by a MOE. From each MOP, a set of lower-level MOPs or, finally, specific data requirements are derived. The data sources are what can actually be measured. A complete hierarchical structure relating performance information from issue statements and MOEs, down to the smallest data elements of MOPs, is formulated in an operational test and evaluation plan. Moreover, the rules for aggregating the data and measures from lower levels to form the higher levels are also specified.

A similar approach should be used for human performance measurement in complex test settings. Although difficult, a hierarchy of data and performance measures can be constructed that clearly shows how elements at lower levels should be combined to form the next higher, broader, and more abstract level of performance representation.

Obtaining Aggregate Weightings

Empirical Weighting of Performance Measures

The neo-classical approach to determining a measurement structure for human performance is to perform a statistical analysis of a large set of discrete, automatically recorded data elements. A global measure can be defined by empirically deriving the weights applied to each measure that best predicts some criterion measure. A discriminant analysis, for example, can be used to determine how much each of several measures of elementary behaviors contributes to a more general outcome measure. Kelly et al. (1979) used this approach to determine what kinds of performance distinguished between successful and unsuccessful pilots in aerial dogfights. Sixty-seven variables were measured for each aircraft during 405 runs in the U.S. Air Force Simulator for Air-to-Air Combat (SAAC), Luke Air Force Base, Arizona. An empirically determined composite of 13 of the measures was found to accurately discriminate between high and low skill pilots. These included mean absolute vertical speed, number of times the throttle was in the idle position, mean speed brake deflection, mean absolute roll rate, and percent time out of opponents view.

One of the problems with empirically derived composite measures, i.e., difficulty of interpretation, is evident from this short list. The nature of the components can speculatively be interpreted as maneuvering and energy management variables (Polzella & Reid, 1987), but another study would be required to validate this hypothesized interpretation. Another, more serious difficulty, is that it is not known if the same measures, with the same weights, would apply to other forms of aerial combat, other simulators, or to actual aircraft. That is, without another large study, the measures cannot be applied confidently beyond the SAAC.

Subjective Weighting of Performance Measures

Another way of determining the weights for the components of a performance hierarchy is to use the judgment of subject matter experts. Lane (1986) describes two general procedures for accomplishing this – the "bid system," credited to Toops (1944), and *policy capture*. The bid system essentially gives expert raters a fixed amount of points to allocate over the set of measures to be combined. The distribution of points reveals the relative importance of each measure. Policy capture techniques, also known as polymorphic representation (Hoffman, 1960), are so called because the procedure consists of analyzing a series of global decisions to determine the factors of importance in the decision and the relative weight of each factor. This approach has an advantage over the bid system in that it is the (decision) performance of the judges, not what the judges say, that determines the weighting of the factors.

Application of the Policy Capture Technique

The policy capture technique can easily be extended to human performance measurement by presenting experts with made-up scores for the established set of measures that are said to come from performance by different individuals or crews. The experts are asked to assign a number on some scale to overall performance. Given several instances of hypothetical sets of scores, the contribution of each measure within the set can be determined.

This technique could be used, for example, to determine an overall performance measure for an armored vehicle mechanic. Suppose that experts have determined that important measures of a tank mechanics skill are (a) time to diagnose a fault, (b) adherence to prescribed repair procedures, (c) dexterity, and (d) maintenance of repair records. Experts are given a set of performance scores on these measures for (hypothetical) mechanics 1 through 15. Based solely on the scores, each expert assigns a rating number, say on a scale from 1 to 100, indicating the judged overall quality of the mechanic. From analysis of the overall judgment scores, the relative importance of each component can be determined. A more detailed, numerical example of this technique is given in the next section of this chapter.

Using a computer and standard statistical software, or specialized programs, greatly simplifies the process of developing and using a human performance measurement hierarchy. At least one commercial software package, Expert87 (Magic7 Software, Los Altos, California), for small computers is available that facilitates the policy capture process. Once a hierarchy of performance measurement is established and entered into the computer, Expert87 presents the expert with a series of sets of scores of performance measures on a common scale. The expert enters the overall rating for each presentation. From the entered data, the program computes the weights for each component performance measure. Expert87 software supports aggregation of the weights at different levels of the hierarchy, so a complete and complex performance measurement system, culminating with an overall figure of merit, can be established.

Expert87 provides tools for examining each expert's data for consistency and conformity to the group average. The performance weight data from the individual experts can be normalized and combined by simple averaging, or other formulas depending on the characteristics of the experts' data and the intentions of the user.

Relying on the judgment of experts to establish the components and structure of a human performance measurement hierarchy and to provide the aggregation weights for components of the performance measurement hierarchy has several advantages. First, experts are likely to identify, almost completely, the important components of the behavioral process and characterize the components in terms of automatically measurable and humanly observable events. When a performance measurement system is based on the premise that automatically recorded data are the only choices for the elements, it is likely to be incomplete and of low usefulness.

Second, component weights from experts, using the policy capture procedure described above, will generalize to a range of contexts because experts' ratings are based on a broad range of situations they have experienced in the past. On the other hand, empirically derived performance weights are sensitive to the specific conditions under which they were obtained.

Third, once the hierarchy is established, different sets of experts can be used to establish different weighting sets for different purposes or changing circumstances. For example, the weights would likely shift within a performance measurement system for crewmembers of a self-propelled howitzer if tactical importance changed from an emphasis on frequent and rapid redeployment to sustained rates of fire over long periods. Extracting weighting information from additional experts can be done more quickly and economically than running another study to empirically derive performance weights.

Use of Video Recording

Videotaping with sound is an important form of test documentation. It is the most complete and accessible record of the events and the most enlightening medium to support briefing the results. Video and audiotape recordings provide a means to revisit the test and extract additional performance data, obtain data in greater detail, or log context factors events that can help interpret the performance data.

Videotape with sound recording may be the best or only source for some of the required observational data. Although data collection requirements are defined ahead of time, in reality not all things that should be measured can be anticipated ahead of time. Also, during any test involving complex performance there is a practical limit to how much a single observer can assimilate at one time.

Techniques have recently been developed for using video for post-hoc analysis of operational tests and exercises (Shaffer, 1989). Events that could not otherwise be measured in the field can be classified, time-stamped, and entered into a data base for summary and analysis similar to other data. Finally, it should be noted that the reliability of the observational data can be checked by having more than one expert observer review the videotapes.

HUMAN PERFORMANCE MEASUREMENT MODEL

In any moderately realistic environment there is little chance of obtaining useful human performance information if the effort is structured as a scientific experiment. There will be too little data, and the conditions too variable, to obtain statistically reliable, automatically recorded data. Also, these data do not capture the essence of complex performance when people are engaged in meaningful tasks because system output data does not reveal human performance data.

A Human Performance Measurement Model is illustrated in Figure 15-3. The approach to employing the model described in detail here is particularly useful in complex settings, such as operational field tests and training exercises, and in large-scale simulations where a test structured as a scientific experiment will not work. The goal is to obtain the best information possible consistent with the constraints. Generally the information goal is to determine how well people are performing relative to known or assumed criteria. Guides for applying the model and a specific example follow.

Guides for Applying the Model

Develop a Hierarchical Performance Structure

At the beginning, the purpose of the performance measurement system should be stated explicitly and the information requirements defined. Using experts in the relevant performance area, develop a hierarchical performance structure. Start at the top and work downward until the bottom elements represent something that can be measured or unambiguously observed. Review the hierarchy with more experts who were not involved in the original development and modify as necessary. More than one review may be necessary. The bottom tier of the hierarchy defines the data collection requirements. Decide how the data will be summarized and the actual format and labeling for tables and graphs. This serves as a check to ensure that all required performance and context information will be obtained.

Use Expert Judgments to Determine the Weights for Aggregating Measures

Use a policy capture technique with experts to determine and develop the weights for each component of the performance measurement hierarchy. Examine each expert's data. Eliminate the data of experts whose weightings deviate substantially from the group. Form an average of the weights by straight averaging or other consensus building techniques.

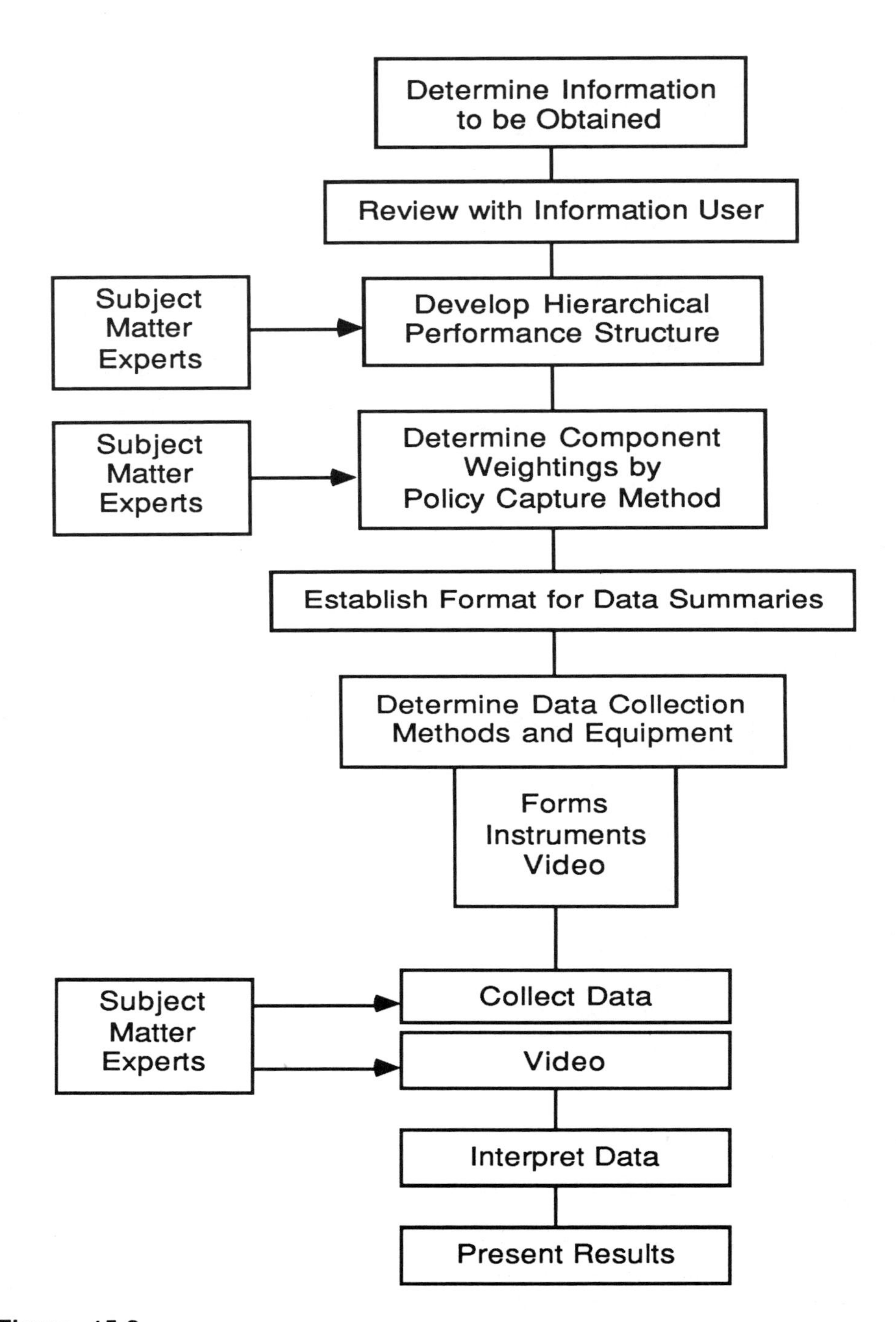

Figure 15-3
Performance Measurement Process Flow

Use Expert Observation and Judgment to Obtain Performance Data

A pragmatic approach to human performance measurement is to make extensive use of the ability of experts to observe and rate performance. Use automatically recorded data only when it captures a relevant behavioral event, a system outcome, or an important context element. Prior to actual data collection, the reliability of the observational data should be verified by comparison of successive measurement sets for controlled conditions. Inter-observer reliability should be verified by comparison of measures between observers. During data collection from the test or videotape, use more than one expert observer if there is a significant judgmental component to the observation.

Applying the Human Performance Measurement Model: An Example

The performance measurement procedures described above are being implemented by the author. The application is a series of studies to determine the effects of various crew-systems and cockpit configurations on mission performance for a U.S. Army, light, reconnaissance and attack helicopter. This work is being performed at the U.S. Army Crew Station Research and Development Branch at the National Aeronautics and Space Administration, Ames Research Center, Moffett Field, California.

Crew Station Research and Development Facility

The Branch operates a sophisticated helicopter simulator, the Crew Station Research and Development Facility (Voorhees et al., 1989; Henderson, 1989). The primary purpose of this simulator is to support research on aircraft-aircrew interactions.

The simulator includes an out-the-window visual system, a speech recognition and production system, and reconfigurable multifunction displays to support one or two crewmember cockpits. Using three separate control stations, up to 8 additional helicopters can participate in simulation exercises requiring multiaircraft team tactics. In addition, the Crew Station Research and Development Facility also incorporates a realistic, interactive, ground-based air defense threats, e.g., Surface to Air Missiles (SA-7, SA-9), Anti-aircraft guns (ZSU-23/4), and tanks (T-72, T-80).

Learning the Hard Way – Objective Measures Not Significant

During 1988 and 1989, the Crew Station Research and Development Facility was being gradually brought up to its full potential to support complex simulations. During that time, a few preliminary studies on stabilization methods for helmet-mounted displays of Forward-Looking, Infra-Red (FLIR) sensor imagery. These studies followed the laboratory research model. Performance measures consisted of automatically recorded variables plus pilot ratings. The results were predictable; almost none of the objective data was statistically reliable or meaningful. The subjective ratings were consistent and showed clear differences between conditions. The lessons learned from these experiences will influence future work. The next series of studies are deliberately planned to take advantage of subjective measures obtained from pilots and expert observers.

General Performance Measurement Hierarchy

The first step was to develop a general performance hierarchy for light combat helicopter crew performance to tie together performance measures from all the studies in the planned series. The principal goal of the performance measurement hierarchy is to determine how well the crewmembers perform with specific crew interface designs. The general performance hierarchy is shown in Table 15-6. It consists of six major performance categories that in turn comprise more specific elements that are formed with observational or measurement data. This hierarchy was developed to assess overall crew system interactions and is developed only to two levels. For studies concentrating on specific subsystems or tasks, the hierarchy is modified by deleting irrelevant categories and expanding others to the point that the lowest level item is measurable or observable.

The original hierarchy was drafted by a group of four U.S. Army aviators and reviewed and modified by two subsequent groups of four to six aviators. The key concern was to not omit any plausibly useful mission performance categories. *Experts can ignore performance factors they consider irrelevant, but they cannot add to the hierarchy once it is established and used.* To add to the hierarchy would require starting over to reestablish weights for each factor.

Realistic Simulation Required for Air-to-Air Tracking

One of the planned studies is on air-to-air tracking of a maneuvering target. The object is to determine the performance of head-mounted sight and a hand-controlled sights for both current sight systems and improved designs. Baseline performance will be measured in the Crew Station Research and Development Facility. Alternative sight control designs will be developed and tested preliminarily in a laboratory environment. After fifteen months the improved sight systems will be implemented in the Crew Station

Table 15-6
Crew Station Research and Development Facility Aircrew Performance Measurement Hierarchy

AIRCREW MISSION PERFORMANCE EFFECTIVENESS

GROUND TARGET/THREAT INTERACTIONS
- Exposure time during engagement
- Engagement time
- Time exposed to AD threat
- Detected by AD threat

AIR TARGET/THREAT INTERACTIONS
- Maintains sight of target
- Out-maneuvers target
- Target engagement
- Distance target engaged

PILOTING AND NAVIGATION
- Timing accuracy
- Position accuracy
- Energy management (fuel use)
- Damaging ground and tree strikes

COMMUNICATIONS
- Optimal number of communications
- Optimal total communications time
- Responds to incoming communications
- Complete and correct message content
- Correct frequencies and callsigns

BATTLE RESOURCE MANAGEMENT
- Use of aircraft resources
- Use of team resources
- Use of support resources
- Execution of mission change

WORKLOAD
- Pilot workload ratings
- Delay in movement
- Tasks deferred
- Tasks skipped
- Crew coordination errors

Research and Development Facility. The government sponsoring agency (the decision makers) are interested in the differences in performance between sight-control types for current and improved designs.

For the test of crew systems, the government sponsor requires full-mission, realistic simulations for the baseline and final tests. Five, ten-hour days of simulator time have been allocated for data collection (following one week of pilot training). Eight U.S. Army pilots will serve as participants. Each run is expected to require 90 minutes, including preparation and debriefing time. Approximately thirty-two test runs, six per day over five days, can be completed. Several targets will be encountered during each mission.

A list of objective performance data variables to be automatically recorded from the simulator has been developed. The set of variables primarily capture time to perform certain actions and errors of omission and commission. Additionally, many "traditional" variables will also be recorded. These include such things as position, altitude and airspeed of all aircraft, aircraft heading and attitude, fuel remaining, weapons fired, targets hit, and ground strikes and tree strikes.

Performance Hierarchy for Air-to-Air Tracking

The general performance measurement hierarchy has been extended to measuring crew performance in air-to-air tracking between maneuvering helicopters in the Crew Station Research and Development Facility. Note that performance factors listed in Tables 15-7 through 15-10 are derived from the "Air Target/Threat Interactions" subfactors shown in Table 15-6. Table 15-7 shows a draft expansion of the performance hierarchy for tracking air targets. Note that data requirements will be a mix of observational and automatically recorded data. In at least two instances, both objective and subjective data will be required. For example, "Reaction Time to Target Maneuver" will probably require an expert observer to note during videotape review when the target began a maneuver and when the subject began to respond. Since both the target and pilot will be moving most of the time, the problem is to determine when a definite, specific type of action began. This is easy for an observer to do and very difficult to do automatically. Establishing the time interval is, of course, a trivial task directly readable from the videotape or obtainable from the test control computer.

In some cases, both objective and subjective measures could be used for some aspects of performance. "Time Threat Maintained in Weapon Sight" is listed in Table 15-7 as a potentially objective measure. However, it may be that even the objective timing data may be too variable to usefully distinguish between the tracking methods. Substantial variability can be expected due to the targets maneuvering – a context effect. The plan is to have other pilots act as observers to monitor and rate "time target maintained

Table 15-7
Hierarchical Performance Factors and Subfactors for Air-To-Air Target Tracking

AIR TARGET/THREAT INTERACTIONS	POTENTIAL DATA SOURCE Objective		Subjective
Maintains sight of target	X		
Time threat maintained in weapon sight			X
Lead tracking of threat			X
Steadiness of tracking			X
No overshoot of target when slewing sight			
Out-maneuvers target			
Maintains target in forward field of view			X
Reaction time to target maneuver	X	(both)	X
Progress to closure on target			X
Target engagement			
Time to visually acquire target after LOS established			X
Time to bring target within weapon sight	X	(both)	X
Time to engage threat within constraints	X		
Distance target engaged			
(No sub-factors)	X		

in sight." Direct observation of display repeaters and latter review of videotape will be used. The expert observers will be able to take into account the trial to trial differences in the tracking difficulty posed by the target. Their observations, expressed on a zero to ten rating scale, will reflect the relative amount of time the pilot kept the target in sight, corrected for target effects. Observers may not be as accurate as a clock for timing but they are good filters. The loss in timing precision could be offset by reduced variability due to target maneuver difficulty.

Policy Capture Used to Obtain Performance Weights

The following steps have yet to be accomplished. However, to finish the example, hypothetical weights and data are shown to illustrate carrying the performance measurement process through to completion.

Expert U.S. Army aviators, with some experience in field tests of simulated air combat, provided weights by a policy capture technique using a commercial software package designed for this purpose, such as Expert87. Each expert is shown a series of graphic displays that represent system performance examples that consist of different combinations of hypothetical scores on the previously established performance dimensions. Figure 15-4 shows the contents of one of the graphic displays. The expert's task is to assign to the sighting system an overall quality score, on a one to one hundred scale, based on the impression by the different performance scores. Although the hypothetical scores only range from 1 to 10, the expert may use numbers from 1 to 100 to allow finer resolution of the responses. The programs essentially use a correlation method to determine the weight ascribed by the expert to each of the factors. To develop the correlation values, a number of sets of scores, each representing a different combination of factor values, are presented to the expert to assign an overall quality score. Depending on the number of factors in each set, from 9 to 36 sets are shown to the expert to develop a correlation score for each factor.

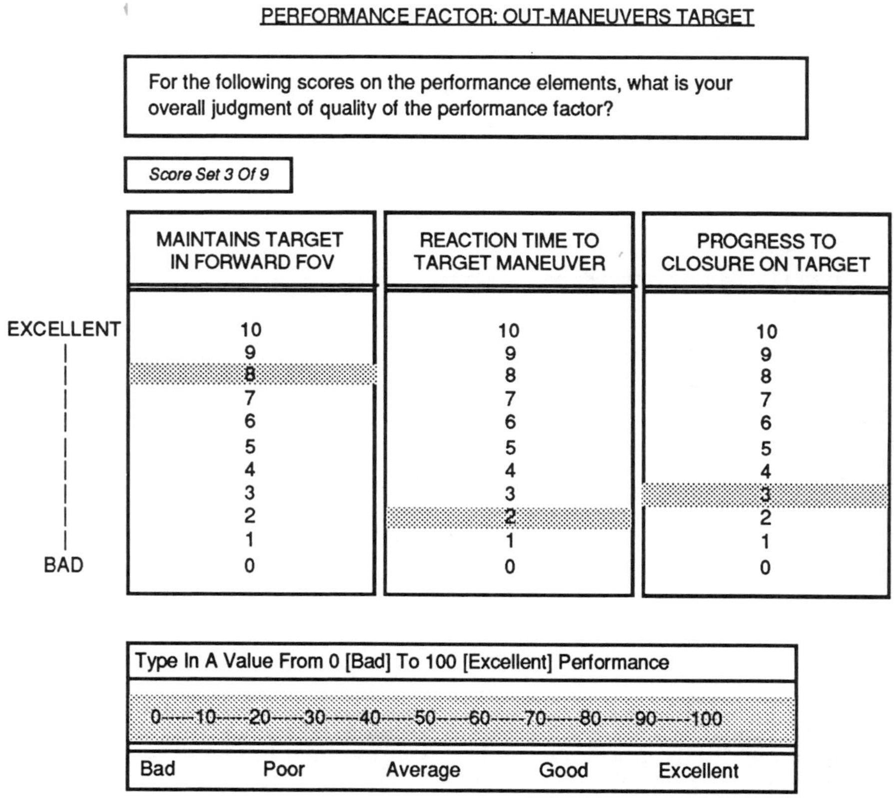

PERFORMANCE FACTOR: OUT-MANEUVERS TARGET

For the following scores on the performance elements, what is your overall judgment of quality of the performance factor?

Score Set 3 Of 9

	MAINTAINS TARGET IN FORWARD FOV	REACTION TIME TO TARGET MANEUVER	PROGRESS TO CLOSURE ON TARGET
EXCELLENT	10	10	10
	9	9	9
	8	8	8
	7	7	7
	6	6	6
	5	5	5
	4	4	4
	3	3	3
	2	2	2
	1	1	1
BAD	0	0	0

Type In A Value From 0 [Bad] To 100 [Excellent] Performance

0-----10-----20-----30-----40-----50-----60-----70-----80-----90-----100

Bad Poor Average Good Excellent

Figure 15-4
Example of Graphic Display Used to Capture Expert's Relative Weighting of Performance Subfactors. (Expert Would Be Presented with a Series of These Displays with Different Combinations of Scores).

By a statistical method known as multiple regression, the relative weights of each performance factor and subfactor is determined. Table 15-8 shows (hypothetical) performance factor and subfactor weights that are derived from experts using the policy capture method. The relative importance of each subfactor as an element of the factor is determined first. Then the relative importance of each factor as an element of overall performance is established by the same procedure. Note that within each factor, the subfactor weights add to 100 percent. Also, the sum of the weights of the top-level factors add to 100 percent.

Table 15-8
Hierarchical Performance Factor and Subfactor Weights for Air-to-Air Target Tracking (Values are Hypothetical)

AIR TARGET/THREAT INTERACTIONS	Subfactor Weights	Factor Weights
Maintains sight of target		0.38
Time threat maintained in weapon sight	0.44	
Lead tracking of threat	0.21	
Steadiness of tracking	0.23	
No overshoot of target when slewing sight	0.12	
Out-maneuvers target		0.22
Maintains target in forward field of view	0.56	
Reaction time to target maneuver	0.34	
Progress to closure on target	0.10	
Target engagement		0.20
Time to visually acquire target after LOS established	0.65	
Time to bring target within weapon sight	0.18	
Time to engage threat within constraints	0.17	
Distrance target engaged		0.20
(No subfactors, weight = 1.00)	1.00	

Note: Factor and subfactor weights are determined independently by the policy capture method.

Weighting and Aggregation of Performance Measures

Table 15-9 contains a set of (hypothetical) crew performance data for each of the two air-to-air tracking sight systems. Regardless of the source of the data, objective or subjective, the data are summarized and transformed to a common, ten-point scale to allow aggregation and weighting of different types of data and measures.

Table 15-10 shows the application of the weighting factors to the performance data. Performance scores for the four main factors as well as the overall score for each of the two sights can be compared. In the data shown, Sight System Two allowed better overall composite performance (64.85 percent vs. 60.55 percent) although Sight System One was superior on the "Maintains Sight on Target" factor (26.30 percent vs. 21.62 percent). A final review by the decision maker is likely to be necessary to determine whether sub-factors have been weighted properly. If weights were determined properly in the first place, then the conclusion is clear – performance is better with the first sighting system.

Table 15-9
Objective and Subjective Performance Scores for Two Sights Transformed to Scales Ranging from Zero to Ten (Values are Hypothetical)

	Score for Sight One	Score for Sight Two
Maintains sight of target		
Time threat maintained in weapon sight	8	5
Lead tracking of threat	9	6
Steadiness of tracking	5	5
No overshoot of target when slewing sight	3	9
Out-maneuvers target		
Maintains target in forward field of view	7	8
Reaction time to target maneuver	4	4
Progress to closure on target	9	5
Target engagement		
Time to visually acquire target after LOS established	4	8
Time to bring target within weapon sight	3	6
Time to engage threat within constraints	7	8
Distance target engaged		
Distance	6	7

Table 15-10
Performance Scores Determined by Applying Subfactor and Factor Weights to Performance Data (Values are Hypothetical)

SIGHT SYSTEM ONE		
Subfactor Weighted Scores (%)	Factor Weighted Scores (%)	
0.44 x 8 x 10 = 35.20% 0.21 x 9 x 10 = 18.90% 0.23 x 5 x 10 = 11.50% 0.12 x 3 x 10 = 3.60% Subscore Sum = 69.20%	0.38 x 69.20% = 26.30%	Maintains sight on target
0.56 x 7 x 10 = 39.20% 0.34 x 4 x 10 = 13.60% 0.10 x 9 x 10 = 9.00% Subscore Sum = 61.80%	0.22 x 61.80% = 13.60%	Out-maneuvers target
0.65 x 4 x 10 = 26.00% 0.18 x 3 x 10 = 5.40% 0.17 x 7 x 10 = 11.90% Subscore Sum = 43.30%	0.20 x 43.30% = 8.66%	Target engagement
1.00 x 6 x 10 = 60.00% SCORE = 60.00%	0.20 x 60.00% = 12.00%	Distance target engaged
SUM OF SCORES FOR FACTORS	60.55%	SIGHT PERFORMANCE

SIGHT SYSTEM TWO		
Subfactor Weighted Scores (%)	Factor Weighted Scores (%)	
0.44 x 5 x 10 = 22.00% 0.21 x 6 x 10 = 12.60% 0.23 x 5 x 10 = i1.50% 0.12 x 9 x 10 = 10.80% Subscore Sum = 56.90%	0.38 x 56.90% = 21.62%	Maintains sight on target
0.56 x 8 x 10 = 44.80% 0.34 x 4 x 10 = 13.60% 0.10 x 5 x 10 = 5.00% Subscore Sum = 63.40%	0.22 x 63.40% = 13.95%	Out-maneuvers target
0.65 x 8 x 10 = 52.00% 0.18 x 6 x 10 = 10.80% 0.17 x 8 x 10 = 13.60% Subscore Sum = 76.40%	0.20 x 76.40% = 15.28%	Target engagement
1.00 x 7 x 10 = 70.00% SCORE = 70.00%	0.20 x 70.00% = 14.00%	Distance target engaged
SUM OF SCORES FOR FACTORS	64.85%	SIGHT PERFORMANCE

The approach to measurement delineated by the example does more than just provide a rank order of performance quality. It also provides a complete audit trail to show the basis for the conclusion that Sight System Two allowed better performance by the crew. Moreover, the relative contribution of the performance components is evident. Lastly, this procedure always results in an outcome if the requirement for statistical significance is waived and statistical tests are not applied to the low-level data. If statistical significance (reliable differences in the performance on some variable among the alternative conditions) is required, the hierarchical structure can still be used if the non-significant variables are assigned the same score, such as the average across alternatives. This, of course, reduces the bases on which a decision may be made, but at least it leaves up to the data collector or user whether statistical significance and or completeness of performance data are absolute requirements.

CONCLUSION

Improved and more comprehensive methods to measure and evaluate human performance in operational settings, with better resolution, is essential to MANPRINT goals. "For MANPRINT to become institutionalized, human performance data will need to be included in underlying combat models and better human performance methodologies for interfacing the combat models with the primary decision-making tools will need to be developed" (Booher, 1988).

Human performance measurement should be done in the system development cycle and certainly during the final operational test. The latter is the best source of data to predict both human and system performance on the battlefield. If human performance is not measured during operational test and evaluation, it probably never will be measured in a way useful to understanding the contribution of the soldiers' performance to system effectiveness and suitability.

The only practical approach to human performance measurement in complex environments such as operational test and evaluations to make better use of subjective measures based on observational data and appropriate use of automatically recorded data. Techniques are available now, and better ones should be developed, to support the collection of observational data and their aggregation in a hierarchical structure to relate elementary but meaningful performance events to overall performance of individuals, crews, and systems.

REFERENCES

Army Regulation 602-2 (1990). *Manpower and personnel integration (MANPRINT) in the materiel acquisition process*. Washington, DC: Department of the Army.

Booher, H. R. (1988). Progress of MANPRINT-The Army's human factors program. *Human Factors Society Bulletin, Vol. 31 (12),* pp. 1-3.

Cooper, G. E., & Harper, R. P. (1969). *The use of pilot rating in the evaluation of aircraft handling qualities* (NASA Technical Note TN-D-5153). Moffett Field, CA: National Aeronautics and Space Administration, Ames Research Center.

Henderson, B. W. (1989, November 27). Simulators play key role in LHX contractor selection. *Aviation Week and Space Technology*, pp. 34-37.

Hoffman, P. (1960, March). The paramorphic representation of clinical judgment." *Psychological Bulletin*, Vol. 57 (2), pp. 116-131.

Kelly, M. J., Wooldridge, L., Hennessy, R. T., Vreuls, D., Barnebey, S. F., Cotton, J. C., & Reed, J. C. (1979). *Air combat maneuvering performance measurement* (NAVTRAEQUIPCEN IH 315/AFHRL-TR-79-3). Williams Air Force Base, AZ: Flying Training Division, Air Force Human Resources Laboratory.

Lane, N. E. (1986). *Issues in performance measurement for military aviation with applications to air combat maneuvering* (NTSC TR-86-008). Orlando, FL: Naval Training Systems Center.

Muckler, F. A. (1977). "Objective" vs. "subjective" measurement: Selecting measures. *Proceedings of Symposium on Productivity Enhancement: Personnel Performance Assessment in Navy Systems* (pp. 169-178). U.S. Navy Personnel Research and Development Center.

Polzella, D. J., & Reid, G. B. (1987). Multidimensional scaling of simulated air combat maneuvering performance data. *Proceedings of the Fourth International Symposium on Aviation Psychology*. Columbus, OH.

Reising, J. M., & Calhoun, G. L. (1982, October 25-29). Color display formats in the cockpit: Who needs them? *Proceedings of the Human Factors Society 26th Annual Meeting* (pp. 446-449). Seattle, Washington.

Rosnow, R. L., & Rosenthal, R. (1989). Statistical procedures and the justification of knowledge in psychological science. *Journal of American Psychologist, Vol. 44, No. 10.*

Semple, C. A., & Cross, B. K.,III. (1982). The real world and instructional support features in flying training simulators. In G. L. Ricard, T. N. Crosby, & E. Y. Lambert (Eds.), *Workshop on Instructional Features and Instructor Operator Station Design for Training Systems* (NAVTRAEQUIPCEN IH-341). Orlando, FL: Naval Training Equipment Center.

Shaffer, M. (1989, July/August). Using video for empirically validated task analysis (EVTA) of system-human interaction and performance. *MANPRINT Bulletin, Vol. IV (I)*, pp. 12-13.

Toops, H. A. (1944). The criterion. *Educational and Psychological Measurement, Vol. 4*, pp. 271-297.

Voorhees, J. W., Bucher, N. M., Gossett, T., & Haworth, L. A (1989, March-April). The Crew Station Research and Development Facility.

Army Research, Development and Acquisition Bulletin, pp. 22-25.
Wherry, R. J., Sr. (1957). *Contributions to correlation analysis.* Orlando, FL: Academic Press.

PART IV

SOURCES OF USER-CENTERED TECHNOLOGY

Implementation of the MANPRINT philosophy in industrial or Government organizations depends on the availability of (1) cost-benefit techniques and human performance data which are translatable into language considered relevant to an organization and its product line, and (2) experts who can act as system integrators along with trained and motivated people who work directly on products and/or product requirements.

The experts and most of the people-oriented producers of technology will come from the nation's college and university school system at, both the graduate and undergraduate levels. As such, the nation's schools will need to learn about and develop coursework appropriate for the new philosophy. The nation's human performance and engineering data bases provide most of the available relevant technical information. University, governmental and industrial research and technology facilities feed these data banks.

Two advanced concepts cover the status of MANPRINT relevant data bases. Van Cott in *Human Performance Engineering Data Bases* , Chapter 16, presents a synopsis of the Air Force *Engineering Data Compendium: Human Perception and Performance.* The relevance of this monumental work to the MANPRINT philosophy of defining the human in system terms and methods of increasing the usability of available human performance data into systems engineering design processes is illustrated.

Haas and Laine in Chapter 17 describe *National Human Performance Data Banks.* They recognize an important aspect of MANPRINT is the contribution of human performance to total system performance. Information on how human performance varies as a function of individual differences, conditions, systems configurations, training time and methods is important to developers and designers of systems. They discuss the difficulties of developing such data bases, as well as the state-of-the-art for existing data sources, primarily in the Department of Defense, the Department of Energy, and National Aeronautical Space Agency.

Muckler and Seven in *National Education and Training,* Chapter 18, presents the problem being studied by the National Academy of Sciences regarding supply and demand of National MANPRINT expertise. Recognizing insufficient knowledge, skills and abilities in the current national pool to meet any dramatic increase in demand for MANPRINT expertise, they discuss the pros and cons of various proposed strategies. Strategies include increases in graduate level courses in human factors, university centers of excellence, industrial training courses, and human performance design curricula for engineering undergraduates.

In *Meeting the Challenge,* Chapter 19, Boff comments on the current state of knowledge as it relates to MANPRINT. He notes that current national awareness of problems in total system effectiveness can translate to meaningful actions and results throughout the country. There are numerous challenges remaining but with MANPRINT as a blueprint for change he is confident that future prospects for a renaissance in systems integration is indeed bright.

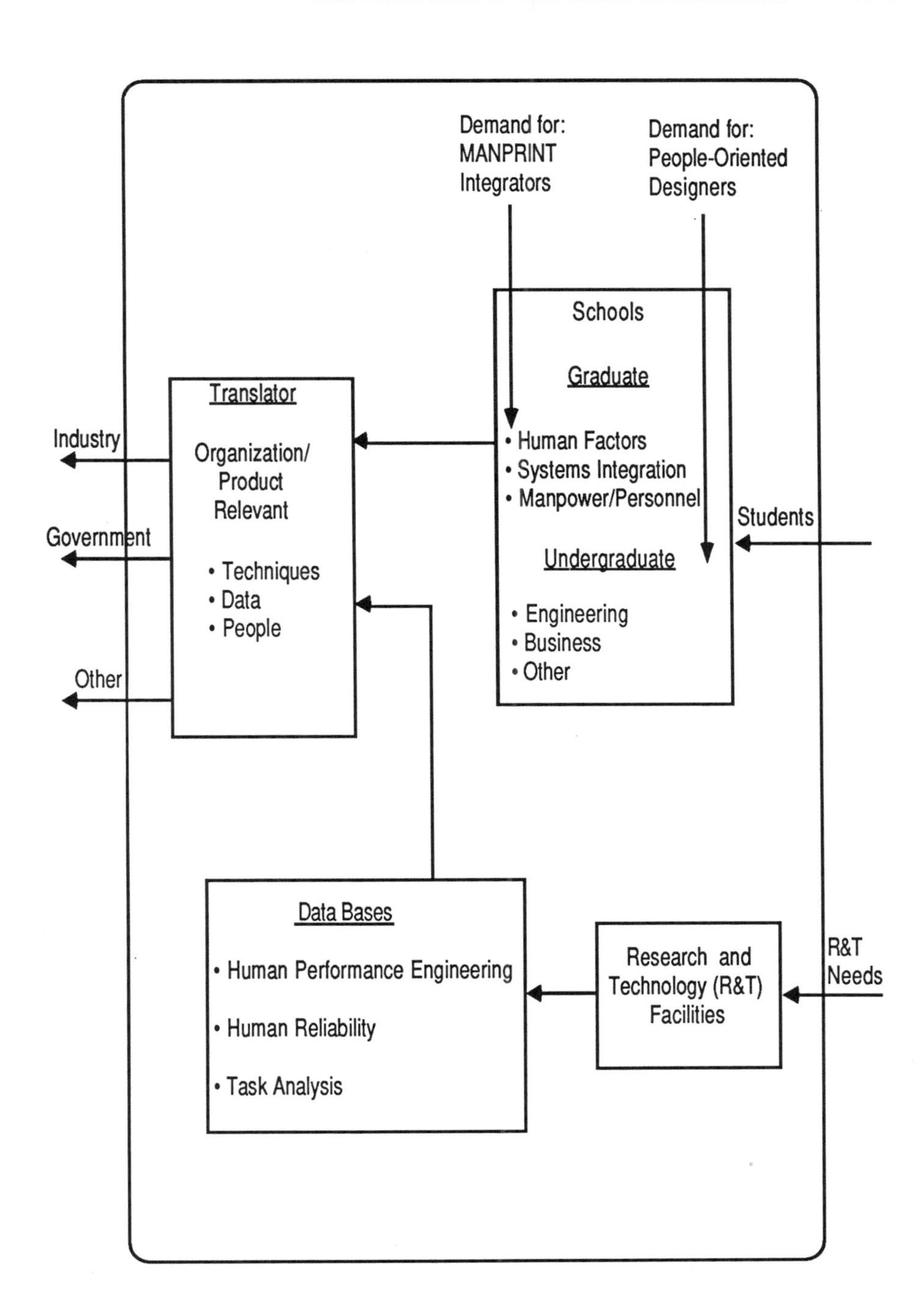

Part IV, Figure 1
Sources of User-Centered Technology

CHAPTER 16

HUMAN PERFORMANCE PRINCIPLES, DATA AND DATA SOURCES

Harold P. Van Cott

ABSTRACT

If human factors considerations are to be incorporated into system design, then designers need to be able to easily access reliable information in a usable form about human capabilities and limitations. Over the years many attempts have been made to meet this need. Two of the most recent are the *Engineering Data Compendium: Human Perception and Performance* and the *Crew Systems Ergonomics Information Analysis Center (CSERIAC)*. This chapter examines the evolution of a series of attempts to package human factors information into increasingly user-friendly design support systems for use with MANPRINT and other human-system integration programs. All of these systems have as their goal the lowering of a long-standing barrier to the use of human factors principles and data by engineers.

INTRODUCTION

A designer in search of information on human performance is like a child suffering from malnutrition amid a plentiful food supply. His condition is not one of starvation for there is a staggering amount to ingest. What is required is a menu from which to select only that information that satisfies specific needs.

The amount of published research applicable to the human factors design of systems has steadily increased, doubling along with the rest of the scientific and technical literature every decade. As early as 1960, the amount of human factors literature had become so large, scattered and unwieldy to access and use that it was hardly an effective aid to design. Recognizing that the prevailing mechanisms for identifying, assessing and using published material were inadequate to the task, the Human Factors Society, the professional focus for the discipline, conducted a thorough examination of the existing information system, pinpointed its inadequacies and suggested a series of remedies. Several of these proposals (Ronco, 1963) are worthy of note because they foreshadow some of today's strategies being adopted to resolve the feast or famine dilemma faced by the designer:

> (1) Greater commitments . . . to report information promptly and explicitly as circumstances permit . . .
> (2) That documentation centers continue to undertake the individually impossible task of acquiring and organizing the existing literature. . .
> (3) The support of these systems for the dissemination of relevant information by a variety of techniques (e.g., critical reviews, annotated bibliographies, and data sheets)
> (4) Periodic review of the accrued wisdom in the field, not only to decide what is valuable and should be perpetuated, but also to decide what is worthless and ought to be thrown out.[1]

Some of the earliest of these attempts to cope with the explosion of human factors information were the *Handbook of Engineering Data* (Tufts College, 1952); the *Handbook of Instructions for Aerospace Personnel Subsystem Design* (United States Air Force, 1966); the *Human Engineering Guide to Equipment Design* (Van Cott & Kinkade, 1972); the *Bioastronautics Data Book* (Parker & West, 1973); various versions of *MIL-STD-1472 - Human Engineering Design Criteria for Military Systems, Equipment and Facilities* (United States Army Missile Command); the *Anthropometric Source Book, Volumes 1-3* (Webb Associates, 1978); and *A Standard Ergonomic Reference Data System* (Van Cott, Kramer, Pezoldt, Porter, Fried, Fechter, & Persensky, 1978).

But even the most widely acclaimed of these attempts to put designers in direct touch with human factors information had less influence on the design process than their technical content would suggest. One explanation for this state of affairs is that like most engineers, designers rely mainly on their immediate colleagues, technical and professional journals and vendor supplied materials to meet their information needs and seldom consult other information sources (Allen, 1977). This observation is consistent with that made by Meister and Farr (1966) who earlier had found, in a well-known study of the utilization of human factors information by designers, that "Designer subjects do not possess or read human factors handbooks."

Another hypothesis that should be examined before these disheartening conclusions can be fully accepted is that the degree to which an information source is used is a function of the extent to which it has been *human factored* or made *user friendly*. Perhaps designers don't use human factors information because the information they need has been difficult to find, interpret and use. In this vein, Boff, Calhoun, and Lincoln (1984) stated that the needs of the user must be taken into account in the presentation of information. This, they said, could be done in two ways: (a) by enhancing the communication of information through attention to presentation format, style, terminology and level of treatment, and (b) by facilitating access to specific information by means of cross-references, indexes, summaries, glossaries, and other user aids. To these it should be added that the information must be credible, must be accompanied by guidance that indicates when it can and should not be used, should be current, and should identify additional sources of information. These are sound principles, many of which were developed long ago for the design of displays. Although it is unfortunate that these principles have been delayed in application to the human factors knowledge base, now that they are being applied, human factors information can be part of the designer's box of tools and its use can become institutionalized in the system acquisition process.

STATE-OF-THE-ART TECHNIQUES FOR INFORMATION ACQUISITION

By the mid-'80s techniques had begun to improve for the designer because human factors principles had been applied in the design of a variety of new human factors information systems. Today's system designer has access to a wide variety of state-of- the-art techniques for acquiring and locating human factors information. If recent interest in human factors by engineers is any indication, these techniques will help transform that interest into an increased use of human factors information.

This section reviews some of the most recent techniques for presenting human factors information. All of them were developed under the Integrated Perceptual Information for Designers Project (IPID), a multiagency program supported by the Air Force, Army, Navy, and the National Aeronautics and Space Administration (NASA) and managed by the Armstrong Aerospace Medical Research Laboratory at Wright-Patterson Air Force Base. Developments of particular interest that are reviewed here are:

- The *Handbook of Perception and Human Performance*
- The *Engineering Data Compendium: Human Perception and Performance*
- The *Compendium* on Compact Disc
- The Crew System Ergonomics Information Analysis Center

The IPID Project, and the innovations in information access that are its products, came about largely through the creative and energetic leadership of Kenneth R. Boff of the Armstrong Aerospace Medical Research

Laboratory and the federal agencies and private sector organizations in the United States and abroad whose support he enlisted.

Handbook of Perception and Human Performance

The first of these IPID innovations, the *Handbook of Perception and Human Performance* (Boff, Kaufman & Thomas, 1986), was an attempt to consolidate within a single reference work research findings on sensation, perception, cognition and human performance that had been judged by subject matter experts to be critical to the design of displays and controls. Detailed and extensive in coverage, this 2,700 page handbook drew on hundreds of research scientists and literature sources throughout the world for its preparation.

While this two volume work was not aimed at a design audience, it does serve the vital function of linking a designer to an information source through the intermediary of a human factors specialist qualified to act as a bridge between design and the often bewildering research literature.

Section I of the *Handbook* provides a general perspective on the theory and methods employed in research on sensory processes and perception. It covers psychophysical measurement, human information processing and its organization, and computer graphics.

Sections II, III and IV summarize what is known about the basic visual process, the vestibular system, kinesthesia, and audition, and then relates these fundamentals to the perception of space and motion and the location and organization of objects in space and time in relation to the observer.

Section V, on human information processing, covers auditory information processing, speech perception, visual information processing, and motor control.

Section VI presents information on what is known about the organization of perception and cognition, on object and event perception, spatial filtering and visual form perception, theoretical approaches to perceptual organization, mental imagery and visual functions and computational approaches to vision.

Section VII, on human performance, covers the topics of human performance in real-world contexts such as the effects of control dynamics on human performance, monitoring and supervisory control behavior, work load and its assessment, and the effects on human performance of environmental stress, fatigue and circadian rhythms. It concludes with a description of an engineering model of human performance.

The *Handbook* is a well indexed, heavily referenced archival work acclaimed by some as a monumental work. Its utility, however, has yet to be established since it was published in 1986, scarcely four years prior to this writing. Nevertheless, the process of developing and organizing the *Handbook* provided a solid foundation for a second, even more innovative information service, the *Engineering Data Compendium: Human Perception and Performance* (Boff & Lincoln, 1988a).

The Engineering Data Compendium

In the introduction to an article announcing the development of the *Engineering Data Compendium*, Lincoln and Boff (1988) said:

> In this technological age when the failure of highly complex human-interfaced systems can threaten multimillion-dollar investments and vital services as well as public safety, it is critical to assure that system demands are matched to the capabilities and limitations of their human operators. Yet the wealth of available information on human sensory, perceptual, and performance characteristics is seldom given any systemic consideration in the design of control and display systems.

The *Compendium* was designed for the engineer, already overloaded with tasks competing for attention, who is aware that the costs in time and effort of attempting to locate, sift, and use information about human characteristics are now unacceptably high. Human performance information is scattered in thousands of different sources. Written by researchers, mainly for other researchers, this information is difficult to locate and once obtained to distill, evaluate and apply.

Under the sponsorship of the Integrated Perceptual Information for Designers (IPID) program, the *Compendium* was an attempt to overcome these barriers. It involved a systematic way of consolidating widely scattered research findings into a single, usable reference data source and of presenting these findings in a manner that is intended to make it easier for non-specialists in human factors, such as a designer, to interpret and apply them.

Data Consolidation

Using the information gathered for the *Handbook* as a starting point, a procedure was developed for consolidating human performance data that would assure that the data selected for inclusion would meet four important criteria:

- Reliability – the data must have been collected using procedures that adhere to scientific standards.
- Representativeness – the data must consist of the most recent and best information about a given topic.

• Generalizability – the data must be applicable to situations and issues beyond the situation and conditions under which they were obtained.

• Relevance – the information must be relevant to helping answer the specific questions raised by design problems facing practitioners.

Making decisions based on these criteria required a mix of expertise ranging from subject matter experts and human factors specialists to designers. This called for a multistage review process. The first stage used those who were knowledgeable about the data and how to interpret it. Subsequent reviews by human factors specialists and members of the design community examined relevance and applicability.

Reviewing the Literature

In consolidating data for the *Compendium*, it was recognized that the scope of potentially worthwhile information had to be traded off against the level of detail needed to make it useful. This required a narrowing of focus to a thorough and intensive treatment of human sensation and perception, information processing, and performance and excluded information on other topics such as anthropometry, safety and accident prevention.

Although the 40 research subareas covered by the *Handbook* provided primary source material for the *Compendium*, 7 additional applied research areas that were judged to be of sufficient technical maturity to provide credible information were added. These areas were:

- Information coding and information portrayal
- Target acquisition
- Controls
- Person-computer dialogue
- Attentional directors and warning indicators
- Human performance reliability
- Vibration and the display of information

Data Selection

Each literature review was carefully evaluated by specialists familiar with the subarea that it covered in order to identify items of information with potential value for system design. Preference was given to choosing data functions, tables and graphic material that illustrate the properties of human sensation, perception and performance, and around which succinct entries could be prepared. The following types of information were given special consideration:

• Basic and parametric data on visual and auditory functions and error rates in operating controls.

• Models and quantitative laws such as models of visual target detection, the power law of practice, and operator control models. To be acceptable a model or law had to provide a way of interpolating or extrapolating data and relating it to a given application, and it had to have a well-documented domain of consistently reliable application.

• Principles and nonquantitative laws that convey important rules that can be used in design in accordance with given conditions and situations.

• Background information that will help a user to understand and interpret models and performance data such as the geometry of retinal image disparity and rudimentary anatomy and physiology of the visual system.

Reviewers also placed special emphasis on information from the literature that would provide a basic understanding of a topic and cautions on the applicability and generality of information being presented. When appropriate, brief tutorials on topics such as signal detection theory and psychophysical methods were added to enhance the reader's understanding of the relevance and importance of information.

Only data and information from research on the normal, adult population was reviewed. Information on children, clinical human and animal populations were excluded as being irrelevant.

Text Preparation

Each candidate item slated for inclusion in the *Compendium* was condensed into a format that included a summary of the proposed data item, key identifying elements such as the citation to the original reference source, and copies of the figures and tables on which the item was to be based. Each of these candidate items was then evaluated by at least three subject matter experts who rated the item for its applicability, representativeness and overall desirability.

Candidate items that passed this screening evaluation were then put into the standard format used for inclusion of items in the *Compendium*, but before final selection, they were reevaluated by subject matter experts, human factors specialists and design engineers knowledgeable in the subject area. This additional review assessed the understandability, technical accuracy and validity of the material and suggested additional key references and cautions on the use of the information selected. The final review and screening provided additional assurance that data in the *Compendium* would be accurate and useful.

Data Presentation Format

A major goal of the IPID project was to develop a standard and structured method for formatting and presenting information that would avoid the difficulties encountered in much of the research literature: wordiness, overemphasis on background and supporting material, the omission of detail necessary to understand and apply data, and too much theory and conjecture. To overcome these shortcomings, several principles were adopted that governed the presentation of human performance data:

- Data should be presented as figures or tables where possible.
- Information units should be independent and self-contained (i.e., give all of the information necessary to understand and interpret a given item of data).
- The format should make it easy for a user to locate desired data without being distracted by extraneous, irrelevant information.
- Emphasis should be given to quantitative information of direct use to designers, adding only that theory or other supporting material to permit its applicability to a design issue to be assessed by a user.
- Writing style should be simple, clear and concise, avoid jargon and unfamiliar terms or, when terms were judged to be unfamiliar but essential, to define them fully.

Figure 16-1 (Boff & Lincoln, 1988b) is an example of the standard format used together with definitions of all of the format elements. This format helps a user in several ways. It allows specific items to be located rapidly. It supports different types and levels of inquiry; for example, the entry title in the header of the entry is a brief statement of what the topic is about, while the General Description below is a more detailed summary of the topic. Even greater detail can be found under other headings.

In its entirety, the *Engineering Data Compendium* consists of 1,138 data entries that incorporate about 2,000 figures and tables. These entries are organized into 12 major sections that cover the sensory acquisition of information, perceptual processes, reading and speech intelligibility, attention and allocation of resources, reaction time and manual control, the effects of environmental stressors, and control and display interfaces.

User Aids

A common failing of many scientific and technical reference works is that they offer the user brief and shallow indexes, limited cross referencing and inadequate bibliographies. The *Compendium* has none of these faults. Its extensive and in-depth indexes with over 2,000 top-level headings and 10,000 items in all, thousands of cross-references, and extensive listing bibliographies all facilitate the ability of a user to locate information on specific topics.

Data entries in the *Compendium* are divided among three loose-leaf volumes of approximately 800 pages each. Each of the 12 topical sections is provided with its individual set of user aids: color-coded marginal tab dividers labeled with section number and title; a logic diagram that shows a taxonomy of the hierarchical organization of the section and facilitates the rapid location of items on given topics. Behind each main section tab divider is a table of contents, a key word listing, and a glossary of terms used in the section.

A separate *User's Guide* (Boff & Lincoln, 1988b) provides an expanded table of contents, a high-resolution general index keyed to *Compendium* entry numbers and a novel design checklist. The design checklist consists of a set of questions about human performance that were chosen for their relevance to system design. These questions, typical of those that a designer might ask, are organized into categories keyed to a listing of equipment related factors. For example, an answer to the question "With optically unaided viewing (naked eye), about how many degrees from straight ahead may the eye rotate with comfort?" is to be found under Entry No. 1.207.

The use of loose-leaf binders for the three volume *Compendium* allows it to be expanded and updated. The volume binders are innovative. The swing-hinge ring design allows pages to lie flat when the binder is open and makes it easy and quick to remove and reinsert pages for copying or other purposes.

Recipients of the *Compendium* are registered in a data base so that they can be notified of supplements and updates and surveyed for their experiences in its use.

Future Plans

Future directions for the *Compendium* include the addition of sections that relate to control and display design on learning, memory, decision making and problem solving, and continuing research to identify additional ways of bringing human factors information to bear on the design of human-equipment systems. As its authors assert:

> It is our hope that the *Compendium* – serving essentially as a sophisticated set of stimulus materials – will be used as a research tool by the IPID program and others, not only to refine methods of presenting human performance data to practitioners but also to address the broader issues of how human factors data are used in the design process and how to encourage and expand this use.

The *Compendium* On Compact Disc

In 1987 work began under the IPID project on the development of a compact disc version of the *Compendium* (Glushko, Weaver, Coonan, &

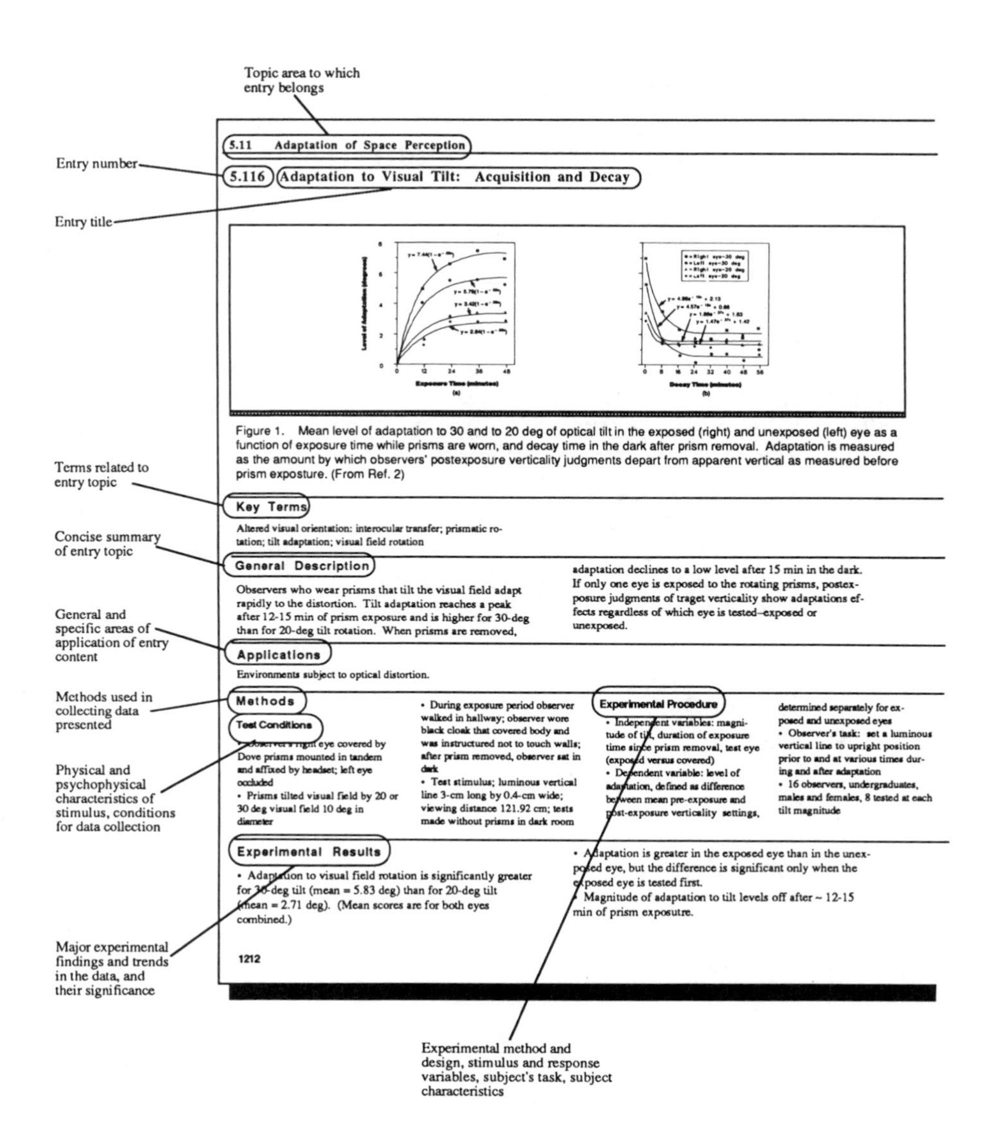

5.11 Adaptation of Space Perception

5.116 Adaptation to Visual Tilt: Acquisition and Decay

Figure 1. Mean level of adaptation to 30 and to 20 deg of optical tilt in the exposed (right) and unexposed (left) eye as a function of exposure time while prisms are worn, and decay time in the dark after prism removal. Adaptation is measured as the amount by which observers' postexposure verticality judgments depart from apparent vertical as measured before prism exposture. (From Ref. 2)

Key Terms

Altered visual orientation: interocular transfer; prismatic rotation; tilt adaptation; visual field rotation

General Description

Observers who wear prisms that tilt the visual field adapt rapidly to the distortion. Tilt adaptation reaches a peak after 12-15 min of prism exposure and is higher for 30-deg than for 20-deg tilt rotation. When prisms are removed, adaptation declines to a low level after 15 min in the dark. If only one eye is exposed to the rotating prisms, postexposure judgments of traget verticality show adaptations effects regardless of which eye is tested–exposed or unexposed.

Applications

Environments subject to optical distortion.

Methods

Test Conditions

- Observer's right eye covered by Dove prisms mounted in tandem and affixed by headset; left eye occluded
- Prisms tilted visual field by 20 or 30 deg visual field 10 deg in diameter
- During exposure period observer walked in hallway; observer wore black cloak that covered body and was instructured not to touch walls; after prism removed, observer sat in dark
- Test stimulus; luminous vertical line 3-cm long by 0.4-cm wide; viewing distance 121.92 cm; tests made without prisms in dark room

Experimental Procedure

- Independent variables: magnitude of tilt, duration of exposure time since prism removal, test eye (exposed versus covered)
- Dependent variable: level of adaptation, defined as difference between mean pre-exposure and post-exposure verticality settings, determined separately for exposed and unexposed eyes
- Observer's task: set a luminous vertical line to upright position prior to and at various times during and after adaptation
- 16 observers, undergraduates, males and females, 8 tested at each tilt magnitude

Experimental Results

- Adaptation to visual field rotation is significantly greater for 30-deg tilt (mean = 5.83 deg) than for 20-deg tilt (mean = 2.71 deg). (Mean scores are for both eyes combined.)
- Adaptation is greater in the exposed eye than in the unexposed eye, but the difference is significant only when the exposed eye is tested first.
- Magnitude of adaptation to tilt levels off after ~ 12-15 min of prism exposutre.

1212

Figure 16-1
Standard Format for *Engineering Data Compendium*
(Source: Boff and Lincoln, 1988b; Figure Source: Redding[2])

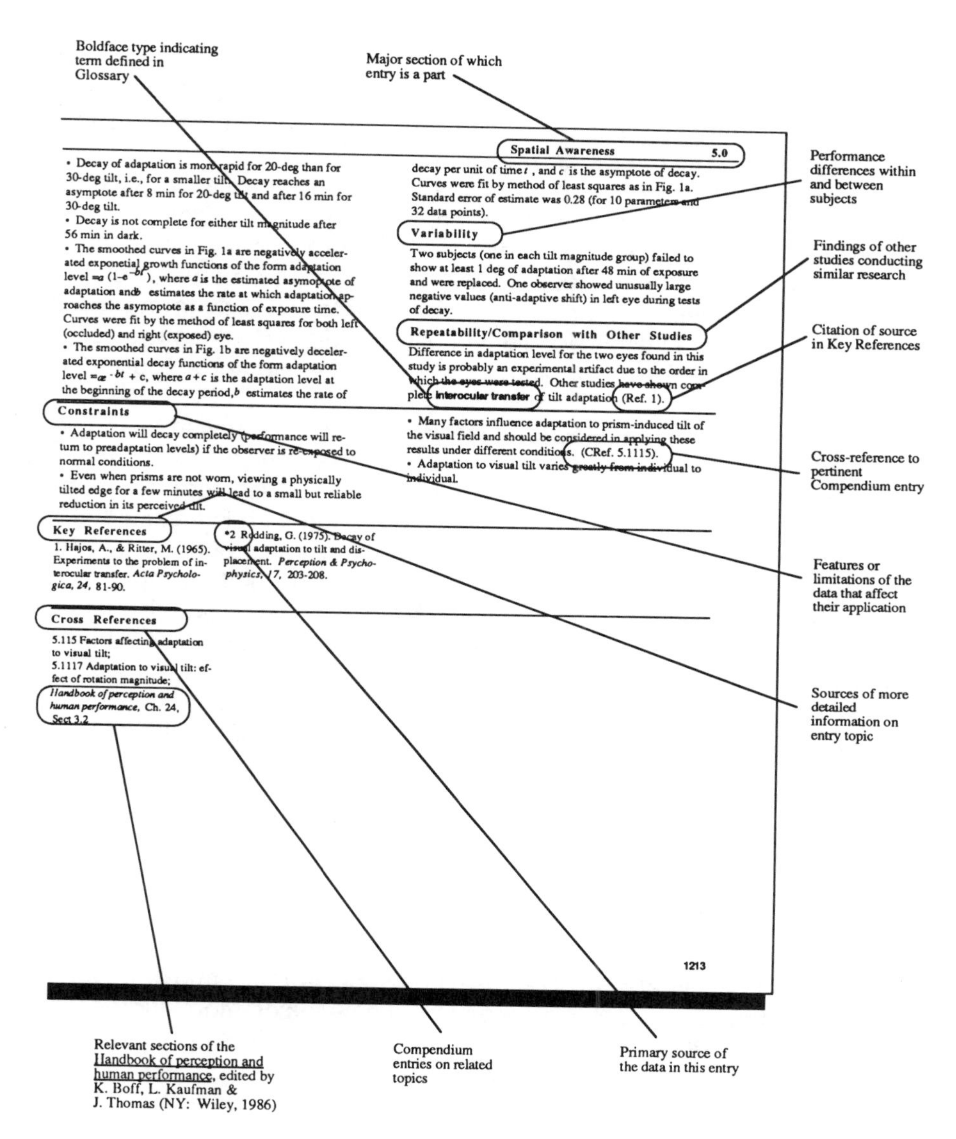

Figure 16-1 *(continued)*

Lincoln, 1988). The disc version, designed to enhance the use of the hard copy document or to be used by itself, employs hypertext features to enhance and supplement the Table of Contents, Index, cross-references and other parts of the printed *Compendium*.

The compact disc contains all of the text and graphics of the printed version and provides a user interface that is designed to exploit the logical, hierarchical structure and links within that structure that were built into the hard copy edition. This interface stresses browsing by a non-specialist of the Table of Contents and Index. It is a more effective way for a novice to locate pertinent material than would be possible with a Boolean logic search system since the latter assumes considerable domain knowledge by one who uses it.

The initial display presented to a user who wishes to browse the Table of Contents contains only the highest level of the hierarchical table; namely, a list of the 12 major topical areas into which the text is organized but not sections or subsections below this level. When an item is selected at this top level, it expands to show the levels below it. To help users who want to browse, the bottom of the display contains "function buttons" that enable them to find a word in the access data base that is being browsed. The FIND function, for example, lets the user browse somewhere other than the beginning by starting with the word located by the FIND button.

The flexibility and precision of searching for and displaying numerous elements and combinations of elements in the *Compendium* data base offer the compact disc user an opportunity to exploit in rich and interesting ways its logic and nesting structure.

The Crew Station Ergonomics Information Analysis Center

For a number of years the Defense Logistics Agency and the Defense Technical Information Center have operated a series of Information Analysis Centers (IACs) which function as gateways to domains of scientific and technical information of interest to the Department of Defense (DoD) and the defense community. Staffed by scientists and engineers who are experts in their fields, these IACs provide their users with specialized engineering, technical and scientific analytical services and products. Today there are over 20 IACs in operation concerned with an array of domains ranging from plastics, coastal engineering, tactical weapons and control, metals, ceramics, and survivability and vulnerability information.

The newest of these IACs to be established is the Crew Station Ergonomics Information Analysis Center (CSERIAC) formed in September 1988 at the United States Air Force Armstrong Aerospace Medical Research Laboratory, Wright-Patterson Air Force Base, Ohio. CSERIAC will provide a gateway to information in the domain of crew system ergonomics.

Information Centers vs. Information Analysis Centers

Individuals who wish to obtain copies of documents and specialized bibliographies produced at government expense on human factors topics have long been able to obtain them through the Defense Technical Information Center (DTIC) or the National Technical Information Service (NTIS). Both store and provide hard copies of these materials on request. While services such as these serve a valuable function, they place the burden of deciding what documents and bibliographic materials will meet a specific need on the requestor. This may be a difficult task for users who have little or no expertise in human factors and ergonomics and may result in search products that are not useful.

IACs, on the other hand, provide a range of services that go beyond the basic functions of document storage, retrieval and the production of specialized bibliographies. Because they are staffed with experts in subject matter domains, IACs are able to conduct tailor-made complex searches and do analyses and interpretations for users that cannot be performed by DTIC or NTIS. In short, the new CSERIAC facility will acquire, digest, analyze, evaluate, synthesize, store, publish and provide advisory and other related services to its users.

CSERIAC is operated for the Department of Defense by the University of Dayton Research Institute. Its staff consists of analysts who are professionals in crew system ergonomics.

The CSERIAC Information Domain

The ergonomic information accessible through CSERIAC encompasses the interaction of humans with equipment in aerospace, land and aquatic environments and the human and equipment design factors that can enhance or degrade crew performance and well-being in real-world military situations. The central focus of CSERIAC is on human characteristics, limitations, physiological needs, performance capabilities, and body dimensions, strength and biomechanics. Users of CSERIAC will be able to obtain information, analyses and expert support on issues related to the design and engineering of equipment used by or supporting military crews. Examples include visual and auditory displays, automated and intelligent aiding systems, manual and voice controls, equipment that supports the health, survival and escape of crew members.

What CSERIAC Will Do

CSERIAC will provide a broad array of products and services using different formats and media. Products to be developed will include handbooks and databooks, state-of-the-art reports, research directories, critical literature reviews, assessments of technologies, abstracts and indexes, current awareness bulletins and training materials. CSERIAC services will consist of support for the revision and development of the ergonomic aspects of military standards and specifications, development of individualized bibliographies, and the implementation and maintenance of computer-based models of human performance. Services will also include sponsored symposia, workshops and short courses for scientists and engineers on important issues and developments in the field of crew system ergonomics. CSERIAC will not duplicate the functions of DTIC and NTIS. Its function is information analysis and not archival storage.

The CSERIAC Survey

In order to confirm the need for the nature of potential CSERIAC users, a survey of crew system information users was conducted in 1985. Questionnaires were mailed to 3,705 individuals in industry and the DoD likely to be interested in crew system ergonomics information. Of the 1,022 questionnaires returned, 829 were amenable to analysis. Eighty-seven percent of these respondents indicated that a central source of crew system ergonomic information was needed and 67 percent indicated that an IAC was a suitable way to meet this need. A full 78 percent expressed a willingness to pay for CSERIAC services. In the aggregate, respondents estimated that they would make approximately 4,000 requests for information services. Background data on the respondents showed that 79 percent were in research and development, management or design, while 97 percent are users of crew systems ergonomics information.

Using CSERIAC's Services

CSERIAC supports the Department of Defense and other federal agencies and their contractors. Within DoD security guidelines and policy, it will also support academic institutions and industrial organizations in the United States and abroad. CSERIAC will offer a full range of information analysis products and services to all qualified users. A steering committee consisting of representatives of the military services and NASA will help to ensure that CSERIAC provides a balanced program of support to the communities that will benefit from its support.

Mention was made earlier that CSERIAC will be the custodian for models of human capabilities and limitations. This will extend CSERIAC's ability not only to provide information and analysis but also to exercise data in a dynamic model of an operator. Thus, data which by itself might be relatively useless when viewed statically can become more meaningful and useful when exercised in conjunction with a dynamic model.

An example of such a model is COMBIMAN (COMputerized BIomechanical MAN-Model), a three-dimensional model of a pilot which can be used to evaluate the physical "fit" of pilots to existing or projected crew station designs. COMBIMAN will perform four types of analysis: fit analysis, visibility analysis, reach analysis, and strength for operating conditions analysis. A user of COMBIMAN is able to specify the body size and body segment proportions of a hypothetical pilot using a 35-segment link system that corresponds to the human skeletal system. This man-model can then be used to evaluate a crew station. The crew station and pilot model are three-dimensional and can be rotated for viewing at any angle on a two-dimensional display screen. A practitioner may use COMBIMAN by providing CSERIAC with specifications for the crew station to be evaluated. CSERIAC will then create a three-dimensional representation of the crew station and exercise the model given pilot characteristics (body size, proportions and visual tasks) provided by the user.

Also accessible through CSERIAC is CREW CHIEF (the computer graphics simulation of an aircraft maintenance technician), another model that can be exercised to test the feasibility of performing maintenance tasks in an existing or proposed maintenance environment. Unlike COMBIMAN which models an individual in a seated position, the human model of CREW CHIEF is mobile.

CONCLUSION

This chapter reviewed the need to put designers in touch with useful, credible information on the human factors/ergonomic aspects of crew system design. It examined attempts to cope with the explosion of research information on human factors and concentrated on a number of recent innovations in selecting, packaging and interfacing designers and practitioners with the best and most applicable of that information for use in design.

The innovations described here were based on the premise that many of the human factors principles that historically have guided the design of displays are applicable to the design of information systems. These principles were put into effect in the design of each innovation. While all of the innovations are too new to have baseline data on their use, utility and durability, it is hoped that they will help reduce the barriers to using human factors information that have so long plagued systems design and compromised system effectiveness and safety.

NOTES

[1] From *Human Factors*, Vol. 5, 1963. Copyright 1963 by the Human Factors Society, Inc., and reprinted with permission.

[2]From *Perceptions and Psychophysics,* volume 17, pp. 203-208, reprinted by permission of Psychonomic Society, Inc.

REFERENCES

Allen, T. J. (1977). *Managing the flow of technology: Technology transfer and the dissemination of technological information within the R&D organization.* Cambridge, MA: The MIT Press.

Boff, K. R., Calhoun, G. L., & Lincoln, J. E. (1984). Making perceptual and human performance data an effective resource for designers. *NATO Workshop Panel IV. Weapon system development process and technology transfer.* Shrivenham, England: Royal College of Science.

Boff, K. R., Kaufman, L., & Thomas, J. (1986). *Handbook of perception and human performance* (Vols. 1-2). New York: Wiley.

Boff, K. R., & Lincoln, J .E., (1988a). *Engineering data compendium: Human perception and performance* (Vols. 1-3). Wright-Patterson Air Force Base, OH: Armstrong Aerospace Medical Research Laboratory.

Boff, K. R., & Lincoln, J .E., (1988b). *User's Guide: Engineering data compendium: Human perception and performance* . Wright-Patterson Air Force Base, OH: Armstrong Aerospace Medical Research Laboratory.

Glushko, R. J., Weaver, M. D., Coonan, T. A., & Lincoln, J. E. (1988). "Hypertext engineering": practical methods for creating a compact disc encyclopedia. *ACM Conference on Documenting Processing Systems*, 11-19.

Lincoln, J. E., & Boff, K. R. (1988). Making behavioral data useful for system design applications: Development of the Engineering Data Compendium. *Proceedings of the Human Factors Society 32nd Annual Meeting.* Anaheim, CA.

Meister, D., & Farr, D. (1966). *The utilization of human factors information by designers* (Contract No. NONR-4974-00O). Washington, DC: Office of Naval Research, Engineering Psychology Branch.

Parker, J. F., Jr., & West, V. R. (Eds.). (1973). *Bioastronautics Data Book* (2nd Edition) (NASA SP-3006). Scientific and Technical Information Office, National Aeronautics and Space Administration.

Ronco, P. G. (1963). A bibliography and overview of human factors reference works. *Human Factors, 5,* 549-568. Tufts College (1952). *Handbook of engineering data.* Medford, MA: Tufts College.

United States Air Force (1966). *Handbook of instructions for aerospace personnel subsystem design* (Report No. AFSCM 80-3). Washington, DC: Andrews Air Force Base.

United States Army Missile Command. *MIL-STD-1472 - Human engineering design criteria for military systems, equipment and facilities.* Redstone Arsenal, AL.

Van Cott, H. P., & Kinkade, R. G. (1972). *Human engineering guide to equipment design* (2nd edition). Joint Army-Navy-Air Force Steering Committee: U.S. Government Printing Office.

Van Cott, H. P., Kramer, J. J., Pezoldt, V. J., Porter, L. G., Fried, C., Fechter, J. V., & Persensky, J. J. (1978). *A standard ergonomics reference data system: The concept and its assessment* (NBSIR-77-1403). Washington, DC: National Bureau of Standards, Institute for Applied Technology.

Webb Associates (Staff of Anthropology Research Project), (1978). *Anthropometric source book, 1-3*. Washington, DC: National Aeronautics and Space Administration.

CHAPTER **17**

NATIONAL HUMAN PERFORMANCE DATA BANKS

Paul M. Haas
Rudy Laine

ABSTRACT

Practice of MANPRINT requires methods to assess human performance in existing systems and to predict human performance in future systems. The data needed to support the required analysis will be extensive, varied, and highly user-dependent. A national human performance information center that would collect and disseminate data widely to meet user community needs could be extremely valuable in implementing user-centered design approches. This chapter discusses some of the problems inherent in developing such a national information center. Then, by highlighting a sample of existing data banks, it illustrates that successes have been achieved in overcoming these problems, and that many of the important elements of a national human performance information center for MANPRINT already exist. A needs/feasibility study is recommended to define the requirements for and assess existing resources for a national human performance information center.

INTRODUCTION

Human performance and the importance of accounting for it as an integral part of system performance are central concepts of MANPRINT philosophy. It will be necessary to predict human performance simultaneously and integrally with the prediction of overall system performance, at a comparable level of detail, as the system progresses from concept through demonstration, full-scale development and production/development. An essential feature of the MANPRINT program, therefore, is the development, adaptation, validation, and use of analytic methods to aid designers in making these predictions of human performance. Equally important is providing the data to support these methods. Centralization of data and data-related services in a national data information center will permit widespread user access to data that can be used to implement MANPRINT methods on a national scale.

There are, of course, a variety of problems associated with the establishment, maintenance, and utilization of a national information center of the scope required for MANPRINT. This chapter addresses the more significant problems associated with human performance data and with the definition and establishment of a national information center. Additionally, it provides summary information on selected data banks that currently exist within the Department of Defense (DoD) and the nuclear power industry. The summary descriptions describe the types of data and data base services that are appropriate for MANPRINT as well as illustrating successes in overcoming many of the identified problems. Finally, the primary needs for a national information center to support MANPRINT analyses are identified as well as a possible strategy for initial development from existing data sources.

It should be noted that nuclear power industry and Department of Defense data bases were selected primarily because these are the data bases most familiar to the authors. No attempt has been made to critically review other existing data bases such as those in aerospace, transportation, etc., which may be as applicable as those selected. It is believed, however, that the combination of DoD and nuclear industry data bases are representative of human performance data stored throughout the United States.

CHALLENGES: PROBLEMS OF HUMAN PERFORMANCE DATA BANKS

MANPRINT practitioners' data requirements can be expected to be highly varied. Some of the major factors determining those needs are the:

- basic purpose/type of analysis
- level of detail required
- phase of system acquisition
- functional role of the analyst/data-user
- specific methods, techniques, and tools employed
- equipment involved and the design of the human-equipment interface

A national human performance information center will have to apply significant resources and expertise to meet these varied needs. In addition, there are both methodological and practical problems associated with data collection, processing, and dissemination that need to be overcome. Some are peculiar to human performance data and some are inherent problems of broadscope, multiuser data bases. Some of the more significant problems/issues relevant to MANPRINT are discussed below.

Human Performance Taxonomies and Data Base Structure

Uniqueness of Human Behavior

Human performance is unique. From a practical standpoint, predictions about human performance are made based on the level of imprecision that can be tolerated in the predictions. In the design process, the degree of tolerance is based on the norm and expected range of human performance. Precise predictions are not required of a specific human performance event.

However, reasonable predictions about human performance are needed as early as possible in the system life cycle. An obvious problem with making these predictions during the design phase is that the system is not available for testing. Consequently, data are extrapolated from predecessor systems, from simulations, laboratory data, simple theoretical models, subjective estimates – any sources considered applicable.

But what data are applicable? Human performance can be affected by numerous variables such as those associated with the:

- specific task
- equipment and human-equipment interface design
- procedures
- operational environment
- system process behavior
- training and experience
- time-dependent individual psychological and physical factors
- interaction with team members
- management and organizational behavior

Certainly not all of these variables are critical to a specific performance of every task or to every analysis associated with a particular task. But how can the critical variables affecting a particular performance observation be determined?

Categorizing Human Performance and Tasks

If data are pooled to produce statistical estimates of quantitative parameters, or even to draw qualitative conclusions about human performance, it is necessary to make assumptions about which variables significantly affected the observed/measured performance. Similarly, acknowledgement is required that some degree of similarity exists between the conditions of the observation and the conditions (modeled) in the system being designed. The degree of similarity required, or the acceptable threshold of dissimilarity, will vary with the particular analysis. However, it must be recognized that there are limitations in the ability to precisely represent the variables that will be important in the operational system of interest.

An associated issue is the specification of the basic level of human performance to be addressed, i.e., the definition of a task. The definition of a unit of human performance as a "task" or a "subtask" is completely arbitrary, dependent on the analysis being performed. But in order to compare and contrast human performance in different instances, it is necessary to be able to do so at comparable levels of behavior. It would be desirable to collect and process data at multiple levels and permit users to aggregate or disaggregate data as they choose, but complete flexibility is not feasible; and the result may not be meaningful if human performance is taken out of context.

The Taxonomic Approach

One of the ways to deal with the issues of uniqueness associated with human behavior and the problems of describing and predicting human performance is to develop taxonomies – listings of labels that classify observed behavior and associated variables. A human performance taxonomy, in effect, defines what is "similar" or "different" about various human performances. Implicitly, taxonomies are defined to discriminate between important differences in performance characteristics. While the definition of the taxonomy is, in a sense, arbitrary, it is important to be explicit and consistent when combining information about human performance obtained from different observations. (Note that in reference to a human performance taxonomy, the taxonomy or taxonomies in general will have to address differences and similarities in equipment because of the interdependence of human tasks and equipment.)

Impact on Data Base Design

Decisions, therefore, about the selection of the taxonomies to structure data within a human performance information center are critical to the ultimate usability of the information. The taxonomy (i.e., data structure) influences virtually all data bank functions – data collection, processing, storage, retrieval, dissemination, etc. To accommodate many users and many data sources, it is desirable to remain as comprehensive and as flexible as possible. But increased comprehensiveness and flexibility translates to increased complexity and increased cost; not just initial development cost, but also operational cost and even some negative impact on the usability. Moreover, the decisions about taxonomy have to be made early in the design/development of the information center. If existing sources are to be incorporated into the center, it is necessary to reconcile differences in the respective taxonomies.

All of these design decisions involve compromises. Conceptually, each aspect of human performance is unique such that no taxonomy can perfectly distinguish or discriminate among every performance. Practically, however, data from different performances can be combined to generalize and make very effective predictions of human performance. Trade-offs would be necessary in the design of the information center to accommodate the breadth of data sources and the user needs within resource constraints and practicality of operations.

Data Collection/Source Problems

Human performances data will be available from a variety of sources. Each source will have its advantages and disadvantages which must be considered in the development and use of a national information center.

Experiments

Controlled "laboratory" experiments provide useful data for both research purposes and for practical applications. However, discussion of problems and methods associated with human performance experiments is not within the scope of this chapter. It is noted, however, that there is a need for advances in fundamental methods for applied cognitive science, i.e., approaches for obtaining data on human cognitive performance.

Field Data

Human performance can be extracted from "historical" records of operating experience. These records typically are not designed for human performance analysis; therefore, it will be unusual to find field data records that provide specific information on critical human performance variables. Although data extraction will be tedious, useful data can be extracted. From these records, the face validity will be high because the data are extracted from practical operating experience.

Increasingly, industry and government organizations are conducting operational systems/environmental data collection specifically designed to obtain human performance data. Over time, these efforts will alleviate some of the problems encountered in extracting data from historical records. There are problems such as the impact of data collection efforts on operations and the natural resistance to self-reporting of errors. But there are notable examples of very successful operational systems. Perhaps the best known is the Aviation Safety Reporting System (Newton et al., 1985) used by the aviation industry. The Human Performance Evaluation System (INPO, 1987) used by nuclear utilities, which is described later in the section on *Successes* and was modeled after the Aviation Safety Reporting System (ASRS), is another example.

Simulator Studies

An approach which has some of the advantages of both experimental and field data collection is the use of simulators in controlled exercises. Data collection on full-scope training simulators has received considerable attention in the nuclear industry (see, for example, Beare et al., 1984; Beare et al., 1985; Kozinsky et al., 1984;). Many of the variables present in the real world are represented well in a simulator. And, experimental control is greater than for field data collection. A major advantage of simulators is the ability to obtain data relevant to performance under accident conditions that would be impossible or unsafe to produce in an operating system.

Programmatic Problems

In addition to the technical issues above, there are programmatic problems having to do with organizational interfaces and, of course, resources that need to be addressed.

Organizational Boundaries

Implementation of a national human performance information center will require crossing many traditional organizational boundaries – research, operations/maintenance, design/engineering, training, and others. Quality data cannot be obtained without cooperation. Information center management must be sensitive to the needs and constraints of each organization. Each group will have to "buy in" to the system. In general, this means that there will have to be some demonstrable benefit for all. In dealing with corporate boundaries, mechanisms need to be developed for handling proprietary or competition-sensitive information. And, of course, there need to be provisions for handling classified information.

Resource Requirements

Besides the normal funding issues for any major program, there are specific resource issues pertinent to a national human performance information center. In particular, there is a need for sustained commitment, and there are special staffing issues.

Perhaps more important than initial funding for a national information center is the long-term commitment that is required. An information center of this magnitude is unlikely to payoff in the short term. There are inherent time lags in its development. It may take several years to apply research results and "lessons learned" from operating experience to improve existing systems and/or design new systems. The true value of the information center is often not apparent until these long-term benefits begin to show up in improved performance. It is difficult for management and sponsoring organizations to maintain this long-term view in the face of annual budget battles and high-visibility, short-term demands.

Staffing of a national human performance information center also presents some challenges to management. A rather broad spectrum of professional and support staff are required for system operation – human factors/psychology professionals, engineering and system specialists, computer systems and library/data base specialists, various data base support personnel, managers, statisticians, and others. Many of these are needed only on a part-time basis. Typically, this means that one individual must perform several functions, or that staffing is shared with other organizations. Further, the job demands change as the system matures. Early on, creativity, organizational and interpersonal skills, and breadth of experience are most important; later, attention to detail and specific in-depth technical skills may become more important. All of these issues must be addressed by the management of the data base system to assure adequate staffing within budget constraints.

SUCCESSES: SOME EXISTING HUMAN PERFORMANCE DATA BANKS

Human performance data banks, as repositories of human factors information, can be grouped into two major categories: *macro* and *micro* (Johnson, 1987). In this context, macro human performance data banks store information concerning the design or adaptation of all manpower, personnel, training, and organizational components of the system. In contrast micro human performance data banks concentrate on information about the design of the interface between the hardware, the person and/or the design of specific tasks. Other macro/micro human performance data bank differences are shown in Table 17-1.

This chapter describes six human performance data banks in the nuclear power industry and Department of Defense as examples of successful data banks that can be useful to design engineers concerned with MANPRINT. These data banks and their organizations affiliations are shown in Table 17-2.

Table 17-3 provides a synopsis of the data categories contained in each of the data banks. A detailed discussion of each of the data banks follows:

Table 17-1
Human Performance Data Bank Differences

	MACRO	MICRO
Domain Association	Manpower, personnel, training	Human factors engineering, system safety, health hazards
Change Impact	Dramatic	Incremental
Orientation	Organizational performance	Individual task performance

Table 17-2
Data Banks and Their Organizational Affiliations

	DATA BANK	ORGANIZATION
MICRO	Human Performance Evaluation System (HPES)	Nuclear Power
	Nuclear Computerized Library for Assessing Reactor Reliability (NUCLARR)	Nuclear Power
	Crew System Ergonomics Information Analysis Center (CSERIAC) (See chapter on Human Performance Engineering Data Bases)	Department of Defense
MACRO	Training and Performance Data Center (TPDC)	Department of Defense
	Materiel Readiness Support Activity (MRSA) MANPRINT Data Base	Department of Army
	Project "A" Data Base	Department of Army

Table 17-3
Data Categories of National Data Banks

DATA CATEGORIES	HPES	NUCLARR	CSERIAC	TPDC	MSRA MANPRINT	PROJECT "A"
Human Performance	X	X	X	X		X
Human Performance (Error)	X	X	X			
Hardward Reliability		X			X	
Human Factors Engineering			X			
Manpower			X	X	X	
Personnel			X	X	X	X
Training				X	X	X
System Safety			X		X	
Health Hazards			X		X	

Micro Data Banks

Human Performance Evaluation System (HPES)

The Institute of Nuclear Power Operations (INPO) has developed the Human Performance Evaluation System (HPES) for use by the nuclear power industry (INPO, 1987). Its mission is "improve human reliability in overall plant operations by reducing human error through correction of the conditions that cause the error." Actually, HPES is not a data base in the conventional sense. INPO does not serve as a central processing and distribution organization and data are not available to anyone other than the participating utilities. INPO's role is primarily as a facilitator. Information is produced and maintained by the individual utilities. The system is described here to explain the general process, content and approach as an example of a system that is succeeding in obtaining data on human performance from an operational environment and providing feedback from that data to improve system design, operation and safety. While the specific data are not available and may not be of interest to MANPRINT users, the concepts and approach are very applicable.

The program is modeled after the Aviation Safety Reporting System (ASRS) developed and used by the aviation industry. ASRS is administered by National Aeronautics and Space Administration (NASA) and endorsed by the Federal Aviation Administration (FAA). HPES is a non-punitive program for reporting, analyzing and disseminating information on human performance in nuclear plant operations.

HPES has three fundamental elements:

- a method for nuclear utilities to identify human error situations (events, potential events, conditions that could lead to human error, etc.)
- plant coordinators trained in a root cause analysis method who perform/guide the analysis of information reported
- an information feedback system for sharing human performance experience with other participants.

According to INPO, the HPES program augments and supports line management's function of managing human performance. It is said to strengthen plant team relationships, increase error reporting, and correct underlying causes of events rather than just symptoms, frequently before an actual event occurs. In addition, INPO feels that the program fosters greater employee satisfaction due to fewer task errors and to employee participation in solving identified problems.

Functionally, there are four basic kinds of activities involved in HPES operation: reporting, analysis, corrective action, and feedback. These four functions and the flow of information through HPES are illustrated in Figure 17-1.

Key personnel and their responsibilities for implementing HPES at an operating nuclear plant are:

- line management – organizes the plant HPES program and establishes coordination with other plant activities; introduces the program coordinator and additional part-time evaluators; ensures the use of program results to resolve the underlying causes of human performance
- reporters – all workers and managers reporting human error events to the program coordinator
- program coordinator – a specially trained worker who analyzes reported events, determines their causes, recommends corrective actions to line management, and provides feedback to reporters
- evaluators – individuals specially trained to assist the program coordinator, normally on a part-time basis, in the evaluation of reported human performance problems.

INPO provides training to plant coordinators and evaluators on how to conduct objective evaluations of events involving human error. The evaluators' and coordinators' plant-specific knowledge is not only essential

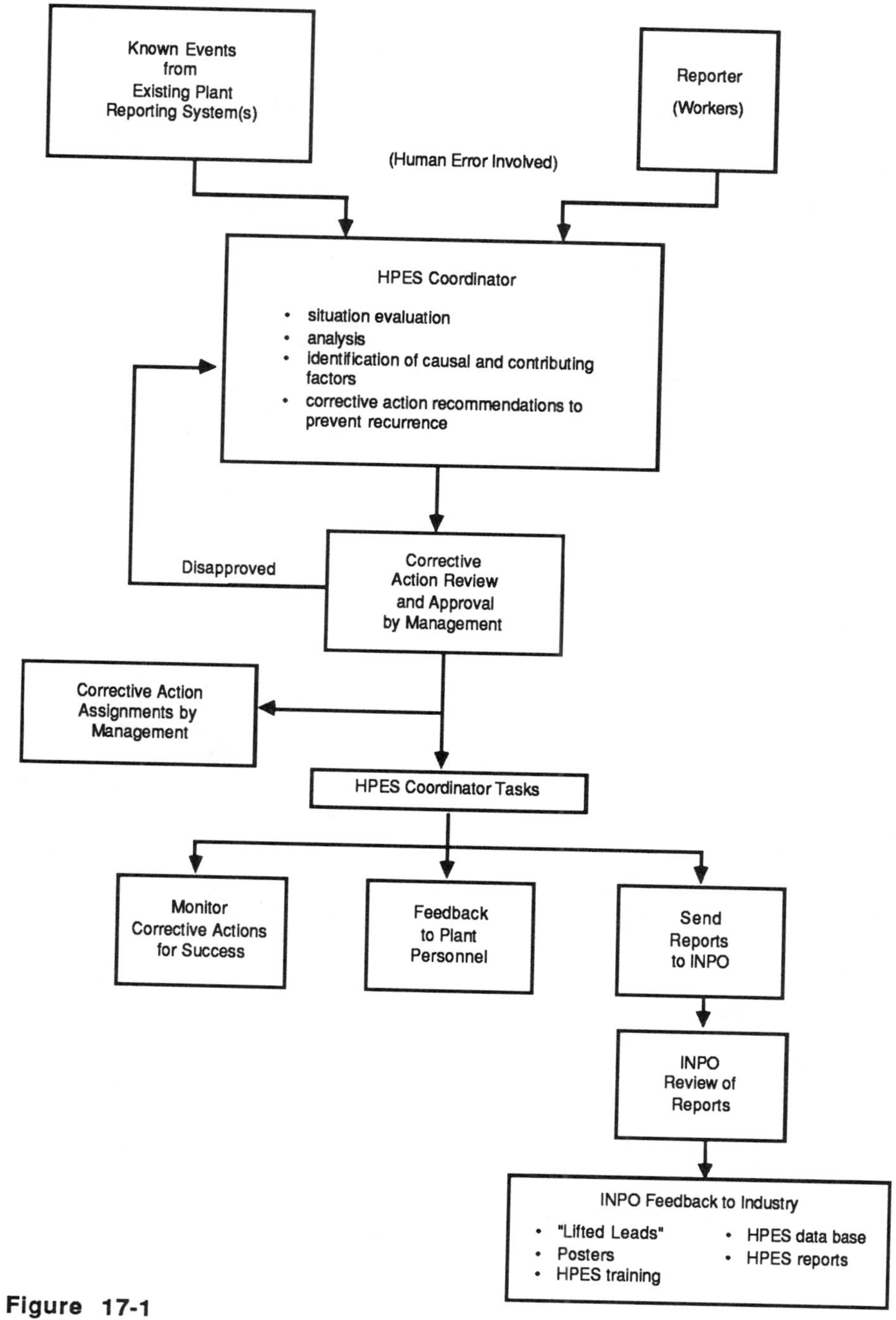

Figure 17-1
HPES Key Personnel and Information Flow

to event analysis, but for recommending viable solutions to identified problems. INPO provides a broad range of support which includes:

- utility-requested assistance in implementing the program
- assistance in screening reports based on industry-wide experience
- entry of report data into the human performance data base
- routine data base analysis and feedback to participating utilities and the industry
- a quarterly newsletter focusing on human performance events at member utilities
- inclusion of a special category, "Human Performance Information Exchange," in the Nuclear Network, and industry-wide information system developed and maintained by INPO
- sponsorship and organization of meetings for coordinators to discuss lessons learned and new developments and sponsorship of workshops on issues related to human performance
- annual review of program methodology and effectiveness

Nuclear Computerized Library for Assessing Reactor Reliability (NUCLARR)

The Nuclear Computerized Library for Assessing Reactor Reliability (NUCLARR) is an automated data base management system used to process, store, and retrieve human and equipment reliability data for nuclear power plants (Gertman et al., 1988). It is in a ready-to-use format for probabilistic risk assessment (PRA), and the human reliability analysis (HRA) included in PRA. NUCLARR was developed by the U.S. Nuclear Regulatory Commission (NRC) to provide the risk analysis community with a repository of human error and hardware failure rate data to support various analytical techniques used to assess nuclear power plant risk. The system is managed by staff at the Idaho National Engineering Laboratory (INEL), operated by EG&G, Idaho, for the Department of Energy.

The major components of the NUCLARR system are:

- NUCLARR Clearinghouse – functions as the primary interface and point of contact for interaction with data suppliers and users; also responsible for developing and administering a quality control program for processing of data by the Human and Hardware Reliability Analysis Group (HHRAG).
- HHRAG – responsible for all data input and data base maintenance functions; reviews/screens data in source documents for suitability prior to entry into the system; develops recommendations for system improvements.
- HHRAG Review Committee – provides technical direction to the HHRAG, data entry clerks, and software engineers to integrate any

approved changes into the system; includes members from NUCLARR project team and outside organizations; conducts quality assessment of the data stored in the system and reviews the Data Manual prior to distribution.

• NUCLARR Computer System – runs software and hardware for automated data base management functions of storing, processing, and retrieving data; includes applications software, computational algorithms, and system "housekeeping" functions. NUCLARR is an IBM-PC based system accessible directly by the user with menu-driven software developed by INTEL. Human error probability (HEP) data can be supplied in ASCII data files directly readable by standard software packages such as SPSS, SAS and dBIII.

Data may be submitted for entry by individuals or organizations. Only data that has been screen by the HHRAG is entered into the system. The basic human performance data entry is a quantitative value of human error probability (HEP) in the form of a point value, probability distribution, or ratio of errors to estimated opportunities. Basic sources of data on human performance include:

- Field data
- Training simulator data
- Laboratory data
- Consensus expert judgment
- Subjective data
- Simulation modeling data
- Analytic data

Individual entries typically contain the following items:

- Task statement
- Failure mode (omission, commission)
- Data type (recovery considered or not)
- Human error probability (mean, number errors, number opportunities, upper and lower confidence bounds, etc.)
- Plant code
- Performance time to complete task
- Time available to complete task
- Scale values for performance shaping factors (PSFs) – stress, experience, procedure, training, etc.
- Data source information
- Event type (pre-initiating event, planned action, recovery action outside of procedures)

HEP data are organized and classified in the three-level hierarchy illustrated in Figure 17-2:

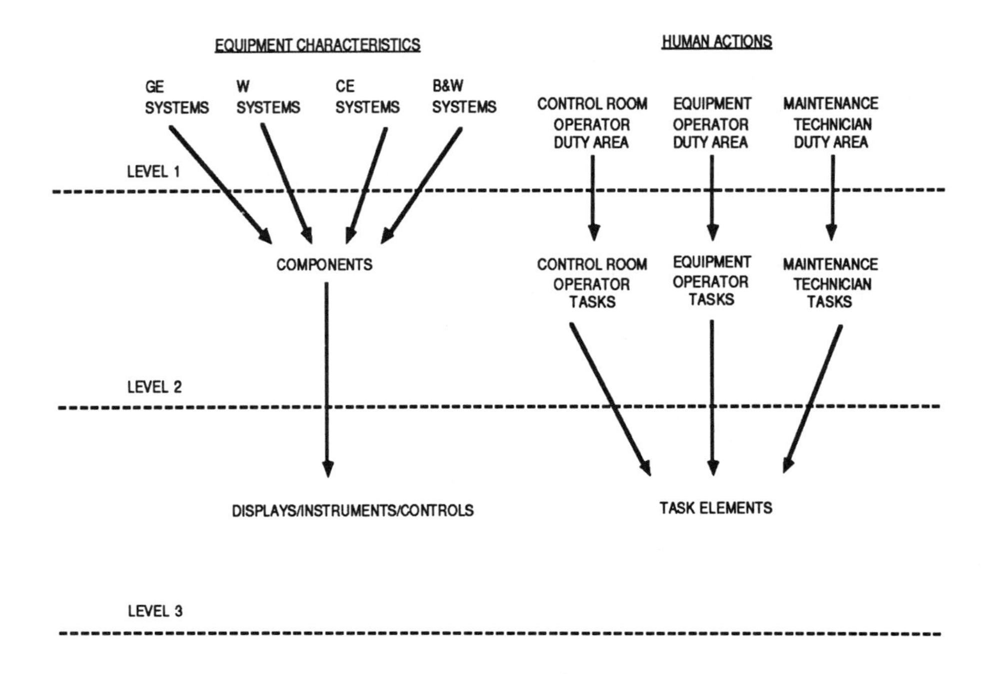

Figure 17-2
NUCLARR System Taxonomy and Organization

- Level 1: Actions are classified by nuclear power plant system and by duty areas that summarize the objective of the human interaction with the system, e.g., control room operator monitors the reactor recirculation system. (System Level)
- Level 2: The task involves one or more actions that change or determine the state of a plant component, e.g., equipment operator starts a motor-driven pump. (Component Level)
- Level 3: The task is a single action directed at a specific control, instrument, or display device used to operate or maintain equipment in the plant or to communicate with other plant personnel, e.g., an operator reads a meter or positions a J-handle switch. (Displays/Instruments/Controls Level)

This hierarchical taxonomy permits the user to choose the appropriate level of detail for the particular analysis.

The basic structure of the NUCLARR system is a series of two-dimensional matrices combining the different levels of the equipment characteristics category with the specific levels of human actions category. A total of 16 separate and distinct matrices are used. Equipment characteristics are listed by rows, and human actions are listed by columns for each matrix. Users can conduct menu-driven "descriptive searches" (basically automated searches with limited user flexibility) or "ad hoc searches" (more flexible, user-controlled searches). A sample tree structure for searching the data base is shown in Figure 17-3.

Crew System Ergonomics Information Analysis Center (CSERIAC)

CSERIAC, as discussed by Van Cott (Chapter 16), is one of over 20 Information Analysis Centers (IACs) (established by the Department of Defense (DoD) and the Defense Technical Information Center [DTIC]) that provide services that go beyond the basic document retrieval functions of DTIC and other repositories of technical information. The mission of an IAC is to acquire, digest, analyze, evaluate, synthesize, store, publish, and provide technical services concerning available worldwide scientific information and engineering data. The type of information covered by each IAC encompasses a clearly defined content area or specialized field of significant DoD interest or concern. Crew System Ergonomics (CSE) information focuses on human and equipment characteristics that effect

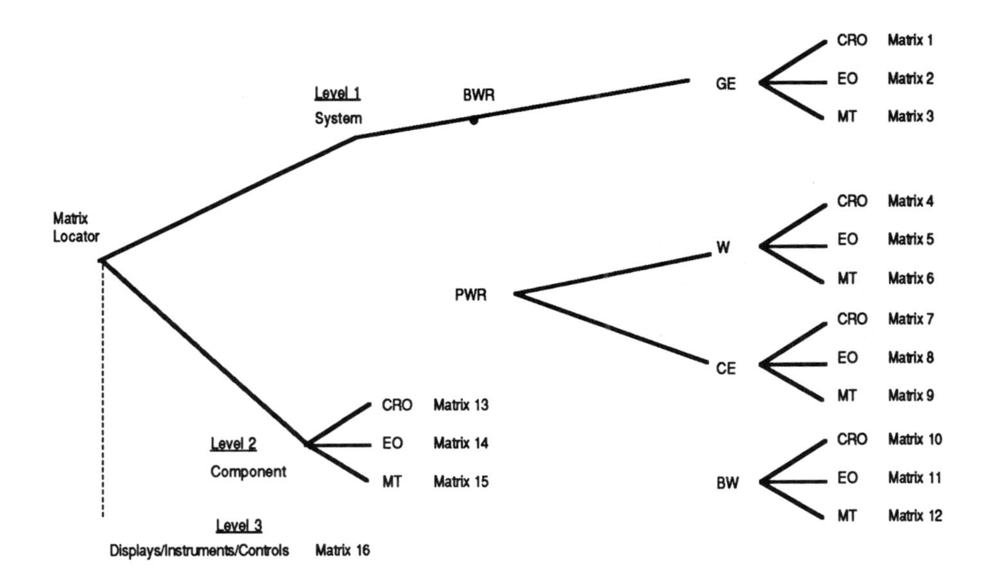

Figure 17-3
Sample Tree Structure for NUCLARR

crew performance and well being in military situations and activities. In other words, CSE information is the interaction of humans with equipment in aerospace, land, and sea environment. CSE data include scientific and technical information concerning human characteristics, abilities, limitations, physiological needs, performance, body dimensions, biomechanical dynamics, strengths, and tolerances.

The objective of CSERIAC is to support DoD requirements for incorporating crew system ergonomics into the design and operation of military systems. Its mission is to provide ergonomic information analysis services to support research, design, and development of space, air, surface, and subsurface crew systems.

In order to fulfill its mission to provide ergonomic information analysis services to designers and engineers, CSERIAC must first gain access to both old and new sources. Ultimately CSERIAC will establish links with primary sources including national and international scientific and engineering data collections as well as repositories with relevant holdings. Major sources of CSE information include professional journals, technical reports, books, and standards. In addition, information generated during test and evaluation in the development of systems provide a wealth of "lessons learned" information and data. This information is generally available only to either the requesting or performing organizations. CSERIAC plans to collect, catalogue, and make available CSE test and evaluation reports and the information contained within.

CSERIAC is hosted by the Armstrong Aerospace Medical Research Laboratory (AAMRL), located at Wright-Patterson Air Force Base, Ohio, and is operated for the DoD by the University of Dayton Research Institute. Although designed primarily to support DoD, government organizations, and United States contractors, CSERIAC is also available to other users. Typically, these would include academic and corporate users (domestic and international), with limitations based upon DoD security guideline policies (e.g., militarily critical information) and the resources of the center. For additional information and to be place on the mailing list, contact: CSERIAC Program Office, AAMRL/HE/CSERIAC, Wright-Patterson AFB, OH 45433-6573.

Macro Data Banks

Training and Performance Data Center (TPDC)

The Training and Performance Data Center (TPDC) serves as the DoD focal point for training and performance data. As a centralized repository, TPDC facilitates information exchange within the defense training community, among other defense agencies, throughout the public and private sectors, as well as various international activities. As specified in its charter, TPDC's mission is to provide training and job performance information and tools/methodologies to the Office of the Secretary of Defense, the individual military services, and to the defense training community to assist in the resolution of large broadly based training issues.

The scope of the TPDC effort is reflected by its organizational structure, five divisions focusing on all aspects of training and job performance.

• Training Division – responsible for both formal and on-the-job training.

• Collective and Joint Training Division – "basic unit and above" dimensions of training.

• Reserve Integration Division – responsible for the unique and increasingly important training mission of the reserve forces (e.g., United States Army Reserve and National Guard).

• Performance and Technology Division – is concerned with performance measurement systems and performance data and technologies used in training, such as computer based instruction and interactive video disk.

• Equipment Integration Division – provides significant assistance to MANPRINT practitioners since it orients specifically on the equipment acquisition (procurement) cycle, the logistics community, and the total training subsystem of weapon systems, training devices and other equipment.

The MANPRINT association with TPDC is primarily through the Equipment Integration Division which orients its data analysis on the linkages among the Military Occupational Specialties (MOS), the weapons systems and the training and performance aspects of the unit. An important aspect of the TPDC data base is that in addition to linkages just mentioned, the identification of MOS/system relationships provides the "key" (link) for accessing relevant information in other data bases (e.g., the Army Training Requirements and Resource System (ATRRS), the Enlisted Master File (EMF), etc.). In other words, the MOS provides a "key" not only to data at TPDC but also to all personnel data bases elsewhere. Similarly, these linkage of MOS to weapons systems provides the "key" to access related weapon systems data (Reliability, Availability and Maintainability [RAM]).

Two MANPRINT related projected developed by TPDC are described below:

CROSSWALK (Urban et al., 1987). The first, entitled CROSSWALK, is an automated "lookup" table that links equipment to MOSs responsible for operation, maintenance and support or, in turn, can link operator and maintainer MOSs to each item of equipment with which they are associated (see Figure 17-4). The significance of CROSSWALK is that it provides an automated means to access relationships between 400 plus MOSs (both current and archival) and over 7,000 end items (i.e., tanks, engines). It provides the opportunity through FOOTPRINT, described below, to obtain a

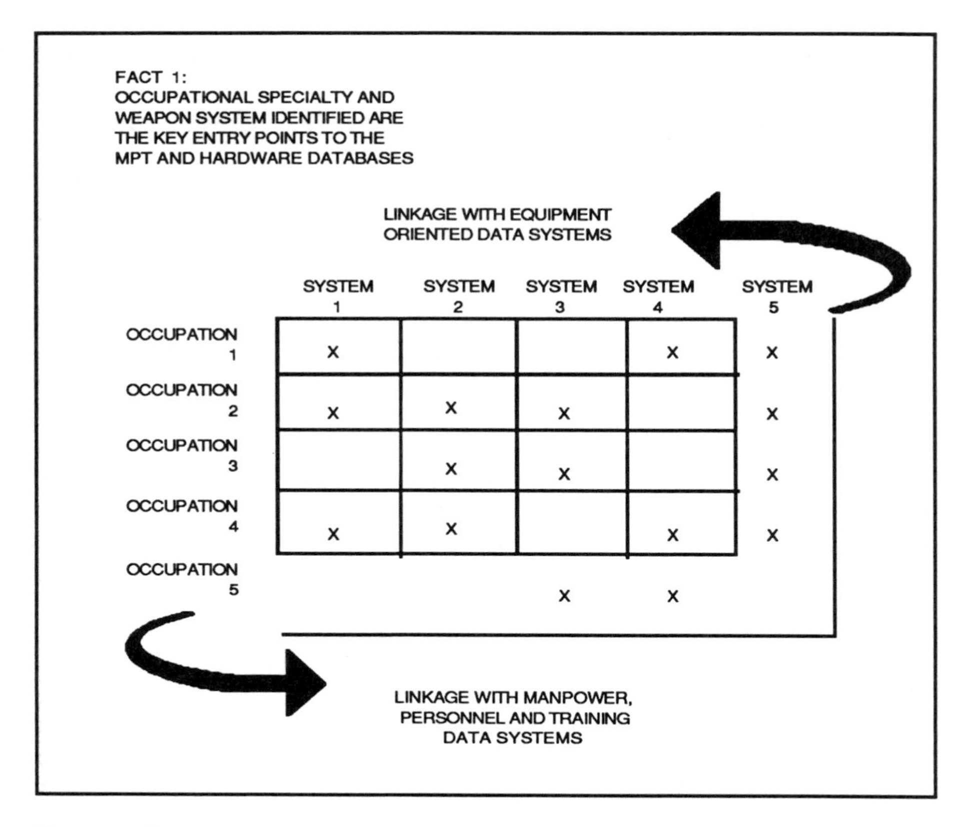

Figure 17-4
CROSSWALK

weapon system view of manpower, personnel and training (MPT) data or an MOS oriented view of weapon system data. CROSSWALK provides the "key" to accessing FOOTPRINT data. Both data bases are updated on a semi-annual basis.

FOOTPRINT (Hoffman et al., 1987). When integrated with CROSSWALK, FOOTPRINT allows MPT information to be aggregated for an existing item of equipment (predecessor), a baseline comparable (notional) system, or a new system (see Figure 17-5). The reports generated by FOOTPRINT provide historical and projected performance indicators and force structure and MPT trends of each MOS required to operate, maintain and support the item. The performance indicators provide a six year historical perspective of aptitude, education, experience and retention. Force structure includes data on current authorizations (requirements) versus operating strength (on-hand) and projected authorizations. The training profiles identify type, length and amount of training, by MOS, associated with the item of equipment.

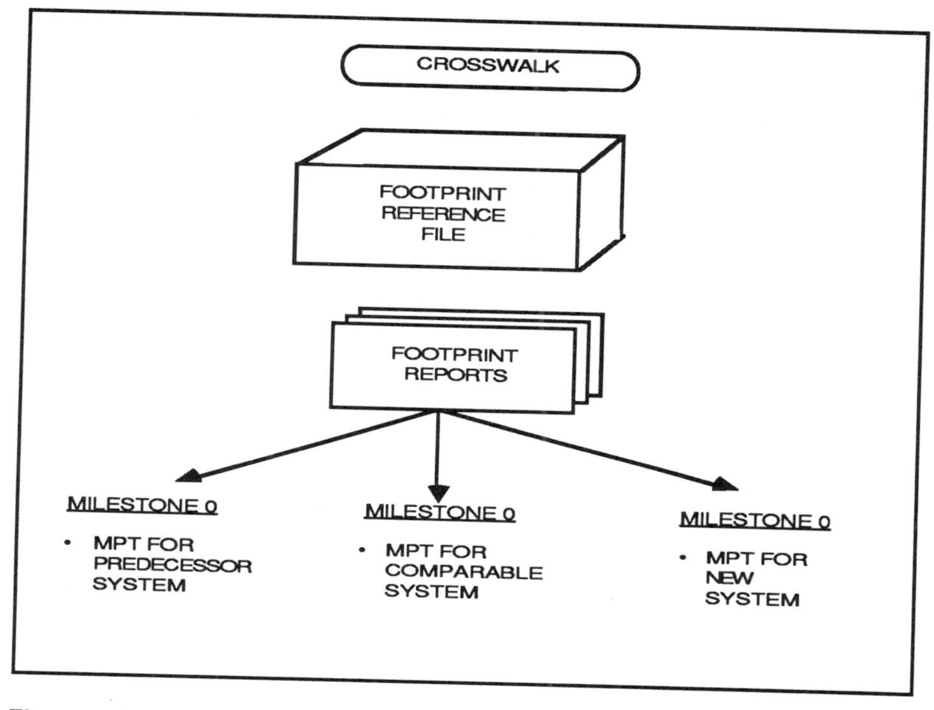

Figure 17-5
FOOTPRINT

Addition information concerning these projects is available from the Department of Defense Training and Performance Data Center, 3280 Progress Drive, Orlando, FL 32826-3229.

Materiel Readiness Support Activity (MRSA) MANPRINT Database

The objectives of the U.S. Army Materiel Readiness Support Activity MANPRINT Data base is to provide weapon and tactical system program information. The six domains of MANPRINT are contained in the data base as well as reliability and maintainability data (which has a direct correlation to manpower, personnel and training requirements) extracted from Logistics Support Analysis Records and Sample Data Collection. Unlike the TPDC data base, which is oriented primarily toward the MANPRINT macro elements (manpower, personnel and training), the MRSA data base encompasses both the macro and micro elements as well as reliability and maintainability data. However, the MRSA data base will contain only 60 plus items of equipment versus the 6,000 plus items of equipment in the TPDC data base. While manpower, personnel and training data is resident in both data bases, the richness of data for each of these domains varies significantly because of the intended use for the data. Differences in both quantity (i.e., number of systems) and richness of data between these two data bases begins to highlight the need for the establishment of an integrated national human performance information center that can both store and facilitate extraction of as much relevant human performance data as is available from other sources. This would establish a much enhanced human performance data baseline from which all personnel involved in the design and development of equipment could begin.

The MRSA MANPRINT database is structured around three basic files: (1) the military occupational specialty (MOS) file, (2) the end item file, major items of specific equipment (e.g., M-1 Tank, Bradley Fighting Vehicle), and (3) the baseline comparison system (BCS) file. The MOS file provides authorization and on-hand strengths, operator and maintainer specific tasks, physical performance tasks, and aptitude data for all MOSs. The end item file contains system specific data on reliability and maintainability, human factors issues, which includes operational impact, recommended corrective actions, and points of contact for additional information. It also contains the same categories of information relative to system safety and health hazard issues. The BCS file enables the user to create and modify baseline comparison systems for the purpose of analysis.

Currently, users of the system are restricted to government and defense industry contractors. Additional information may obtained from the U.S. Army Materiel Command, Materiel Readiness Support Activity, Lexington, KY 40505.

The Project "A" Database

In 1980, Congress mandated that each of the military services begin research directed at determining whether the Armed Services Vocational Aptitude Battery (ASVAB) is a valid, reliable, and fair test for use in selecting, classifying, training and assigning recruits into military occupational specialties (Campbell et al., 1988). The U.S. Army Research Institute (ARI), under the sponsorship of the Deputy Chief of Staff for Personnel, initiated in 1982 a nine year research program to address the Congressional mandate. The objectives of the project are to:

1. Validate ASVAB against both existing and project-developed performance measures which includes both Army-wide performance measures based on rating scales and direct hands-on measures of MOS specific task proficiency.
2. Develop and validate new selection and classification measures against existing and project-developed performance measures.

3. Validate intermediate criteria such as training performance, as predictors of later criteria, such a first-tour job performance, so that more informed decisions could be made on reenlistment and retention.

4. Examine the relative effectiveness of alternative selection and classification procedures in terms of their validity and utility for making operational selection and classification decisions in the Army.

Three major initiatives resulting from the Project "A" are:

1. Improvements in the use of ASVAB Aptitude Area (AA) Composites to determine eligibility for enlistment in particular Army MOSs. Although the majority of Army MOSs were assigned to the appropriate Aptitude Area Composites, the research has enhanced the validity of these composites and has resulted in recommendations for realignment of MOSs within existing composites and changes to the ASVAB subtests of the Mechanical Maintenance Composite.

2. Development of a test battery for predicting tank gunnery performance based on spatial and psychomotor tests. In an initial sample, these tests showed strong predictive validity for scores on the Unit Conduct of Fire Simulator, a tank gunnery simulator. These findings have also been replicated on several independent samples of anti-tank gunnery performance.

3. Development of a new test battery which supplements the ASVAB by measuring more of the motivational and experiential attributes not previously measured by the ASVAB. The results of the battery, Assessment of Background and Life Experience (ABLE), indicate that soldiers receiving high scores will have higher leadership potential, lower attrition rates, and fewer disciplinary problems.

Longitudinal Validation (LV), the third phase of Project "A," was started in 1986. The initial phase involved the administration of the new battery of Project "A" selection tests to approximately 55,000 soldiers in 21 MOSs at all Recruiting and Reception Stations throughout the United States. At the end of the Advanced Individual Training or equivalent, each solider in the LV initial phase was administered the training knowledge test for his or her MOS and a set of rating scales. These soldiers were tracked in their first-tour duty assignments and were administered a set of MOS-specific and Army-wide job performance measures. Second-tour job-performance measures are planned for the LV sample in 1991.

The Army Research Institute has created a Longitudinal Research Database (LRDB) to serve as a permanent storehouse for all Project "A" empirical data. This database is extremely important to the Army in analyzing accession policies, establishing standards for enlistment and reenlistment, predicting attrition, linking initial training to job performance, and linking entry level characteristics to overall performance.

THE NEED: AN INTEGRATED NATIONAL HUMAN PERFORMANCE INFORMATION CENTER

The data bases and information sources described above in *Successes* are representative of only a small portion of the commercial nuclear power industry and the Department of Defense efforts relative to macro and micro human performance data needs applicable to MANPRINT. As indicated previously, this chapter does not provide a comprehensive assessment of all data needs and data sources. The discussion of these representative data bases should, however, indicate that the task of collecting, processing, and distributing the necessary human performance data for MANPRINT is such a formidable one that it will require significant and dedicated resources at a national level. And, from the discussion above on *Challenges* , it is apparent that development and operation of such a center will require resolution of a variety of inherent problems. However, the successes of the data bases previously described show that these inherent problems can be and are being resolved.

If this is so, what more needs to be done? The real need is greater concentration on integration of existing sources, distribution of information, and assistance to the user in applying the data. Bringing together the existing sources of information, processing, combining, distilling data into formats needed by varied users, and distributing data to users promptly and in an easy-to-apply form are critical functions that can best be accomplished by a dedicated national data center.

For many years, human factors specialists have been acutely aware that the sparsity of data on human performance has limited the ability of the field to contribute significantly to quantitative systems analysis. To some degree, particularly in the area of human cognitive performance, this is still true. However, the data banks and information sources discussed above demonstrate that there has been considerable advances over the past decade in the compilation and processing of existing information and, indeed, increases in the amount of basic data that are available. It is reemphasized that the data banks and information sources discussed are a sample, not a comprehensive summary, of the data stores that are available. Clearly, there is a need to continue to develop the body of raw data/information on human performance even though a significant amount currently exists. The most urgent need now, however, is to evaluate, process and disseminate the information, efficiently and effectively.

An integrated National Human Performance Information Center that supports MANPRINT needs and the needs of the nation for similar information may be the appropriate means to meet this end. State-of-the-art information system technology, now, makes such a national center feasible. The combination of relational data bases, improved communication capabilities, and intelligent software to aid user access can greatly enhance the ability to overcome the basic complexity/flexibility problems cited above in *Challenges*. As a result of these technological improvements, large quantities of data, quite different in form and content, can be searched, merged, and processed rapidly within reasonable costs.

The needs, feasibility and requirements for establishing a national information center should be examined more carefully, as well as the capabilities of existing systems to fulfill the requirements. Specific recommendations should be made on the basis of such a study. It appears that some elements of the system may already be in place and that an effective initial system can be brought on-line with software that would enable users to access existing sources through a flexible interface.

A cursory examination of the selected data bases suggests the need for an overall system architecture with two general categories – macro and micro human performance data. At this point, the desirability or feasibility of creating an architecture that would integrate these two categories of data is not clear nor is it clear that the state of knowledge concerning human performance is such that the mapping between the two could be established. Consequently, the illustrative top-level architecture in Figures 17-6 and 17-7 shows separate macro and micro nodes. These high-level descriptions are intended primarily to illustrate that key elements of each node already exist and to suggest that the definition of interfaces would be an initial step toward providing integrated data to a National Human Performance Information Center.

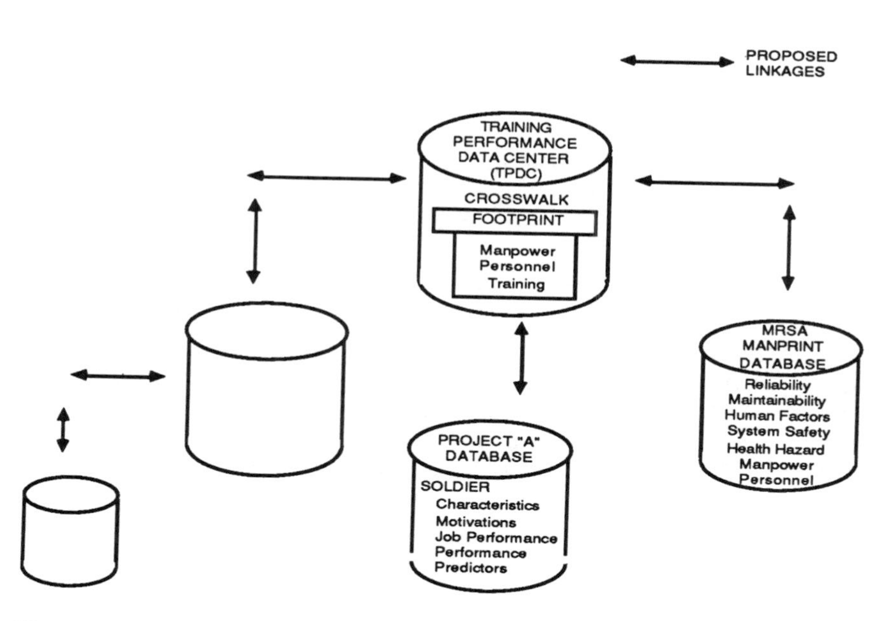

Figure 17-6
Macro Human Performance Node

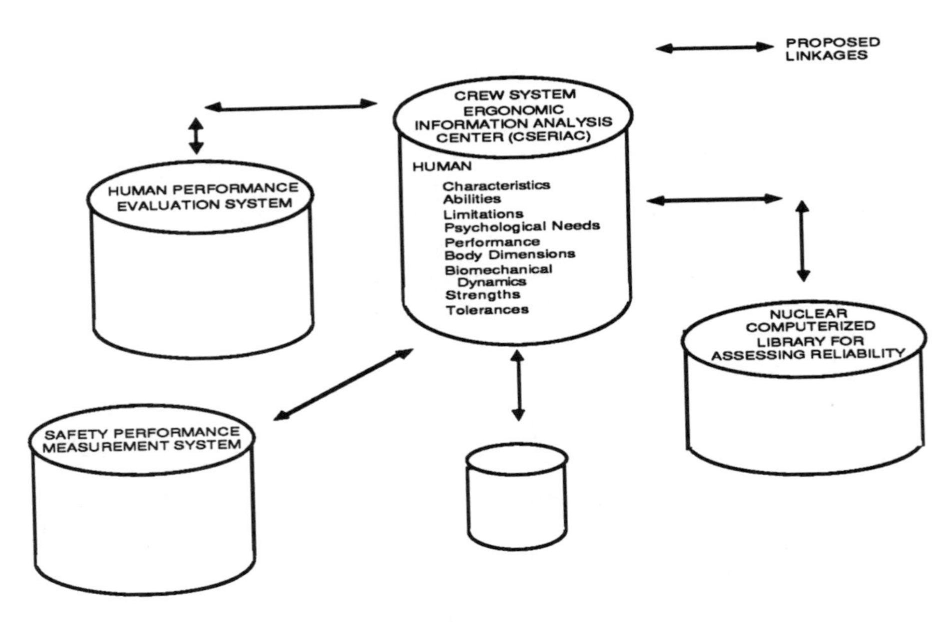

Figure 17-7
Micro Human Performance Node

Macro Human Performance Node

A proposed architecture for the macro human performance node, Figure 17-6, assigns the responsibility for accessing all of the available macro human performance data banks to the Training Performance Data Center (TPDC). In turn, TPDC would then process and transmit the data to the National Human Performance Information Center.

TPDC was selected as the most likely node for macro human performance data based on the fact that it currently has or is establishing access to other macro human performance data banks discussed. Therefore, TPDC could use CROSSWALK as the "key" to access and extract all available macro human performance data in the other data banks from either an equipment or a Military Occupational Specialty perspective. Additionally, TPDC is in the process of establishing a CROSSWALK/ FOOTPRINT data bank for the U.S. Air Force and has been seeking access to the data base at the U.S. Army National Training Center, a source of human performance data under simulated wartime conditions. It could reasonably be expanded to any commercial macro data centers in existence or be expanded to cover other non-military systems of national importance. One candidate may be the complex air traffic control system currently under development. Based on the above, TPDC currently appears to be the best positioned to assume the role as the macro human performance node.

Micro Human Performance Node

Similar to TPDC, the proposed architecture for the micro human performance node, Figure 17-7, assigns responsibility to the Crew Station Ergonomic Information Analysis Center (CSERIAC) for accessing, processing and transmitting all micro human performance data to the National Human Performance Information Center.

CSERIAC was selected as the micro human performance node based on the significance, from a MANPRINT perspective, of the human performance data currently available from this source. Unlike TPDC, CSERIAC does not have established interfaces with the other micro human performance sources discussed and in order to perform its nodal functions, CSERIAC would have to establish the necessary interfaces with these and other significant micro human performance data sources.

National Human Performance Information Center

Although attempts to establish a National Human Performance Information Center may encounter the same or similar obstacles identified above, the need far outweighs the obstacles. Currently available data bases and information sources coupled with state-of-the-art software and communication interfaces could provide the basis for a rudimentary architecture dedicated to providing the latest available information on human performance.

It is recommended that a national advisory committee be formed to study and direct a feasible approach for creating a national human performance information center. The purpose of the center would be to make available to all participating individuals or organizations an equal and readily accessible baseline of human performance information. The capabilities of industry as well as industrial professional organizations should be examined for contribution and participation in this undertaking. It is well known that development of human performance data tends to be primarily a government or academic activity because of the universality of application and the difficulty for industry independent research and development (IR&D) to acquire a proprietary or competitive advantage by investing in human performance data collection. However, the INPO HPES approach based on the Aviation Safety Reporting System may provide a preliminary model for a national information center. Conceivably, all of the nation's industries which build or operate complex systems could (like the nuclear power industry) submit data to and draw information from a center (like INPO). Through IR&D programs, data could be developed but placed into a national center for retrieval. A formal study of the needs, existing resources, selection options, and resource requirements should be conducted. The structure suggested in this chapter is only one of many possibilities.

Consideration, for example, should be given to incorporating the center under an existing industrial organization such as the National Security Industrial Association, The Army Defense Preparedness Association, or a national library system like the Library of Congress.

REFERENCES

Beare et al. (1984). *A simulator-based study of human errors in nuclear power plant control room tasks* (NUREG/CR-3309). Washington, DC: Nuclear Regulatory Commission.

Beare et al. (1985). *The effects of supervisor experience and assistance of a shift technical advisor (STA) on crew performance in control room simulators* (NUREG/CR-4280). Washington, DC: Nuclear Regulatory Commission.

Campbell et al. (1988). *The Project A approach to describing job proficiency.* Presented at the 96th Annual Convention of the American Psychological Association, Atlanta, GA.

Gertman et al. (1988). *Nuclear computerized library for assessing reactor reliability (NUCLARR) Volume 1: Summary description* (NUREG/CR-4639). Washington, DC: Nuclear Regulatory Commission.

Hoffman et al. (1987). *FOOTPRINT: One small step for MPT.* Presented at the 9th Interservice/Industry Training Systems Conference, Washington, DC.

INPO (1987). *Human performance evaluation system* (INPO 87-007). Atlanta, GA: Institute of Nuclear Power Operations.

Johnson, E. M. (1987). *The role of man in the system design process: The unresolved dilemma.* In W. B. Rouse and K. R. Boff (Eds.), System design. New York: North Holland.

Kozinsky et al. (1984). *Safety-related operator actions: Methodology for developing criteria* (NUREG/CR-3515). Washington, DC: Nuclear Regulatory Commission.

Newton et al., (1985). *Nuclear power safety reporting system: Implementation and operational specifications* (NUREG/CR-4133). Washington, DC: Nuclear Regulatory Commission.

Urban et al. (1987). *Directory of Defense Training and Performance Data Center databases and extract files* (TPDC 88-001). Orlando, FL: Defense Training and Performance Data Center.

CHAPTER **18**

NATIONAL EDUCATION AND TRAINING

Frederick A. Muckler
Sally A. Seven

ABSTRACT

To bring about MANPRINT objectives, it is essential that there be adequate numbers of trained generalists, specialists, researchers, and practitioners. In order to better understand MANPRINT education and training issues, this chapter considers the kind of skills and knowledges needed to conduct the MANPRINT effort; examines some of the current institutional systems that educate and train many of the specialties of MANPRINT; and reviews some of the relevant MANPRINT information available for training in technical handbooks and introductory textbooks. An existing source of future manpower supply is the present human factors academic training programs from psychology and engineering. What is currently taught in these programs is exemplified by a brief look at three very strong programs at the Ohio State University, the University of Southern California, and Virginia Polytechnic Institute and State University. Several training-related issues are highlighted: (1) What will be the aggregate demand for trained professionals in the future? (2) Is it possible, or even desirable, to institute certification and licensing procedures? (3) What cost-effective provisions can be made for continuing education? (4) Would the creation and sustenance of centers of excellence enhance this technology? (5) How can the training need of the non-specialist be satisfied?

MANPRINT PERSONNEL

To improve system integration and operation through the application of the MANPRINT approach, trained personnel capable of achieving specific technical MANPRINT objectives for systems are needed. At issue in this chapter is how the national education and training "system" can provide trained MANPRINT personnel. In a sense, MANPRINT issues such as manpower supply and demand, personnel quality, and training will be applied to the execution of MANPRINT programs.

Supplying Trained MANPRINT Personnel

Trained MANPRINT personnel will have to come from somewhere. In considering where MANPRINT people will come from and what they will be like, there are several questions of direct interest:

1. What will MANPRINT professional personnel need to know?
2. From where will MANPRINT personnel come in the national education and training system?
3. What changes might have to be made in current specialist sources to be more effective in MANPRINT programs and projects?
4. Will curricula changes be required?
5. What will be the MANPRINT manpower demand requirements in the future?
6. Can the current national education and training system supply sufficient manpower to meet future MANPRINT needs?
7. What kinds of training can be provided for those who will not be MANPRINT technical specialists but who should have some awareness of MANPRINT objectives in system design and operation?

These questions, and others, will be considered in this chapter with tentative answers to some but certainly not all the questions. The main concern will be: Can the amounts and kinds of skilled technical personnel that can make the MANPRINT approach a reality be provided?

It is important to start with the reminder that the education and training system must consider trained people for many different organizational and professional settings. MANPRINT and the associated technical areas will naturally require continuing scientific research. So, some part of the manpower supply must be directed to research. Others may find themselves in system design and integration where the requirements for technical information and the way information is used may be rather unique (cf., Muckler, 1969; Rouse, 1986; Rouse & Boff, 1987). Individuals are already conducting versions of MANPRINT investigations in system test and evaluation. MANPRINT people in system operation and maintenance up to and including system obsolescence and decommissioning (e.g., removal of overage nuclear power plants) are also expected to be required. In short, MANPRINT personnel have skills and knowledges useful to any system of which people are a part, and some consideration must be given to their training and to how and where they will practice their profession.

Training to Know What?

Previous chapters of this book have already discussed the six basic MANPRINT domains: human factors engineering, manpower, personnel, training, system safety, and health hazards. From the standpoint of education and training, one immediate question is: What skills and knowledges might be required for these areas which should dictate, in part, what should be taught? One approach to this problem is shown in Table 18-1 where an attempt has been made to define some of the essential and minimum content areas that must be considered for MANPRINT formal training (with assistance from the analyses of Koch, 1980; Snyder & Eggleston, 1984).

Table 18-1 implies that the items listed under each major category are not the total extent of the skills and knowledges for that area, but that they are essential for performance at a minimal professional level. For example, it is very difficult to perform adequately in system design and human factors engineering without some knowledge of task analysis.

Table 18-1
Recommended Content Areas for MANPRINT Training

- HUMAN FACTORS ENGINEERING
  - Psychological and physiological capabilities, limitations and tolerances
  - Mission, function, and task analysis
  - Anthropometric and biometric criteria
  - Display-control task design
  - Workspace requirements and design
  - Organizational design
- MANPOWER
  - Human resources systems predictions
  - Manpower models
  - Personnel models
  - Assignment models
  - Training models
- PERSONNEL
  - Skills, knowledges, and abilities analysis (SKA)
  - Personnel selection
  - SKA/training tradeoffs
  - Personnel quality and performance predictions
  - Motivations, incentives, and performance
- TRAINING
  - Human learning and transfer of training
  - Training requirements and needs analysis
  - Instructional system design
  - Training media and devices
  - Training system evaluation

Table 18-1 *(continued)*
Recommended Content Areas for MANPRINT Training

- SYSTEM SAFETY
  - System reliability analysis
  - Human error analysis
  - System safety planning
  - Safety training
- HEALTH HAZARDS
  - Environmental stressors identification
  - Psychological stressors identification
  - Designing for health and safety
  - Personal protection and equipment
  - Controlling workplace hazards
  - Product reliability and liability
- BASIC METHODS
  - Literature retrieval
  - Handbooks and data bases
  - Experimental and quasi-experimental methods
  - Statistics and quantitative methods
  - Computer methods and applications
  - Man-machine system models and simulations
  - System and performance measurement – estimation and evaluation of performance
  - Cost-benefit analysis
  - Engineering approaches to system design
  - Sociotechnical system design

Human Factors Engineering

Human factors engineering means primarily those activities concerned with the design of jobs, tasks, and related equipment. Further, interest here is not limited to single task or single person job design but rather expands into multiple person complexes. This extension is covered under the general category of "organizational design" as described, for example, by Hendrick (1987). As will be shown later, this expansion exceeds that normally covered in texts of human factors engineering.

Manpower

The term "manpower" in Table 18-1 covers a large number of concerns about the total manpower demands of a given system and/or the aggregation of manpower demands across all systems for a given organization. The objective of these concerns is twofold. On the one hand, we are often told that "people are our most important resource." On the other hand, it also seems to be true that "people are our most expensive resource." Because of the cost of manpower, it is essential to try to predict how many people will be needed by systems and how much they will cost. This concern is translated into the use of quantitative prediction models. The structure of Rostker (1984) is followed in defining four general classes of models – manpower, personnel, assignment, and training. In all cases, these models attempt to predict system manpower, personnel, and training demands; supplies and costs at total system levels; and, where possible, to achieve less costly and more optimum solutions to the use of people in systems.

Personnel

For every system and for every organization, the people who come to those systems and organizations come with certain attributes. In particular, they come with certain skills, knowledges, and abilities (SKA) which would be useful to measure and use in system operation. Further, SKA data are essential to the selection of system personnel. In general, although not without exception, the greater those skills, knowledges, and abilities, the less the training resources that might be needed. Thus, it is important that ways be available to make SKA/training trade-offs so that better training can be provided. Finally, people must also bring a set of desires to participate in system performance (a property here termed "motivation"). A great deal of attention has been given to the control and manipulation of motivation.

Training

Billions of dollars are spent each year for system training, on-the-job training, and formal education programs and institutions. There is some reasonable concern that training be effective and place a minimum possible demand on resources. All training systems must be evaluated in terms of the principles of human learning and transfer of training and must be designed with respect to specific, well-defined training requirements and needs analysis. Table 18-1 is a gross, first-cut look at the training needs for MANPRINT personnel training. Unfortunately, at the present time, there appears to be much emphasis on high technology and very expensive media (e.g., interactive videodisc, multimedia training, full-mission simulation) which may be unnecessary to meet actual training requirements (cf., Swezey, Perez, & Allen, 1988). These are often "solutions" looking for a training problem and may be used because they "look good" rather than because they are training effective. Because of this problem and because some good-looking systems actually do not train well, it is important that *all* training systems be subjected to some form of training system evaluation.

System Safety

Predictions of potential problems in system safety begin with quantitative predictions of system reliability. It is important to predict what parts of the system will tend to fail, when they will fail, and what the consequences of those failures might be. So much is made of "human error" in system failure that all system design must consider what errors humans might make and what the possible results of these errors might be. As Rouse has described in this book (Chapter 8), zero error is probably not a reasonable design objective, and he states: "Instead, systems should be designed so that they are error tolerant in the sense that errors can occur without leading to unacceptable consequences." This concept represents a major approach to system safety planning. There are some who believe that systems should be designed so that failures and errors cannot occur, but there is much doubt that such a result can be obtained by system designers (cf., Petroski, 1985). Thus, preparation for system failures and safety problems, as well as training in safety for all systems personnel, is essential.

Health Hazards

With respect to human health hazards, the intensity of the search for environmental hazards is increasing. For example, the appearance of toxic substances or gases at the workplace needs to be anticipated. From other disciplines such as physiology and psychology, there has also been radically increased concern for psychological hazards or workplace stressors (cf., Smith, 1987). If physical stressors imposed by system safety problems are combined with environmental and psychological stressors, a serious and very difficult problem in design for health and safety may develop. On the one hand, better design can help reduce the probability of health and safety hazards; but, on the other hand, personal safety training and, where necessary, personal protection and equipment (cf., Moran & Ronk, 1987) can help prepare these professionals. When all else fails, entering the domain of product liability and the legal system – a world for which training should be provided – may be expected.

Basic Methods

The last category in Table 18-1 concerns basic methods and techniques with which the MANPRINT professional should be acquainted (see also Howell, Colle, Kantowitz & Wiener, 1987, for a different perspective). As in all sciences and technologies today, there is a continuing outpouring of relevant literature, handbooks, and data bases (cf., Boff, Kaufman, & Thomas, 1986; Boff & Lincoln, 1988) with which the practitioner and researcher should be familiar. In all of the areas above, there can be need for empirical investigations which would require skill both in experimental and quasi-experimental methods (cf., the classic Cook & Campbell, 1979). Increasingly, human tasks are being permeated by computer techniques, and the human-computer interface is becoming perhaps the most important single type of human task. The computer even aids in designing human-computer tasks, either by rapid prototyping (Helander, 1988) or by the use of quantitative models of the human operator (Williges, 1987).

Finally, the use of MANPRINT in system design and integration represents an investment of resources. One can reasonably ask the question: What is the potential return on this investment? One way of answering that question is through cost-benefit analysis (Corlett, 1988). This involves quantitative techniques from economics by which a prediction can be made of the potential return on investment from human factors, ergonomics, and MANPRINT techniques. There are some suggestions that these techniques can show and bring about outstanding returns on investment, but the modeling and measurement complexities can be so great that model applications can create more problems than they cure. Corlett (1988, page 102) has suggested that "To leave cost-benefit analysis to accountants, to economists, even to ergonomists, is to build in a cause for failure."[1] In this one area alone, then, this very complex technique could require very considerable education, training, and experience.

Considering Table 18-1 from the standpoint of education and training, several implications seem apparent. First, there are many traditional disciplines that could be involved such as economics, operations research, statistics, physiology, engineering, industrial design, computer sciences, management and organizational development, system safety, and psychology. MANPRINT thus becomes a cross-disciplinary area from the standpoint of education and training and requires the participation of many disciplines.

Second, this immediately raises the question of what discipline, if any, should assume primary responsibility for MANPRINT education and training. As discussed in the next section, current educational practice suggests a number of possible alternatives within traditional institutional settings.

Third, it seems doubtful that there will be the possibility of producing, at either the undergraduate or graduate level, a trained MANPRINT generalist. There is simply too much to know and learn. Although an objective would be to expose students to all MANPRINT domains, a curriculum that could encompass all the content areas in Table 18-1 appears to be overwhelming. Therefore, some interdisciplinary division of training patterns is likely. What can be provided in the future will depend in part on the existing institutional structure in education and training, which is discussed next.

HUMAN FACTORS AND MANPRINT TRAINING

The emergence of MANPRINT as a recognized, relatively coherent realm of expertise is comparatively recent. There is no easy way to determine the number of practitioners who would be considered, or who would consider themselves, "MANPRINT experts." However, there is a closely allied field with a long-standing concern for most of the issues encompassed under the MANPRINT umbrella. That field is human factors, a field from which information on trained manpower can be obtained.

Traditional Sources of Supply

As long ago as 1916, a conference held at Ohio State University encouraged "the engineer to understand the human factors with which we will have to deal." In 1938 that university became the first to offer a human factors degree program. That program was offered by the Psychology Department as an emphasis within Experimental Psychology. In the intervening years, a number of other psychology departments began human factors programs, and they have become a major source of human factors professionals.

A second major academic home of human factors programs is engineering. In 1953 a human factors focus was established in the Department of System Engineering at the University of Pennsylvania, a program that continues to produce human factors specialists. Human factors programs are now offered by a number of other engineering departments as well. Industrial engineering departments have proven particularly compatible with human factors training and are now the second major source of such training in the United States and Canada according to data extracted from a recent survey (Sanders & Smith, 1988).

In the absence of specific data pertaining to MANPRINT as a whole, the best estimate of the available supply of MANPRINT manpower will be based on data pertaining to the availability of human factors professionals emerging from graduate degree programs. Dependence on such data for the purposes of these discussions is not meant to imply that these are the only possible sources of well-trained, accomplished, and productive MANPRINT professionals; some other potential paths will be discussed later.

The Human Factors Society (HFS) periodically issues a directory of human factors graduate programs offered in the United States (Sanders & Strother, 1982; Sanders & Strother, 1985) and most recently the United States and Canada (Sanders & Smith, 1988). Data in the following discussions are taken from these directories since they provide information on which institutions, and which departments within these institutions, offer human factors programs; on how many degrees are granted in these programs; and on how many students apply to each program, are accepted into it, and are actively enrolled there.

Program Diversity

One of the things reflected in the HFS directories is the diversity of nomenclature used in referring to the field. Human factors, human factors engineering, ergonomics, engineering psychology, human factors psychology, industrial/organizational psychology, and man-machine-environment systems all appear in program titles. The preferred terms differ from time to time, from place to place, and from person to person, but none challenges the underlying reality that this is a field which concerns itself with the man-machine-system-environment interface, and that field is, by whatever name, essentially and necessarily multidisciplinary.

It is the need to integrate various disciplines that accounts for the diversity of academic sponsors that human factors enjoys (or suffers from, depending on one's views). Of the 65 human factors programs listed in the 1988 HFS directory, 30 are in engineering departments, 27 are in psychology departments, 4 are offered jointly by an engineering and psychology department, and 4 are offered by departments other than engineering or psychology. Table 18-2 lists the specific department names and the number of programs associated with them. Seven institutions have separate programs under two different departments; they have been tabulated separately in the table.

The number of human factors degree programs has gone from 1 in 1938 to 65 in 1988. During the early years of these programs and continuing through 1970, an approximate equality was maintained between the number of programs in psychology departments and the number in engineering departments. In the early 1970s, most of the new programs were begun by engineering departments (Sanders, 1982). As Table 18-3 indicates, this imbalance has diminished during the present decade and engineering and psychology are once again approaching a parity with respect to the number of human factors programs they offer.

In subsequent discussions, data on the joint programs are not tabulated separately until the 1988 data. Prior to that, joint programs were categorized as primarily engineering or psychology, depending on the background of the students and other criteria, and their data combined as appropriate. Also, some institutions (five in 1982 and 1985, seven in 1988) offered separate programs in two different areas. As in Table 18-2, their data were tabulated separately.

Program Output

Some programs offer master's degrees only, others offer doctorates, and many programs offer both. Data compiled from the 1988 HFS directory of human factors graduate programs (Sanders & Smith, 1988) were combined with summaries of the earlier periods (Sanders, 1982, 1985) and are presented in Table 18-4. As the table shows, there are approximately three times as many master's degrees awarded as there are doctorates. That ratio appears fairly stable. The balance between engineering and psychology

Table 18-2
Current Human Factors Graduate Training Programs, 1988
United States (59 Programs) and Canada (6 Programs)

Academic Area	Department Name	Number of Programs	
ENGINEERING			**30**
	Industrial Engineering	14	
	Industrial and Systems Engineering	4	
	Industrial and Management Systems Engineering	2	
	Industrial Engineering and Operations Research	2	
	Industrial and Operations Engineering	1	
	Mechanical and Industrial Engineering	1	
	Engineering	1	
	Engineering and Computer Science	1	
	Engineering and Applied Science	1	
	Mechanical Engineering	1	
	Systems Engineering	1	
	Systems Design Engineering	1	
PSYCHOLOGY			**27**
	Psychology	24	
	Engineering Psychology	1	
	Human Factors	1	
	Human Biology and Psychology	1	
JOINT			**4**
	Industrial Engineering and Psychology	1	
	Industrial Engineering and Psychological Science	1	
	Psychology and Mechanical Engineering	1	
	Psychology and Mechanical-Industrial Engineering	1	
OTHER			**4**
	Management Sciences	1	
	Operations Research	1	
	Design and Environmental Analysis	1	
	Kinesiology	1	
	TOTAL		**65**

Table 18-3
Number of Human Factors Program Offered

DIRECTORY YEAR PERIOD COVERED	1982 1978-1980	1985 1982-1984	1988 1985-1987
PROGRAM AREA			
Engineering	27	28	30
Psychology	18	19	27
Joint	4	5	4
Other	0	5	4
TOTAL PROGRAMS	**49**	**57**	**65**

departments as degree sources is more nearly equal at the doctorate level than at the master's level, where engineering shows a consistent numerical advantage.

Although the overall number of degrees granted continues to increase with each new survey, the rate of increase appears to be slowing. Whereas the 1985 data show an increase of 26 percent over 1982, the increase from 1985 to 1988 is under 10 percent. Several limiting factors could be at work. The capacity of the training institutions, the interests of the student population, or the demands of the marketplace may all play roles in the number of human factors degrees being earned. Next, the number of openings available in human factors graduate training programs and at the level of student applications for those positions as indicators of the first two factors will be examined. The demand for those students after they are trained, and the degree to which industrial, governmental, and commercial institutions actualize that demand by funding jobs for human factors specialists, may be more important limiters. That support may also be the final determiner of whether or not the MANPRINT approach is actually implemented because training such specialists is clearly not enough. Their expertise must also be put to use in the many ways discussed earlier in this chapter.

Table 18-5 presents data relevant both to the capacity of the present human factors training system and to the available supply of students. It shows enrollment figures by program area at three points during this decade – 1980, 1984, and 1987. For both 1980 and 1984, students in joint programs were assigned either to engineering or psychology, so the decrease in student enrollment that engineering human factors programs experienced in the middle of the decade would have been made up by 1987 had the students in joint programs been distributed as they had been before. Growth in the number of students enrolled in the psychology human factors programs continued throughout the decade. When all program areas are taken together, the percentage increase in student enrollment from 1984 to 1987 was more than four times as great as the increase from 1980 to 1984.

Table 18-4
Number of Human Factors Degree Granted
(Sources: Sanders and Strother, 1982, 1985; Sanders and Smith, 1988)

DIRECTORY YEAR	1982	1985	1988
PERIOD COVERED	1978-1980	1982-1984	1985-1987
MASTER'S DEGREES			
Engineering	265	320	282
Psychology	118	159	149
Joint			37
Other			42
TOTAL	**383**	**479**	**510**
DOCTORATES			
Engineering	52	78	74
Psychology	62	67	75
Joint			19
Other			6
TOTAL	**114**	**145**	**174**
TOTAL DEGREES			
Engineering	317	398	356
Psychology	180	226	224
Joint			56
Other			48
TOTAL DEGREES (ALL AREAS)	**497**	**624**	**684**

Table 18-5
Human Factors Students, Active and Potential, 1980s

PROGRAM AREA	Engineering	Psychology	Joint	Other	TOTALS
Enrollment					
1980	448	352			**800**
1984	419	420			**839**
1987	421	444	79	81	**1025**
Openings					
1980	N/A	N/A			**300+**
1984	N/A	N/A			**N/A**
1987	194	183	35	29	**441**
Applications					
1980	429	314			**743**
1984	602	451			**1053**
1987	374	620	74	47	**1115**
Acceptances					
1980	175	109			**284**
1984	343	126			**469**
1987	232	150	29	19	**430**
Acceptance Rate					
1980	41%	35%			**38%**
1984	57%	28%			**45%**
1987	62%	24%	39%	40%	**39%**
	Engineering	Psychology	Joint	Other	TOTALS

N/A = not available

Applications to engineering programs show a 38 percent decline from 1984 to 1987, while applications to psychology programs show a 37 percent increment during the same period. Despite an increase in the total number of openings, applications continued to stay well ahead of places to accommodate them. Only three engineering programs had fewer applicants than openings, and one of those programs listed its number of openings as "infinite." (That program, however, accepted only eight students in 1987.)

With respect to individual programs, acceptances matched openings (in number) in nine engineering and ten psychology programs. Five engineering and five psychology programs accepted more students than the number of openings they listed. Eleven engineering and ten psychology programs had more openings than they filled. In general, a lack of applicants does not seem to be limiting the growth of human factors graduate training programs.

The acceptance rate in engineering programs is increasing gradually, and that in psychology programs is decreasing. Fewer than one of every four applicants to psychology programs is accepted. Whether student quality, program capacity, or some combination of these or other factors accounts for the rejection rates cannot be established from these data, but the number of unfilled openings in psychology programs together with the high rejection rate suggests at least a mismatch between student preparation or characteristics and the expectations of the institutions.

Prospects for Future Growth

Whether in terms of number of programs, number of degrees granted, or number of students enrolled or applying, continuing growth in human factors graduate training programs is evident. Different forms of growth should be distinguished, whether in looking at the past or forecasting the future. Koch (1980) surveyed psychology human factors programs and reported an anticipated increase of 13.75 percent per year in the rate of graduate applications for the subsequent five years.

Overall, the growth in psychology human factors programs did not keep pace with that prediction, but it would be wrong to conclude too quickly that the estimates Koch reported were incorrect. If only the schools that contributed data to Koch's (1980) survey are considered, and if the projection is carried out to seven years rather than five so that a comparison can be made with the available 1987 data, those selected psychology programs came very close to fulfilling Koch's expectations. The average number of applications in 1987 for the specific subset of programs surveyed by Koch matched the prediction for the 1986 level of applications to the first decimal (31.43 actual mean vs. 31.45 predicted mean). Whether or not growth will continue at that rate probably depends more on demand than on any other single factor.

Thus, one form of growth is the growth in size of existing programs, and it appears that such growth can be predicted fairly accurately. A second type of growth is an increase in the number of institutions and departments offering human factors training. The tables and discussion in this chapter are based on a combination of both types of growth. Growth in the quantity and quality of student applicants is yet another type of growth, one that may produce pressure for an increase in numbers of programs as well as their student capacity.

Looking at the history of human factors programs during the 1980s indicates that there has been growth in each of these areas and that there is a potential for continued and even greater increases if the demand for graduates increases. The present level of production of professionally trained entry-level human factors practitioners is over 200 per year, but that is certainly not the ceiling.

Human Factors/MANPRINT Curricula

Ideally, the curricula of each of the human factors programs could be compared with the MANPRINT content areas and subareas listed in Table 18-1. From these comparisons, an estimate could be made of the comprehensiveness of the programs with MANPRINT as the criterion. Unfortunately, practical realities intrude. Such a comparison could start with the categorization of all required and elective courses offered by each program. Most programs listed in the HFS 1988 directory of programs (Sanders & Smith, 1988) noted the titles of their required courses, but as Sanders (1982, page 4) himself commented, "We attempted to categorize the courses, but after several hours of effort, realized this was futile. Never have so few found so many creative ways of saying 'human factors'."[2]

A comparable problem was encountered in simply trying to judge whether or not MANPRINT areas were covered on the basis of listings by title in the directory and in other more extensive descriptions of curricula of a subset of the programs. The inferences required in going from titles to judgments about course content are substantial.

While such a comprehensive approach to characterizing program content proved impractical here, some comments with respect to the treatment of MANPRINT areas of these programs are appropriate. Certainly a program requiring one or fewer human factors courses cannot be expected to cover even one of the MANPRINT areas very thoroughly, and the 1982 survey showed that 58 percent of the psychology human factors programs fit that description (Sanders, 1982). There has been improvement. The 1988 listing shows that only 4 of the 65 programs require one or fewer courses, and another 3 state that the curriculum is custom designed. Thus, 7 programs (11 percent), counting all areas, now have no specific course requirements. Another 3 list requirements only in terms of units. The fewer the requirements, the less possible it is to specify any core of knowledge possessed by a nominal human factors graduate.

Not surprisingly, the one MANPRINT area best represented in the curricula is the one listed first in Table 18-1, i.e., human factors engineering. It is difficult to know what the content of a course is from the title alone, but courses such as "Human Factors," "Human Factors Principles and Applications," and "Human Performance" will clearly cover material relevant to the first MANPRINT area. Other courses such as "Psychology of Office Design," "Human Physical Capabilities," and "Ergonomics in Work Design" are more specific and thus more reliably relatable to specific subareas of Table 18-1, but they are also less widely offered.

The other content area listed in Table 18-1 that is well represented in the curricula of most of the human factors programs is the final one, basic methods. That is not to say that all of its listed subareas are well covered by most programs, but only that most programs do provide courses to introduce the student to the statistical and experimental principles that they need to understand the field and its literature and, in some cases, to contribute to its growth. Computer methods, models, and applications are also the subject of a variety of courses now available.

If human factors engineering and basic methods are the areas best covered by the curricula of the human factors programs as a group, manpower, the second listed area, seems to be the most neglected. None of the listed courses could reasonably be inferred to provide training in the concepts or the models important for making manpower evaluations and predictions. Most human factors graduate training courses were as silent about personnel concerns as they were about manpower. However, there are courses sprinkled here and there called "Motivation," "Learning and Motivation," "Principles of Psychological Testing," and "Psychometric Theory" indicating that there were elective courses dealing with the third MANPRINT area available in some programs, and in one case "Personnel Psychology" was a required course.

Training, the fourth MANPRINT area, is not a major concern of most programs, but as with the area of personnel some elective courses are available. One in particular, "Training and Skill Acquisition," gave promise of providing a link between two areas, personnel and training.

System safety was quite well represented among the elective offerings. There were courses in "Human Performance and Accident Causation," "Industrial Safety," "Safety Systems Engineering," "Safety and Work Physiology," "Safety Engineering," and just "Safety." One program is associated with an Institute of Safety and Systems Management and has several research laboratories available to its students (e.g., a Safety and Industrial Hygiene Laboratory and a Head Protection Laboratory). While not generally offered by the human factors training programs, a focus on safety is apparent at several institutions.

Health hazards, a closely allied MANPRINT area, is not represented by as wide a range of offerings as is system safety, but many of the issues treated in some of the courses cited above can be expected to impinge on health issues too. In addition, there are electives available on "Industrial Hygiene" and "Environmental Hygiene Engineering." While not typical, such courses indicate that health and safety are specific concerns of some of the graduate human factors training programs.

For a closer look at the MANPRINT coverage provided by some specific programs, three institutions were used – one in the East, Virginia Polytechnic Institute and State University; one in the Midwest, Ohio State

University; and one in the West, the University of Southern California. The latter two institutions offer dual programs in two separate departments. Thus, for each of five programs, the course requirements and the elective courses available as options were related to the MANPRINT areas listed in Table 18-1.

Table 18-6 reflects these relationships. Liberal interpretations of available descriptions were used, but an attempt was made to avoid obscuring differences which did seem to exist. These summaries are not meant to indicate that one program or another is better than the rest, but simply to indicate how programs very reasonably categorized as human factors programs differ in their emphasis and in their coverage of the various MANPRINT areas.

As Table 18-6 indicates, and as noted earlier, requirements tend to focus on basic methods and the first MANPRINT area, human factors engineering. Some of the material subsumed under those areas may also provide background and preparation relevant to the other areas, but if a specific

Table 18-6
Relationship of Selected Programs to MANPRINT Areas

MANPRINT Area	Ohio State University		University of Southern California		Virginia Polytechnic Institute and State University
	ISE	Psych	HF	ISE	
Human Factors	X	O	X	X	X
Manpower					
Personnel	O		O		
Training			O	O	X
System Safety	O		O	O	O
Health Hazards			O	O	O
Basic Methods	X	O	X	X	X

KEY *X = Required*
O = Optional
ISE = Industrial and Systems Engineering

course requirement or available elective did not relate specifically to that area, it is not marked. The comments made about human factors curricula in general pertain also to the programs cited in Table 18-6. That table sums up the strengths and weaknesses of human factors curricula as they relate to MANPRINT content areas. The areas that need augmentation if a MANPRINT foundation is to be adequate are obvious.

Human Factors Texts

In addition to programs and curricula, another way to look at what information is provided for training is in terms of source books and texts. That is, what information is being provided for the general training and technical audiences in the source books and texts published in human factors?

The training content areas, as defined in Table 18-1, are shown again in Table 18-7. In 1987 a major source book for human factors was published, Salvendy's (1987) *Handbook of Human Factors*. This book has some 1,874 pages with 68 technical chapters. It is clearly intended as a comprehensive summary of the field with detailed technical reviews in each of the 68 chapters. In Table 18-7, Salvendy is listed first in the "Authors" columns to serve as a basis for relative comparison with the training content areas and with the human factors texts.

To represent human factors texts, six texts from the past decade have been selected: Huchingson (1981), Oborne (1982), Kantowitz and Sorkin (1983), Wickens (1984), Sanders and McCormick (1987), and Adams (1989). Texts have been selected that represent both the United States and foreign points of view (with Oborne from England) and a variety of approaches from a more basic point of view of engineering psychology (e.g., Wickens, 1984) to a highly applied human factors engineering point of view (e.g., Huchingson, 1981).

It should be said immediately that no comparative "evaluation" of the texts is intended. They were written for different purposes and for different audiences. All seem purposely restricted in length to stay within reasonable space limitations for texts. Huchingson (1981, page 30) includes a typical qualification in particularly excluding selection and training of personnel in favor of design-oriented material. And Wickens (1984, page 3) states specifically: "Designing machines that accommodate the limits of the human use is the concern of a field referred to as human factors. The field is a very broad one – broader than this book, which is focused specifically on designing systems that accommodate the information-processing capabilities of the brain."[3] But these differences are useful because they help to indicate the kinds of information a variety of authors present for the field of "human factors" and presumably part of MANPRINT as well.

Based on Table 18-7, a number of comments can be made: If the training content subareas are at the least a minimal representation of this critical

Table 18-7
Training Content Areas and a Sample of Sources

TRAINING CONTENT AREA	AUTHORS*						
	S (1987)	H (1981)	O (1982)	K/S (1983)	W (1984)	S/M (1987)	A (1989)
• HUMAN FACTORS ENGINEERING							
– Psychological and physiological capabilities, limitations and tolerances	X	X	X	X	X	X	X
– Mission, function, and task analysis	X	X				X	X
– Anthropometric and biometric criteria	X	X	X	X		X	X
– Display-control task design	X	X	X	X	X	X	X
– Workspace requirements and design	X	X	X	X		X	X
– Organizational design	X						
• MANPOWER							
– Human resources systems predictions							
– Manpower models							
– Personnel models							
– Assignment models							
– Training models							
• PERSONNEL							
– Skills, knowledges, and abilities analysis (SKA)	X			X	X		
– Personnel selection	X						X
– SKA/training tradeoffs		X					
– Personnel quality and performance predictions	X		X				
– Motivations, incentives, and performance	X						
• TRAINING							
– Human learning and transfer of training	X				X		X
– Training requirements and needs analysis	X						
– Instructional system design	X						
– Training media and devices	X	X					X
– Training system evaluation	X						X
• SYSTEM SAFETY							
– System reliability analysis	X			X			X
– Human error analysis	X	X	X	X	X	X	X
– System safety planning	X					X	
– Safety training	X	X					
• HEALTH HAZARDS							
– Environmental stressors identification	X	X	X	X		X	
– Psychological stressors identification	X	X	X	X	X		
– Designing for health and safety	X	X	X				
– Personal protection and equipment	X	X					
– Controlling workplace hazards	X	X	X				
– Product reliability and liability	X	X		X			X
• BASIC METHODS							
– Literature retrieval							
– Handbooks and data bases	X						
– Experimental and quasi-experimental methods	X	X	X			X	X
– Statistics and quantitative methods							
– Computer methods and applications	X	X		X	X	X	X
– Man-machine system models and simulations	X	X		X	X	X	
– System and performance measurement – estimation and evaluation of performance	X	X	X			X	
– Cost-benefit analysis			X				
– Engineering approaches to system design	X					X	X
– Sociotechnical system design	X						

* S = Salvendy (1987); H = Huchingson (1981); O = Oborne (1982); K/S = Kantowitz and Sorkin (1983); W = Wickens (1984); S/M = Sanders and McCormick (1987) ; A = Adams (1989)

MANPRINT technical area, then Salvendy (1987) certainly covers a very large number of them. The only major exception is in the manpower domain. And there are three secondary omissions: SKA/training trade-offs, literature retrieval, and cost-benefit analysis. Additionally, there are many more things in Salvendy (1987) beyond those content areas listed in Table 18-7.

None of these seven texts shows any apparent interest in the manpower area. This is surprising since the results of human factors engineering job design can have a tremendous impact on system human resource requirements in terms of the quantity and quality of personnel as well as system training. Indeed, it is in the manpower area that one can show major cost savings in total systems (cf., Corlett, 1988). Unfortunately, this omission in the texts is another example of what appears to be a disconnect between the human factors and the manpower technical communities.

The texts included in Table 18-7 tend to cover personnel variables only insofar as they directly interact with human engineering design problems. As just noted in the case of Huchingson, most of these authors are principally interested in system design variables. The same general comment may be made of the training area. But here there is sometimes a design concern with training devices, since the design of a training device is as much a design problem as the task for which the operator is being trained.

The texts principally concentrate on three major areas: human factors engineering, system safety, and health hazards. The latter two are not surprising because of three major trends over the years. First, for a long time there has been a fundamental interest in the study of human error. Second, a major initial impetus to starting the field of human factors engineering was the problem of environmental stressors – noise, temperature, and acceleration problems have plagued many operator tasks for decades. Third, many specialists are being called upon to participate as expert witnesses in product liability suits and investigations.

Human factors has traditionally concentrated on the single operator/ maintainer job design problem. It is not that the practitioners in the field were unaware of teams, groups, and organizations, but that there were perhaps more than sufficient problems in the design of each workstation and job to justify that concentration. At the same time, some researchers were concerned with larger units and organizations, so that organizational design and the larger sociotechnical system became the focus of interest (Hendrick, 1987). This trend most naturally moved into problems of organizational development and assessment and theories and practice of organizational structure and process. Table 18-7 shows that Salvendy (1987) is responsive to this part of the technology but that none of the texts are. It might be useful in the future for human factors text writers to include some consideration of both organizational and manpower problems.

As shown in Table 18-7, there is substantial interest in basic methods which reflects the fact that human factors has been concerned with the sophistication, reliability, and validity of its methods from the very beginning of the field. One of the most positive aspects of this field has been the constant concern with sound scientific methodology, particularly where very complex phenomena must be faced.

At some future time, it will be desirable to have a text or source book covering all the MANPRINT content areas listed in Table 18-7. It is apparent from this table that none of the seven books covers all MANPRINT concerns adequately. These books were not intended to. Authors of future human factors texts might consider expanding their texts to encompass a broader MANPRINT base. Such texts would be useful for the classroom student, the researcher, the practitioner, or the interested and involved non-specialist.

RELATED EDUCATION AND TRAINING ISSUES

To this point, the possible sources of trained manpower, what they might have been trained for, and from where they might have come have been the concern. On the other side of the manpower equation is the issue of predicting future manpower demands. That is, how many trained people will be needed in the future? In addition, there is the question of what knowledges, skills, and abilities employers will demand from future employees. But first the concern is simply the problem of estimating quantity of demand.

Manpower Demands

Predicting the future for anything is difficult, and apparently generating valid predictions of labor demands and requirements is extraordinarily hard. So much depends upon the fundamental status of an economy about which many make predictions but few seem to be right. With respect to demands for technically trained workers, much depends upon advances in technology development. For example, the development and application of the digital computer over the past two decades has created a demand for completely new areas of technological skills most of which were quite unexpected 35 years ago (cf., Muckler, 1987). Specific technological initiatives (e.g., the manned space program) can create a very high demand for workers trained in the problems of human-system interactions. As a directly applicable example, Booher (1988) has predicted a significant increase in demand for human factors/MANPRINT trained personnel as a function of the expansion of the MANPRINT programs in the U.S. Army.

For the past 30 years there has been a unique increase in the demand for highly skilled scientific and engineering personnel. The human factors community has experienced the same type of rapid rise. For example, in 1962 the Human Factors Society had slightly fewer than 1,000 members; by 1982, it had grown to over 3,000 members. Klemmer (1989) has extrapolated the growth figures of the Society, and he has estimated 100,000 members by the year 2022.

Further, if the 1980s are any indication, it would appear that the labor market is able to absorb the labor supply in human factors. Sanders (1982) estimated that there would need to be 150-200 jobs per year to absorb graduating students and that this demand (Sanders, 1982, page 4) ". . . presents a continuing challenge to the profession to provide employment opportunities for these new professionals."[4] It would appear that the demand has met the supply.

What might be very useful is a major national survey to estimate both future demand and supply for human factors/MANPRINT professionals. In 1988, the Committee on Human Factors of the National Academy of Sciences/National Research Council initiated such a survey (Van Cott, 1988). The results will be available in 1990 and are intended to answer the following questions:

> How many human factors professionals are there in the United States? How many will be needed in the future? Can existing educational and training programs meet the projected need? What specialized knowledge and activities are wanted by employers of human factors professionals? Are these expectations being met by the present professional work force and by graduates of academic human factors degree programs? What are the important career problems and ethical issues confronting the profession?[5]

Data providing some answers to these questions will be very useful.

Finally, while the concentration here has been on the formal human factors programs in the United States and Canada, mention could be made of the fact that there are many potential sources of manpower supply beyond those specific programs. Some indication of this comes from the Human Factors Society directory of members and is summarized in Table 18-8. The variety of possible academic specialties is apparent. Should the demand increase radically, there are several possible sources of increased supply. However, if the demand should increase, advanced notice is really desirable so that the supply sources can expand to meet that demand. It might take, for example, from two to five years to produce a major increase in graduates. Demand may fluctuate rapidly, but supply of highly skilled personnel is not, and perhaps cannot be, immediately responsive.

Certification and Licensing

Once a labor pool of highly skilled professionals is established, a way of indicating to the public that they are (1) skilled and (2) skilled in something

Table 18-8
Educational Background of HFS Membership from Human Factors Society Directory, 1988, page 263

ACADEMIC SPECIALTY	Percent
Psychology	45.6
Engineering	16.2
Human Factors/Ergonomics	6.3
Industrial Design	2.5
Medicine/Physiology/Life Sciences	1.9
Education	1.6
Business Administration	1.4
Computer Science	0.9
Other	11.6
Students	11.8
Unspecified	0.2

specific may be inevitable. The issue is certification and/or licensing of human factors/MANPRINT professionals where "certification" usually is the weaker form of regulation and restricts the use of a professional title. "Licensing," on the other hand, usually implies some specific kind of skill level and requires some form of examination.

The basic purpose of certification and licensing is said to be the protection of the public. The practice goes back at least 3,000 years to Hammurapi and the Babylonians (Oates, 1979, pp. 180-183, on the practice of medicine). It should be understood that this is an extraordinarily controversial subject, and there are those who argue that certification and licensing protects no one except guild members and may operate to the detriment of the public as well as practitioners (cf., Danish & Smyer, 1981; Gross, 1978; Koocher, 1979; Wiens & Menne, 1981; Herbsleb, Sales, & Overcast, 1985). It has been an issue of major concern to human factors practitioners (cf., Siegel, 1980; Blanchard, 1985). Perhaps the best that can be done here is to describe some of the difficult issues raised by certification and licensing.

- It is possible to apply the process either to individuals or organizations. Normally, it is the individual who will be certified or licensed. An alternative would be blanket approval for all those who graduate from an accredited educational program. What agency would do the accreditation has been the subject of disagreement.

- For individuals making their initial entry into a field, there is the question of what kinds of requirements should be established for either certification or licensing. For example, some fields require a formal, written, and often very rigorous, examination. Another alternative would be an oral examination of some form. Further, it may probably be assumed that some minimum level of experience and/or education would be required.
- There must be some written set of standards or ethics for what constitutes good and bad practice. There should be some form of statement as to what constitutes "harmful" activity. The American Psychological Association has a detailed code of ethics and defines at length acceptable and unacceptable behavior. The Human Factors Society has also recently ratified a statement of ethics.
- There have been many questions as to who should perform the certification and licensing. Paramount is the legal desire of the state. If the state deems the profession potentially harmful to the public, the state will probably provide formal mechanisms for certification and licensing. Even then, the state rarely does so without the approval and assistance of professional societies and organizations. In some cases, there will be both professional and public representation. For many professions (e.g., medicine) the state provides certification and licensing which may create some problems in jurisdiction when practicing in different states (cf., Henderson & Hildreth, 1965) or in standardization of requirements.
- One useful question is: What do the people in the field want? Rarely are current practitioners asked, but in one case they were. Siegel (1980) queried human factors professionals. His findings are interesting. First, current professionals preferred certification over licensing, principally on the practical grounds of the investment required to do licensing. Second, a master's degree was felt to be a sufficient level of education for certification (even then 10 percent of the respondents felt that degrees were either not required or "immaterial"). Third, there was little agreement on the most appropriate education background. Fourth, practitioners felt that experience should be substituted for formal education where appropriate. Perhaps the last two items reflect, in part, the fact that several major contributors to the field of human factors have not come from traditional educational backgrounds.
- A final issue is how frequently an individual should be assessed for his or her competency. For some, initial assessment is sufficient. But there has been a growing social feeling that more frequent assessments might be useful to see if competence is maintained. There appears to be no particular agreement on what form of assessment might be best. One could require formal, written and/or oral examinations. Another method could be the submittal of performance and achievements to an assessment board. A third way could be a requirement to return to an educational institution at certain prescribed times for certain required number of courses. The latter raises the issue of continuing education.

Continuing Education

The assessment of individual competency at any time in a career assumes that the individual will have resources for sustaining lifetime career competency from initial entry into the profession to retirement. The assumption can reasonably be made that the technical areas listed in Table 18-1 will continue to show improvement in knowledge and method. Furthermore, the assumption can be made that unless the individual keeps aware of the enhancements in the field, he or she will suffer professional obsolescence.

There is some empirical evidence on the decay of professional skills, at least in engineering (cf., Zelikoff, 1969; Dalton & Thompson, 1971). It has been estimated that the half-life of the engineering baccalaureate in 1955 was about seven to eight years, in 1960 about five years (Zelikoff, 1969; Kaufman, 1978). Assuming a linear model as the best fit to these two data points, all engineers after 1970 should have returned to the classroom immediately after graduation. That is, by the time they had graduated in 1970, half of what they had learned was no longer applicable. One would expect that there are actually three critical factors in obsolescence: (1) the rate of change of the technology, (2) the rate of forgetting of the individual professional, and (3) the work the professional is doing – hopefully, he or she is maintaining as well as increasing competency while performing.

There are many alternatives for providing continuing education: reading books and journals, attending conventions and workshops, audio-visual programs such as cassettes or videotapes, university short courses, in-house courses at the place of employment, and formal university courses. There are at least four general problems that might be considered:

First, should the professional be allowed to select what he or she thinks is the educational need and the mechanism, or should the alternatives be dictated? If the latter, by whom?

Second, how does one evaluate the effectiveness of the alternatives? One common measure is the student's evaluation of the instructor and the course effectiveness. Surely, some additional measures are desirable.

Third, do any or some combination of these alternatives work in terms of job performance? With respect to engineering skills, Kaufman (1978) compared the relative effectiveness of two kinds of training, in-house versus formal university graduate courses, to on-the-job performance ratings of 110 engineers in research and development (R&D), applied development, and manufacturing organizations. The only significant effect was in favor of formal graduate courses, provided the engineer was in an R&D group. In-house courses were at best "promising" with respect to future job performance ratings. Moderating all of this is the attitude of the organization towards continuing education. It seems probable that the organization must be supportive and must provide resources if any form of continuing education is to be worthwhile.

Fourth, and last, there seems to be no consistent responsibility shown by any organization to provide the kind of quality continuing education material that people would need. This could be very expensive, particularly if high technology training media were used. It is hoped that responsibility will be assumed by some organization or combination of organizations in the future.

Centers of Excellence

One particular source for good continuing education materials of all kinds might be human factors/MANPRINT centers of excellence. As such, these do not exist today. However, one could imagine regional centers of excellence specializing in advancing the state of the art, providing a wide range of educational resources from short courses to formal course programs, and running frequent seminars and colloquia open not only to professionals but to interested non-professionals. They could provide major forums of debate where significant policy and technical issues could be discussed publicly and in-depth. And, as just noted, they could serve as coordinated centers for continuing education with a very large variety of educational options. It is important that they not be limited to traditional academic teaching methods only.

When one looks at the many education and training programs now in existence (Sanders & Smith, 1988) which have previously been examined, many potential candidates – based on programs, longevity, size, and staff – could be listed for roles as centers of excellence. One could consider, but certainly not limit it to, the following places: Ohio State University, University of Illinois, University of Southern California, Virginia Polytechnic Institute and State University, Georgia Institute of Technology, and Old Dominion University. These universities and their faculties have made major contributions to human factors and closely related fields.

One might suggest also that a center of excellence would be very much strengthened by a close association with regional industry and, to the degree it would be appropriate, governmental agencies. Rather than being a part of a traditional university department, the center of excellence might best, organizationally and administratively, be a part of a university foundation or institute. It is also assumed that there could be interuniversity ties within the region. The aim is to concentrate all possible resources at a focal point without regard to institutional setting with the objective of really serving the profession and the public.

The first question most likely will be: Who pays for all of this? While the federal government is a natural candidate, there could well be major regional interests that could supply substantial sources of funding. Indeed, for the purposes of the center of excellence, multiple sources of income may be the safest financial approach. Dependence upon a single source or just a few sources of funding, however well-intentioned, may not be successful;

note the case of Microelectronics and Computer Technology, the University of Texas, and the state of Texas (as described in Williams, 1988). Here the entire political, university, and industrial communities were involved, a combination which could not overcome a major swing downward in the fundamental economy of the region.

"Training" for the Non-Specialist

The basic focus of this chapter has been formal education and training as expressed centrally around the academic setting. But there are other possible "training" requirements and needs that center around those who would interact with human factors professionals. There are a number of other communities with which interactions are essential and for whom some degree of "training" could be worthwhile.

Traditionally, the human factors professional can be seen as a bridge between the user (operator/maintainer) and the system and technology. In a sense, the human factors professional represents the user in seeing that the design is appropriate to the user. It is important, therefore, to "train" the user not only in his or her tasks but also in what the field can and cannot provide. While national advertising is increasingly using such words as "ergonomics" and "human factors," it is doubtful that the general public has any specific understanding of those terms.

Another major community with which the human factors practitioner must interact is the general engineering community. In system design, for example, there will be many engineering disciplines represented, and it is essential that they be "trained" as to the contribution of the human factors/MANPRINT activity. Even within the discipline there is considerable disagreement about the understanding of design engineers, system designers, and engineering program management as to the contributions of human factors/MANPRINT.

Finally, there is the higher-level of leadership and management that must often make basic commitments to the allocation of resources. These individuals rarely understand the technology or the terminology, and a great deal of translation is necessary. Furthermore, as mentioned above, leaders will often decide allocation on the basis of some kind of estimates of return on investment. It is critical, therefore, that some kind of answer be provided to the question: What return will they get from any investment in education and training?

In modern industrial and sociotechnical systems, it is impossible for a discipline to exist without essential interaction with others, many of whom may not understand what the discipline is trying to accomplish. From one point of view, it is the responsibility of the discipline to "educate and train" those other communities in which they exist. Certainly no one else has a greater vested interest in seeing it done and done well.

NOTES

[1]From Corlett, E. N. (1988). Cost-benefit analysis of ergonomic and work design changes. In D. J. Oborne (Ed.), *International reviews of ergonomics: Current trends in human factors research and practice* (Vol. 2) (p. 102). London: Taylor & Francis. Reprinted with permission.

[2]From the *Human Factors Society Bulletin*, Vol. 25, No. 11, 1982). Copyright 1982 by the Human Factors Society, Inc., and reprinted by permission.

[3]From Wickens, C. D. (1984). *Engineering psychology and human performance.* Columbus: Charles E. Merrill. Copyright Scott, Foresman and Company.

[4]From the *Human Factors Society Bulletin*, Vol. 25, No. 11, 1982). Copyright 1982 by the Human Factors Society, Inc., and reprinted by permission.

[5]From the *Human Factors Society Bulletin*, Vol. 31, No. 12, 1988). Copyright 1988 by the Human Factors Society, Inc., and reprinted by permission.

REFERENCES

Adams, J. A. (1989). *Human factors engineering.* New York: Macmillan.

Blanchard, R. (1985). Certification of human factors specialists. *Human Factors Society Bulletin, 28*(3), 1-2.

Boff, K. R., Kaufman, L., & Thomas, J. (Eds.). (1986). *Handbook of perception and human performance: Volume I, Sensory processes and perception; Volume II, Cognitive processes and performance.* New York: John Wiley & Sons.

Boff, K. R., & Lincoln, J. E. (Eds.). (1988). *Engineering data compendium* (3 volumes). Wright-Patterson AFB, OH: Harry G. Armstrong Aerospace Medical Research Laboratory.

Booher, H. R. (1988). Progress of MANPRINT–The Army's human factors program. *Human Factors Society Bulletin, 31*(12), 1-3.

Cook, T. D., & Campbell, D. T. (1979). *Quasi-experimentation: Design & analysis issues for field settings.* Chicago: Rand McNally.

Corlett, E. N. (1988). Cost-benefit analysis of ergonomic and work design changes. In D. J. Oborne (Ed.), *International reviews of ergonomics: Current trends in human factors research and practice* (Vol. 2) (pp. 85-104). London: Taylor & Francis.

Dalton, G. W., & Thompson, P. H. (1971). Accelerating obsolescence of older engineers. *Harvard Business Review, 49*(5), 57-67.

Danish, S. J., & Smyer, M. A. (1981). Unintended consequences of requiring a license to help. *American Psychologist, 36*(1), 13-21.

Gross, S. J. (1978). The myth of professional licensing. *American*

Psychologist, *33*(11), 1009-1016.
Helander, M. (Ed.). (1988). *Handbook of human-computer interaction*. Amsterdam: North-Holland
Henderson, N. B., & Hildreth, J. D. (1965). Certification, licensing, and the movement of psychologists from state to state. *American Psychologist*, *20*(6), 418-421.
Hendrick, H. W. (1987). Organizational design. In G. Salvendy (Ed.), *Handbook of human factors* (pp. 470-494). New York: John Wiley & Sons.
Herbsleb, J. D., Sales, B. D., & Overcast, T. D. (1985). Challenging licensure and certification. *American Psychologist*, *40*(11), 1165-1178.
Howell, W. C., Colle, H. A., Kantowitz, B. H., & Wiener, E. L. (1987). Guidelines for education and training in engineering psychology. *American Psychologist*, *42*(6), 602-604.
Huchingson, R. D. (1981). *New horizons for human factors in design*. New York: McGraw-Hill.
Kantowitz, B. H., & Sorkin, R. D. (1983). *Human factors: Understanding people-system relationships*. New York: John Wiley & Sons.
Kaufman, H. G. (1978). Continuing education and job performance: A longitudinal study. *Journal of Applied Psychology*, *63*(2), 248-251.
Klemmer, E. T. (1989). Where will we be 65 years from now? *Human Factors Society Bulletin*, *32*(1), 7-8.
Koch, C. G. (1980). A survey of graduate education in human factors/engineering psychology. *Human Factors Society Bulletin*, *23*(2), 1-3.
Koocher, G. P. (1979). Credentialing in psychology: Close encounters with competence? *America Psychologist*, *34*(8), 696-702.
Moran, J. B., & Ronk, R. M. (1987). Personal protective equipment. In G. Salvendy (Ed.), *Handbook of human factors* (pp. 876-894). New York: John Wiley & Sons.
Muckler, F. A. (1969). Human factors research on weapon systems project teams. *Human Factors*, *1*(4), 28-31.
Muckler, F. A. (1987). The human-computer interface: The past 35 years and the next 35 years. In G. Salvendy (Ed.), *Cognitive engineering in the design of human-computer interaction and expert systems* (pp. 3-12). Amsterdam: Elsevier.
Oates, J. (1979). *Babylon*. London: Thames and Hudson.
Oborne, D. J. (1982). *Ergonomics at work*. Chichester: John Wiley & Sons Ltd.
Petroski, H. (1985). *To engineer is human: The role of failure in successful design*. New York: St. Martin's Press.
Rostker, B. D. (1984). Human resources models: An overview. In W. P. Hughes, Jr., (Ed.), *Military modeling* (pp. 187-200). Washington, DC: Military Operations Research Society.
Rouse, W. B. (1986). On the value of information in system design: A

framework for understanding and aiding designers. *Information Processing and Management, 22*(2), 217-228.

Rouse, W. B., & Boff, K. R. (Eds.). (1987). *System design: Behavioral perspectives on designs, tools, and organizations.* New York: North-Holland.

Salvendy, G. (Ed.). (1987). *Handbook of human factors.* New York: John Wiley & Sons.

Sanders, M. S. (1982). Human factors graduate education programs: The state of the union. *Human Factors Society Bulletin, 25*(11), 1-4.

Sanders, M. S. (1985). Human factors graduate education: An update. *Human Factors Society Bulletin, 28*(12) 1-3.

Sanders, M. S., & McCormick, E. J. (1987). *Human factors in engineering and design* (Sixth Edition). New York: McGraw-Hill.

Sanders, M. S., & Strother, L. (1982). *Human Factors Society directory of graduate human factors programs in the U.S.A.* Santa Monica, CA: The Human Factors Society.

Sanders, M. S., & Strother, L. (1985). *Directory of human factors graduate programs in the U.S.A.* Santa Monica, CA: The Human Factors Society.

Sanders, M. S., & Smith, L. (Eds.). (1988). *Directory of human factors graduate programs in the United States and Canada*. Santa Monica, CA: The Human Factors Society.

Siegel, A. I. (1980). Certification/licensing of human practitioners in industry. *Human Factors Society Bulletin, 23*(10), 1-3.

Smith, M. J. (1987). Occupational stress. In G. Salvendy (Ed.), *Handbook of human factors* (pp. 844-860). New York: John Wiley & Sons.

Snyder, H. L., & Eggleston, R. G. (1984). *Human Factors engineering in the Air Force, Task IV: Education, training and career development* (AFSC-TR-80-8). Washington, DC: Air Force Systems Command.

Swezey, R. W., Perez, R. S., & Allen, J. A. (1988). Effects of instructional delivery system and training parameter manipulation on electromechanical maintenance performance. *Human Factors, 30*(6), 751-762.

Van Cott, H. P. (1988). National Academy of Sciences to study human factors professionals. *Human Factors Society Bulletin, 31*(12), 5.

Wickens, C. D. (1984). *Engineering psychology and human performance.* Columbus: Charles E. Merrill.

Wiens, A. N., & Menne, J. W. (1981). On disposing of "straw people": Or an attempt to clarify statutory recognition and educational requirements for psychologists. *American Psychologist, 36*(4), 390-395.

Williams, F. (Ed.). (1988). *Measuring the information society.* Newbury Park, CA: SAGE Publications.

Williges, R. C. (1987). The use of models in human-computer interface design. *Ergonomics, 30*(3), 491-502.

Zelikoff, S. B. (1969). On the obsolescence and retraining of engineering personnel. *Training and Development Journal, 23*(5), 3-15.

SUGGESTIONS FOR FURTHER READING

Salvendy, G. (Ed.). (1982). *Handbook of industrial engineering.* New York: John Wiley & Sons.

Salvendy, G. (Ed.). (1987). *Handbook of human factors.* New York: John Wiley & Sons.

Sanders, M. S., & Smith, L. (Eds.). (1988). *Directory of human factors graduate programs in the United States and Canada*. Santa Monica, CA: The Human Factors Society.

Rouse, W. B., & Boff, K. R. (Eds.). (1987). *System design: Behavioral perspectives on designers, tools, and organizations.* New York: North-Holland.

CHAPTER **19**

MEETING THE CHALLENGE: FACTORS IN THE DESIGN AND ACQUISITION OF HUMAN-ENGINEERED SYSTEMS

Kenneth R. Boff

ABSTRACT

The design of effective military systems requires consideration of the skills, capabilities and fundamental performance limitations of end-user personnel involved in operations and maintenance over the life of the system. This concept, culminating in MANPRINT, is an important coalescence of system human engineering philosophy and concerns that have been evolving since World War II. Achieving a satisfactory match between system capabilities and the performance characteristics of users presumes the availability of requisite knowledge and the means to act upon it in the course of system design. The major challenges to broad acceptance and implementation of this concept lie in the ability to ensure an adequate supply of: (1) *Useful and Usable Data Resources*. Applicable MANPRINT data are needed that can be traded off against equipment variables; (2) *Effective Methods and Media*. Reliable means are needed to support the retrieval and application of people-oriented resources in the context of system acquisition and design. Also needed are methodologies and metrics for the test and evaluation of "MANPRINTed" system designs; and (3) *Sophisticated and Motivated Users*. Designers, design management and system acquisition personnel must believe that consideration of MANPRINT factors is "valuable" to the design of effective systems. Designing for the inclusion of these elements vitally depends on influencing individual, organizational and regulatory variables that jointly influence and support the design and acquisition processes.

INTRODUCTION

Previously developed material systems have not performed in the field as designed or desired because they were not designed with adequate considerations of the performance capabilities and limitations of their operators, maintainers, and support.

Lieutenant General R. M. Elton, U.S. Army (1986)

Despite spectacular technological advances in controls, displays, and information handling, the effectiveness of military systems remains inextricably linked to the performance capabilities of human operators and maintainers. Though few would argue with this assertion, there have been deep and long standing difficulties in translating this awareness into practice in the design and acquisition of military systems. Innumerable examples, data and prospective solutions germane to present system difficulties, abound in the post World War II military human factors literature. Yet problems in equipment operability, trainability, maintainability and safety stemming from deficits in human-systems integration are pervasive (General Accounting Office, 1981).

Steadily mounting concerns over the cost and readiness consequences of these problems have moved the Department of Defense (DoD) and other government agencies to take actions to improve the human factors effectiveness of the equipment and systems for which they are responsible (Elton, 1986; Johnson, 1987; Boff, 1987a; Office of Technology Assessment, 1988; Gentner, 1988; Booher, 1988). The fact that this "awareness" has suddenly broadened beyond human factors practitioners and their handlers to policy makers is the result of a timely confluence of social, economic and political developments:

- *Increased Public Sensitivity and Concern.* In recent years, in-depth press coverage of catastrophic incidents in air traffic management (Office of Technology Assessment, 1988), nuclear power plants (Rubinstein & Mason, 1979), etc., have raised the consciousness of the nation to the consequences of human error associated with system design and personnel training.
- *Higher Stakes in Military Systems Acquisition and Operations.* Military systems are increasing in complexity and sophistication. Correlated with this are increased skill and work demands on operators and system maintainers parallelled by a dramatic growth in acquisition, training, and maintenance costs (Boff, 1987b). Hence, both the risks and consequences of low system effectiveness have become unacceptably high. Investigations identifying human error as the prime causal factor in over 50 percent of military system failures have enlisted the attention and concern of legislators and DoD managers (General Accounting Office, 1981).
- *Difficulties in Achieving and Sustaining Adequate Force and Skill Levels of Personnel.* With declining birth rates and enhanced competition for the best and brightest, the prospective pool of "quality" military recruits is

shrinking. Barring compulsory military service, the projections are that future personnel will be fewer in number and of potentially lower capability, aptitude or competence with concomitant effects on the operational effectiveness and maintenance of military systems (Malone & Waldeisen, 1986; Malone, Kirkpatrick, & Kopp, 1986).

• *Negative Public Image of Military System Acquisition.* Throughout the 1980s, the barrage of critical press releases and follow-up coverage of congressional reviews of procurement practices and policy have collectively yielded a negative public perception of the costs and effectiveness of military systems (Comeau, 1984; Luttwak, 1984; Coates & Kilian, 1985; Stubbing & Mendel, 1986). DoD's sensitivity to this notoriety, coupled with publicized estimates that human factors related costs over the life cycle of current military systems will exceed 60 percent of total costs (Malone & Waldeisen, 1986) have contributed to the inevitability of intense DoD managerial attention to system human factors.

• *Leadership Responsive to the Need for Reform in Military Systems Acquisition.* Historically, human factors has had a weak constituency within the decision making power structure (Lane, 1987). Recently, however, senior managerial support and involvement, as evidenced by Army Staff advocacy in the creation and sustenance of the MANPRINT Program, has boosted the consideration of human factors throughout system design, acquisition and operations (Elton, 1986; Spurlock, 1988).

The cumulative effect of these factors has been a heightened awareness throughout DoD of the need to reduce the risks and consequences of poor system effectiveness stemming from ill-considered human factors in design. This has, in turn, led to more open-minded consideration and, in the case of the U.S. Army, complete organizational embracement of an integrated systems concept in which operability, trainability, maintainability and safety of system personnel are considered interdependent with the fundamental design of system hardware and software.

MANPRINT: A BLUEPRINT FOR CHANGE

Since World War II when it was determined that operational deficiencies in bombing, artillery targeting, submarine sonar detection and aircraft were associated with poor equipment design, training and personnel selection, the military has played the lead role in nurturing human factors to its present state of professional maturity (Christensen, 1964, 1988). It is, therefore, ironic that attempts by the military (United States Air Force Systems Command, 1967, 1969; Department of the Army, 1968) and other civilian agencies to institutionalize the integration of human factors considerations within the systems acquisition, design, and engineering process have, for the most part, met with little success (Chatelier, 1987; Booher, 1988). Though some of the blame may lie with the nature of human factors per se, it is evident that it can also be attributed to the basic attitudes and training of personnel involved in acquisition and design, support limitations in the design environment, and constraints imposed by the system acquisition and design processes (Rouse & Boff, 1987a; Boff, 1987a, 1987b, 1987c, 1988).

In 1982 the Army Research Institute, by studying Army weapons systems, determined what could be done to improve integration of manpower and training in system development. Based on their findings, the Army inaugurated the MANPRINT Program in 1984 to overcome past problems and improve human performance and equipment reliability (Elton, 1986; Booher, 1988; and elsewhere in this volume). Unlike past efforts at integrating human factors in equipment design, MANPRINT became Army policy requiring a wide spectrum of actions on multiple fronts during the system acquisition process. The policy both provided the reasons and the process for integrating separate manpower, personnel, training and other human factors requirements as a total system. Advocacy began at the top with the Vice Chief of Staff of the Army and was enthusiastically amplified by MANPRINT representatives strategically placed throughout the Army's acquisition process. Civilian, military, and contractor personnel engaged in research and development (R&D), acquisition, and operations are indoctrinated and trained in the MANPRINT process and methodology.

Being the most significant, broadly implemented effort of its kind to date, the existence of MANPRINT has helped raise organizational awareness throughout DoD and civilian agencies. The Air Force, anticipating a multi-billion dollar savings from implementing a similar process, established in 1988 a formal program named IMPACTS (Integrated Manpower, Personnel, and Comprehensive Training/Safety). The U.S. Air Force's goals are "to integrate human-centered disciplines of MPT [manpower, personnel, and training], safety and human engineering to support development of mission capable systems that can be safely operated, maintained and supported in present and future operational environments at the lowest life cycle cost and with the people who will be available" (Gentner, 1988). The Navy, building on its earlier HARDMAN (Hardware vs. Manpower) program, continued to refine its methodology for comparing manpower, personnel, and training requirements of planned systems to existing baselines that have been implemented in the acquisition of systems by the Department of the Navy (Council, 1986; Lane, 1987).

In December 1988, the Secretary of Defense formally embraced the manpower, personnel, training and safety (MPTS) concept. DoD Directive 5000.53 (Manpower, Personnel, Training and Safety in the Defense System Acquisition Process) was approved establishing MPTS criteria that must be addressed by all DoD components in cooperation with industry.

> The Department of Defense shall maximize the operational effectiveness of all systems, whether being procured initially or

> being refurbished, by ensuring those systems can be effectively operated, maintained, and supported by well qualified and trained people. To do so, human capabilities and limitations must be fully considered early in the systems design process. Such MPTS concepts, requirements and goals shall be developed in a consistent manner, communicated to industry, evaluated in contract proposals, and weighed positively and substantially as criteria for source selection.
>
> DoD Directive 5000.53

There can be little doubt that these events signal a serious, and indeed historical, organizational commitment to MPTS or MANPRINT goals. However, while this commitment is a vital prerequisite to meaningful actions and results, it is unlikely, by itself, to be sufficient. Full consideration of human performance capabilities and limitations within the systems acquisition process, as directed, presumes the availability of requisite knowledge and the means to act upon it in the course of system design. In other words, a technological basis is needed on which to make and implement good system human factors decisions. In turn, the multitude of players distributed over government/industrial organizations and agencies which influence decisions in the design of military systems must *value* this basis as beneficial to the design of effective systems. Finally, the effort and costs/risks associated with integrating human factors in design must be minimized by increasing the availability of design support that is compatible with standard or accepted practice. In effect, the enabling of these conditions constitutes a complex human factors design problem nested within the broader issue of the improvement of "total system" human engineering.

THE HUMAN FACTORS OF DESIGN

The general goal of system design is to conceive a system whose form and function fulfill a set of defined needs and requirements within prescribed cost, schedule and material constraints. The underlying complexities involved in the military system acquisition and design process have been widely subjected to dissection and scrutiny from an array of perspectives. Regardless of perspective, the process of system design ultimately reduces to some dependence on iterative human intervention to solve problems, assess options, and make decisions. As behavioral processes, design problem solving and decision making are, in part, subjective in nature and dependent on the experience of the decision maker, the situational context and constraints, and the nature of the information that is sought and utilized.

Contemporary approaches to integrating MANPRINT into systems acquisition and design decision making must take account of past failures of human factors to be naturally assimilated by the process and the practitioners of design. To succeed, a strategy is needed to motivate system designers to seek and use MANPRINT information. In essence, this requires some understanding of the "human factors *of* design" which includes: (a) the nature of system design terms of the technical, organizational, political and economic conditions and constraints under which military systems design decision making occurs; (b) the nature of designers in terms of their basic skills, inclinations and limitations as integrators of information and arbiters/architects of design decisions; and (c) the nature and perceived value of prospective design information. Collectively, these factors should determine the effectiveness with which MANPRINT resources are likely to be integrated into design decisions.

Nature of System Design

Design Process

Much has already been documented in the literature on the general logic and formalisms associated with the process of design (Pahl & Beitz, 1984; Meister, 1987). Likewise, many accounts exist of the apparent and hidden complexities associated with the acquisition and design of military systems (Department of the Navy, 1986). In formal descriptions, system design is frequently characterized as an orderly, hierarchical process. However, in reality, it involves many iterative steps, stages and procedures dependent on the sometimes chaotic control and communications of multiple individuals and organizations with their respective skills, biases and inclinations.

The process of acquiring new systems (i.e., not "off the shelf") in DoD typically encompasses technical design activities in addition to accounting and procurement actions necessary to justify and facilitate the purchase of military systems. The acquisition process is governed by DoD policies and regulations broadly concerned with auditing and ensuring compliance with a defined continuum of steps, stages, procedures, specifications and standards in the design, test, and production of a product/system (Packard Commission, 1986). While these instructions and directives provide a procedural framework for design, they do not attempt to technically guide the design of systems. Though these "ungoverned" technical design activities can "either make or break a project" (Department of the Navy, 1986), the latitude of control by DoD acquisition managers is generally constrained to ensuring "correctness of process" to assure that everything "goes by the book" (Packard Commission, 1986; Rouse & Boff, 1987b).

DoD systems acquisition and design is also a social and politically mediated process in which systems engineering criteria are not the sole objective determinants of outcome. Indeed, the far reaching implications of military acquisition on the nation's economy and security often requires that source selection decisions be made on the basis of competing social or political agendas such as those intended to foster equal economic opportunity (Luttwak, 1984; Stubbing & Mendel, 1986).

Design Decision Making

Fundamentally, system design is a cumulative function of a multitude of design decisions and trade-offs bounded by a regulatory framework and dependent on the information and experience factored into them. The interdependence among the myriad factors which contribute to the design of complex systems makes it difficult to predict the influence of any single factor, bit of information or decision on a given design. Likewise, it is nearly impossible to objectively attribute all the root sources of inspiration or information for any given original design decision. In contrast, the pressures of limited time and resources typical in system design drive designers to bias decisions and trade-offs toward reduction of uncertainty and risk. As a result, most new designs tend to be adaptive (i.e., accommodating a known solution to a changed requirement) or variant (i.e., in which parameters such as timing or size of a known solution are varied without changing the basic design) as opposed to "original" which may depend on untested approaches or new technology. Hence, the selection of appropriate baselines – a proven system or subsystem analogous to the one under development – will generally account for the largest portion of variance in the effectiveness of a given design. Therefore, choosing good baselines reduces the risk to a given system's effectiveness.

The implication of this fundamental dependency on prior designs as baselines is that it is unlikely that additional information will be sought beyond that viewed as satisfactory to meet requirements. In other words, if MANPRINT is not embedded in the baseline design, then it is unlikely to be invoked unless specifically required and paid for. The lack of existing MANPRINT system baselines for designers to draw upon will continue to be a limiting factor to MANPRINT applications in new systems.

The design decision process (simply schematized in Figure 19-1) is best represented as a subjective integration of information resources and personal experience which is, in turn, constrained by limitations of available time and resources. At different times, single individuals or individuals working together as a team may be involved. It is an iterative process, recurring at all stages and levels (i.e., component through system levels) of a given design. Effective design, therefore, involves a skillful blending of past baselines with new decisions and trade-offs, counterbalanced to minimize risks to achieving predefined functionality within material, cost, and schedule constraints.

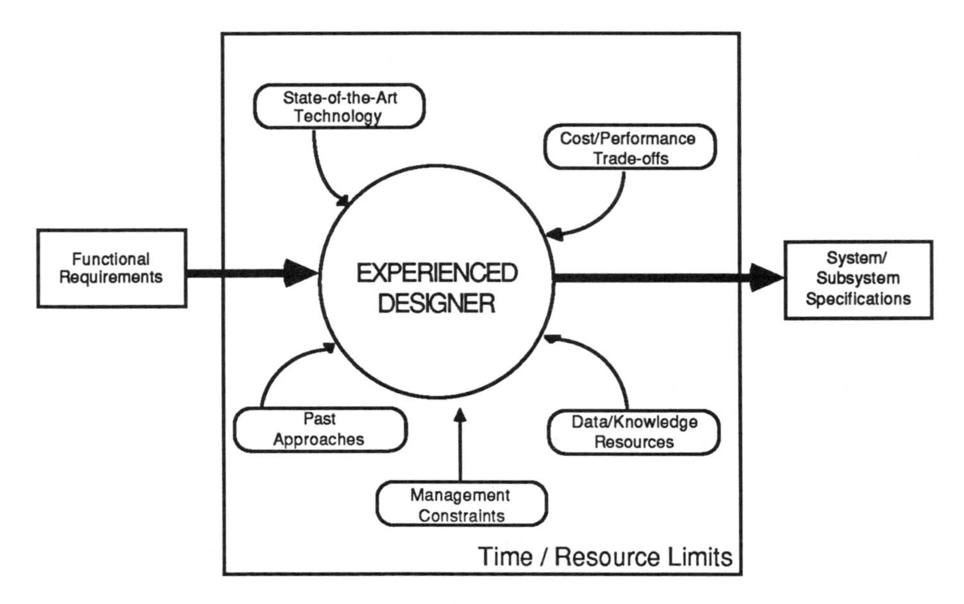

Figure 19-1
Simplified Characterization of the Design Decision Process

Given that effective design is a function of the information factored into design decisions, then decisions made without consideration of potentially leveraging information may be suboptimal. Depending upon their impact on system function, these decisions may collectively undermine design effectiveness. In a series of studies, Allen (1977) has convincingly shown that well over 90 percent of the information factored into technical decisions by engineers already resides, at the time it is sought, in personal files or in the files of trusted colleagues (i.e., "the technological gatekeeper"). Therefore, raising the efficiency with which information is considered and factored into design decision making should, by inference, raise the probability of design effectiveness. Ironic as it seems, enhancing design effectiveness by improving designer access and utilization of design relevant information may be hampered by the fact that designers are already deluged by too much potentially relevant information competing for their time and attention (Boff, 1987a). Given these circumstances, a strategy is needed which raises the competitive value of MPTS data for the designer's limited attentional resources.

Support Environments

The past few years have witnessed a revolutionary increase in engineering capability to do efficient equipment trade-off analyses requiring modeling and manipulation of complex three-dimensional forms in space and time. Computer-aided design (CAD) has rapidly moved from support of drafting and blueprinting functions to aspects of system performance analysis. With CAD use growing at an approximate rate of 30-50 percent per year, it is estimated that by 1995, 80 percent of all designs will employ some form of CAD support (Kulp & Coppola, 1987). Obviously, it is both critical and urgent that efforts be directed at integrating MPTS data and models, as may exist, to enable their accessibility and implementation via design support media common with that deployed in equipment design.

Major hurdles must be overcome if this goal is ever to be met. One problem is that the state of the art in modeling of domains germane to MANPRINT, such as reliability, maintainability, inspectability, accessibility, repairability, operability, personnel availability and trainability is, at best, poorly understood. This availability of human performance models of potential value to computer-aided engineering was assessed in a recent National Research Council study (Elkind, Card, Hochberg, & Huey, 1989) which concluded that existing models have been insufficiently validated against human performance capabilities; "usable" models are not within "immediate reach"; numerous gaps exist between the functions that can be modeled; and "little has been done to integrate models of diverse components into a coherent reality that works together." Another significant problem is that new and innovative forms of computer-aided design will be needed to aid the integration of MPTS in the design of military systems. Cody (1988) suggests these should emphasize support of "problem formulation" and "synthesis and composition" (e.g., aids in generating conceptual design solutions). In any case, major investments are needed to shape the emerging direction of new design support technologies in favor of aiding human-system integration.

The Nature of Designers

Who Is the Designer?

The design of complex military systems typically involves large numbers of individuals, usually distributed over many organizations, who make the decisions which determine the form and functionality of a given design. Though one might expect that designers constitute an easily recognizable, titled groups of professionals, it is the case that design decision making in military systems acquisition typically involves many individuals who neither identify with the role or responsibilities of designers (Rouse & Boff, 1987b). Needless to say, this poses significant difficulties in maintaining the accountability of an evolving design and supporting a process in which key participants do not explicitly acknowledge their responsibilities. Recognition of the fact that the design of complex military systems is dependent on effective communication and decision making among organizationally and geographically distributed individuals and groups (including end-users, acquisition specialists, design management and engineers, research and development technicians, marketing specialists, etc.) is vital to any strategy directed at influencing the outcome of this process. All participants who can strongly influence elements in the acquisition and design of systems must be identified, educated and made accountable.

Designer Bias and Inclinations

A significant obstacle to institutionalizing MANPRINT in the system acquisition and design processes can still be the negative attitude of many engineers and managers regarding human factors. Some of the misconceptions harbored by designers (Meister, 1987) include:

• Humans are sufficiently flexible to compensate for design inefficiencies.
• The system will buffer or compensate for the effects of personnel deficiencies (e.g., error).
• Human factors inputs are not effective in predicting or overcoming human system mismatch problems.
• The cost of including human factors considerations is excessive.
• Good engineering practice already considers the role of personnel in the system, therefore, specialized human factors engineering efforts are not needed.

In sum, the perception of many designers and their organizational management is that the *costs* for implementing MANPRINT are too high, the *usefulness* of MANPRINT design resources are too low, and the probable *gains* resulting from consideration are insignificant. Meister argues that as a result of this negative attitude towards human factors, "otherwise mandatory directives are often evaded or fulfilled half-heartedly." Hence, instituting widespread use of MANPRINT in system design will require persuading or motivating design decision makers to overcome ingrained biases in the use of human factors information.

Cost/Value Considerations in the Use of Information

As discussed earlier, design is a process characterized by decision making, driven by requirements, and fueled by information that is sought and used on the basis of its anticipated utility in making decisions, fulfilling requirements or meeting system goals (Ozog, 1979; Epstein, 1982; Rouse & Rouse, 1983; Rouse, 1986; Sage, 1987). Given the serious constraints of time and resources typically associated with the design of complex military systems, decision making is, by necessity, biased towards minimizing costs and maximizing benefits. Costs refer to undesirable consequences with respect to either resource utilization or the achievement of system performance goals. Benefits are the gains or advantages to resource or performance goals resulting from a given decision. Risk is the potential for costs or lost benefits and typically represents a major source of uncertainty in any new complex system design.

The benefits and costs associated with a design decision depend on the "usefulness" and "usability" of the information used. Information is "useful" to a given decision if its use promotes advantage. Therefore, it is likely that decisions made without benefit of leveraging information will be suboptimal. The use of (or failure to use) information may also exact liabilities or "costs," in terms of added uncertainty and effort needed to acquire, interpret, or apply it in a given situation. Information may be inaccessible or represented in such a manner that it is difficult to interpret or apply it. In other words, the worth or value of potentially useful information may be corrupted by the costs associated with its "usability." Figure 19-2 illustrates the underlying benefit-to-cost relationship between the usefulness and usability of information and predicts the region for which information should have optimal value or worth. This relationship is consistent with contemporary expectancy value models (Feather, 1982) and provides a basis for anticipating the likely behavior of system designers as consumers of technical information. For instance, designers are unlikely to seek or use high-cost (low usability) information of low usefulness, whereas low-cost (highly usable) information of high usefulness should have a higher likelihood of use.

In the absence of quantifiable measures of value, information tends to be sought and used based on expectancies of its usefulness and usability in a given context. Simply stated, information not believed to be useful is not likely to be used. Given the negative attitude that many designers and their organizational managements have regarding *costs* of implementing human factors (Meister, 1987), it is reasonable to assume that the "perceived" value of this information will be biased lower than its potential value. Based on an extensive review and evaluation of primary source literature on human performance research, Boff (1987a, 1987b, 1987c) concluded that the generally low perceived value of this information by practitioners was often a consequence of its poor usability. Fully consistent with the relationship shown in Figure 19-2 these findings suggest that the high costs and risks of using this information will, in the mind of the designer, outweigh its potential benefits or usefulness to the user. Hence, information that is not usable, or perceived not to be usable, may not be considered useful.

In the systems acquisition and design environment, the use of technical information may be naturally motivated by its perceived value or mandated by regulation. The required use of standards, guidelines, source data, and

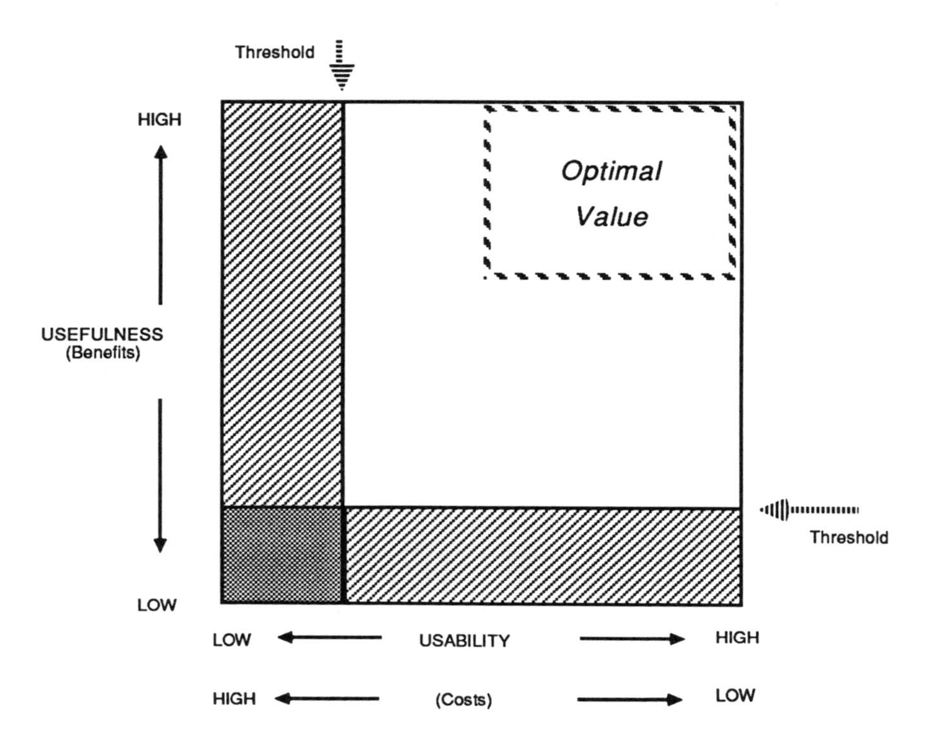

Figure 19-2
Shows the Regions in Which the Use of Information is Likely to Have its Greatest Value or Worth

predetermined specifications are instances of regulation (i.e., where perceived value may not necessarily be a determinant in its consideration or use). In general, information requirements that are regulated are doubtlessly intended to insure the use of data/information of value that may otherwise be ignored. However, success in regulating the use of information, not generally *perceived* as having value, will ultimately depend on the consequences of non-use, the probability of detection of non-use (i.e., policing and inspection) and the anticipated costs of usability.

The cost-benefit perspective on the value of information sets the stage for presenting the critical challenge to the institutionalization of MANPRINT. While a necessary prerequisite to their use, the mere existence of human factors data, models or methodology is, in and of itself, insufficient. To be used, these resources must be positively valued in terms of their usefulness and usability while the consequences associated with their use or non-use must be clearly evident or predictable to the user. Therefore, an optimal strategy for the eventual integration of MANPRINT in systems acquisition and design must depend on ensuring (a) *availability* of reliable and useful human factors data, information or techniques; (b) *usability* of information resources (i.e., where corresponding costs and risks of use are low); (c) a *high perceived value* of this information among anticipated users; and (d) *consequences* of the use or non-use of MANPRINT information resources that are predictable and measurable. To effectively meet this challenge requires knowledge about the baseline availability of useful MANPRINT information, its usability, the needs and inclinations of prospective users and the characteristics of the support and regulatory environment in which system design occurs.

IMPLICATIONS FOR THE "VALUE" OF MANPRINT IN THE DESIGN OF MILITARY SYSTEMS

It is beyond the scope of this chapter to outline the wide range of manpower, personnel, training and human engineering tools, methods, models, and data that exist (for further details and examples, see the chapters in Part III and the other chapters in Part IV of this volume). Suffice it to say that the impact of these human factors resources on human-system design has been far less than their availability or content would suggest. One reason is that they are conceptually and physically scattered; that is, they vary widely with respect to contextual or theoretical frames of reference (Boff, 1987c), in their relevance to practical design concerns (Meister, 1987), and in their consistency with contemporary design resources and culture. While a number of techniques exist that are of potential value to analysis in system development (Table 19-1; National Research Council, 1983), there is almost a "complete absence of information" about their effectiveness or utility (Meister, 1983). In addition, MANPRINT data and methodological resources are distributed over a wide range of hard copy sources (e.g., journals, periodicals, and government and industrial reports) and computer-based media. As a result, extracting, let alone applying, these resources can be a daunting challenge for the most seasoned professional. The temptation to exploit the massive scientific research literature rapidly vanishes in the face of the risk and frustration associated with identifying usable information from this low signal-to-noise data mass. As a result, the high overhead and risks involved in use of human factors research findings often pose an unreasonable barrier for the practitioner (Boff, 1987c).

CONCLUSIONS: CHALLENGES AND PROSPECTS

As discussed at the outset of this chapter, the MANPRINT concept of a system design process that takes full account of the human's role in training, operations and maintenance has deep historical roots. What is especially

Table 19-1
Generally Known Applied Methods Categorized by Purpose

ANALYSIS	*DATA COLLECTION*
System Analysis	Activity Analysis
Function/Task Analysis	Time Lapse Photography
Information Analysis	Real Time Film/Video Recording
Scenario Analysis	Direct Observation
Workload Analysis	Physiological Recording
Time-Line Analysis	Quantitative Performance Recording and Analysis
Operational Sequence Analysis	
Failure Mode Analysis	
Fault Three Analysis	*PREDICTION*
Link Analysis	The Human Error Rate Procedure (THERP)
Function Allocation	
Anthropometric Analysis	Data Store
Decision Analysis	Human Operator Simulator (HOS)
Display Evaluation Index	Control Theory
	Accuracy Theory
	Predetermined Time Analysis
IDENTIFICATION OF NEEDS	Readability Indices
Critical Incident Technique	
Surveys/Questionnaire	*EVALUATION*
Accident Investigation	Test Plan Evaluation
Interviews/Group Technique	Simulation
Definition of User Population	Mock-Ups
	Walk Throughs
	Check Lists
	Ratings

novel, however, is a broadbased awareness of the causes and consequences of deficiencies in contemporary military systems and a growing consensus regarding the nature of the action that must be taken. It is the thesis of this chapter that the enhancement of human-system integration in the design of systems is itself a human factors design problem.

System design is essentially a decision making process by which designers interpret and attempt to meet defined requirements within time, resource, management, social and political constraints. Though decision making is abundantly treated in the behavioral science literature, it remains poorly understood in the context of system design. Nonetheless, it is reasonable to assume that the nature of design decisions depends on the experience and capability of the decision maker and the nature of the information utilized. Because designers are human beings, they are subject to human limitations. They have biases and inclinations; they make errors; they have limited capability to deal with conceptual complexity; etc. Moreover, this behavior is bounded by serious constraints in time and resources which limit parametric consideration of all the information which may relate to a needed decision. Therefore, the seeking and utilization of information by designers is guided by the need to reduce risk with respect to system cost and performance. Given these conditions, information must compete for a share of the designer's constrained attentional resources. The basic challenge to influencing this process is to raise the competitiveness of human factors for the designer's attentional resources. What is needed are strategies which raise the perceived value of MANPRINT by lowering the perceived costs of its usability in the decision making process while also raising perceived gains of its use with respect to design objectives.

Significant progress has already been made by MANPRINT efforts in both government and industry (see Chapters 3, 4, 5, and 10). Nonetheless, considerable challenges remain in achieving institutionalization of this concept throughout the entire Defense acquisition, R&D, and equipment manufacturing communities.

Sustaining Broadbased Awareness of the Benefits

Aggressive maintenance of public and professional awareness of the need and accrued benefits of a MANPRINT perspective (i.e., keeping the "window of opportunity" open) is essential to its institutionalization and long-term success. System design is, by necessity, a conservative process that will tend to resist change which can increase uncertainty or add risk. During the period that it will take for the "natural value" of total systems integration ideas to permeate and be widely assumed by entrenched levels of acquisition and design middle management, it is vital that MANPRINT continue to be actively competitive for "public attention," support, and resources.

Raising Awareness of Roles and Responsibilities in Design Decision Making

Individuals and organizations which contribute to, make, or implement decisions at all levels of system acquisition and design are the key agents for integrating MANPRINT into that process. The challenge arises from the fact that key players ranging from end-users who become involved in the definition of need, to personnel who initiate or approve engineering changes or retrofittings, often do not recognize their roles, responsibilities, and the consequences of their decisions on system effectiveness. These individuals must be identified, educated, and made accountable to meeting MANPRINT system objectives.

Raising the Perceived Value of Human Factors Among Design Decision Makers

It is obviously vital that individuals and organizations engaged in design decision making be motivated to seek and utilize MANPRINT because of its "perceived value" to achievement of system design objectives. The principal challenge, once these individuals are identified and accountable, is to influence their attitudes, expectations and capabilities. Given the availability of an adequate supply of useful MPTS data resources, active measures must be taken to assure:

1. Low cost-low risk usability of MANPRINT design resources, in terms of their accessibility, interpretability and applicability.
2. Wide dissemination and sensitization on the value and design application of MANPRINT design resources. For long-term success, this strategy must strive to seed a well-positioned cadre of "technological gatekeepers" (Allen, 1977) who can *credibly* aid the process of total system design from within the peer group.
3. Regulation of the value of MANPRINT by raising the costs or consequences of non-use (i.e., introducing significant penalties for non-compliance with directives, regulations and standards). However, to be effective, this strategy raises new challenges that must be met. For instance, the regulatory authority must possess the means or measures to monitor or test for compliance.

Indeed, understanding the key variables contributing to variance between perceived and actual values and "costs" associated with usability of information by system designers should be a principal area of concern for human factors researchers and practitioners.

Assumption of Control Over Supply and Demand

To raise perceived value, MANPRINT must ensure an adequate supply of useful and usable MANPRINT design resources that is commensurate with demand. Demand may be a natural outcome of the recognized need and search for useful information or be a response to consequences of non-use that are predictable, measurable, and significant. If the supply of useful/usable MANPRINT design resources does not increase to meet increased demand, then both system designers and regulatory authorities bowing under the pressure to continue fielding systems will likely continue to rely on the only information sources available. The extent to which all parties accept to these compromises will reduce demand on supply of needed MPTS data resources. The danger is that these compromises will quickly become entrenched modes of responding to MANPRINT directives and objectives. Therefore, it is vital that demand not be raised beyond the feasible limits of the design community to effectively respond.

Building the Supply of MANPRINT Data Resources

The challenge exists to ensure the existence and availability of an adequate supply of useful and usable human factors design resources. These include data, models, methodology, and metrics to support system design planning, decision making, and test and evaluation. Two logical means are available to raise the supply of these design resources: (1) stimulation of the direction and objectives of new R&D, and (2) selective harvesting of scattered design resources of potential value (i.e., data from the existing R&D literature). Neither option is, however, without some risk. Collecting new data or developing new methodologies is, at best, a costly process with highly uncertain outcomes. With respect to MANPRINT objectives, the problem of acquiring useful data through new R&D is exacerbated by the historical separation of the government laboratories into specialties for training, human engineering, manpower and personnel, physiological, and biomedical. On top of this, of course, are all the equipment engineering specialties. As a result, a common R&D environment where synthesis across these specialty areas might be explored or validated does not presently exist. Furthermore, such integration is simply not affordable, in terms of time or resources, during the course of system design. An integrated MANPRINT technology base is urgently needed by the design community to support new regulatory demands.

Harvesting the Existing Research Investment

Since World War II, the federal government has invested huge sums in behavioral research which has potential value for exploitation as human factors design resources (Wulfeck & Zeitlin, 1962; Boff, Calhoun, & Lincoln, 1984). Following a massive, labor intensive review and distillation of thousands of archival publications (as well as government and industrial reports), Boff and Lincoln (1988) concluded that a substantial volume of research studies exists that possess potential value and relevance to practical applications (See Van Cott, this volume, for a detailed description of the product of this effort). The key problems they encountered were the low signal-to-noise ratio of useful data (i.e., relevant to system design applications) in the research knowledge base as a whole and low usability (i.e., accessibility, interpretability and applicability) of those data classified as potentially useful. While these problems provide a basis for understanding why this research investment remains relatively unexploited by the process or practitioners of system design, it would be imprudent to "toss the baby out with the bath water." The challenge is to institutionalize the process of exploiting the existing investment of research findings and accelerate the transition and validation of this technology as a vital source of MANPRINT design resources.

Measurement of Gain in MANPRINT System Effectiveness

A basis is urgently needed for assessing the effectiveness of total system performance with respect to human factors considerations (e.g., determining obvious deficiencies, evaluating whether the system performs to standards, determining the contribution of variables that affect system performance, etc.). This is obviously vital for (1) detecting non-compliance with MANPRINT directives in the course of system acquisition and acceptance testing, and (2) demonstrating a return on investment in terms of measurable enhancement of system effectiveness resulting from implementation of MANPRINT. Systematic feedback on and analysis of the performance of operational systems is essential to building good human factors baselines and setting performance standards. The present lack of human performance standards, for example, reduces evaluation of the adequacy of human performance to intuitive calls of judgment (Meister, 1987), often mediated by individuals without appropriate training or experience.

Developing an Enhanced Integrated Design Support Capability

The final challenge is to invest in shaping the future of design. There can be little doubt that emerging design support system technologies (e.g., computer-aided design/computer-aided manufacturing/computer-aided engineering) are forcing a redesign of the design process. Future support systems will enhance the designer's capability in areas such as problem formulation, synthesis and composition, consequence finding and analysis, rapid prototyping and system evaluation (Rouse & Cody, 1987; Cody, 1988). In order for MANPRINT to be successfully integrated into the designer's repertoire in the future, MANPRINT must play a visible role in driving the direction of design support technologies.

ACKNOWLEDGEMENT

This chapter immensely benefited from many helpful comments and suggestions offered by colleagues. In particular, I am grateful to William Cody of Search Technology, Inc., Georgia, and Sarah Swierenga of Logicon Technical Services, Inc., Ohio.

REFERENCES

Allen, T. J. (1977). *Managing the flow of technology: Technology transfer and the dissemination of technological information within the R&D organization.* Cambridge, MA: The MIT Press.

Boff, K. R. (1987a). Matching crew system specifications to human performance capabilities. *Proceedings of the 45th NATO AGARD Guidance and Control Panel Symposium.* Stuttgart, Germany: NATO Advisory Group for Aerospace Research and Development.

Boff, K. R. (1987b). Designing for design effectiveness. *The design, development and testing of complex avionics systems*. Las Vegas, NV: NATO Advisory Group for Aerospace Research and Development.

Boff, K. R. (1987c). The tower of babel revisited: On cross-disciplinary choke points in system design. In W. B. Rouse & K. R. Boff (Eds.), *System design: Behavioral perspectives on designers, tools and organization* (pp. 83-96). New York: North Holland.

Boff, K. R. (1988). The value of research is in the eye of the beholder. *Human Factors Bulletin, 1*(6), 1-4.

Boff, K. R., Calhoun, G. L., & Lincoln, J. (1984). Making perceptual and human performance data an effective resource for designers. *Proceedings of the NATO DRG Workshop (Panel 4).* Shrivenham, England: Royal College of Science.

Boff, K. R. & Lincoln, J. (1988). *Engineering data compendium: Human perception and performance* (4 Volumes). Wright-Patterson AFB, OH: Armstrong Aerospace Medical Research Laboratory.

Booher, H. R. (1988). Progress of MANPRINT: The Army's human factors program. *Human Factors Bulletin, 31*(12).

Chatelier, P. R. (1987). Psychology or reality. In W. B. Rouse & K. R. Boff (Eds.), *System design: Behavioral perspectives on designers, tools and organization* (pp. 97-102). New York: North Holland.

Christensen, J. M. (1964). *The emerging role of engineering psychology* (AMRL-TR-64-88). Wright-Patterson AFB, OH: Aerospace Medical Research Laboratory.

Christensen, J. M. (1988). The human factors profession. In G. Salvendy (Ed.), *Handbook of Human Factors* (pp. 3-16). New York: John Wiley and Sons.

Coates, J., & Kilian, M. (1985). *Heavy losses: The dangerous decline of American defense.* New York: Viking.

Cody, W. J. (1988). *Recommendations for supporting helicopter crew system design* (Contract No. DAAD05-87-M-L584). Norcross, GA: Search Technology, Inc.

Comeau, L. (1984). *Nuts and bolts at the Pentagon: A spare parts catalog.* Washington, DC: Defense Budget Project, Center on Budget and Policy Priorities.

Council, G. S. (1986). Early consideration of human factors in weapon system design. *Proceedings of the Human Factors Society 30th Annual Meeting*, 1290-1293.

Department of the Army (1968). *Man-material systems - Human factors engineering program* (AR 602-1). Washington, DC: U.S. Government Printing Office.

Department of the Navy (1986). *Best practices: How to avoid surprises in the world's most complicated technical process: The transition from development to production* (NAVSO P-6071). Washington, DC: U.S. Government Printing Office.

Elkind, J. I., Card, S. K., Hochberg, J., and Huey, B. M. (1989). *Human performance models for computer-aided engineering.* Committee on Human Factors, National Research Council. Washington, DC: National Academy Press.

Elton, R. M. (1986). MANPRINT (Manpower and Personnel Integration). *Proceedings of the Human Factors Society 30th Annual Meeting*, 905-907.

Epstein, B. J. (1982). An experimental study of the value of information. *OMEGA*, *10*(3), 249-258.

Feather, N. (1982). *Expectations and actions: Expectancy value models in psychology.* New Jersey: Erlbaum Associates.

Gentner, F. C. (1988). *USAF Aeronautical Systems Division's Model Manpower, Personnel and Training Organization – An update*. Military Testing Association.

General Accounting Office (1981). *Effectiveness of U.S. forces can be increased through improved weapon system design* (PSAD-81-17). Washington, DC: U.S. Government Printing Office.

Johnson, E. M. (1987). The role of man in the system design process: The unresolved dilemma. In W. B. Rouse & K. R. Boff (Eds.), *System design: Behavioral perspectives on designers, tools and organizations* (pp. 159-174). New York: North Holland.

Kulp, B., & Coppola, A. (1987). Unified life cycle engineering. In W. B. Rouse & K. R. Boff (Eds.), *System design: Behavioral perspectives on designers, tools and organizations* (pp. 187-197). New York: North Holland.

Lane, N. E. (1987). *Evaluating the cost-effectiveness of human factors engineering* (Final Report PO6509; Draft manuscript). Alexandria, VA: Institute for Defense Analyses.

Luttwak, E. N. (1984). *The Pentagon and the art of war.* New York: Institute for Contemporary Studies/Simon & Schuster.

Malone, T. B., Kirkpatrick, M., & Kopp, W. H. (1986). Human factors engineering impact on system workload and manning levels. *Proceedings of the Human Factors Society 30th Annual Meeting*, 763-767.

Malone, T. B., & Waldeisen, L. E. (1986). MANPRINT applications for design of advanced armored weapons systems. *Proceedings of the Human Factors Society 30th Annual Meeting*, 753-757.

Meister, D. (1983). Are our methods any good? A way to find out. *Proceedings of the Human Factors Society 27th Annual Meeting*, 75-78.

Meister, D. (1987). Systems design, development and testing. In G. Salvendy (Ed.) *Handbook of Human Factors* (pp. 17-42). New York: John Wiley and Sons.

National Research Council, Committee on Human Factors (1983). *Research needs for human factors*. Washington, DC: National Academy Press.

Office of Technology Assessment (1988). *Safe skies for tomorrow: Aviation safety in a competitive environment*. Congress of the United States. Washington, DC: U.S. Government Printing Office.

Ozog, S. (1979). On the value of information. *Journal of the American Society for Information Science*, 310-315.

Packard Commission (1986). *Interim report by the President's blue ribbon commission on defense management*. Washington, DC: U.S. Government Printing Office.

Pahl, G., & Beitz, W. (1984). *Engineering design*. New York: Springer-Verlag.

Rouse, W. B. (1986). On the value of information in system design: A framework for understanding and aiding designers. *Information Processing and Management, 22*, 217-228.

Rouse, W. B., & Boff, K. R. (1987a). *System design: Behavioral perspectives on designers, tools and organizations*. New York: North Holland.

Rouse, W. B., & Boff, K. R. (1987b). Designers, tools and environments: State of knowledge, unresolved issues, and potential directions. In W. B. Rouse & K. R. Boff (Eds.), *System design: Behavioral perspectives on designers, tools and organizations* (pp. 43-63). New York: North Holland.

Rouse, W. B., & Cody, W. J. (1987). On the design of man-machine systems: Principles, practices, and prospects. *Proceedings of the 10th World Congress of the International Federation of Automatic Control*. Munich, West Germany.

Rouse, W. B., & Rouse, S. H. (1983). Human information seeking and design of information systems. *Information Processing and Management, 20*, 129-138.

Rubinstein, T., & Mason, A. F. (1979). The accident that shouldn't have happened: An analysis of Three-Mile Island. *IEEE Spectrum*, 33-57.

Sage, A. P. (1987). Information systems engineering for distributed decisionmaking. *IEEE Transactions on Systems, Man and Cybernetics, 17*(6), 920-931.

Spurlock, D. L. (1988). View from the top: An interview with the Honorable Delbert L. Spurlock, Assistant Secretary of the Army for Manpower and Reserve Affairs. *MANPRINT Bulletin*, *3*(3). Washington, DC: Headquarters, Department of the Army (DAPE-MR).

Stubbing, R. A., & Mendel, R. A. (1986). *The defense game.* New York: Harper & Row.

United States Air Force Systems Command (1967). *Handbook of instructions for aerospace personnel subsystems design* (AFSCM 80-3). Washington, DC: Andrews Air Force Base.

United States Air Force Systems Command (1969). *Personnel subsystems* (AFSC DH 1-3). Washington, DC: Andrews Air Force Base.

Wulfeck, J. W., & Zeitlin, L. R. (1962). Human capabilities and limitations. In R. M. Gagne (Ed.), *Psychological principles in system development* (pp. 115-156). New York: Holt.

CONCLUSION

CHAPTER **20**

MANPRINT AS THE COMPETITIVE EDGE

Harold R. Booher
William B. Rouse

THE MANPRINT GOAL

Many of the chapters of this book identify tragic and costly problems with modern technology that could have been avoided with proper human factors technology application. However, if we survey consumers, designers, managers, and bureaucrats, nearly all will agree that they would like safe and well-performing systems. None wants technology that will maim, poison, or fall out of the sky on the user. None wants systems that cannot be operated, or create demands for such high job skills that the only alternative is unconscionably expensive training programs. As Lieutenant General Allen K. Ono, the Deputy Chief of Staff for Personnel for the U.S. Army, has so succinctly stated, "No one is purposely trying to design equipment that the soldiers can't use."

But these same individuals when asked what they will pay for systems designed for safety and ease of use, will, when pressed, state "I want a safe and well performing system at a price I can afford." In all cases, economics rules. The consumer wants both safe air transportation and affordable air fares. Deregulation, by allowing greater competition among airlines, reduces air fares, which increases number of passengers and routes. This, in turn, increases air traffic volume making air safety more difficult to maintain. If the price of seats goes up to pay for more safety, passengers stop flying or go to cheaper air lines.

Similar arguments are found with manufacturers of equipment. "If I develop a safer and easier to maintain automobile, but it costs more, my competition gets my business, because the consumer will not pay more for added safety and maintenance features." In weapon system procurements, earlier arguments were, "If I add MANPRINT at my expense because the government doesn't require it, my competition will win the contract because of lower price." What everyone wants is greater safety and usability at an affordable price.

Human factors as a discipline has urged, however, that investment in early stages will be returned many times over in the long term, through reduced operations and maintenance costs (Price, Chapter 6), through reduced waste and higher productivity (Booher & Fender, Chapter 2), and by avoiding catastrophe (Price, Chapter 6; Rouse, Chapter 8; Boff, Chapter 19). These are good arguments, provided an organization has a long term vision and rewards individuals accordingly. But, decision makers tend to be rewarded for meeting near term costs and schedules, not for long-term savings.

There are situations, however, where organizations are caused to apply human factors technology regardless of cost. One situation occurs from changes in government regulations (as with nuclear power plant and commercial air safety) or changes in procurement policy (as with the Department of Defense in weapon systems procurement). Since in these instances all competitors' costs can rise together, cost considerations are lessened. Any cost increases are passed to the consumer (or the tax payer).

In another situation, organizations are forced to emphasize human factors to gain a competitive advantage in the global market place. When the Japanese started producing quality automobiles (more reliable, more comfortable, and just as safe) for lower costs, the American automobile industry had no choice but to change or continue to lose business. As Booher and Fender in *Total Quality Management and MANPRINT* (Chapter 2) discuss, at least one automobile manufacturer has made serious attempts to face the problem of improving product quality without raising prices beyond what the market will pay, with very favorable results.

The ultimate goal of MANPRINT is to encourage both government and industry to seek human factors technology as one of the primary foundations of future competition. Applied as a quality management philosophy, MANPRINT can aid in competitive positioning. This can happen in two major ways. First, MANPRINT can serve as a driver for advancing technology. Second, MANPRINT can assure better utilization of job skills across the spectrum of personnel aptitudes. MANPRINT technology stimulates both meaningful jobs for the employee and good business practice for the employer by matching jobs to skills and aptitudes from the widest possible range; from the seriously handicapped to those with exceptionally adept psychomotor and physical skills; from those with learning and cultural disadvantages to the highest of cognitive abilities.

MANPRINT STATE OF THE ART

The contributions by the book's authors as well as those by the advisory panel and the many others from government, industry, and academia provide a good representation of MANPRINT experience. A very useful result of this book has been the bringing together of the MANPRINT state of the art from several perspectives. Using the four parts of the book, the highlights of these perspectives are discussed as they to apply the MANPRINT objectives of advancing technology, enhancing personnel utilization, and increasing product safety.

Organization/Management Context

MANPRINT has both management and technical components. The chapters of Part I provide the context for the management component within the larger organizational framework and for specific project management. To achieve a competitive edge with MANPRINT, organizational leaders must first understand and believe in its benefits, next they must require it be utilized, then reward those who apply it and when necessary, punish those who ignore it.

Booher and Fender (Chapter 2) describe the importance of top management in supporting the wide, sweeping philosophy of MANPRINT and in understanding its context with the Total Quality Management (TQM) philosophy currently being offered to American management. Any philosophy which hopes to achieve as much as promised by MANPRINT and TQM must embrace a common vision for top management to focus upon and must employ measures of success in the language of the decision makers.

TQM is supported by many major industries and the Department of Defense, but there is still more talk about TQM than substantive action. In a recent small survey (72 industry people) using Booher and Fender's test for evaluating TQM maturity, only 3 percent of those surveyed considered their organization "mature," whereas a full 65 percent considered their organization "closed minded." TQM should be watched carefully. It's ideas are good and compatible with MANPRINT, but is so broad in its sweep that the entire industrial culture of an organization must change. MANPRINT is much smaller and directed. Progress can be more easily measured and controlled. MANPRINT can rise with TQM, but is not dependent upon it for demonstrable progress.

Both MANPRINT and TQM have characteristics which encourage institutional change. Blanchard and Blackwood in *Change Management Process* (Chapter 3) describe the various principles inherent in any major institutional change and recognize how extremely difficult it is to make lasting changes. But institutional changes can and should be made when their objectives are sound. MANPRINT and TQM support change to cause improved products to be placed in the hands of the user. Many of our governmental and industrial organizations require a major overhaul if they are to move from the "technology-for-technology-sake" orientation to ones that recognize the importance of people (as consumers and operators of technology) being the dominant focus for the future competition. As

Blanchard and Blackwood point out, MANPRINT and other productivity enhancements cannot be successfully implemented without some fairly fundamental "changes to organizational structuring, policies, and procedures and to the values and norms of groups and individuals."

The examples given by Blanchard and Blackwood apply to how a government organization (the Army) institutionalizes change. The examples of Mittler, Hewitt, and Vehlow in *Management Integration Methods* (Chapter 4) and Shafer in *MANPRINT in a Systems Engineering Organization* (Chapter 5) demonstrate institutional change in industry. The place of emphasis depends upon organizational type. When the government is the buyer, it must present a clear signal to industry that human factors technology is important. Industry, on the other hand, need not change its entire institution to respond favorably to the Army's requirement. Many of the MANPRINT success stories from industry relate to projects (or in the case of IBM, an entire division) but do not entail changing an entire organization.

By leading the reader step by step through the acquisition process, Shields, Johnson, and Riviello describe *The Acquisition Decision Process* (Chapter 10) and show how MANPRINT questions and issues can be addressed early in the process. In this way, those who wish to gain a competitive advantage through MANPRINT need not wait until later stages when design influence will be far less.

To be a successful MANPRINT manager, at least two ingredients are needed to insure a competitive edge. First is human factors expertise. There is nothing in the MANPRINT approach which suggests reducing the need for qualified human factors specialists. Muckler and Seven in *National Education and Training* (Chapter 18) identify sources of expertise and tell why such skills will be at a premium in the future. The second ingredient is the knowledge and use of the various tools, techniques and data sources needed by designers and facilitators to conduct MANPRINT. Guides to these are found throughout the book and detailed in Booher and Hewitt, Chapter 12. Armed with the philosophy, training, and techniques of MANPRINT, it is possible for managers to articulate human factors principles, require their application, and reward engineers for successful designs.

Current applications of the MANPRINT philosophy are most visible in Army programs, but the philosophy is spreading to the other services (Shields, Johnson, & Riviello, Chapter 10; Boff, Chapter 19), to other agencies, like the Federal Aviation Agency (Booher, 1990a), and other countries (e.g., United Kingdom and West Germany).

Blanchard and Blackwood also provide a clear understanding of the principles of change management, the pitfalls to avoid, along with sufficient guidance so that any organization can develop a good change management plan and be able to assess progress with confidence. The lessons learned from the Army should prove valuable, especially to other government or large scale industrial organizations. Blanchard (1989) has provided a change management model which should be helpful, but it has not been validated. It is clear that MANPRINT needs industrial change to realize its benefits. This is especially true if the competitive edge due to MANPRINT is what is desired. Attempts to introduce a MANPRINT-like approach which does not have the domains represented nor the leadership absolutely dedicated to the idea are likely to fail. The importance of mid-level managers is also brought out. While top level managers provide the vision, requirements and rewards, it is the mid-level which "represents the marketing and promotional force for change".

In *Management Integration Methods*, Mittler, Hewitt, and Vehlow show much can be done to accomplish systems integration through applying the principles and techniques of modern managerial procedures. This is a difficult task, especially in the "definition and conceptual phase where the cost/effective opportunities of engineering design are realized" because of the inherent uncertainty of dynamic design. The authors identify eight principles (communication, physical proximity, horizontal processing, commitment, decision documentation, flexibility, feedback mechanisms and measure of effectiveness, and design decision influence) which are inherent to quality integration. They describe a very promising tool developed by McDonnell Douglas called the MANPRINT Design Analysis Technique (MDAT). It is used to maximize MANPRINT influence during design and provides an automated method to manage system design analyses from the various functional disciplines. Something like MDAT is greatly needed to aid in managing complex systems development. Any organization with modern interactive computer capability can easily develop a similar system.

Few organizations have current formal systems for monitoring institutional change, and cost benefits analyses are largely non-existent. Shafer (Chapter 5) notes at IBM that the specific terms of cost-effectiveness parameters for MANPRINT are "difficult to assess and largely unknown." Blanchard and Blackwood note the generally poor research base for organizational change decisions. They conclude, ruefully, that there is "no basis for explaining organizational behavior in a manner that can be understood or tested empirically" and state that a whole "new vision" is needed for the research community "if the technology of organizational change and change management is to be advanced." Conducting research on what produces lasting institutional effects is extremely difficult. There are too many factors outside the control of the experimenter to use the typical laboratory approach. Also, it may take years to observe effects of institutionalization. A large scale validation of Blanchard's model would be beneficial, but unless funded by a large corporate or governmental organization, is unlikely to be done. Any new vision in organizational management research should probably move away from the specifics of organizational structure itself and move toward a systems viewpoint where controlled studies and multivariant models of change methodology might be classified by effectiveness in clearly defined situations. Studies of a smaller nature can and should be done as suggested by the TQM philosophy of Deming and Taguchi. Improvement in the process can be measured and validated quite quickly by small scale experiments. Also "macro" research on methods to enhance communication among domains and methods to validate and document MANPRINT influence on design quantitatively are needed.

User-Centered Design Advances

It is with the latest tools and concepts that the direct influence of human factors technology can be made on systems and design engineering. Yamada (1989) shows the Japanese have moved to human technology for the competitive edge. The U.S. has not done so, but the "New Reality" trends (Table 20-1) suggest a need for a fresh appraisal of how we invest in

Table 20-1
The "New Reality"- Trends Affecting U.S. Science and Technology Investment (Source: Singley, 1990)

Trend	Direction
• Potential for European War	↓
• Potential for Low Intensity Conflict	↑
• Lethality of Third World	↑
• Role in Drug Wars and Treaty Compliance Monitoring	↑
• Global Economics Influence	↑
• Global Technical Competition	↑
• U.S. Defense Industry Base and Long-Term Research and Development	↓
• Supply of Quality U.S. Citizens Scientists, Engineers, and Technicians	↓
• Defense Budget	↓

our science and technology base for the future. Price (Table 6-3) lists the 22 critical technologies for Defense submitted in 1989 to the Senate and House Armed Services Committees. As Price notes, HFE and other MANPRINT domains are relevant to many if not all of them. To develop a system using these technologies, people will have to be integrated into the design. Some, like "Machine Intelligence/Robotics," directly imply human factors technology.

The state of the art for human factors design concepts is reflected in Part III in a number of ways. Price with his emphasis on *Conceptual System Design and the Human Role* and Rouse with his *Concepts for Error Tolerant Systems* discuss fundamental issues that systems and design engineers must consider early in the conceptual stages of system design. How should functions be allocated between people and machines? As Price and Rouse both point out, there are some functions definitely best left to people (e.g., response flexibility and creativity in unforeseen events, authority in decision making) others which are best left to machine (repetitive, tedious tasks) and others which either can do as well depending on the circumstances. There are some (like high-speed estimation of risk) that *neither* does well. In most instances, Price concludes "allocations will have an acceptable solution across a range of automation and people, and the final design may be based on utilitarian and cost considerations."

Methodology to minimize human errors is at the heart of human factors technology. Price discusses numerous examples of equipment induced human error. There are those instances where the operator makes a mistake (e.g. hits the wrong button) but where a better design would have made such a mistake unlikely. Price also shows that the greatest potential today for human error having negative effects on system performance is in microelectronics and computers. In this case it is designer error, not operator error, that is the source of the problem. Programmer errors arise faster than they can be eliminated simply because of the immense complexity of programming required in modern sophisticated systems. Because potential for error is never totally eliminated, Rouse argues for designers to take this into consideration. They should design error tolerate systems so when errors do occur, consequences will be manageable.

Computer-Aided Ergonomic Design Tools (McDaniel & Hofmann, Chapter 7) allow the resolution of human factors issues both by human factors specialists-and design engineers. Expert systems can allow the engineers to design-in human factors considerations along with hardware/software design factors. McDaniel and Hofmann describe some highly useful tools (COMBIMAN, CREW CHIEF) which are already in use as well as the Virtual Mockup which pushes the state of the art into animated 3-D model with complex, multiple human figures to aid in task visualization. The challenge for the future in ergonomic design tools is probably closest to what Yamada means by human technology. The research here is currently driving the entire field in computer-aided displays. According to McDaniel and Hofmann, the goal is to achieve a fully integrated dynamic ergonomic model.

More and more the functions allocated to people are cognitive ones, and the demand for higher aptitude people is coming at the very time when demographics for the next decade show they are in short supply. Automation tends to drive the demand for aptitudes even higher. It is extremely important, then, that people's talents are not wasted by wrong job assignments, and that equipment be designed to reduce the demand for high aptitudes. This is not easy, especially since workload (cognitive or otherwise) is difficult to assess and predict. *Workload Assessment and Prediction* (Hart & Wickens, Chapter 9) provides the kind of guidance needed to help the designer, first, to ask the right question and, then, to select the relevant workload measurement method. They conclude that workload can be measured with a considerable degree of accuracy, although workload prediction is a much more difficult and less precise task. Considerable research emphasis is currently being directed to the workload issues; Hart and Wickens provide excellent background for either the practitioner or the new researcher in this field.

The importance of this part of the book to our theme "the competitive edge" is the recognition that human factors technology is a high leverage investment. Not only are major costs avoided, as in the prevention of disaster and accidents, or in reducing unnecessary manpower, personnel, and training costs; but human factors technology is at least as important as any other technology in assuring a system will actually perform in the operational world as expected.

Systems Integration Methodologies

For all the talk of the importance of human performance factors influencing design, there is little that can be done by decision makers in an informed manner if human factors issues are not routinely raised, discussed and resolved throughout all stages and levels of systems integration. For complex systems to be used in complex environments, years of planning and analysis must be undergone before a system idea becomes a system reality. In all instances, decisions must be made considering as many facets as can be possibly imagined. The Department of Defense employs extensive procedures and analyses to enable decisions to be made considering thousands of product and cost variables. MANPRINT, in *System Integration Methodologies*, shows how people factors are woven into every *other* process and analysis used to reach system decisions. The methodologies of MANPRINT make it possible to institutionalize human factors. They allow the advocacy for human factors to rise above the personal inclination of system designers and producers.

Using *The Acquisition Decision Process* of the Defense of Department as their model, Shields, Johnson, and Riviello (Chapter 10) show how necessary it is to first get the decision makers to ask the right questions. If questions about the cost of manpower and training associated with the purchase of a major new system are not asked before a decision is made, how can they be given information that has relevance to their decision?

Shields, Johnson, and Riviello set the framework for an organization to include human performance along with the complexities of technical, managerial, and economic concerns. Through government policy outlined in Department of Defense Directive 5000.53 and Army Regulation 602-2, it is possible to focus an entire organization towards decisions that blend people data into its cost and performance data.

Creating a decision making framework which includes people variables need not be limited to the Department of Defense. Although consumers who use automobiles, TVs, and washing machines do not write design specifications for new products, they make choices that do influence design decisions. Shields, Johnson, and Riviello remind us that buyers in large industry and service companies become wiser to the integration of people as they invest in high-technology machinery and systems. There is nothing to prevent industry from writing specifications for new computer-assisted manufacturing systems or computer-based satellite-distributed training systems or in evaluating alternative candidates that address questions like: Will the size and quality of the work force be adequate to operate, maintain, and support the total system?; or how sure can we be sure that operation, maintenance, or support of the system will not result in safety or environmental health hazards?

A very critical issue for MANPRINT is how early can human factors be considered part of quantitative prediction methodology? Parry, Collins, and Van Nostrand *(Complex Environment Models in Systems Integration, Chapter 11)* discuss the most complex of all known models for decision aids – those coming from wargames and combat modeling. These models are used by military planners and can influence whole fields of technology 20 to 30 years before intended system application. It is through decisions this far in advance that MANPRINT can have the most dramatic influence on systems design – by influencing the very technology that is to be explored and advanced. Again, even though military models are just beginning to incorporate human performance factors, it is not too early to consider ways to apply large-scale complex modeling concepts to non-military decision making. Econometric models for policy formulation or corporate long-range planning, for example, could both profit from MANPRINT.

Of all the chapters in the book, *MANPRINT Tools and Techniques* (Booher & Hewitt, Chapter 12) comes closest to being a "how to" methodology for the MANPRINT practitioner. Without laying out step-by-step how to go through any specific MANPRINT tool or guide, it is a comprehensive description of the state of the art. As a "door to currently existing MANPRINT domain techniques and MANPRINT integration tools," it should provide the reader with enough information on how and where to get started on a project that requires MANPRINT. The wealth of tools and techniques already in use coupled with those soon to be available, make it possible to produce high quality answers to both performance and cost issues for all acquisition stages and all MANPRINT domains. There is no longer any technical reason for human factors technology to be ignored on the argument that it lacks a method to predict performance effectiveness. It also is encouraging to know that any project which applies MANPRINT has sufficient technology to show an immediate and measurable performance contribution (value-added). There is still much needed research and development in all related areas of MANPRINT as is documented at some length in this chapter. As with any advancing technology, each new development stimulates new ideas, suggests new areas of research, and opens a host of new applications.

Two chapters of *Systems Integration Methodologies* concentrate on the training domains. Training deserves special attention in the MANPRINT philosophy because after personnel selection (reliance on people's aptitudes and abilities), the most important trade-off is equipment design versus training. In fact, training is the one domain in which system designers and managers expect to make up for any design limitations. Oneal (Chapter 13), in *Integration of Training Systems and Analyses*, shows how the error of this kind of thinking can result in unexpected and exorbitant costs for training courses, facilities, and devices. Moreover the high training investment does not assure performance can be raised or maintained at the appropriate level. Training analyses through MANPRINT help to insure that training is not the backup to poor design, but rather, whatever training is required has been carefully planned as part of the total system performance package.

Rouse (Chapter 14) with *Training and Aiding Personnel in Complex Systems* presents what may be the most central trade-off to consider in an overall human resources investment strategy – particularly in issues of automating complex operations tasks or maintenance diagnostics. When speaking of trade-offs within "human resources," one needs to restrict the definitions of *aiding*. In one sense, aiding can be defined so broadly as to be nearly indistinguishable from equipment design. For example, the purpose of the FLIR (Forward Looking Infrared Radar) technology is to enhance operator visual performance well beyond his natural capability. Can that be considered aiding in the sense of a computer-aided diagnostic tool, or maintenance technical manual for troubleshooting? In the former sense, every new technological advancement that replaces functions previously assigned to the operator could be considered an "aiding" technology. This issue is primarily one of how best to enhance human performance in complex systems once it has been decided to invest in a new hardware/software technology. Whenever that decision entails an assumed increase in human performance over some baseline, the next decision is how best to augment the human performance (a) through providing knowledge and skills to the operating and maintenance personnel or (b) augment performance directly – diagnostic tools, troubleshooting guides, task simplification, or automation.

All of the analysis methodologies are sophisticated ways of trying to predict, plan for, and execute a systems design which when finally produced and placed in the hands of users will perform as intended. Although our predictions may be quite accurate, no one knows for sure whether the system will really perform until it is in an operational environment. Confidence can be greatly increased with a test and evaluation program appropriate to each stage of development, but the problem Hennessy in *Practical Human Performance Testing and Evaluation* (Chapter 15) brings forcefully to light is how disappointing human performance testing and evaluation results tend to be. His comments are important to the overall context of *MANPRINT: An Approach to System Integration* with its focus on product quality and organizational effectiveness. First of all, testing is not inexpensive. MANPRINT has bridged the first gap which is to continually test and evaluate with the user "in-the-loop." But Hennessy challenges us to go further. To advance human technology, or to understand the variations of total system performance in complex environments, we need to be able to assess and predict realistically the human portion of the man-machine interface.

Several of the authors (Price; Parry, Collins, and Van Nostrand; Hennessy, for example) discuss decision makers tendency to misapply or overapply new technology – especially computer technology. They warn, as Deming, that any action taken in ignorance (in attempts to improve system integration processes when the variables influencing the process, are not known) is likely to make the situation worse. The most frequent violation of this rule discussed throughout the chapters is "throwing" technology (high priced automation or training facilities) at the problem of rising MPT costs. In seeking ways to gain a competitive edge, it is important not to make major technology-driven decisions that force an organization out of the competitive range altogether.

Part III gives two other examples of where technology capability is starting to determine how decisions are made about complex systems development. One example from, Parry, Collins, and Van Nostrand, shows that the capability now provided by high resolution computers, analytical models, and engineering simulation gives decision makers the ability to select new technologies and candidate systems without first building and testing a prototype. In this instance, the technology appears sufficiently advanced that the use of high technology contributes to overall cost effectiveness, particularly if human performance is part of the formula. Hennessy, however, with his example, shows just the opposite. He concludes the blind adoption of computer technology in the performance testing world of complex environments is providing less useful information on human performance decisions today than the highly subjective methods of the 1970s.

Sources of User-Centered Technology

Van Cott in *Human Performance Principles, Data and Data Sources* (Chapter 16) states "If human factors considerations are to be incorporated into system design, then designers need to be able to easily access information in a usable form about human capabilities and limitations." This is a fundamental principle to be applied at the engineering design level and, fortunately, as described in his chapter and that of Haas and Laine *National Human Performance Data Banks* (Chapter 17), highly reliable and accessible data are available. Access to high quality human performance data coupled with the *System Integration Methodologies* now make it possible to overcome perhaps the most difficult barrier to the use of human factors principles and data. An organization which wishes to implement the MANPRINT philosophy, not only to meet government requirements but also to enhance its competitive standing, can now do so using a language readily understood throughout the organization. The concepts of *micro* and *macro* (Hendrick, 1986; Johnson, 1987; Booher, 1990b) are useful in describing human performance data sources for the various MANPRINT domains.

Haas and Laine describe *micro* human performance data banks as those which concentrate on storing "information about the design of the interface between the hardware, the person, and/or the design of specific task." This fits best with the HFE, system safety, and health hazards domains. Together, Van Cott (Chapter 16) and Haas and Laine (Chapter 17) identify some of the better sources for the micro domains including the *Engineering Data Compendium: Human Perception and Performance;* the *Crew Systems Ergonomics Information Analysis Center* (CSERIAC); the *Human Performance Evaluation System* (HPES); the *Aviation Safety Reporting System* (ASRS); and the *Nuclear Computerized Library for Assessing Reactor Reliability* (NUCLARR). On the other hand, *macro* human performance data banks store information concerning the design or adaptation of all manpower, personnel, training, and organizational components which relate to systems integration. The most fully developed sources for the macro MANPRINT domains today are the *Training and Performance Data Center* (TPDC) with FOOTPRINT and CROSSWALK, the *Materiel Readiness Support Activity*, and the *Army Research Institute Project A Data Base*. The availability of these highly useful data sources makes it possible to design systems which not only perform safely and effectively, but which match personnel skills and aptitudes to jobs associated with the system operation.

The future challenge given by Haas and Laine of creating a National Human Performance Information Center, could greatly benefit the national industrial base. Although development and implementation would require wide cooperation across industry and government, such an operating center could raise the entire technology baseline for all participating industries.

Unlike basic human performance data, which is plentiful, MANPRINT specialists who can act as system integrators, equipment designers, or guide others in applying the MANPRINT philosophy are at a premium. The shortage of engineers and scientists as a national problem (see Table 20-1) may be the most serious of all facing the United States industrial base and technology leadership (Singley, 1990). But those with academic degrees in human factors engineering are fewer in number than any engineering field and fewer still actually end up doing design work.

Muckler and Seven, in *National Education and Training,* survey the strongest existing MANPRINT science and technology personnel source (human factors academic training programs in psychology and engineering) and highlight some very crucial issues – again from a national point of view – that need to be addressed along with the more encompassing "reality" of declining scientists and engineers. Muckler and Seven note that demand for human factors has generally met the supply (we have not been overproducing) but carrying out the MANPRINT philosophy described in this book on a national level would create a demand far outstripping the supply. If we had to pick a single most concern with the overall institutionalization of MANPRINT, it would be the potential gap of "too few professionals" being filled by self-appointed experts. Expanding existing human factors academic programs, establishing centers of excellence, and some kind of MANPRINT certification and licensing are potential ways to address the gap. But there will never be enough MANPRINT specialists to work shoulder to shoulder with every design engineer, every senior and technical manager, every training equipment developer, every maintenance and supply technician. What is more desperately needed is a reduction in human errors which flow throughout any organization from failure to focus on people and total process quality. MANPRINT's contribution in that regard will be the continued emphasis of its role and interface with TQM.

Greater dependency will have to be place on the non-specialist. A goodly number of the various MANPRINT tools and techniques are being developed for the non-HFE specialist. But perhaps the most attractive concept of all is better education of undergraduate engineers. If every new engineer started with an appreciation of critical human performance variables associated with design, the potential for design induced human error might drop to a level where the professional human factors expert could control the remainder.

Boff in *Meeting the Challenge* (Chapter 19) concludes that the major problems to broad acceptance and implementation of the MANPRINT concepts presented in this book lie in three arenas: (1) useful and usable data resources; (2) effective methods and media; and (3) sophisticated and motivated users. The contributors to this volume have all in one way or another provided valuable information in these arenas to aid the reader who wishes to participate in this new adventure. But Boff's twist on the perceived "value" being central to producing sophisticated and motivated users within the systems design and acquisition communities is a most critical insight. He says, "Designer, design management, and system acquisition personnel must believe that consideration of MANPRINT factors is "valuable" to the design of effective systems." To make MANPRINT factors valuable to these individuals, they must compete favorably for designers time and resources in reducing overall design uncertainty and risk.

Many designers and their organizational managers have a negative attitude regarding *costs* of implementing human factors both in terms of their personal time to get useful and usable information and in terms of the ultimate system performance. In addition to increasing the perceived value of human factors among design decision makers, he presents many helpful suggestions to bring about a broader-based, positive attitude to human factors technology:

- Sustaining public and professional awareness of MANPRINT benefits.
- Raising decision makers' awareness of their roles and responsibility toward MANPRINT issues and consequences.
- Insuring the supply of useful/usable MANPRINT design resources match increased demand.
- Building an integrated MANPRINT technology base.
- Harvesting existing research investment.

ECONOMIC IMPACT

The adoption of the MANPRINT approach to system integration has, to date, been primarily motivated by government regulatory and contractual requirements. On Army programs where the experience of MANPRINT is the greatest, a company competing for a major system contract must satisfactorily show how it will meet MANPRINT requirements, or it will not be ranked as favorably as those who do. This can result in MANPRINT making the difference in who receives the contract award. This is a common way for the Federal Government to institute change in line with its policies, ranging from defense to environmental protection to affirmative action.

This approach is necessary to communicate clearly the value of human factors technology to industry. Our personal experience and the studies of others (Perrow, 1984, for example) conclude that industry is slow to change practices related to issues of public health and safety without regulatory urging. In military procurements, industry tends to change practices when convinced that the change will increase their overall competitive posture.

MANPRINT has taken an approach to help industry become convinced that change is in their best interests. The priority of MANPRINT is made clear in the procurement documents, in its source selection policy, and in its final selection of contractors. Success stories told by industry are those who won or lost contracts because of MANPRINT. But the message of this chapter is that MANPRINT can do far more for industry than just meeting government requirements. It is now possible to show cost benefit examples from MANPRINT implementation which suggests MANPRINT should be viewed as a good investment. By investment is meant a traditional dollar and cents point of view, rather than from regulatory, contractual, or legal liability perspectives.

Referring to Table 20-2, a number of MANPRINT cost/benefits can be described. The results of the T-800 engine development is a prime example because the benefits came without additional cost to the government. When industry accepted the challenge of adding MANPRINT as a requirement at no added cost, it had to change its organization and design approach (See Mittler, Hewitt, & Vehlow, Chapter 4). The impressive

Table 20-2
MANPRINT Cost Benefits

System	Government Cost	Benefit
T-800 Engine	None	• Easier Maintenance • Fewer Maintainers • Higher Reliability
AVENGER	$300,000	• $61,000,000 Cost Avoidance, or • Increased Warfighting Capability
LOS-F-H	$800,000	• $80,000,000 Cost Avoidance, or • Increased Warfighting Capability
ATHS	$1,500,000	• $1,000,000 Per Annum Manpower Cost Avoidance, and • Increased Warfighting Capability
M151A2 Jeep Rollover Protection System	$20,000,000	• Damage Reductions: $285,000 (Fiscal Year 87) to $21,000 (Fiscal Year 89) * Non-Fatal Injuries: 55 (Fiscal Year 88) to 1 (Fiscal Year 90-First 6 months)

reduction from 134 tools for organizational maintenance of the predecessor engine to only 6 tools on the T-800 was only one of the resultant features. Fewer maintenance personnel are needed now, and the highest area aptitude requirements were also reduced. But what really impresses design engineers and corporate presidents is the indication of higher reliability for engine performance itself. How can this be? It comes, as in so many technology advances, from unforeseen and synergistic results. Because MANPRINT and Integrated Logistics Support had such a major influence on T-800 design for maintenance, it also helped to enhance machine performance life. For example: (a) The engine has fewer and simpler parts, no gaskets or o-rings, and shorter fuel and lube lines (Overall reliability is increased because of fewer parts to fail); (b) Parts are placed for easy access. A maintenance person does not have to remove good parts to look at other parts; (c) Maintenance induced component failures are reduced, (For example, better electrical connector pin arrangements increase the component reliability figure).

The results are shown in Table 20-2 (the Pedestal Mounted Stinger [PMS] known as Avenger and the Line of Sight Forward Heavy [LOS-F-H]). Studies were done by the Army Research Institute to identify where human factors errors would degrade systems performance (probability of successful detection, identification, tracking, and firing) for the Forward Area Air Defense (FAAD) System. The PMS study, for example, showed how to increase probability of successful operation from .816 to .918, a 10% increase. This translated into a $61,000,000 cost avoidance if additional equipment were procured to offset the projected system degradation. Similarly, $80,000,000 cost avoidance was shown for the LOS-F-H. The investment by the government for these two studies was a little over $1 million. The cost avoidance was $140 million.

The ATHS (Airborne Target Hand Over System) is a good example of what can happen when all the MANPRINT domains are considered together. Product improvement was required on the ATHS to decrease pilot workload and allow a more rapid firing of the Hellfire missile from the Apache helicopter. This change made a large bundle of electrical wires no longer functional. Decision makers were reluctant to incur the cost to remove the wire ($1.5 million). However, MANPRINT analysis projected that the problem of troubleshooting with so many "dead" wires throughout the aircraft would require another $1 million per year for additional electricians. MANPRINT made the trade-off feasible for overall manpower cost savings to the Army. Another serendipitous result was a reduction of 16 pounds of aircraft weight allowing either better fuel consumption or increased weapon load.

The final example in Table 20-2 is the reduction in damage and number of soldier injuries because of rollover protection added to the jeep. In this case, the cost to the government is not a negligible one. The cost is well invested not only because of injury and damage reduction, but also in lives being saved. If these installations were made as part of the original design, however, the cost would have been a fraction of the figure shown.

These success stories are not the results of carefully controlled studies or detailed cost benefit analyses. These are all examples of benefits to the government from requirements made of industry. The natural question is: Isn't somebody paying for all these added advantages? Even though there is little or no added cost to the government in some of these examples, what is the MANPRINT cost to industry? This is a very difficult question to answer. There is, of course, some expense associated with doing MANPRINT. But we have asked this question of industry in each of the above cases and in several others. To date, none has done a complete analysis of what MANPRINT costs and what specific influence this investment has on design and ultimately on system performance.

Some data are beginning to be collected, however, on these Army programs. Estimates for the cost of MANPRINT range from zero to 8 percent of total design costs (in cases of zero cost for MANPRINT to industry, the estimate is based on the fact that there would have been some costs associated with the domains like human factors engineering, system safety, and training support anyway, so the added cost for some coordination activity is negligible). An 8 percent estimate is high unless it also includes the costs associated with Integrated Logistics Support. It is not unusual to hear figures of 2 to 4 percent of design costs assigned to "MANPRINT" for all the domains on a major program. But on these same programs, MANPRINT has been given anywhere from 15 to 40 percent weight on design decisions. However the industry cost benefit studies come out, the investment appears to be a factor of 10 smaller than the importance given to design decisions. MANPRINT, even without government requirements, should begin to make good business sense. What other investment promises so much measurable return in such a short time?

There is clearly a need, however, for better quantitative data to show that adopting MANPRINT as a corporate methodology, first, results in higher quality, better performing products and systems; and, second, affects sales and profits directly and positively by these improvements. Further, there is almost no information to guide the relative investment of MANPRINT classified by type of product or system. Moreover, how should any particular MANPRINT investment get proportioned among its domains and other technology? Figure 20-1 shows a Human Resources Investment model for which investment evidence is beginning to be assembled (Rouse & Cacioppo, 1989). Human resource variables, measures, and relationships of interest have been identified. Attempts have also been made to identify data sources for the variables and the relationships among variables. It was found that data are generally available for determining variable ranges from a variety of industries, but that data on relationships among variables are severely lacking.

Although assembling the evidence for a full human resource investment model will be difficult, it is believed to be a feasible task. Currently under

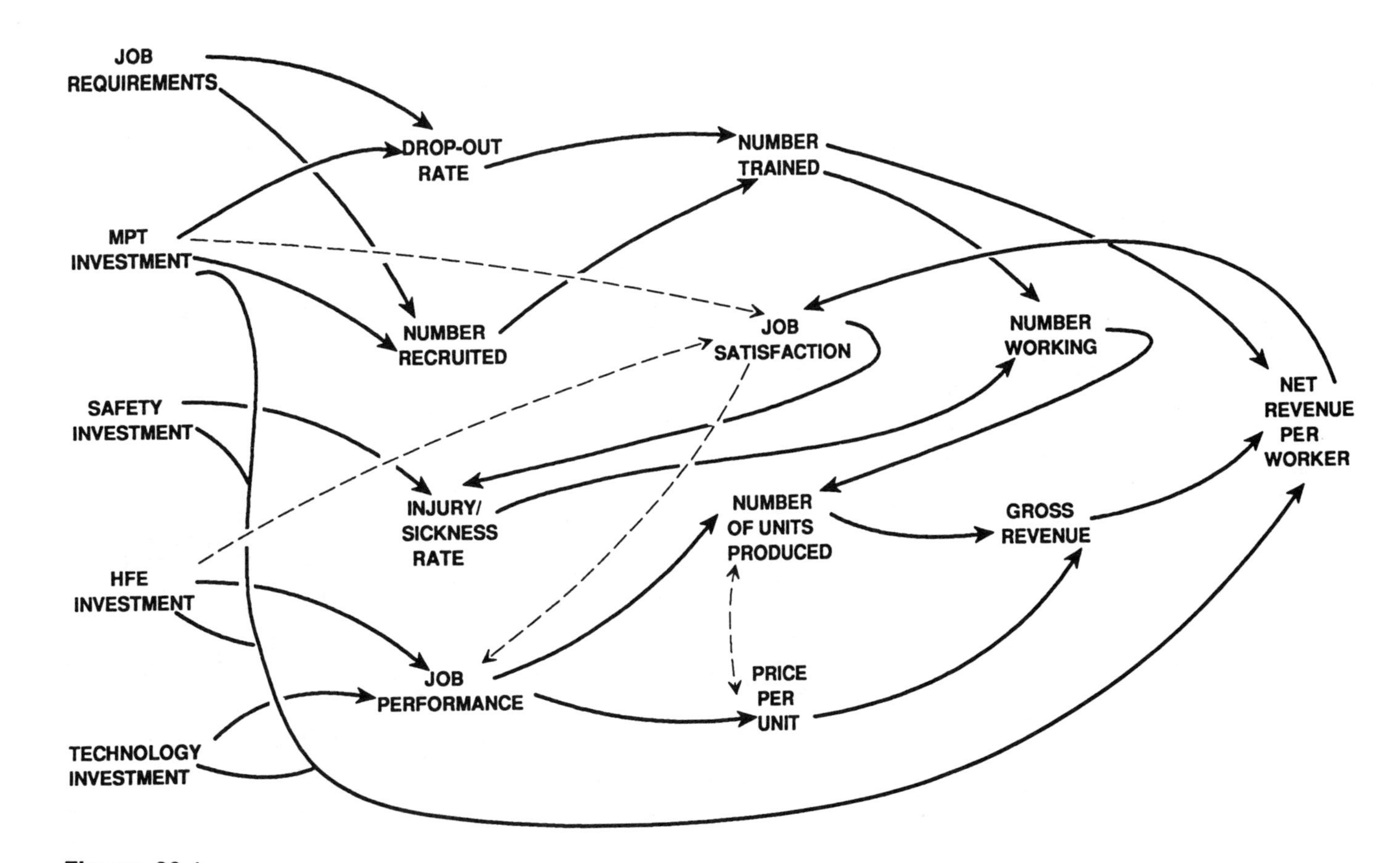

Figure 20-1
Human Resources Investment Model

consideration is the development of an interactive software package for modeling the impact of human resource investments among key variables shown in Figure 20-1. A validated model would allow corporate decision makers to identify what investment is required, what is the payback period, and what is the return on investment. This information could be used not only by business people and industrialists, but their bankers and stockholders as well.

ACROSS CULTURES

Interest in MANPRINT should maintain a global perspective. U.S. foreign trade with Pacific rim countries now accounts for over 33 percent of our foreign trade and this percentage is growing. In contrast, our foreign trade with European countries accounts for less than 25 percent and this percentage is decreasing. The percentage of world GNP for under-developed countries is also growing (currently 25 percent).

Thus, U.S. foreign trade will become more and more dominated by non-western and developing countries. What does this mean for the future of MANPRINT? Will any fundamental changes to MANPRINT implementation be needed?

At the very least, these trends mean that human resource issues identified via MANPRINT will have to be resolved with consideration for the cultural context within which they occur (Rouse, 1990). For example, in some countries, preservation of traditions and creation of jobs may have more importance than performance and productivity.

Beyond the ways in which issues are resolved, MANPRINT may have to evolve in its fundamental philosophy. For example, MANPRINT is currently packaged in a manner consistent with the usual adversarial nature of U.S. business practices, where every requirement is an explicit and legally enforceable part of a written contract. However, in non-western and/or underdeveloped countries, these types of business practices are not necessarily acceptable. A cooperative and non-adversarial approach may be an important element of success in these cultures.

The types of cross-cultural issues that we have raised are difficult to understand and resolve fully. Consequently, such issues have only recently received concerted attention (e.g., Martin et al., 1990). Nevertheless, once MANPRINT is cast in economic terms and adopted by business and industry, the next concern of these companies will be how they can apply these ideas to their international operations. If MANPRINT continues to open new doors in technology innovation and show profitability in human resource investment, along with safer and more efficient system performance, it may well become one of the nation's foremost competitive philosophies leading the U.S. and its industrial base into the 21st century.

REFERENCES

Blanchard, R. E. (1989). *Conceptual model for assessing MANPRINT's organizational effectiveness* (Report No. BM 106-4). San Diego, CA: Behavior Metrics.

Booher, H. R. (1990a, February 14). Application of MANPRINT to air traffic control. *Human factors in air traffic control, Air Traffic Control Association Symposium*, Arlington, VA.

Booher, H. R. (1990b). MANPRINT implications for product design and manufacture. *International Journal of Industrial Ergonomics.* (to appear)

Hendrick, H. W. (1986). Macroergonomics: A conceptual model for integrating human factors with organizational design. In O. Brown, Jr., & H. W. Hendrick (Eds.), *Proceedings of the Second International Symposium Human Factors in Organizational Design and Management* (pp. 467-478). Amsterdam: Elsevier.

Johnson, E. M. (1987). The role of man in the system design process: The unresolved dilemma. In W. B. Rouse and K. R. Boff (Eds.), *System design.* New York: North Holland.

Martin, T., Kivinen, J., Rijnsdorp, J. E., Rodd, M. B., & Rouse, W. B. (1990). Appropriate automation - Integrating technical, human, organizational, economic, and cultural factors. *Proceeding of IFAC World Congress.*

Perrow, C. (1984). *Normal accidents: Living with high-risk technologies.* New York: Basic Books, Inc.

Rouse, W. B. (1990). Human resource issues in system design. In N. P. Moray, W. R. Ferrell, & W. B. Rouse (Eds.), *Robotics, control, and society.* London: Taylor & Francis.

Rouse, W. B., & Cacioppo, G. M. (1989). *Prospects for modeling the impact of human resource investments on economic return.* Washington, DC: Department of the Army, Office of the Deputy Chief of Staff for Personnel.

Singley, G. T. (1990, February 27). *Army tech-based master plan, Executive breakfast seminar.* Washington, DC: Pentagon.

Yamada, S. (1989, June 1). Paradigm shift in product/service development. *Japanese business strategies at turning point.* New York: NRI Forum International.

AUTHOR INDEX

SUBJECT INDEX